TREATISE ON ANALYTICAL CHEMISTRY

A comprehensive account in three parts

PART I

THEORY AND PRACTICE

PART II

ANALYTICAL CHEMISTRY OF INORGANIC
AND ORGANIC COMPOUNDS

PART III

ANALYTICAL CHEMISTRY IN INDUSTRY

TREATISE ON ANALYTICAL CHEMISTRY

PART I
THEORY AND PRACTICE

VOLUME 9

Edited by I. M. KOLTHOFF
School of Chemistry, University of Minnesota

and PHILIP J. ELVING
Department of Chemistry, University of Michigan

with the assistance of ERNEST B. SANDELL
University of Minnesota

WILEY–INTERSCIENCE

a division of John Wiley & Sons, Inc., New York–London–Sydney–Toronto

TREATISE ON ANALYTICAL CHEMISTRY

PART I
THEORY AND PRACTICE

AUTHORS OF VOLUME 9

J. H. BADLEY   R. W. KING

C. E. CROUTHAMEL   G. H. MORRISON

H. L. FINSTON   WILLIAM SEAMAN

VINCENT P. GUINN   R. K. SKOGERBOE

R. R. HEINRICH   F. H. STROSS

Authors of Volume 9

J. H. Badley

Shell Development Company, Emeryville, California, Chapter 101

C. E. Crouthamel

Chemical Engineering Division, Argonne National Laboratory, Argonne, Illinois, Chapter 96

H. L. Finston

Department of Chemistry, Brooklyn College of the City University of New York, Brooklyn, New York, Chapters 94 and 95

Vincent P. Guinn

Department of Chemistry, University of California, Irvine, California, Chapter 98

R. R. Heinrich

Chemical Engineering Division, Argonne National Laboratory, Argonne, Illinois, Chapter 96

vii

R. W. King
Sun Oil Company, Research and Engineering Division, Marcus Hook, Pennsylvania, Chapter 99

G. H. Morrison
Department of Chemistry, Cornell University, Ithaca, New York, Chapter 100

William Seaman
American Cyanamid Company, Bound Brook, New Jersey, Chapter 97

R. K. Skogerboe
Department of Chemistry, Cornell University, Ithaca, New York, Chapter 100

F. H. Stross
Shell Development Company, Emeryville, California, Chapter 101

PART I. THEORY AND PRACTICE

CONTENTS—VOLUME 9

SECTION D-6: Radioactive Methods

Part I
Section D-6

Chapter 94

RADIOACTIVE AND ISOTOPIC METHODS OF ANALYSIS: NATURE, SCOPE, LIMITATIONS, AND INTERRELATIONS

By H. L. FINSTON, *Department of Chemistry, Brooklyn College of the City University of New York, Brooklyn, New York*

Contents

I. INTRODUCTION

The phenomenon of radioactivity presents to the analyst a number of specific properties, characteristic of a particular radionuclide and, consequently, of any particular element for which there is a radioisotope. These properties include the type and energy of the radiation, alpha, beta and gamma, the half-life, and the decay scheme. The fact that a given nuclide may be radioactive does not, in any way, affect its chemical properties before radioactive emission takes place, and after detection its fate is of no consequence. Radionuclides of suitable half-lives exist in all the elements except a very few, for example, oxygen, nitrogen, helium, lithium, and boron; and even of these, oxygen, lithium, and boron have been determined by activation techniques.

The sensitivity inherent in radioactivity methods is due to the fact that as few as several thousand radioactive molecules can be detected. The practical working limits are fixed by the half-life and nature of the radiations; for example, ^{210}Po yields 35 disintegrations per second per milliliter in $10^{-12}M$ solution, and ^{140}La may also be studied at a concentration of $10^{-12}M$. Other advantages for the analyst are the possibility of nondestructive analysis and also of in-line analysis, since the radioactivity of an aliquot or of the whole sample can be determined without altering the composition of the sample.

In addition to the methods of analysis dependent on the tracer characteristic of radioactive material, the ionizing radiations themselves are also of widespread utility in analysis. Radiation thickness gages have their basis in the fact that nuclear radiations are attenuated to a predictable amount by a certain mass of material. Beta gages employ ^{90}(Sr-Y), which emits low-energy particles for gaging exceptionally thin materials. Gamma radiation has been employed for measuring the thicknesses of more dense material such as metal pipe (14) and the concentrations of high-atomic-number elements in solution (5). Neutron attenuation has been applied for the analysis of elements such as boron and cadmium which have high probabilities for capturing a neutron (28).

Another characteristic phenomenon of neutrons and nuclear radiation which has found application in chemical analysis is back scatter. Neutron scattering is by now a well-known technique for oil-well logging and for the determination of moisture in a wide variety of substances (18,19). The basis for this technique is the high scattering cross section of hydrogen. The back scatter of beta particles has been demonstrated to follow a simple linear function within periods of elements corresponding to the noble-gas configuration (24). Both beta scattering and neutron scattering have been applied to the analysis of hydrocarbons (23,8). Alpha-particle scattering has been used in the determination of some of the low-atomic-number elements; oxygen and carbon, in particular, have very large scattering cross sections for the 6.1-MeV alpha particles from ^{242}Cm (29).

The ability of beta rays to excite characteristic fluorescent X-rays in a target element affords yet another application of ionizing radiation to chemical analysis. A pure beta emitter such as ^{147}Pm or ^{90}Sr may be placed in an aluminum block, a magnetic field imposed to deflect scattered beta rays, and the characteristic X-rays of various target elements detected with a proportional counter. This technique has been applied to the determination of plating thicknesses (10); under optimum conditions the characteristic X-rays of the plate increase with thickness, and, simultaneously, the characteristic X-rays of the backing decrease.

A third analytical technique which has achieved widespread application in the short span of two decades is activation analysis, which consists in the formation of an artificially radioactive isotope by exposure of an element to a flux of energetic charged particles or neutrons. The versatility of the technique is inherent in the fact that there are approximately 1000 known artificially produced radioisotopes; and with fluxes of 10^{12} neutrons/cm^2-sec available in nuclear reactors, sensitivities for most elements are in the range of 10^{-16} to 10^{-11} g. The specificity is excellent since each radionuclide is characterized by its radiations and half-life.

The relationship between analytical chemistry and nuclear science has been a symbiotic one, dating back to the very origins of nuclear research. Within 2 years after the initial discovery of radioactivity by Becquerel in 1896, the Curies applied the technique of fractional crystallization to the separation and identification of radium and polonium. The hypothesis of nuclear transformation initially proposed by Soddy and Rutherford in 1902 had its origins in the researches of Sir Richard Crooke, who used solvent extraction (ether extraction), complexation (uranyl carbonate soluble in excess carbonate), and fractional crystallization for the first separation of uranium and its daughter products. By 1930, Rutherford, Chadwick, and Ellis had summarized and categorized the analytical techniques for the preparation of naturally occurring radioactive sources. Precipitation techniques were described for the separation of radium sulfate, thorium oxalate, and radiothorium (^{228}Th) hydroxide with thorium carrier from mesothorium (^{228}Ra). Electrochemical methods, including electrodeposition at controlled potentials and chemical plating, were described for the preparation of radium C (^{214}Bi), radium B (^{214}Pb), radium F (^{210}Po), radium D (^{210}Pb), and radium E (^{210}Bi) sources. Volatilization methods were more briefly described and summarized as follows: "B bodies volatilize more readily than A bodies and the A bodies more readily than the C bodies."

The naturally occurring radioisotopes were soon applied by von Hevesy and his co-workers to the study of the phenomena of self-diffusion and exchange of atoms between compounds having a common atom. They evolved a very sensitive method for studying the diffusion of lead.

Lead containing thorium B (^{212}Pb) was pressed into contact with a thin foil of inactive lead just thick enough to stop all the alpha rays from the thorium C (^{212}Bi) daughter; as diffusion progressed, the alpha activity appeared and increased as measured through the foil.

During the 1930s, Fermi and his co-workers irradiated uranium with slow neutrons and observed a number of new activities which they attributed to transuranium elements. In 1938, Hahn, Strassman, and Meitner, convinced that these new activities were transuranics, attempted a chemical characterization of the new elements and were led to an early conclusion that a reaction of the type U (n, 2α) occurred, leading to radium. Subsequent attempts to confirm this conclusion proved the inseparability of the activity from barium carrier and led to the well-known conclusion that fission occurred, which was announced by Niels Bohr at a meeting of the American Philosophical Society in Washington late in 1938.

The tremendous impetus given to the study of nuclear science by the Manhattan Project is obvious and well known, but it is perhaps appropriate to point out that the chemical procedures devised for the isolation of fission products (7) were adaptions of procedures found in standard analytical textbooks. The procedures were so modified that most of the precipitations and dissolutions could be performed in 50-ml centrifuge tubes, and centrifugation was extensively applied for rapid separation of solid and solution phases. With the single exception of separation by recoil, the chemical operations performed in radiochemistry in general are the standard techniques of analytical chemistry: precipitation, volatilization, solvent extraction, and ion-exchange chromatography.

In the years following the conclusion of the Plutonium Project, analytical chemistry has continued to make major contributions to the techniques applied in radiochemistry. The usefulness of metal chelates in radiochemical separations has long been recognized (7), and the application of reagents which form insoluble compounds with metal cations is commonplace. A number of aminopolycarboxylic acids, known as complexones, for example, ethylene-diaminetetraacetic acid (EDTA), have the special property of forming water-soluble chelates with a large number of metallic ions; and since the order of chelation varies for the different cations and is a function of pH, the complexones afford a technique for exercising a selective masking action in precipitation reactions. The formation of a precipitating ion or radical by the gradual hydrolysis or oxidation of a parent compound, referred to as "precipitation from homogeneous solution" (11), eliminates the occurrence of concentration gradients and consequently reduces the coprecipitation and occlusion of foreign ions. In many cases, radiochemical procedures take advantage of the occlusion and coprecipitation which accompany the formation of gelatinous precipitates in order to "scavenge" a solution of unwanted activities. However, there are often situations in which the scavenging process is undesirable, and a homogeneous technique should be applied. The techniques of electroanalysis, such as deposition at a mercury cathode and controlled potential electrolysis, also represent methods capable of extensive application in the field of radiochemistry.

II. RADIOACTIVITY

A. DECAY AND GROWTH

The observation by Crookes in 1900 that the uranium-free carbonate which precipitated from a uranium solution had many times the activity of the uranium fraction was followed by similar separations of more active fractions from both uranium and thorium solutions by Rutherford and Soddy. They called the separated substances uranium X and thorium X, respectively, and noted that these activities decreased rapidly, while the residual activities of the mother liquors soon rose to their former values. In 1903 Rutherford and Soddy published their conclusions that nuclear transformation occurred, that is, an element changed into another element by the emission of radiations, and that the phenomenon was a subatomic process. The growth and decay curves of the separated fractions from uranium and thorium were explained by the assumption that the parent atom, uranium (or thorium), decays, giving rise to a radioactive daughter, uranium I (or thorium I), which in turn undergoes radioactive decay. The decay rates were observed to correspond to a mononuclear process.

The decay rate is proportional to the number of decaying atoms present, that is,

$$\frac{-dN}{dt} = \lambda N \tag{1}$$

where N is the total number of atoms present, dt is the time interval, and λ is the decay constant. Rearranging and integrating between time limits, where $N = N_0$ at $t = 0$, we obtain

$$\int_{N_0}^{N} \frac{dN}{N} = \int_{t=0}^{t} - \lambda \, dt \tag{2}$$

or

$$[\ln N]_{N_0}^N = [-\lambda t + c]_{t=0}^t \tag{3}$$

Thus

$$\ln \frac{N}{N_0} = -\lambda t \tag{4}$$

or

$$N = N_0 e^{-\lambda t} \tag{5}$$

Equation (5) agrees with the observed exponential decay of activity in uranium X and thorium X, and leads to the growth curves observed for the respective uranium and thorium residues.

The decay constant, λ, is characteristic of the radioactive species and has the dimensions of reciprocal time. The characteristic rate of decay is, however, more conveniently expressed in terms of the half-life, $t_{1/2}$, which is the time required for half of an initially large number of radioactive atoms to decay. Thus, if $t_{1/2}$ is the half-life of the radioactive species,

$$\frac{N_0}{2} = N_0 e^{-\lambda t_{1/2}} \tag{6}$$

or

$$t_{1/2} = \frac{\ln 2}{\lambda} = \frac{0.69315}{\lambda} \tag{7}$$

The rate of increase in the number of radioactive daughter atoms is $\lambda_P N_P$, while the rate of decrease of radioactive daughter atoms is $\lambda_d N_d$ where the subscripts p and d refer, respectively, to parent and daughter atoms. The rate of change of daughter atoms is thus given by

$$\frac{dN_d}{dt} = -\lambda_d N_d + \lambda_p N_p = \lambda_p N_{p0} e^{-\lambda_p t} - \lambda_d N_d \tag{8}$$

which can be arranged as follows:

$$\frac{dN_d}{dt} + \lambda_d N_d = \lambda_p N_{p0} e^{-\lambda_p t} \tag{9}$$

Multiplying each term by $e^{\lambda_d t}$ yields a perfect differential,

$$e^{\lambda_d t} \frac{dN_d}{dt} + e^{\lambda_d t} \lambda_d N_d = \frac{d}{dt} N_d e^{\lambda_d t} = \lambda_p N_{p0} e^{(\lambda_d - \lambda_p)t} \tag{10}$$

When the condition that the daughter activity $N_d = 0$ at $t = 0$ is imposed, integration yields

$$N_d e^{\lambda_d t} = \frac{N_{p0}\lambda_p}{\lambda_d - \lambda_p} [e^{(\lambda_d - \lambda_p)t} - 1] \tag{11}$$

or, upon rearrangement,

$$N_d = \frac{N_{p0}\lambda_p}{\lambda_d - \lambda_p} (e^{-\lambda_p t} - e^{-\lambda_d t}) \tag{12}$$

When the half-life of the parent is much longer than the half-life of the daughter, that is, $\lambda_p \ll \lambda_d$, then $e^{-\lambda_p t} \simeq 1$ and $\lambda_d - \lambda_p \simeq \lambda_d$; thus

$$N_d = \frac{N_{p0}\lambda_p}{\lambda_d} (1 - e^{-\lambda_d t}) \tag{13}$$

This corresponds exactly to the buildup of daughter activities in separated uranium and thorium, which represent very long-lived parent activities. At equilibrium the daughter activity is equal to the parent activity,

$$\lambda_d N_d = \lambda_p N_p \tag{14}$$

and such a case represents "secular equilibrium," wherein the parent and daughter activities decay with the half-life of the parent after equilibrium is reached.

An analogous situation is presented by the production of a radioactive substance by any steady source such as a nuclear reactor or accelerator. If we replace $N_{p0}\lambda_p$ with R, the rate of production, the activity of the radioactive substance is given by

$$\lambda N = R(1 - e^{-\lambda t}) \tag{15}$$

The term $(1 = e^{-\lambda t})$ may be referred to as the "saturation factor." When t is very long in comparison with the half-life, the maximum activity is attained, which is equal to R; one-half the maximum is obtained when t equals one half-life, three-fourths is obtained after two half-lives, etc.

We have so far considered the case of a very long-lived parent decaying to a radioactive daughter with a much shorter half-life, such that the parent undergoes negligible decay during many half-lives of the daughter. It is also of interest to note the case of "transient equilibrium," in which the parent is longer lived than the daughter ($\lambda_p < \lambda_d$) but does undergo significant decay. Referring again to the general equation (12) and making the approximation $(e^{-\lambda_p t} - e^{-\lambda_d t}) \simeq e^{-\lambda_p t}$, we have

$$N_d = \frac{\lambda_p}{\lambda_d - \lambda_p} N_{p0} e^{-\lambda_p t} \tag{16}$$

Since $N_p = N_{p0} e^{-\lambda_p t}$, the ratio of parent to daughter activities at any time after equilibrium has been reached is given by

$$\frac{N_p}{N_d} = \frac{\lambda_d - \lambda_p}{\lambda_p} \tag{17}$$

A statistical approach to the law of radioactive decay was taken by E. von Schweidler in 1905. He assumed, as for any random phenomenon, that the probability p for an atom to disintegrate in the time interval Δt depends only on the length of the time interval. It is independent of any past history and also of any present circumstances of the atom. For short time intervals, the probability is just proportional to Δt; thus

$$p = \lambda \, \Delta t \tag{18}$$

The probability that an atom does not decay in time interval Δt is

$$1 - p = 1 - \lambda \, \Delta t \tag{19}$$

If this is now compounded, the probability that an atom does not decay in n intervals is

$$(1 - \lambda \, \Delta t)^n \tag{20}$$

Setting $n \, \Delta t$ equal to t, the total time, we can rewrite equation (20) as follows:

$$\left(1 - \lambda \frac{t}{n}\right)^n \tag{21}$$

The probability that an atom is unchanged after time t is the limit of this quantity as $n \to \infty$ (Δt becomes very small).

If we now recall that

$$\lim_{n \to \infty} \left(1 + \frac{x}{n}\right)^n = e^x \tag{22}$$

probability (19) becomes

$$e^{-\lambda t} \tag{23}$$

Considering now a large group of radioactive atoms with an initial number N_0, we find that the fraction remaining unchanged after time t is given by

$$\frac{N}{N_0} = e^{-\lambda t} \tag{24}$$

where N is the number remaining unchanged after time t. This brings us back to equation (5).

The appearance of daughter products other than uranium X and thorium X was noted by Rutherford and Owens, who isolated and characterized a radioactive emanation (^{220}Rn) from thorium. They further observed that the emanation appeared to "induce" radioactivity in materials exposed to it; the "induced radioactivity" had a half-life of 10.64 hr, regardless of the material exposed, and was later identified as the radioactive decay product of radon, ^{212}Pb. Almost simultaneously the Curies published a paper about a radioactivity induced in the presence of radium, but they were as yet unaware of any radium emanation. Subsequently Rutherford and Soddy found that the emanation from both thorium and uranium could be condensed and was chemically inert. With the help of A. G. Grier they were able to distinguish between alpha and beta decay and so could identify five successive transformations in radium. Their suggestion that helium was an end product of uranium and thorium decay was verified when Ramsay and Soddy in London identified the spectrum of helium in condensed emanation.

Pierre Curie, after an initial reluctance to accept the transformation theory, found three successive products of radium emanation with half-lives of 2.6 min, 21 min, and 28 min, which he called radium A, B, and C, respectively. Early in 1905, Rutherford had this chain of transformation worked out, and by 1907 the descent of radium was verified by Boltwood. The concept of isotopy advanced by Soddy in 1910 and his classification of the then known radioelements in the periodic table was followed by Fajan's rules of radioactive decay, which stated that an atom upon emitting an alpha particle was transmuted to an atom with atomic number of 2 less and a mass of 4 less; the emission of a beta particle was a transmutation to an element of next higher atomic number.

The products of ^{238}U, ^{235}U, and ^{232}U are shown as members of a decay series in Fig. 94.1. Since the mass numbers change only in alpha decay and by 4 mass units, they can be arranged in three series which correspond to the approximate mass formulas $4n + 2$, $4n + 3$, and $4n$.

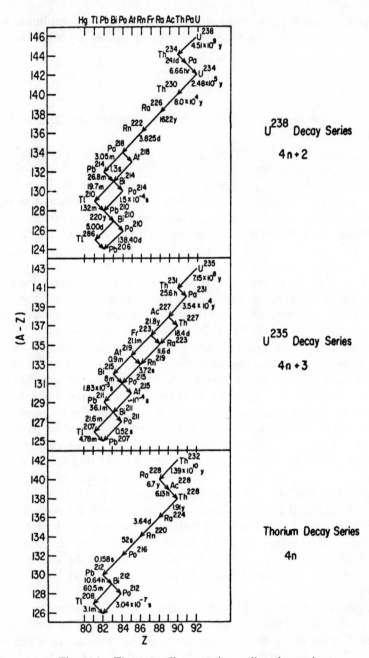

Fig. 94.1. The naturally occurring radioactive series.

B. THEORY OF SUCCESSIVE TRANSFORMATIONS

If we consider a decay chain in which successive transformations occur, that is, $A \rightarrow B \rightarrow C \rightarrow$ etc., then at any time t we may have P, Q, R, etc., atoms of the respective nuclides. As a practical case we will consider the changes that occur in the active deposit produced when a plate is exposed to a constant amount of radon and then removed. It is necessary to consider three special cases: (*1*) when the time of exposure is extremely short compared with the period of the changes (long half-life); (*2*) when the time of exposure is so long that the amount of each of the products has reached a steady limiting value (equilibrium); and (*3*) any time exposure.

Case 1. Assuming that only A is initially present, we are required to find the number of atoms P, Q, R, S of A, B, C, D, . . . , respectively, that are present after time t.

Thus $P = P_0 e^{-\lambda_1 t}$, and dQ, the increase in the number of atoms of B per unit time, is the number formed by the decay of A minus the number lost by the decay of B into C.

$$\frac{dP}{dt} = \lambda_1 P \tag{25}$$

$$\frac{dQ}{dt} = \lambda_1 P - \lambda_2 Q \tag{26}$$

$$\frac{dR}{dt} = \lambda_2 Q - \lambda_3 R \tag{27}$$

Substituting for P in equation (26), we obtain

$$\frac{dQ}{dt} = \lambda_1 P_0 e^{-\lambda_1 t} - \lambda_2 Q$$

which is similar to equation (9) and can be integrated in the same way.
Upon integration

$$Q = P_0(ae^{-\lambda_1 t} + be^{-\lambda_2 t}) \tag{28}$$

where $a = \lambda_1(\lambda_2 - \lambda_1)$.
Since $Q = 0$ when $t = 0$, $b = -\lambda_1/(\lambda_2 - \lambda_1)$; thus

$$Q = \frac{P_0 \lambda_1}{\lambda_2 - \lambda_1} (e^{-\lambda_1 t} - e^{-\lambda_2 t}) \tag{29}$$

If we now substitute this value of Q into equation (27),

$$R = P_0(ae^{-\lambda_2 t} + be^{-\lambda_2 t} + ce^{-\lambda_3 t}) \tag{30}$$

where

$$a = \frac{\lambda_1\lambda_2}{(\lambda_2 - \lambda_1)(\lambda_3 - \lambda_1)}; \qquad b = \frac{\lambda_1\lambda_2}{(\lambda_1 - \lambda_2)(\lambda_3 - \lambda_2)};$$

$$c = \frac{\lambda_1\lambda_2}{(\lambda_1 - \lambda_3)(\lambda_2 - \lambda_3)}$$

The method of solution of the general case of n products has been given by Bateman; the amount of the nth product, $N(t)$, at any time t is

$$N(t) = C_1 e^{-\lambda_1 t} + C_2 e^{-\lambda_2 t} + \cdots + C_n e^{-\lambda_n t} \tag{31}$$

where
$$C_1 = \frac{\lambda_1\lambda_2 \ldots \lambda_{n-1}P_0}{(\lambda_2 - \lambda_1)(\lambda_3 - \lambda_1) \ldots (\lambda_n - \lambda_1)}$$

$$C_2 = \frac{\lambda_1\lambda_2 \ldots \lambda_{n-1}P_0}{(\lambda_1 - \lambda_2)(\lambda_3 - \lambda_2) \ldots (\lambda_n - \lambda_2)}$$

etc.

Case 2. A primary source supplies A at a constant rate; and some time after equilibrium has been reached, the primary source is removed. We are required to find the amounts of A, B, etc., remaining at any subsequent time, t.

In this case (equilibrium), the number, n_0, of atoms of A deposited per second from the source equals the number changing to B per second, and of B changing to C, etc. Thus

$$n_0 = \lambda_1 P_0 = \lambda_2 Q_0 = \lambda_3 R_0 \tag{32}$$

The values of Q, R, etc., at time t after removal of the source are given by equations of the forms (28) and (30) for a short exposure. When we remember that the initial condition is

$$P = P_0 = n_0/\lambda_1, \quad Q = Q_0 = n_0/\lambda_2, \quad R = R_0 = n_0/\lambda_3$$

it follows that

$$P = \frac{n_0}{\lambda_1} e^{-\lambda_1 t} \tag{33}$$

$$Q = \frac{n_0}{\lambda_1 - \lambda_2} \left(\frac{\lambda_1}{\lambda_2} e^{-\lambda_2 t} - e^{-\lambda_1 t} \right) \tag{34}$$

$$R = n_0(ae^{-\lambda_1 t} + be^{-\lambda_2 t} + ce^{-\lambda_3 t}) \tag{35}$$

where

$$a = \frac{\lambda_2}{(\lambda_2 - \lambda_1)(\lambda_3 - \lambda_1)}$$

$$b = \frac{\lambda_1}{(\lambda_1 - \lambda_2)(\lambda_3 - \lambda_2)}$$

$$c = \frac{\lambda_1\lambda_2}{\lambda_3(\lambda_1 - \lambda_3)(\lambda_2 - \lambda_3)}$$

Case 3. A primary source has supplied matter A at a constant rate for any time T and is suddenly removed. We are required to find the amounts of A, B, C at any subsequent time.

Suppose that n_0 atoms of A are deposited each second. After time T, the number of atoms P_T of A present is given by

$$P_T = n_0 \int_0^T e^{-\lambda_1 t}\, dt = \frac{n_0}{\lambda_1}(1 - e^{-\lambda_1 T}) \tag{36}$$

At any time t after removal of the source,

$$P = P_T e^{-\lambda_1 t} = \frac{n_0}{\lambda_1}(1 - e^{-\lambda_1 T}) e^{-\lambda_1 t} \tag{37}$$

Consider the number of atoms $n_0\, dt$ of A produced in the interval dt. At any later time t, the number of atoms dQ of B which result from the decay of A is given by [see equation (29)]

$$dQ = \frac{n_0\lambda_1}{\lambda_1 - \lambda_2}(e^{-\lambda_2 t} - e^{-\lambda_1 t})\, dt = n_0 f(t)\, dt \tag{38}$$

After time T, the number of atoms Q_T of B is given by

$$Q_T = n_0 \int_0^T f(t)\, dt$$

If the body is removed from the emanation after an exposure T, at any time t after the removal the number of atoms of B is given by

$$Q = n_0 \int_t^{T+t} f(t)\, dt \tag{39}$$

Substituting for $f(t)$ from equation (38), we get after integrating

$$\frac{Q}{Q_T} = \frac{ae^{-\lambda_2 t} - be^{-\lambda_1 t}}{a - b} \tag{40}$$

where

$$a = \frac{1 - e^{-\lambda_2 T}}{\lambda_2} \quad \text{and} \quad b = \frac{1 - e^{-\lambda_1 T}}{\lambda_1}$$

III. THE NUCLEUS

The fact that atoms of a radioactive element are transformed into atoms of another element by emission of positively or negatively charged particles led to the obvious conclusion that atoms are made up of positive and negative charges. The atom being neutral, the numbers of positive and negative charges must be equal; and since early experiments showed that the mass of the electron was only about 1/2000 the mass of the proton, it was immediately recognized that practically all of the mass of the atom was associated with the positive charge. Three questions remained unanswered in 1902:

1. How many electrons are in an atom?
2. How are the electrons arranged?
3. How are the positive charges arranged?

The first nuclear model was proposed by Thomson in 1910. It consisted of a sphere of positively charged particles of uniform density, throughout which was distributed an equal and opposite charge of electrons. The diameter of the sphere was assumed to be approximately 1 Å, which corresponds to the diameter of atoms. However, such a structure could exert little influence on an incident charged particle and proved inconsistent with the large angle scattering of incident alpha particles observed by Geiger and Marsden. A consistent model was proposed by Rutherford in 1911, in which the positive charge was centered in a nucleus, and the negative charge was distributed over a sphere of radius comparable with atomic radii.

The following characteristics of the nucleus were recognized:

1. Radii of atomic nuclei $= 10^{-12}$ to 10^{-13} cm.
2. Volumes of atomic nuclei $= 10^{-36}$ cm³.
3. Density of nuclei (even the lightest nuclei) $= 10^{12}$ g/cm³.

A. NUCLEAR PARTICLES

The deviations of atomic weights from integral numbers, plus the fact that these weights were approximate multiples of atomic numbers, led to the hypothesis that nuclei contained extra protons which were neutralized by electrons in the nucleus. This proton–electron hypothesis soon proved inconsistent, however, with observed spins and magnetic moments of nuclei. A further argument against this theory arises from considerations of wave mechanics.

The spectroscopists had observed that the individual components of multiplet lines were split into a number of lines lying very close together—"hyperfine structure." This was attributed to the fact that the nucleus has an angular momentum or "spin." The results of nuclear spin determinations can be summarized in the statement that the spins of all nuclei of odd mass numbers are odd half-integral multiples of $h/2\pi$, while all nuclei of even mass numbers have spins which are zero or integral multiples of $h/2\pi$. The spin of the nucleus is denoted by I, where

$$I = 0, 1, 2, \ldots \qquad \text{if } A \text{ is even,}$$

and

$$I = \tfrac{1}{2}, \tfrac{3}{2}, \tfrac{5}{2}, \ldots \quad \text{if } A \text{ is odd.}$$

Consideration of this rule led to the first breakdown of the proton–electron hypothesis: for example, in Table 94.I the nuclear electrons would be counted as nuclear particles.

The nucleus having both charge and an angular momentum, it also has associated with it a magnetic moment, which led to the second argument against the existence of electrons in the nucleus. The magnetic moment of an electron, defined as a Bohr magneton, is

$$M_e = \frac{eh}{4\pi m_e c}$$

and for a proton the magnetic moment should be $1/1837$ as large, corresponding to the ratio of electron–proton masses, or

$$M_p = \frac{eh}{4\pi m_p c}$$

The actual measured moments of nuclei are considerably less than those expected on the basis of the electron–proton hypothesis.

Wave mechanics requires than an electron existing in a nucleus have a kinetic energy of 60 MeV. Since beta energies greater than 4.8 MeV have never been observed, this constitutes a third argument against the existence of electrons in the nucleus.

TABLE 94.I

Isotope	7^{14}N	48^{odd}Cd	80^{odd}Hg	82^{odd}Pb
Particles	21	Even	Even	Even
Calc. spin	Odd $\tfrac{1}{2}$	0, int.	0, int.	0, int.
Obs. spin	0, int.	Odd $\tfrac{1}{2}$	Odd $\tfrac{1}{2}$	Odd $\tfrac{1}{2}$

A neutral particle had been proposed by Rutherford as early as 1920, but it was not until 1932 that Chadwick found experimental verification for the neutron. In the course of bombarding elements with alpha particles, it was noted that marked transmutations occurred in the case of boron, nitrogen, and aluminum, all elements up to $Z = 19$ yielding protons. Bombardment of boron and beryllium also yielded a highly penetrating particle with zero charge and a mass approximately equal to that of a proton.

B. NUCLEAR FORCES

Once it was established that the nucleus consists of neutrons and protons, attention was directed toward an explanation of the forces which hold a nucleus together. It was obvious, of course, that these forces could be neither simply electrical, since the neutron has zero charge, nor simply gravitational, since such forces would be too weak by many orders of magnitude. Furthermore, the experimental evidence indicated that the forces involved are extremely short range, even shorter than nuclear dimensions.

1. Exchange Force

The first hypothesis that was considered postulated that a proton and neutron can interact by emitting and absorbing an electron–neutrino pair. Such an exchange force appeared quite logical at first, but proved inconsistent with the known binding energies of nucleons in the nucleus. A compatible theory was developed in 1934 by Yukawa, who proposed that a new subnuclear particle, the meson, was transferred between the neutron and proton. The application of quantum mechanics to the electromagnetic field surrounding an electrically charged particle leads to the conclusion that the electrical force is exerted by the transfer of a photon from one charged body to another, as illustrated in Fig. 94.2. Yukawa suggested that nuclear forces are due to a field similar to an

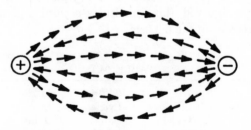

Fig. 94.2. The transfer of a photon from one charged body to another.

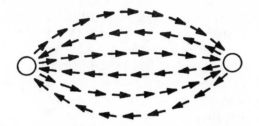

Fig. 94.3. The transfer of a meson between a neutron and proton.

electromagnetic field but involving the transfer of a particle of an entirely different character with a mass approximately 200 times that of an electron (Fig. 94.3). Such a "meson" was indeed observed in cosmic rays by Neddermeyer and Anderson in 1936 (mass = 210 times electron mass).

Further consideration of this theory led to the independent conclusion by Sakata and Inoue in Japan, and Bethe and Marshall in the United States, that a satisfactory explanation of the nuclear forces required a heavier meson, one which would exist in the free state for less than 10^{-8} sec and would transmute into the lighter mu meson. In 1947 the pi meson (mass = 270 electron masses) was observed by Powell, Occhialini, and Lattes, and proved consistent with the theory.

Our current understanding is that every nucleon is surrounded by a meson field, through which it interacts with other nucleons in a manner analogous to the interaction of electrically charged bodies via an electromagnetic field. Mesons have a very evanescent existence, being continually created from the mass of the neutron and absorbed back into it without escaping from the structure of the nucleus. A free pi meson (pion) will change spontaneously into a mu meson (muon) in about 10^{-6} sec, and the muon in turn decays into an electron and a neutrino (Fig. 94.4).

2. Alpha Decay Theory

The positive electrical charge of the nucleus quite obviously presents a Coulombic barrier for the emission or entry of a positively charged

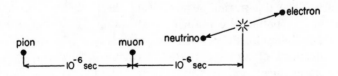

Fig. 94.4. The decay of a free pi-meson (pion).

particle, yet certain anomalies were observed in alpha decay which were not resolved until 1928 by the alpha decay theory of Gamow, Condon, and Gurney. The ranges and energies of alpha particles are determined by the difference between the mass of the parent nuclide and the sum of the mass of the alpha particle plus the mass of the product nuclide ($E = mc^2$). Thus, all alpha particles which are the product of a specific decay process have the same energy or range. This energy has been experimentally related to the decay constant by the Geiger-Nuttall relationship,

$$\log \lambda = A + BE \tag{41}$$

where A and B are experimentally derived constants. The decay probabilities for the naturally occurring alpha emitters vary widely, with half-lives ranging from seconds to thousands of years, yet the energy varies only by a factor of approximately 2.

We can represent the nucleus as a potential well. When an alpha is a considerable distance from the nucleus, it is influenced only by the repulsive force of the nuclear field, and work must be applied to the particle in order to bring it closer to the nucleus, thus increasing its potential energy. When the alpha particle is sufficiently close to the nucleus, however, the short-range nuclear forces of attraction are effective and they overcome the electrical repulsion. Consequently, after a certain potential barrier has been overcome, the force of repulsion changes into a force of attraction, and from that point on the potential energy drops steeply in the direction of the interior of the nucleus.

The case for a particle emerging from the nucleus is exactly the same, but in the opposite direction of course. The alpha particle travels a certain distance from the nucleus with a certain quantity of kinetic energy, and its total energy outside the nucleus is positive since its potential energy vanishes. Figure 94.5 illustrates the nuclear potential. The two horizontal lines a and b indicate the energy with which alpha particles emerge from the thorium C′ and ^{238}U nuclei, respectively. Extrapolating line a, for example, back into the region within the nucleus, we see that the effective energy of the alpha particle is somewhat greater inside the nucleus, given by the vertical c, as compared with d outside the nucleus. The illustration suggests that the particle can bounce from one side of the "potential container" to the other, but classical mechanics offers no possibility for the particle to escape.

The movement of alpha particles is governed, however, by the laws of wave mechanics; and if wave-particle duality is considered, there is an analogy in the total reflection of light at the surface separating

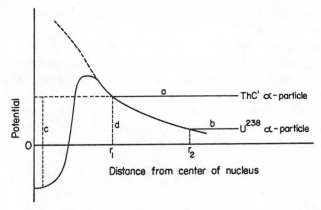

Fig. 94.5.   The nuclear potential well.

two transparent refractive substances, which can explain the escape of an alpha particle. If light falls perpendicularly on one of the short sides of a right-angled triangular glass prism, it will strike the surface of the hypotenuse at 45° and will be totally reflected. If a second prism is placed close to the first, no change occurs as long as the distance is great. In total reflection, however, a little light always seeps through the reflecting surface for a short distance, of the magnitude of the wavelength. The closer the surfaces are, the greater the amount that goes through, and if the surfaces are pressed firmly together no total reflection at all occurs (Fig. 94.6).

In this analogy, the interior of the nucleus is regarded as the first prism, while the potential barrier is the equivalent of the space separating the two prisms. Thus, some waves will always get through the barrier (that part of the potential curve which rises above the horizontal energy level of the particle), and it is quite obvious that the more energetic the particle, the narrower is the potential barrier. The exact mathematical expression derived from these considerations leads to the experi-

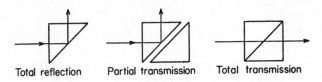

Fig. 94.6.   The analogy between the movement of alpha particles and the total reflection of light.

mentally deduced Geiger-Nuttall law. The leakage through the barrier is the "tunnel effect."

When transmutation is brought about by bombarding a nuclide with a charged particle, the particle must have sufficient energy to overcome the potential barrier, or at least to permit a significant amount of tunneling. The probability for tunneling, however, drops off rapidly with decreasing energy. The height of the potential barrier around a nucleus of charge Z_1e and radius R_1 for a positively charged particle with charge Z_2e and radius R_2 is given by

$$V = \frac{Z_1 Z_2 e^2}{R_1 + R_2} \qquad (42)$$

and R can be approximated sufficiently well by the formula

$$R = 1.5 \times 10^{-13} A^{\frac{1}{3}} \qquad (43)$$

The height of the potential barrier around a given nuclide is proportional to the charge of the bombarding particle, being twice as high for alpha particles as for protons and deuterons. It is actually approximately proportional to $Z^{\frac{2}{3}}$, because the nuclear radius R increases approximately as $Z^{\frac{1}{3}}$.

C. NUCLEAR STRUCTURE

1. The Liquid Drop Model

The mass of a nucleus can be calculated by means of a semiempirical equation of the type first devised by von Weizsaecker in 1935:

$$M(Z, A) = (A - Z)M_n + ZM_H - aA + bA^{\frac{2}{3}} + \frac{cZ^2}{A^{\frac{1}{3}}} + \frac{d(A - 2Z)^2}{A} - \frac{e}{A} \qquad (44)$$

where a, b, c, d, and e are suitable constants, and the last constant, e, is positive for nuclides with an even number of protons and an even number of neutrons, zero for even–odd nuclei and odd–even nuclei, and negative for odd–odd nuclei. The equation can be rearranged to give the binding energy of the nucleus, that is, the difference between the sum of the masses of the nucleons (a collective term for neutrons and protons) and the mass of the nucleus:

$$E_b = [ZM_H + (A - Z)M_n - M(Z, A)] \times 931.4 \text{ MeV} \qquad (45)$$

Thus, the semiempirical equation for the binding energy is

$$E_b(Z, A) = aA - bA^{2/3} - \frac{cZ^2}{A^{1/3}} - \frac{d(A - 2Z)^2}{A} \pm \frac{e}{A} \qquad (46)$$

For mass numbers greater than 80, the following constants yield good agreement:

$$E_b(Z, A) = 14.1A - 13A^{2/3} - \frac{0.6Z^2}{A^{1/3}} - \frac{20(A - 2Z)^2}{A} \pm \frac{125}{A} \qquad (47)$$

The semiempirical binding energy equation is consistent with a liquid drop model of the nucleus if we make the following assumptions.

1. All nuclei have the same density; therefore all the nucleons are the same average distance apart. The radius of a nucleus is

$$R = R_0 A^{1/3} \qquad (48)$$

and R_0 is approximately constant for all nuclei:

$$R_0 = 1.4 \times 10^{-13} \text{ cm} \qquad (49)$$

2. The forces binding nucleons together in the nucleus saturate in a manner analogous to chemical bonds. Nucleons interact only with a limited number of other nucleons which are close to them.

3. The forces between nucleons are independent of charge, that is, they are the same for n–n, n–p, and p–p. In addition, there is the long-range force of Coulomb repulsion for p–p, which is of much smaller magnitude, but in large nuclei the many protons act together to exert Coulomb repulsion.

The individual terms of the binding energy equation will now be discussed in terms of the liquid drop model.

Term aA. According to assumption *2*, each of the A nucleons is bound only to a certain number of its nearest neighbors, and the number does not depend on A. Thus the binding energy is proportional to A and not to $A(A - 1)$, as would be the case if each of the A nucleons could interact with all of the others $(A - 1)$.

Term $bA^{2/3}$. By analogy with the liquid drop, the nucleons at the surface do not have as many close neighbors and are, consequently, less tightly bound. An amount of binding energy corresponding to the number of nucleons at the surface must be subtracted. This will be proportional to the surface area, and thus to R^2 or $A^{2/3}$.

Term $cZ^2/A^{1/3}$. The loss of binding energy due to the coulomb repulsion between the protons must also be subtracted. If the protons are uniformly distributed in a sphere of radius R, the potential energy is proportional to Z^2/R or $Z^2/A^{1/3}$.

Term $d(A - 2Z)^2/A$. $(A - 2Z)$ is the excess of neutrons over protons. Since there is a given n/p ratio corresponding to stability, the equation must include a negative term which increases when the number of either neutrons or protons is greater than that corresponding to the stable ratio.

Term $\pm e/A$. In an even–even nucleus every nucleon can be paired off with another and these nuclei are observed to be the most stable; thus the term is positive. For odd–odd nuclei, generally observed to be unstable, the term is negative, and for even–odd or odd–even it is zero.

The semiempirical binding energy is used to calculate masses, and conversely it can serve to calculate the energy released or required for nuclear transmutation, from the tabulated masses. The equation can also predict the particular mode of beta decay, β^-, β^+, or EC (electron capture), that will occur for a radioactive nucleus.

The liquid drop model has also proved particularly useful in the study of high-energy reactions.

2. The Shell Model

The treatment of the nucleus as a liquid drop in which neutrons and protons essentially lose their identities does not account for the remarkable stability patterns which are discerned for particular combinations of neutrons and protons. It is noteworthy that neutrons and protons show a strong tendency to pair, as indicated by the fact that out of the approximately 1000 isotopes known there are only six stable nuclei containing an odd number of neutrons and protons. There is also strong experimental evidence for a shell structure in the nucleus, as shown by the stability of nuclei with N and Z values of 2, 8, 20, 28, 50, and 82 and also nuclei with 126 neutrons. Calculations of the binding energy, which yield, very accurately, the relative stabilities of the lighter elements, demonstrate that nuclei with the magic numbers 2, 8, and 20 have much greater binding energies than their neighbors; these nuclei correspond, respectively, to helium with 2 protons and 2 neutrons; oxygen with 8 protons and 8 neutrons, which is remarkably stable; and calcium with 20 protons, which has 6 stable isotopes ranging in neutron number from 20 to 28.

The binding energies of the heavier nuclides have not been so accurately determined, and the evidence for stability is based on the number of stable nuclei with a given proton number (isotopes) or neutron number (isotones) and/or the relative abundance of a given nuclide in the universe. Tin with 50 protons has 10 stable isotopes, more than any other element, and is more abundant than any of its neighboring elements. In both calcium ($Z = 20$) and tin the stable isotopes span an unusually large mass range. Lead with 82 protons is the end product of all the naturally radioactive decay chains, and the only case in which more than 5 isotopes occur is for the neutron number $N = 50$, where there are 7.

In keeping with the natural abundance of the isotopes, ^{138}Ba, with 82 neutrons, accounts for 72% of that element, and ^{140}Ce, also with 82 neutrons, represents 88% of the naturally occurring element. Further evidence for the "magic numbers" is the low cross section of these elements for absorption of a neutron. (The cross section, σ, is the probability of a nuclear process expressed in terms of an area, which is derived from the concept that the probability for a reaction between a nucleus and a particle is proportional to the cross-sectional area of the nucleus.)

The spins and magnetic moments of nuclei lead to a description of the nucleus in terms of orbits of single particles. It will be recalled that spins of nuclei with an odd number of particles are all half-integral multiples of $h/2\pi$, even-mass nuclei have spins and magnetic moments of zero or integral multiples of $h/2\pi$, and even–even nuclei have spins and magnetic moments of zero. Two isotopes with the same odd number of protons, but different even numbers of neutrons, will have the same spins, almost the same magnetic moments, and may even have the same kind of excited states. As an example, ^{113}In and ^{115}In, with 49 protons, both have spins of 9/2, and their magnetic moments are approximately the same; thus the extra pair of neutrons in the heavier isotope does not seem to have any effect. Strontium-87, with 49 neutrons, also has a nuclear spin of 9/2. Such considerations have led to the picture of an odd nucleus as a spherically symmetric core containing an even number of protons and neutrons with no spin, around which revolves the last odd particle, whose motion alone determines the spin and magnetic moment of the nucleus.

The nucleons, analogous to electrons in the atom, can assume only certain orbits, which are characterized by four quantum numbers. These are N, l, j, and m, corresponding, respectively, to the total energy, the orbital angular momentum, the total angular momentum (including spin), and the orientation of the orbit with respect to some arbitrary direction. Table 94.II defines the quantum numbers with respect to the

TABLE 94.II
The Four Quantum Numbers[a]

Quantum number			Value of quantum number	
			Allowable	Typical
Approx. total energy of nucleon in orbit	N		Any positive integer	$+1, 2, 3, \ldots\ldots$
Orbital angular momentum (orbit ellipticity)	l		$N -$ (any positive odd integer N)	$N-1, N-3, \ldots$ 1 or 0
Total angular momentum (l + spin)	j		$l \pm \tfrac{1}{2}$	$N - \tfrac{1}{2}, N - \tfrac{3}{2}$ $N - \tfrac{5}{2}, N - \tfrac{7}{2}$
Spatial orientation of orbit with respect to any arbitrary direction	m		All half-integers from $+j$ to $-j$	$N - \tfrac{1}{2}, -(N - \tfrac{1}{2})$ $N - \tfrac{3}{2}, -(N - \tfrac{3}{2})\ldots$

[a] Reproduced by permission of the editor from *International Science & Technology* (6).

shell model and summarizes the rules for finding the values they can assume.

All the allowed orbits for $N = 1$, 2, and 3, derived from the allowed values of Table 94.II, and the total number of orbits for each j, are shown in Table 94.III. Summation of the number of orbits for each allowed j yields the total number of orbits for each shell; thus, as shown in the last row of Table 94.III, there are totals of 2, 6, and 12 orbits, respectively, for $N = 1$, 2, and 3.

Although the energy of a particle in an allowed orbit depends primarily on the quantum number N, there is also a dependence on l, j, and m, so that all orbits corresponding to a given N do not have exactly the

TABLE 94.III
Allowed Orbits in $N = 1$, 2, and 3 Shells[a]

Quantum number		Allowable values				
N =	1	2		3		
$l = (N - 1 \ldots)$ =	0	1		2		0
$j = \left(1 \pm \dfrac{1}{2}\right)$ =	$\dfrac{1}{2}$	$\dfrac{3}{2}$	$\dfrac{1}{2}$	$\dfrac{5}{2}$	$\dfrac{3}{2}$	$\dfrac{1}{2}$
$m = (+j) \ldots (-j)$ =	$+\dfrac{1}{2}$ $-\dfrac{1}{2}$	$+\dfrac{3}{2}$ $+\dfrac{1}{2}$ $-\dfrac{1}{2}$ $-\dfrac{3}{2}$	$+\dfrac{1}{2}$ $-\dfrac{1}{2}$	$+\dfrac{5}{2}$ $+\dfrac{3}{2}$ $+\dfrac{1}{2}$ $-\dfrac{1}{2}$ $-\dfrac{3}{2}$ $-\dfrac{5}{2}$	$+\dfrac{3}{2}$ $+\dfrac{1}{2}$ $-\dfrac{1}{2}$ $-\dfrac{3}{2}$	$+\dfrac{1}{2}$ $-\dfrac{1}{2}$
Total number of orbits with this j	2	4	2	6	4	2
Total number of orbits with this N	2	6		12		

[a] Reproduced by permission of the editor from *International Science & Technology* (6).

same energy. However, orbits with a given N are generally much closer in energy than orbits with different values of N, and there are larger energy gaps between the different shells. It should be noted, however, that the radii of nucleonic orbits, unlike those of electrons, depend only slightly on the total energy N, and all the orbits have approximately the same radius. For $N > 3$ the energy of the orbit is strongly dependent on j, and in some cases it is more appropriate to group an orbit, not with its own value of N, but with the shell corresponding to the next lower value of N (see Table 94.IV).

Before attempting to draw the shell structure for a nucleus it is well to recall, first, the Pauli exclusion principle, which states that there can be no more than one proton and one neutron in each allowed orbit, and, second, the fact that the nucleons in a normal nucleus, like all particles in nature, will go into the lowest available energy states or orbits. Considering some of the low-atomic-number nuclei, and referring to Table 94.IV, we see that all four nucleons in ^{4}He can be accommodated in the two $N = 1$ orbits corresponding to the lowest available energy

TABLE 94.IV
Number of Orbits with Various Values of N, l, $j^{a,b}$

N	1	2		3			4				5					6						7								
l	0	1		2		0	3		1		4		2		0	5		3		1		6		4		2		0	7	
j	$\frac{1}{2}$	$\frac{3}{2}$	$\frac{1}{2}$	$\frac{5}{2}$	$\frac{3}{2}$	$\frac{1}{2}$	$\frac{7}{2}$	$\frac{5}{2}$	$\frac{3}{2}$	$\frac{1}{2}$	$\frac{9}{2}$	$\frac{7}{2}$	$\frac{5}{2}$	$\frac{3}{2}$	$\frac{1}{2}$	$\frac{11}{2}$	$\frac{9}{2}$	$\frac{7}{2}$	$\frac{5}{2}$	$\frac{3}{2}$	$\frac{1}{2}$	$\frac{13}{2}$	$\frac{11}{2}$	$\frac{9}{2}$	$\frac{7}{2}$	$\frac{5}{2}$	$\frac{3}{2}$	$\frac{1}{2}$	$\frac{15}{2}$	$\frac{13}{2}$
Number of orbits with j	2	4	2	6	4	2	8	6	4	2	10	8	6	4	2	12	10	8	6	4	2	14	12	10	8	6	4	2	16	14
Number of orbits in shell	2	6		12			8	22				32					44						58							
Cumulative number	2	8		20			28	50				82					126						184							
Shell	1st	2nd		3rd			4th	5th				6th					7th						8th							

^{a} Reproduced by permission of the editor from *International Science and Technology* (6).
^{b} Vertical heavy lines show the shell groupings. Note that some orbits (e.g., $N = 5$, $j = 9/2$) are grouped in shells of next lower N.

states, but for ^{5}Li, which has 3 protons and 2 neutrons, one of the protons has to be in an $N = 2$ orbit.

To go to somewhat larger nuclei, ^{16}O with 8 protons and 8 neutrons has all of its $N = 1$ and $N = 2$ orbits filled; thus in ^{17}O the extra neutron has to go into an $N = 3$ orbit. The nuclei ^{4}He and ^{16}O represent "closed-shell" nuclei, and ^{5}Li and ^{17}O represent "single-particle" nuclei; both types have particularly simple properties.

As nucleons are added beyond the closed-shell configuration, which is spherically symmetrical, the nucleus at first becomes more and more distorted; but as the next closed shell is approached, it again becomes more symmetrical. If the first nucleon added beyond a closed shell enters an orbit near the horizontal plane, subsequent nucleons will be attracted by it and will also go into horizontal orbits. The nucleons within the closed shell will also be influenced by the added nucleons, with the result that their orbits will be distorted so as to lie closer to the horizontal plane. Thus the whole nucleus will assume the shape of an ellipsoid of revolution. However as more nucleons are added after the horizontal orbits are filled, they will preferentially enter vertical orbits in the same shell rather than horizontal orbits in the next shell, and the nucleus will approach symmetry. The principal regions of distorted nuclei are the regions of magnesium and aluminum, the heavy rare earths, tungsten and tantalum, uranium and plutonium.

Radioactive decay can be understood in terms of the shell model if we consider that a nuclear transformation will occur whenever it is energetically possible, and that the rate depends upon (1) the characteristic rate of the mechanism, (2) the amount of kinetic energy released, (3) the amounts by which the orbits change, and (4) the probability for penetrating the potential barrier. The third factor depends on the number of nucleons which must change orbits and the difference between the initial and final orbits. Obviously, if many nucleons must change orbits the rate must be slower; however, in the usual case the initial and final nuclei differ only in the orbit of one nucleon. Similarly, if there is a drastic change in the orbit of the nucleon, from very circular to very elliptical or vice versa, corresponding to a large change in l, the rate will decrease.

As an example we will consider a nucleus which has a single nucleon in a high-energy orbit when there is available a lower-energy orbit. The nucleon can drop into the lower-energy orbit, and the excess energy will be emitted as electromagnetic radiation. The characteristic rate for such a gamma decay can be very fast ($t_{1/2}$ of the order of 10^{-14} sec);

but if there is a large change in l, that is, if the orbit must change from very elliptical to very circular or vice versa, and if the kinetic energy release is very small, the life-time for gamma emission may be very long. When the excited state has a measurable half-life, which may be as short as 10^{-11} sec or as long as hundreds of years, it is known as a metastable or isomeric state.

There are, however, two other processes competing for de-excitation of the nucleus following such an orbit transition, namely, the emission of internal conversion electrons and pair production. Internal conversion results from the interaction of the nucleus with the extranuclear electrons and is indicated by the ejection of an electron with a kinetic energy corresponding to the energy of the nuclear transition minus the binding energy of the electron. Since the ejection of an extranuclear electron leaves a vacancy in one of the electron shells, and electronic transition occurs to fill the vacancy, giving rise to characteristic X-rays and/or Auger electrons. If the energy of the nuclear transformation is greater than 1.02 MeV (twice the rest mass of an electron), the nucleus may simultaneously create and emit a positron and an electron with kinetic energies that add up to the excitation energy minus 1.02 MeV.

The beta decay process, in which a neutron can exchange roles with a proton or vice versa, affords another mechanism for lowering the energy of a nucleus. If we consider the nuclide ^{16}N, which has 9 neutrons and 7 protons, then, according to Table 94.III, the ninth neutron is in an $N = 3$ orbit. There is, however, an $N = 2$ orbit available for a proton. Thus there is an exothermic reaction, with the neutron in the $N = 3$ orbit replaced by a proton in the $N = 2$ orbit, an electron and neutrino being emitted to conserve the charge and carry off the excess energy. When there is an excess of neutrons in the nucleus, we always observe the emission of an electron; when the situation is reversed, that is, there is an excess of protons, the beta decay process permits the exchange of a proton for a neutron by the emission of a positron or by electron capture. The latter phenomenon is also accompanied by the appearance of X-rays and Auger electrons. Usually the orbit shape is less drastically changed if the beta decay transition corresponds to some intermediate state rather than to the lowest unfilled orbit, and the nucleus then goes to its lowest energy state by a gamma transition. Thus the emission of one or more gamma rays (since, for the same reasons, the gamma transition may be stepwise) is frequently concomitant with beta decay.

Alpha decay is another mechanism for attaining a lower-energy state and is especially favored energetically because the ^{4}He nucleus is very tightly bound, all of its nucleons being in the low-energy, $N = 1$, orbits.

3. Other Nuclear Models

As we have already seen, it has been necessary to construct various, and sometimes apparently contradictory, models of the nucleus, each of which describes certain properties of the nucleus. Three additional models aid in the understanding of nuclear reactions, namely, the optical model, the compound-nucleus model, and the direct-interaction model.

The optical model treats the nucleus as a cloudy crystal ball and the neutron as a wave rather than a particle. The amplitude of the wave at any point is a measure of the probability of finding the neutron at that point, and the frequency of the wave measures the energy of the particle. In a typical scattering experiment in which a beam of neutrons impinges upon a target, such as a thin sheet of iron, some of the neutrons will pass through, some will be deflected by collison with the nuclei of iron atoms, and some will be captured by iron nuclei. The percentage of neutrons scattered or absorbed divided by the number of iron nuclei in the area covered by the beam yields the cross-section area for scattering plus absorption presented by each iron nucleus. For neutron energies of about 15 MeV, each nucleus presents a cross section corresponding to a disk with radius equal to $1.4 A^{1/3} \times 10^{-13}$ cm (the unit 10^{-13} cm is referred to as the "fermi").

An optical model is immediately suggested if we consider that the scattering and absorption of neutrons corresponds to the dimming of the neutron beam in its passage through the target. For neutrons of 15 MeV the iron nucleus is an opaque ball with a radius of 1.4 fermis, blocking part of the neutron beam and casting a sharp shadow. Such considerations also imply that the cross section should be a function of neutron energy; for example, a beam of wavelength much greater than the nuclear radius should not cast a sharp shadow. This is indeed found to be the case. However, when the energy loss of long-wavelength (slow) neutrons is calculated for passage through an assembly of opaque nuclei, it is found that the energy loss is much less than expected. It appears that nuclei are opaque to short-wavelength neutrons, but become increasingly transparent as the wavelength increases. Thus the nucleus looks like a cloudy crystal ball to slow neutrons.

If we consider a beam of light passing through a conglomerate of semitransparent glass balls, some light is lost by refraction and some by absorption; and optical theory permits the calculation of the amount of light lost when the index of refraction and the absorption coefficient are known. In an analogous manner, nuclear theory permits the prediction of the degree of attenuation of a neutron beam of given energy by a group of nuclei of given size. The refractive index and absorption coefficient for nuclear matter can be experimentally observed. Such calculations yield remarkably good agreement with experiment; in general, the cross section decreases with increasing energy, with the appearance of distinct and sharp peaks in certain cases.

The model presents a paradox, however, since it implies that part of the wave energy passes through the nucleus unaltered, in spite of the known strong interaction among nucleons. If, however, we recall from the shell model that each orbit corresponds to a specific energy level, and recall also the Pauli exclusion principle, we recognize that individual nucleons occupy their respective positions because the other positions are filled. Therefore, the force acting on any single nucleon is a group force, that is, all the other nucleons act as a single body. An incoming neutron may thus fail to interact with an individual nucleon and be subjected only to the group force which deflects it from individual contacts. Consequently the neutron may pass through the nucleus as if it were transparent.

The concept of the nucleus as a cloudy crystal ball is consistent with the potential-well model. The cloudy ball signifies a finite probability that a neutron will be absorbed or will fall into the well, and this probability is the coefficient of absorption. Thus it is only absorption, and not scattering, that determines the cloudiness of the nucleus.

The compound-nucleus model and the direct-interaction model attempt to explain the mechanisms by which an incident particle is absorbed by a nucleus. The former assumes that all of the energy of the incident particle is randomly distributed among all the nucleons in the nucleus, including the incident particle. After a lifetime of 10^{-14} to 10^{-18} sec, which is long compared, for instance, to the time for a neutron to traverse the nucleus (10^{-20} to 10^{-23} sec), there may be sufficient energy concentrated on a single nucleon (or group of nucleons) to permit its (or their) escape ("evaporation").

The direct-interaction model encompasses two types of reactions, namely, the "Oppenheimer-Phillips" and the knock-on reactions. The anomalous observation that (d, p) reactions occur at deuteron energies below the Coulomb barrier for penetration of the nucleus, and the much higher cross sections for (d, p) than for (d, n) reactions, are explained by Oppenheimer and Phillips as the result of only part of the incident particle penetrating the nucleus. The deuteron, consisting of a neutron and a proton, is polarized by the Coulomb field of the nucleus. The proton end being repelled, the neutron can approach the nucleus; and since the neutron–proton distance in the deuteron is large, it can penetrate the barrier while the deuteron is still outside most of the Coulomb barrier. The deuteron binding energy is only 2.23 MeV, and thus the nuclear forces acting on the neutron can fragment the deuteron, leaving the neutron inside the nucleus and the proton outside. The opposite reaction, in which a proton can pick up a neutron and leave the nucleus as a deuteron, has also been observed.

The direct-interaction model has also been proposed to explain the experimental observation that for some medium-energy reactions [e.g., Ni(p, α)Co] a finite number of high-energy particles are emitted in the forward direction. This observation is in contradiction with the compound-nucleus model and indicates that direct interactions can also occur. Further evidence for the direct-interaction model lies in the fact that for certain reactions, such as the (p, pn) and (p, p2n), the cross sections as a function of proton energy reach a characteristic maximum value and stay at a plateau, rather than decreasing as other competing reactions become energetically possible.

IV. PRODUCTION OF RADIONUCLIDES

The tools currently available for the production of radionuclides are the various accelerators, nuclear reactors, and neutron sources. The purpose of an accelerator is to produce charged particles of high energy, and in all cases this is accomplished by creating, in one way or another, a very high potential drop through which the particle is allowed to fall. The nuclear reactor is an assembly of fissionable material, usually uranium, and moderator; the self-sustaining fission reaction releases excess neutrons. The common neutron sources consist of an intimate mixture of beryllium and an alpha emitter such as ^{226}Ra, ^{210}Po, or ^{234}Pu and produce neutrons by the reaction ^{9}Be(α, n)^{12}C.

A. ACCELERATORS

In the 13 years following the discovery of nuclear transmutation, the only tools available to the scientist were the naturally occurring alpha particles. Consequently there was a "drive" to develop techniques for the production of higher-energy alpha particles, as well as other accelerated particles. Indeed, the competition between various groups of investigators took on the aspects of a race. Simultaneously Cockcroft and Walton in England worked on a voltage-multiplying rectifier device,

Lawrence at Berkeley was developing the cyclotron, and Lauritsen at the California Institute of Technology was experimenting with a cascade transformer.

Inasmuch as Cockcroft and Walton won the race in 1931 with a machine that produced 0.8-MeV protons, we shall first consider their voltage-multiplying rectifier design (Fig. 94.7). The Cockcroft-Walton generator is fitted with a number of successive electrodes, each one maintained at a higher negative potential than the one preceding it. These serve (1) to divide the overall voltage across the whole device so as to reduce the chance of "flashover," and (2) being hollow cylinders, to confine and focus the beam. The system involves a network of vacuum-tube rectifiers and condensers so connected that each stage doubles the voltage of the preceding stage. Although this type of device has produced energies up to 3 MeV, its greatest application today is as a preaccelerator for very-high-energy devices and as a neutron generator.

For the purpose of continuity we will look next at the Van de Graaff generator, although, chronologically, Lawrence's cyclotron followed immediately. The Van de Graaff is an electrostatic generator consisting of a motor-driven belt made of insulating material, which bodily transports electrical charges to an insulated electrode at the top of a vertically supported column (Fig. 94.8). The belt (A) runs at high speed over the two pulleys, one at the base of the cylindrical insulating column, and the other inside the hollow spherical electrode at the top. At the bottom of the belt a positive charge is imparted by means of a comblike electrode (B). Actually, electrons pass from belt to comb, leaving a positive charge. At the top of the machine, a similar electrode (E) draws

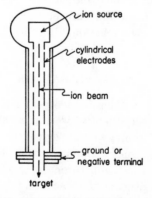

Fig. 94.7. The Cockcroft-Walton generator.

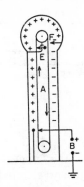

Fig. 94.8.   The Van de Graaff accelerator.

off positive charge, transferring it to the hollow sphere (F) and thus building up a potential on the latter.

Energies as high as 5 MeV have been attained. Most often the machines are filled with gas up to 225 psi, which reduces both the clearances required and the volume of the machine. Their great advantage is that they can deliver charged particles at potentials with any selected value up to their maximum, within $\pm 0.1\%$.

The most recent development in this type of accelerator is the tandem Van de Graaff. In this device, negative ions are produced at ground potential and then accelerated to a high-voltage positive terminal. Within the terminal, the swiftly moving negative ions are stripped of electrons by collision with other gas molecules; they thus become positive ions and receive an additional acceleration from the positive terminal back to ground.

The Cockcroft-Walton and the Van de Graaff achieve high energies essentially by the application of a single high voltage between electrodes. The emergent ions have energy directly equivalent to the applied voltage, and this constitutes the limitation of such machines. When the voltage between electrodes is too great, they will short or spark over.

This difficulty is overcome in a resonant accelerator, the principle of which is to accelerate to a high energy, not in one step, but in a large number of small successive steps. The very first device incorporating this principle as a nuclear device was the cyclotron, but once again we shall defer our discussion and take up first the linear accelerator.

The first system utilizing this principle was devised by R. Wideroe in Germany about 1928. A series of colinear cylindrical electrodes is arranged in an evacuated glass tube as shown in Fig. 94.9. Alternate

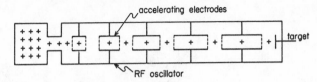

Fig. 94.9. The linear accelerator.

cylinders are connected together; 1, 3, 5, etc., are connected to one terminal and 2, 4, 6, etc., to the other terminal of a high-frequency oscillator. At any instant adjacent electrodes have opposite potentials; and when the oscillation are properly synchronized, a positive ion will be accelerated from one electrode to the next. Once within the electrode, it coasts in this field-free region. The increasing length of electrodes compensates for the increased velocity of the particle; thus it is always in phase.

The use of such machines was restricted to very heavy ion acceleration until the advent of ultrahigh-frequency oscillators during the war (for radar systems). In 1947 Alvarez constructed a linear accelerator in which the drift tubes are enclosed in a metal tube. When excited by an ultra-high-frequency source of power, the structure behaves as a resonant cavity; it oscillates in a mode in which the electric field is longitudinal, pointing first in one direction and then, a half-cycle later, in the opposite direction. The oscillations are so synchronized that the ions are within the electrodes when the field is in the opposite direction.

The cyclotron (Livingston and Lawrence, 1932), which closely followed the Crockcroft-Walton accelerator, consists of a pair of "hollow D electrodes" within a uniform magnetic field (Fig. 94.10). If a particle

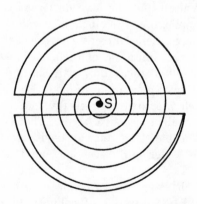

Fig. 94.10. The principle of the cyclotron.

is introduced into this region with some kinetic energy, it will travel in a circle. If, however, the dees are given opposite charges with an electric radio-frequency oscillator, then each time a particle arrives at the gap it will gain energy, provided the voltage is synchronized, by being kicked out of one dee and attracted to the other. As the particle picks up speed, it will describe a larger circle, thus arriving again at the gap in the same time. Thus, by properly synchronizing the electrical field, the total energy acquired can be many times the voltage applied to the dees. The frequency of revolution is

$$f = \frac{eH}{2\pi Mc} \tag{50}$$

Thus, as long as the mass does not change, the frequency is independent of the energy of the particle.

The practical limit of such a cyclotron, however, is ~10 MeV for protons, 20 MeV for deuterons, and 40 MeV for alpha particles. This limitation arises from two factors. First, as the energy increases the mass must increase in accord with the equation

$$E = Mc^2 \tag{51}$$

Second, in order to keep the particles in a median plane and prevent them from straying up or down, the magnetic field is made a little weaker at the outer edges than at the center. Both these factors operate in the same direction, limiting the range of such a cyclotron.

Immediately after World War II, Veksler in Russia and McMillan in the United States suggested that, by increasing the ratio of the field to the frequency during the flight of the particle, the energy could be increased. If we rewrite equation (50), this becomes obvious; thus

$$Mc = \frac{eH}{2\pi f} \tag{52}$$

Multiplying by c, we obtain

$$E = Mc^2 = \frac{eH}{2\pi f}c \tag{53}$$

Such a device is practicable because of the principle of phase stability, which tends to make all particles attain the same final energy; a particle with too much energy will describe an orbit with a larger radius and will arrive at the gap late, thus receiving less energy. A particle with too little energy, conversely, will arrive at the gap early and receive more energy.

Accelerators incorporating this principle are of two forms: (*1*) constant circle, increasing magnetic field; and (*2*) frequency-modulated cyclotron. The limitation of such machines is, once again, the drift of particles out of the median plane, upward or downward; there is also a tendency for a particle to drift outward. These drifts are characterized by a certain wavelength of oscillation, λ. The wavelength of oscillation is given in terms of the radius of the orbit, R, and a magnetic parameter, the gradient index η, which depends on the derivative of the magnetic field:

$$\text{Horizontal: } \lambda = \frac{2\pi R}{\sqrt{1 - \eta}} \tag{54}$$

$$\text{Vertical: } \quad \lambda = \frac{2\pi R}{\sqrt{\eta}} \tag{55}$$

$$\eta = \frac{R \, dH}{H \, dR} \quad (0 < \eta < 1) \tag{56}$$

The problem of drift is solved by alternate gradient focusing; the circumference of the magnet is divided into a number of sectors in half of which the field is such that it is good for vertical focusing and in the other half for horizontal focusing.

B. NUCLEAR REACTORS

The fission of a ^{235}U nucleus produces, on the average, 2.5 neutrons. These neutrons can produce fission in other ^{235}U nuclei, and thus a chain reaction can take place in which an ever-increasing number of neutrons are produced. This process is the basis of the nuclear reactor, in which a chain reaction is allowed to occur under controlled conditions. In order to maintain a chain reaction it is necessary that the number of neutrons lost by (*1*) escape and (*2*) nonfission capture by uranium and other materials be no greater than the number of neutrons produced. The first factor imposes a certain critical size, since, if we assume a regular array of uranium in a matrix, the production of neutrons depends on the volume, whereas the escape of neutrons depends on the surface area. A small size means a large surface-to-volume ratio; the greater the size, the less the probability that a neutron will escape. A further improvement in neutron economy is obtained by surrounding the core of fissile material with a neutron reflector such as graphite and/or beryllium. The loss of neutrons by nonfission capture can be minimized by exploiting the very high ^{235}U fission cross section (580 barns) for thermal neutrons (about 0.03 eV). The spectrum of neutrons from fission ranges in energy from less than 0.05 MeV to more than 17 MeV; but if the uranium is distributed in a lattice of a moderator such as graphite or D_2O, the neutrons will rapidly lose energy by collisions with moderator nuclei and reach thermal energies. At thermal energy, the cross section for nonfission neutron capture is considerably less.

As already mentioned, both graphite and D_2O are suitable moderators for natural uranium. Because of the higher neutron absorption cross section of hydrogen, H_2O can be used only if the uranium is enriched in ^{235}U. In addition to enriched uranium, other fissile materials such as ^{239}Pu and ^{233}U, which is made by the reaction ^{232}Th$(n, \gamma)^{233}$Th and subsequent beta decay to ^{233}U, can also fuel reactors with less efficient moderation. The decreased fission cross section for higher-energy neutrons is compensated for by the increased concentration of the fissile nuclide in the fuel.

A further requirement for a reactor is dissipation of the tremendous heat output. The earliest reactors consisted of an array of natural uranium in a graphite matrix and were air cooled; more recently, similar reactors cooled with carbon dioxide and helium have been operated. There are currently many D_2O-moderated natural uranium reactors in operation, the essential difference among them being the coolant; the D_2O itself, N_2, and H_2O have been used. The swimming pool reactor is a simple, relatively inexpensive, and very useful type of research reactor. It consists of an enriched uranium core suspended in a pool of water which serves as moderator, coolant, reflector, and radiation shield. It is not useful for the production of power but can have thermal neutron fluxes up to 5×10^{13}/cm²-sec. This compares with fluxes greater than 10^{15}/cm²-sec for some of the newer special-purpose reactors.

C. NEUTRON SOURCES

Although several very short-lived nuclides, produced by artificial means or in fission, decay by neutron emission, the only practical neutron sources are various nuclear reactions themselves. The most familiar sources depend on the reaction

$$^9\text{Be}(\alpha, n)^{12}\text{C}$$

and consist of an intimate mixture of an alpha emitter such as ^{226}Ra, ^{210}Po, or ^{239}Pu with beryllium. Such sources, emitting about 10^7 neutrons/sec, are commercially available.

Gamma rays can interact with both beryllium and deuterium to yield neutrons by the reactions

$$^9Be(\gamma, n)\,^8Be \quad and \quad D(\gamma, n)H$$

but these reactions give significantly fewer neutrons.

A much more intense source of neutrons, which is currently available, is based on the reaction

$$^3H(d, n)\,^4He$$

which has a strong resonance at deuteron energies of 100 keV. The most common device of this type consists of a Cockcroft-Walton generator to accelerate the deuterons into a target of tritium adsorbed in titanium. Although such neutron generators suffer from rapid target depletion, they can yield more than 10^{11} neutrons/sec, and targets are both relatively inexpensive and easy to replace.

V. ACTIVATION ANALYSIS

As recently as 1949, G. E. Boyd (31) was able to present a complete survey of all the applications of activation analyses to date, in a rather brief review article containing a total of only 43 references, most of which referred to fundamental nuclear physics and chemistry. The statement was made that "the widespread use of activation analysis appears to hinge in a decisive fashion upon the emergence of machines of relatively modest cost for laboratory production of artificial radioactivity." Needless to say, the ready availability of nuclear reactors throughout the world, and the easy availability of both neutron sources and neutron generators at costs within reason for almost all laboratories, have indeed led to widespread application. A bibliography compiled by the National Bureau of Standards in 1968 (32), with an addendum compiled in 1969 (32), lists 8035 papers on activation analysis.

The basic principle of activation analysis, unchanged from its earliest concepts, consists of placing the sample to be analyzed in a flux of energetic charged particles or neutrons for a sufficient time to produce enough radionuclide product to measure. The rate of production of radioactive atoms is given by

$$\frac{dN^*}{dt} = \varphi\sigma_{act}N - \lambda N^*$$

and integration over the time of irradiation yields the number of radioactive nuclei, N^*, at the end of the irradiation:

$$N^* = \frac{\varphi\sigma_{\text{act}}N}{\lambda}\,(1 - e^{-\lambda t})$$

The flux in number of bombarding particles per square centimeter per second is designated by φ; σ_{act} is the reaction cross section in units of square centimeters per target atom; t is the time of irradiation; and λ is the radioactive decay constant of the product nuclide. The activity, A, in units of distintegrations per second is the product λN^*, and substituting $0.693/t_{1/2}$ for λ yields

$$A = \varphi\sigma_{\text{act}}N\left[1 - \exp\left(-\frac{0.693t}{t_{1/2}}\right)\right]$$

Since there is usually a significant interval between the end of the irradiation and the measurement of the activity, the activity at any time, t', after irradiation is given by

$$A_{t'} = \varphi\sigma_{\text{act}}N\left[1 - \exp\left(-\frac{0.693t}{t_{1/2}}\right)\right]\exp\left(-\frac{0.693t'}{t_{1/2}}\right)$$

In practice, the unknown sample is compared with a known standard exposed under identical conditions (i.e., simultaneously). Thus the sample and comparison standard(s) are subject to the same flux for the same period of time; and since the cross section for any given reaction at a specific particle energy is constant, the relative activities are directly proportional to the respective concentrations of parent nuclide:

$$\frac{A_{\text{unknown}}}{A_{\text{standard}}} = \frac{N_{\text{unknown}}}{N_{\text{standard}}}$$

New developments in reactor design and counting techniques now permit much greater sensitivity and scope for activation analysis. High-flux beam reactors permit irradiations with neutrons of selected and well-defined energies, enabling the analyst to take advantage of resonance neutron capture. The extremely high-resolution lithium-drifted germanium detectors afford a resolution so great that in many cases chemical separation of the desired radionuclide is obviated. Other counting techniques, such as the use of lithium-drifted silicon detectors for counting conversion X-rays and coincidence counting, may also yield great specificity.

REFERENCES

1. Bateman, H., *Proc. Cambridge Phil. Soc.*, **15**, 423 (1910).
2. Bethe, H. A., and P. Morrison, *Elementary Nuclear Theory*, 2nd Ed., Wiley, New York, 1956.
3. Blatt, J. M., and V. F. Weisskopf, *Theoretical Nuclear Physics*, Wiley, New York, 1952.
4. Choppin, G. R., *Experimental Nuclear Chemistry*, Prentice-Hall, Englewood Cliffs, N.J., 1961.
5. Connally, R. E., *Nucleonics*, **17**, No. 12, 98 (1959).
6. Cohen, B. L., *Intern. Sci. Technol.*, November 1963, p. 65.
7. Coryell, C., and N. Sugarman, eds., *Natal. Nucl. Energy Serv.*, IV, Part V, *Radiochemical Studies: The Fission Products*, McGraw-Hill, New York, 1951.
8. Finston, H. L., and E. Yellin, *Anal. Chem.*, **35**, 336 (1963).
9. Friedlander, G., J. W. Kennedy, and J. Miller, *Nuclear and Radiochemistry*, Wiley, New York, 1964.
10. Gatrousis, C., R. Henrich, and C. Crouthamel, "Recent Advances in Counting Techniques," in C. Grouthamel, ed., *Analytical Chemistry (Progr. in Nucl. Energy*, Ser. IX), Pergamon, New York, 1961, pp. 1–80.
11. Gordon, L., M. L. Salutsky, and H. H. Willard, *Precipitation for Homogeneous Solution*, Wiley, New York, 1959.
12. Halliday, D., *Introductory Nuclear Physics*, 2nd Ed., Wiley, New York, 1955.
13. Harvey, B. G., *Introduction to Nuclear Physics and Chemistry*, Prentice-Hall, Englewood Cliffs, N.J., 1962.
14. Jefferson, S., *Radioisotopes, A New Tool for Industry*, Newnes, London, 1957.
15. Kaplan, I., *Nuclear Physics*, 2nd Ed., Addison-Wesley, Cambridge, Mass., 1963.
16. Kramer, A. W., "Nuclear Energy—What It Is and How It Acts" (a continuing series), *Atomics*.
17. Overman, R. T., and H. M. Clark, *Radioisotope Techniques*, McGraw-Hill, New York, 1960.
18. Leveque, P., et al., *Proceedings of The Second United Nations International Conference on the Peaceful Uses of Atomic Energy*, Vol. 19: *The Uses of Isotopes: Industrial Uses*, United Nations, Geneva, 1958, pp. 34–41; Martinelli, P., and H. Rici, *CEA Rept.* 1785, Centre d'Etudes Nucleaires de Saclay, 1960.
19. Martinelli, P., and H. Ricci, *CEA Rep.* 1785, Centre d'Etudes Nucleaires de Saclay, 1960.
20. Mayer, Maria G., *Sci. Am.*, March 1951, pp. 22–26.
21. Mayer, M. G., and J. H. D. Jensen, *Elementary Theory of Nuclear Shell Structure*, Wiley, New York, 1955.
22. Moses, A. J., *Nuclear Techniques in Analytical Chemistry*, Pergamon, New York, 1964.
23. Muller, Doris C., *Anal. Chem.*, **29**, 975 (1957).
24. Muller, Ralph, *Anal. Chem.*, **29**, 969 (1957).
25. Peierls, R. E., *Sci. Am.*, January 1959, pp. 75–82.
26. Rochlin, R. S., and W. W. Schultz, *Radioisotopes for Industry*, Reinhold, New York, 1959.
27. Rutherford, E., J. Chadwick, and C. D. Ellis, *Radiations for Radioactive Substances*, Cambridge University Press, 1930.
28. Taylor, T. I., and W. W. Havens, Jr., "Neutron Spectroscopy and Neutron

Interactions in Chemical Analysis," in W. G. Berl, ed., *Physical Methods in Chemical Analysis,* Vol. III, Academic, New York, 1956, pp. 447–621.
29. Turkevich, A., "Radioisotopes Research, Development and Related Activities," *Quart. Progr. Rept.* No. 11 (June–September 1961), Chicago University Laboratories for Applied Science, TID-14226.
30. Weisskopf, V. F., and E. P. Rosenbaum, *Sci. Am., December* 1955, pp. 75–82.
31. G. E. Boyd, *Anal. Chem.* **21,** 335 (1949).
32. *Natl. Bur. Std. Tech. Note* No. 467, Part 1, issued September 1968; Addendum 1, issued December 1969; U.S. Department of Commerce, Washington, D.C.

Part I
Section D-6

Chapter 95

NUCLEAR RADIATIONS: CHARACTERISTICS AND DETECTION

By H. L. FINSTON, *Department of Chemistry, Brooklyn College of the City University of New York, Brooklyn, New York*

Contents

I. INTRODUCTION

Radioactivity is detected as a consequence of the interactions of radiation with matter, and a first step in understanding the various phenomena involved should be the consideration of wave-particle duality. Anomalies in the wave theory of light, for example, the observation that under certain conditions light exhibits the characteristics of particles, led to the development in the early 1900s of quantum mechanics. A further anomaly was the observation that electrons, heretofore considered as particulate, at times exhibit the characteristics of waves. The recognition of this dualism resulted in the quantum theory, which states that the energy of the wave, the "quantum," behaves in certain respects like a particle, and, conversely, every particle has certain characteristics of a wave, for example, a definite wavelength.

Newton had maintained that light consisted of particles, which he called corpuscles, emitted by the light source. Huygens, on the contrary, held that light consisted of waves, and the experimental evidence of the time supported this theory. Subsequently, however, the experimental evidence required that light be considered particulate (photons), and this was followed by De Broglie's suggestion that matter consisted of waves.

Electromagnetic waves do not have the same physical significance as waves in a fluid; the latter represent the actual physical motion of mass, whereas the former are merely the manifestations of the electric and magnetic fields of which they are constituted. Modern physics recognizes the validity of both the wave and particle aspects of matter and radiation, and can explain all physical phenomena on the basis of one or the other. There are, however, differences between quanta and particles. Particles (alphas, protons, neutrons, electrons) have an existence within the atom, they have a strong tendency to exist after being emitted, and they can travel with relatively slow velocities. Photons, or light quanta, are the product of an energy change within the atom and are created at the instant of the energy transformation. Thus they have no existence within the atom. Furthermore, their existence is evanescent; and since they always travel at the speed of light, their very detection is difficult.

Whenever a photon is emitted or absorbed by matter, a definite quantity of energy is either created or dissipated. This amount of energy, corresponding to the light quantum, is $h\nu$, where h is Planck's constant and ν is the frequency. The momentum is given by the theory of relativity as $h\nu/c$, where c is the velocity of light; and since c/ν is equal to λ, the wavelength of the light,

$$h/\lambda = \text{momentum}$$

which is the De Broglie equation.

The positron, last to be recognized of the radiations emitted by radio-active isotopes, affords a specific example of wave-particle duality. It was first observed in 1932 by C. D. Anderson at the California Institute of Technology in the course of his studies of cosmic rays with a cloud chamber. The tracks Anderson observed could be accounted for only by a particle of electronic mass with unit positive charge. The positron had been previously predicted by Dirac, however, on the basis of mathematical considerations. His relativistic wave equation yielded two sets of solutions, one corresponding to a positive total energy and the other to negative total energy. Thus,

$$E \geq M_0 c^2 \quad \text{and} \quad E \geq -M_0 c^2$$

where M_0 is the rest mass of the electron. Dirac's interpretation was that normally all the negative energy states are filled. Consequently, when an electron is raised from a negative to a positive energy state, the phenomenon should be manifested not only by the appearance of an electron but also by the appearance of a hole in the "infinite sea" of electrons in negative energy states. The hole has the properties of a positively charged particle, otherwise identical with an ordinary electron. The positron, since it is a hole readily filled by an ordinary electron, of which a large number are always available, has a short lifetime. When the hole is filled, the two charges neutralize and annihilate each other, with two gamma quanta of 0.51 MeV each (which corresponds to the rest mass of an electron) emitted at 180° to each other, thus conserving momentum.

II. INTERACTIONS OF NUCLEAR RADIATIONS WITH MATTER

While traversing matter, alpha, beta, and gamma radiations all lose energy by interaction with electrons. Interactions with nuclei are negligible, since the probabilities for such reactions (cross sections) are of the order of "barns," that is, 10^{-24} cm^2, corresponding to the physical cross-sectional area of a nucleus. The quantitative differences in their various modes of interaction are due to the double charge, heavier mass, and almost monoenergetic nature of the alpha particle; the single charge, much smaller mass, and continuous distribution in energy of the emitted electron (from nearly zero energy to the characteristic $E_{\max}$); and the zero electric charge, essentially zero mass, and line spectra of gamma rays.

A. ALPHA PARTICLES

The alpha particles emitted in radioactive disentegrations are easily absorbed in matter and are completely stopped in a few centimeters of air, or by a sheet of paper or a layer of aluminum foil 10/1000 in. thick. Thus it has been possible to determine the ranges of alpha particles, and consequently their relative energies, by varying the distance between an alpha source and a detector, or by varying the gas pressure in a closed vessel containing the source and the detector. The rate of detection of alpha particles as a function of distance between a constant source and a detector remains almost constant up to a certain distance R, the range, and then drops rapidly to zero. A direct and precise determination of energy is afforded by the method of measuring the deflection of a charged particle in a known magnetic field. A charged particle moving in a magnetic field describes an orbit with radius r, as given by

$$r = \frac{MV}{Hq} \tag{1}$$

and transposing gives us

$$V = \frac{q}{M} Hr \tag{2}$$

where H is the magnetic field strength, q the charge of the particle, M its mass, and V its velocity. When H is expressed in gauss, r in centimeters, and q/M in emu per gram, velocity is in centimeters per second. Since the alpha particle is a doubly charged helium nucleus, the charge is given by

$$q = \frac{2 \times 4.8029 \times 10^{-10} \text{ esu}}{2.9979 \times 10^{10} \text{ cm/sec}} = 3.2043 \times 10^{-20} \text{ emu} \tag{3}$$

and its mass is that of the helium atom minus two electron masses, or

$$M = 6.6430 \times 10^{-24} \text{ g} \tag{4}$$

This gives the ratio

$$\frac{q}{M} = 4823.5 \text{ emu/g} \tag{5}$$

and the velocity is given by

$$V(\text{cm/sec}) = 4823.5 Hr \tag{6}$$

Velocities from 1.6×10^9 to about 2.2×10^9 cm/sec can be determined in this way; and since the relativistic mass correction is negligible at these velocities, the kinetic energy is

$$E = \tfrac{1}{2}MV^2 = 2.074 \times 10^{-18}V^2 \text{ MeV} \tag{7}$$

Absolute energies can be calculated from measured ranges by calibrating a range–energy curve obtained by comparison of ranges with energies determined by the magnetic deflection method.

The alpha particle interacts with matter by producing ionization, removing an electron from an atom and leaving an ion pair. The ionization, which can be measured, is related to the energy and is expressed in terms of "specific ionization," that is, the number of ion pairs formed per millimeter of ion path. Although energies, and consequently ranges of different alpha particles, vary significantly, the specific ionization–distance curves are qualitatively the same for any gas and any group of alpha particles, as shown in Fig. 95.1.

The extent of the ionization caused by an individual alpha particle depends on both the number of collisions with molecules in its path and on the way in which the alpha strikes these molecules. Some particles will make more, and others less, than the average number of collisions per unit path; consequently, the exact distance from the source at which the energy is completely dissipated is slightly different for the individual alpha particles in a monoenergetic beam. This effect is called straggling, and because of the uncertainty in the range either an extrapolated or a mean range is used, which is usually expressed in centimeters of air

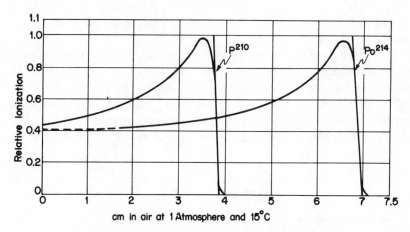

Fig. 95.1. Specific ionization–distance curves for typical alpha particles.

at 15°C and 760 mm Hg. In Fig. 95.1, as the distance from the source increases, the relative specific ionization increases, at first quite slowly and then more rapidly, until it reaches a maximum; thereafter, it drops rapidly to zero. The effect of straggling is noticeable, and the extrapolated range is the point on the abscissa crossed by a tangent to the curve at its inflection point. The derivative curve gives the relative number of particles stopping at a given distance as a function of distance from the source when the unit of the ordinate is chosen so that the area of the differential range curve is unity. Thus all the particles are accounted for. The mean range, R, corresponds to the maximum and is so defined that half of the path lengths exceed this value, and half are shorter.

The average number of electron volts required to produce an ion pair in a given gas can be calculated by completely absorbing an alpha particle of known energy in the gas and measuring the current produced. In air this value corresponds to about 35 eV, which is greater than the energy required only to ionize the atom. The excess energy goes into dissociation of the molecule and excitation of the atoms and molecules. Conversely, the total ionization produced by an alpha particle can be used to calculate its energy.

The early ionization experiments led to the formulation of the empirical Geiger-Nuttall rule, which relates the range of an alpha particle to its energy:

$$\log R = A \log \lambda + B \tag{8}$$

when R is the range, λ is the decay constant, and A and B are constants. A plot of $\log \lambda$ versus $\log R$ for each of the three series of naturally occurring radioelements yields three parallel lines since the value of A is essentialy constant.

A comparison of the characteristics of the natural alpha emitters yields some noteworthy relationships. The energies of the alpha particles emitted by the respective nuclides vary from 4.05 to 8.95 MeV; the half-lives vary from 1.39×10^{10} years to 3×10^{-7} sec; the energies are roughly inversely proportional to the half-lives; and the half-lives vary widely with small changes in energy. These facts are inherent in the Geiger-Nuttall rule.

B. BETA PARTICLES

The distinguishing feature of beta radiation, which sets it apart from both alpha-particle and gamma radiation, is the continuous distribution in energy from nearly zero energy to a maximum E, characteristic of the specific beta transition. This feature, plus the fact that the lower mass of betas permits a greater degree of scatter, results in an exponential absorption for these particles. In general, beta particles, in contrast to alpha particles, have energy less than 4 MeV. However, because of its much smaller mass, a beta with the same kinetic energy will travel much faster than an alpha. At 4 MeV, for example, an alpha has a velocity 0.05 that of light, but a beta will have a velocity 0.995 that of light.

The velocity, or momentum, of beta particles can also be measured by their deflection in a magnetic field. However, since charge-to-mass ratios are much greater for beta than for alpha particles, beta-ray spectrometers require much smaller magnetic fields. The beta particles, because of their much greater velocities, must be treated relativistically. The corresponding formulas for the velocity and kinetic energy are as follows:

$$v = Hr \frac{e}{m_0} \sqrt{1 - \frac{v^2}{c^2}} \tag{9}$$

$$T = m_0 c^2 \left[\frac{1}{\sqrt{1 - (v^2/c^2)}} - 1 \right] \tag{10}$$

where m_0 is the rest mass of the electron, and e is the magnitude of the electric charge. Equation (9) may be rewritten as a function of the ratio v/c:

$$\frac{v}{c} = \frac{Hre}{m_0 c} \sqrt{1 - \frac{v^2}{c^2}} \tag{11}$$

The specific charge of the electron, in electromagnetic units, is

$$e/m_0 = 1.75888 \times 10^7 \text{ emu/g}$$

and the velocity of light is

$$2.99793 \times 10^{10} \text{ cm/sec}$$

Thus

$$Hr \text{ (gauss-cm)} = \frac{m_0 c}{e} \frac{v}{c} \left(1 - \frac{v^2}{c^2} \right)^{-\frac{1}{2}}$$

$$= 1704.5 \frac{v}{c} \left(1 - \frac{v^2}{c^2} \right)^{-\frac{1}{2}} \tag{12}$$

The quantity m_0c^2 is the energy corresponding to the rest mass of the electron, 0.511 MeV. Consequently, the kinetic energy is

$$T = 0.511 \left[\frac{1}{\sqrt{1 - (v^2/c^2)}} - 1 \right] \tag{13}$$

When the beta particles emitted by a radionuclide are studied with a magnetic spectrograph, the deflected beta particles fall on a photographic plate, producing a fogging which extends for a definite distance and then ceases at a characteristic maximum energy. In some cases a superimposed line spectra is observed which corresponds to the emission of secondary electrons. When the photographic plate is replaced by a counter, the instrument becomes a spectrometer, and the number of particles corresponding to a given energy is determined by holding the source and detector in a constant position and varying the magnetic field. A characteristic spectrum is shown in Fig. 95.2. The energies of beta particles, like those of alphas, can be determined by measurement of their absorption. Because of the greater ranges of betas, however, aluminum

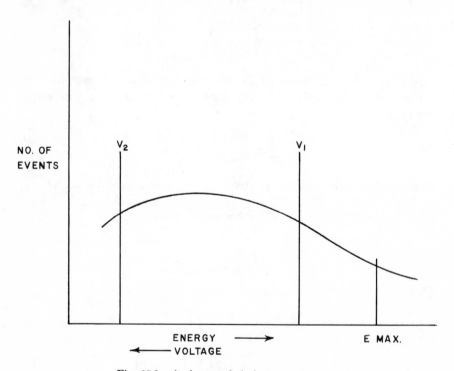

Fig. 95.2. A characteristic beta spectrum.

foils are most often used as the absorbing medium, rather than air. In practice, successively thicker aluminum foils are placed between the beta source and the detector, and the counting rate is plotted on a logarithmic scale versus the thickness of absorber expressed in milligrams per square centimeter. Such an absorption curve decreases nearly linearly over a large fraction of the absorber thickness to a nearly constant value, the background, which may be due to the electromagnetic radiation in the sample itself and in the environment and to cosmic rays. As the absorption curve nears the end, it deviates from its exponential form, and the point where it meets the background is the range. The exponential absorption, which is fortuitous, since it results from the effects of the energy continuum of beta particles and the scatter, can be represented by the formula

$$A(X) = A_0 C \exp - \mu X \qquad (14)$$

where A_0 is the counting rate corresponding to zero absorber, $A(X)$ is the counting rate through absorber thickness X, and μ is the absorption coefficient of the absorber.

The range of a beta particle may be obtained by visual inspection of the absorption curve, or more accurately by one of a number of comparison methods in which the range of beta particles from a particular nuclide is measured in terms of the range of a known standard. These methods do not, however, approach the accuracy of the beta spectrometer for energy determinations.

The interactions of beta particles with matter are considerably more complicated than those of alpha particles. The beta, because of its smaller mass, can lose a large fraction of its energy in a single collision with an atomic electron, resulting in a much greater degree of straggling. Betas are also much more easily scattered by nuclei. The combination of straggling and scattering results in widely different path lengths, even for a monoenergetic electron. Consequently, the range is much less precisely defined for beta particles than for heavier particles.

An additional mechanism for the loss of energy by a beta particle is "bremsstrahlung" (literally, "braking radiation") when it is accelerated by the field of a nucleus. The effect is a continuous spectrum of electromagnetic radiation with a maximum energy equal to the maximum beta energy. Denoting the energy losses by radiation and by ionization as $(dT/dx)_{rad}$ and $(dT/dx)_{ioniz}$, respectively, we can express the ratio of the two losses approximately by

$$\frac{(dT/dx)_{rad}}{(dT/dx)_{ioniz}} = \frac{TZ}{800} \qquad (15)$$

where T is the beta energy in millions of electron volts, and Z is the atomic number of the absorber. The effect is significant for heavy-element absorbers but amounts to only a few per cent for even the most energetic beta emitters with aluminum.

Our understanding of the beta decay process is based on the theory first developed by Enrico Fermi in 1934, in which negatron decay is represented by a neutron in the nucleus being replaced by a proton, with the simultaneous emission of a beta particle and an antineutrino:

$$n \rightarrow p + \beta^- + \bar{\nu} \tag{16}$$

Subsequently, positron decay was represented by a proton in the nucleus being replaced by a neutron, with the simultaneous emission of a positron and a neutrino:

$$p \rightarrow n + \beta^+ + \nu \tag{17}$$

In these reactions, the neutron and the proton are not considered as reacting as free particles, but rather as bound in the nucleus.

In the free state, the neutron mass exceeds the proton mass by 0.782 MeV; thus reaction (16) is not only energetically possible but does indeed occur, with a half-life of 12.8 min. Reaction (17) can occur only within a nucleus; the nucleus as a whole can provide the energy needed.

The Fermi theory of beta emission is analogous to the theory of light emission by atoms since the light is also created at the moment of emission. The well known electromagnetic interaction characterized by the electronic charge is replaced by a new type of interaction, characterized by a new universal constant, the Fermi constant, g, whose magnitude must be determined by experiment. The probability, P, that an electron with momentum between P_e and $P_e \, dp_e$ is emitted per unit time is given by

$$P(P_e) \, dp_e = \frac{4\pi^2}{h} |\psi_e(0)|^2 |\psi_\nu(0)|^2 |\mu_{if}|^2 g^2 \frac{dn}{dE_0} \tag{18}$$

where ψ_e is the electron wave function;

ψ_ν is the neutrino wave function;

$|\psi(0)|^2$ is the probability of finding an electron or a neutrino, respectively, at the nucleus;

μ_{if} represents the matrix element characterizing the transition from the initial to the final nuclear state;

$|\mu_{if}|^2$ is a measure of the amount of overlap between the wave functions of the initial and the final states;

dn/dE_0 the so-called statistical factor, is the density of final states, that is, the number of states of the final system per unit of decay energy with the electron in the specified momentum interval, P_e to $P_e + dp_e$.

Fermi first derived the shape of the beta spectrum as predicted by equation (18), and subsequent integration over all electron momenta from zero to the maximum possible yields the transition probabilities or lifetimes.

Transitions in which both the electron and the neutrino are emitted with zero angular momentum (s-wave electron and neutrino emission) are allowed transitions. The magnitudes of $|\psi_e|^2$ and $|\psi_v|^2$ at the position of the nucleus are certainly much larger for s-wave electrons and neutrinos than for electrons and neutrinos emitted with larger angular momenta. Thus the largest transition probabilities must correspond to s-wave emission. The magnitudes of $|\psi_v(0)|$ and $|\mu_{if}|$ are independent of the division of energy between electron and neutrino; thus the shape of the spectrum is determined entirely by $|\psi_e(0)|$ and dn/dE_0. The term $|\psi_e(0)|$ enters only through the Coulombic interaction between the nucleus and the emitted electron; for a first approximation (valid for low atomic numbers) it can be neglected.

Consider now an infinitesimal interval, dE_0, of the total (electron + neutrino) kinetic energy, E_0. An electron with kinetic energy E_0 has associated with it a neutrino with kinetic energy $E = E_0 - E_e$. For the neutrino of zero rest mass the relation between momentum and kinetic energy is

$$P_v = \frac{E_v}{c} = \frac{E_0 - E_e}{c} \tag{19}$$

Therefore, for a given electron energy, E_e, we have

$$dp_v = \frac{dE_0}{c} \tag{20}$$

The number of neutrino states with neutrino momentum between p and $p + dp$, which is derived from the quantum-mechanical treatment of the particle in a box, is given by

$$\frac{4\pi p_v^2 \, dp_v}{h^3}$$

and is the number of neutrino momentum states associated with a given electron momentum.

The number of electron states in the momentum interval p_e and $p_e + dp_e$ is

$$\frac{4\pi p_e^2 \, dp_e}{h^3}$$

and the number of neutrino states above can be associated with each one of these electron states. Therefore the total number of states of the system in the interval dE_0 with electron momentum p_e and $p_e + dp_e$ is

$$dn = \frac{4\pi p_\nu^2 \, dp_\nu \, 4\pi p_e^2 \, dp_e}{h^3} \tag{21}$$

If we substitute equations (19) and (20) into equation (21),

$$dn = \frac{16\pi^2}{h^6 c^3} \, p_e^2 (E_0 - E_e)^2 \, dp_e \, dE_0 \tag{22}$$

Following customary practice by expressing momentum in units of $m_0 c$ and total energy W (kinetic plus rest energy) in units of $m_0 c^2$, we have

$$p_e/m_0 c = \eta, \quad (E_e/m_0 c^2) + 1 = W, \quad \text{and} \quad (E_0/m_0 c^2) + 1 = W_0$$

Rewriting equation (22) now gives us

$$\frac{dn}{dE_0} = \frac{16\pi^2 m_0^5 c^4}{h^6} \, \eta^2 \, (W_0 - W)^2 \, d\eta \tag{23}$$

From the relativistic equation

$$\eta^2 = W^2 - 1 \quad \text{and} \quad \eta \, d\eta = W \, dW$$

Therefore we can write

$$\frac{dn}{dE_0} = \frac{16\pi^2 m_0^5 c^4}{h^6} \, W(W^2 - 1)^{1/2}(W_0 - W)^2 \, dW \tag{24}$$

Obviously, dn/dE_0 goes to zero at both $W = 1$ and $W = W_0$; thus the characteristic bell-shaped spectra are qualitatively reproduced by the statistical factor.

C. COULOMB CORRECTION

The effect of Coulombic interaction is to decelerate electrons and accelerate positrons. Thus one might expect low-energy particles in a negatron spectrum and fewer in a positron spectrum than so far predicted. This is experimentally observed for ^{64}Cu, which emits both.

D. KURIE PLOTS

Beta-ray spectrometers are employed to analyze beta particles with respect to their momenta; the quantity measured is the number of betas per unit of momentum. For comparison between theory and experiment it is convenient to use equation (24), modified by the Coulomb correction:

$$F(Z, W) = \frac{2\pi X}{1 - \exp(-2\pi X)} \qquad (25)$$

where $X = \pm Ze^2/hv$ (+ for β^-, — for β^+); v = velocity of $\beta\pm$ far from the nucleus; and Z = charge of product nucleus (the atomic number). We then have for the probability of electron momentum between η and $\eta + d0$

$$P(\eta)\ d\eta \propto F(Z, W)\eta^2(W_0 - W)^2\ d\eta \qquad (26)$$

A plot of $[P(\eta)/\eta^2 F(Z, W)]^{1/2}$ versus W should thus yield a straight line for allowed transitions with the intercept on the energy axis at W_0. Extrapolation of Kurie plots (Fermi plots) are the only reliable means for determining beta end-point energies.

E. COMPARATIVE HALF-LIVES

We can now obtain the probability per unit time that an electron in the energy interval W and $W + dW$ is emitted. If we insert the spectrum shape from equation (24) and the Coulomb correction into equation (18), we get

$$P(W)\ dW = \frac{64\pi^4 m_0^5 c^4 g^2\ |\mu_{if}|^2}{h^7} F(Z, W)W(W^2 - 1)^{1/2}(W_0 - W)^2\ dW \qquad (27)$$

Integrating over all values of W from 1 to W_0 yields the total probability per unit time that a particle is emitted, that is, the decay constant, λ:

$$\lambda = \frac{64\pi^4 m_0^5 c^4 g^2}{h^7} |\mu_{if}|^2 \int_1^{W_0} F(Z, W)W(W^2 - 1)^{1/2}(W_0 - W)^2\ dW \qquad (28)$$

It should be remembered that we assumed the matrix element to be independent of energy, which is valid only for allowed transitions. Denoting now the integral as f_0 and combining all our constants, we can write

$$\lambda = \frac{\ln 2}{t_{1/2}} = K|\mu_{if}|^2 f_0$$

The product $f_0 t_{1/2}$, usually denoted $f_0 t$ and loosely called the ft value, is the comparative half-life and should be the same for all transitions with similar matrix elements. The ft value may be thought of as the half-life corrected for the difference in Z and W_0.

Approximate ft values sufficiently accurate for most purposes are given by

$$\log f_{\beta^-} = 4.0 \log E_0 + 0.78 + 0.02Z - 0.005(Z - 1) \log E_0$$

$$\log f_{\beta^+} = 4.0 \log E_0 + 0.79 - 0.007Z - 0.009(Z + 1)(\log E_0/3)^2$$

where Z is the product nuclide, and E_0 is the kinetic energy in millions of electron volts of the upper limit of the spectrum (E_{max}).

F. ELECTRON CAPTURE

This phenomenon corresponds to capture by the nucleus of an electron bound with energy E MeV and the emission of a neutrino of energy E_0 MeV. The calculation of the statistical factor is simpler, since only the neutrino phase space need be considered; no integration over energy is involved. Values of f_{ec} can be approximated by

$$\log f_{ec} = 2.0 \log E_0 - 5.6 + 3.5 \log (Z + 1) \qquad (29)$$

Whenever positron emission is energetically possible, these two processes compete, and since the initial and final states are the same for both modes,

$$\frac{\lambda_{ec}}{\lambda_{\beta^+}} = \frac{f_{ec}}{f_{\beta^+}}$$

G. GAMMA RADIATION

The interaction of electromagnetic radiation with matter is more complicated than the interactions of alphas and betas because of three essentially different mechanisms by which the quantum can lose energy; the dominant mechanism depends more or less on the energy of the quantum. The important mechanism at low energies is photoionization, in which the photon rips an electron off the atom, leaving a positive ion. The excess energy of the photon, that is, the difference between the binding energy of the electron and the energy of the photon, is imparted to the electron as kinetic energy.

Photons of intermediate energy can be scattered by essentially free electrons, that is, electrons with binding energies negligible in comparison to the photon energy; this is the Compton effect. It was observed by Arthur H. Compton that X-rays scattered through an angle underwent an energy loss which depended on the angle. The energy loss is in quantitative agreement with the energy loss derived from the energy–momentum laws. In the following diagram, the electron initially has zero momentum and energy corresponding only to its rest mass, m_0c^2. The photon moving in the x direction has no rest mass, but its moving mass is equivalent to its energy, $M = E/C^2 = h/c\lambda$. The photon has momentum equal to its mass times its velocity, that is, the velocity of light; thus $M_x = (h/c\lambda)c = h/\lambda$. The interaction with the photon also imparts kinetic energy to the electron, giving it a total energy mc^2. Its momentum is mv, and in these equations (Table 95.I) the particle mass is relativistic moving mass.

At energies greater than 1.02 MeV (twice the rest mass of the electron) pair production can occur, that is, a photon disappears and an electron–positron pair is created. The process requires the presence of matter and occurs with greatest probability in the vicinity of a highly charged nucleus, where high electrostatic fields exist. The nucleus with its great mass can absorb almost any amount of momentum and is acted upon by the electrostatic fields to produce an electron–positron pair. Any extra momentum is transferred to the nucleus.

All three processes, photoionization or photoelectric effect, the Compton effect, and pair production are summarized in the following diagram.

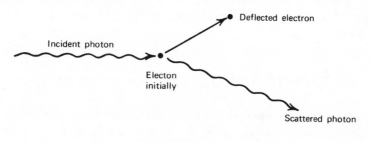

TABLE 95.I

	Incident photon	Scattered photon	Deflected electron
Energy	$E = hc/\lambda$	$E' = h/\lambda$	$E' = mc^2$
Momentum	$M = h/\lambda$	$M' = h/\lambda$	$M = mv$

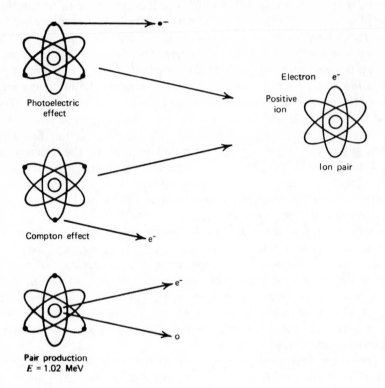

The characteristics of all three modes of interaction are such that 1–2 in. of a high-atomic-number element is required to completely absorb gamma rays. Table 95.II summarizes the characteristics of the different radiations.

III. RADIATION DETECTION AND MEASUREMENT

A discussion of radiation detection might appropriately begin with the photographic method, since the first observation of radioactivity by Becquerel in 1896 resulted from the accidental darkening of a photographic plate.

A. PHOTOGRAPHIC METHODS

Photographic emulsions are still used extensively for measuring the radiation exposure of personnel, for radiography, and for the study of

TABLE 95.II

Characteristics of Different Radiations

Radiation	Type	Charge	Energy range	Path length[a] Air	Path length[a] Solid	Primary mechanism of energy loss	General comments
Alpha	Particle	++	5–8 MeV	cm^s		Ionization excitation	Identical to ionized He nuclei, arises from nucleus
Beta	Particle	—	0–4 MeV	1–10^m	0–1^{mm}	Ionization excitation	Identical to electron, arises from nucleus
Neutron	Particle	None	0–10 MeV	0–100^m	0–cm	Elastic collision with nuclei	Arises from nuclear bombardment
X-ray	Electromagnetic radiation	None	eV–100 keV	–10^m	–cm	Photoelectric effect	Photons arising from atomic electron transitions
Gamma ray	Electromagnetic radiation	None	10 keV–3 MeV	cm–100^m	mm–10^{cm}	Photoelectric effect, Compton effect, pair production	Photons arising from nuclear transitions

[a] Exponential attenuation in the case of electromagnetic radiation.

5443

energetic charged particles. The positron and the meson were first observed by the photographic method. The nuclear emulsions, which are especially made for particle detection, contain a higher concentration of silver bromide grains, much smaller in diameter than those in ordinary film. The grain size can vary from 0.1 to 0.6 μ; and since the sensitivity varies with grain size, different emulsions can be used for particle discrimination. The advantages of nuclear emulsions are inherent in their high density, which results in charged particles being stopped in millimeters, as compared with meters in air; the narrow trails produced in the emulsion, with widths of the order of a micron; and their continuous sensitivity. The emulsions can be arranged in stacks to give a three-dimensional record of the observed events and permit studies of the type performed with cloud and bubble chambers.

B. ION COLLECTION METHODS

The instruments based on the principle of ion collection include the various electroscopes, the ionization chamber, and both Geiger-Müller and proportional counters. Electroscopes, which are now almost of historical importance only, depend on the rate of leakage of electrical charge due to radiation-produced ionization. The gold leaf electroscope, familiar in the elementary physics laboratory, can measure the intensity of a radioactive source as a function of the time it takes to dissipate an induced charge. The Lauritsen electroscope, which operates on the same principle, consists of an air-filled chamber within which is a highly insulated electrode connected to a gold-coated quartz fiber. The image of the fiber is projected onto a graduated scale, which is viewed through the eyepiece of a microscope attachment. The electroscope is charged by a battery, deflecting the fiber, and the time to return across the scale in the presence of ionizing radiation is measured. Beta intensities of the order of 10^3–10^6 particles/min entering the chamber are conveniently measured.

The basic principle of the ionization chamber, the proportional counter, and the Geiger-Müller counter is radiation-induced ionization of a gas and subsequent separation and collection of the ions by an electrostatic field. This principle can be explained by consideration of a chamber consisting of a cathode at ground potential and an insulated anode to which a positive potential can be applied as in Fig. 95.3. If a radioactive source containing alpha, beta, and gamma rays is used, ions produced in the gas will recombine in the absence of any anode potential. As the voltage is increased to tens of volts, alpha particles will produce

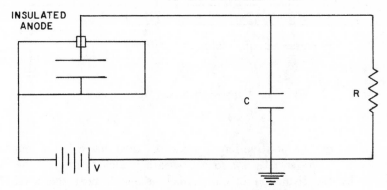

Fig. 95.3. A schematic diagram of an ionization chamber.

sufficient primary ions which can be collected to give a detectable signal with a sufficiently sensitive electrometer. As the voltage is further increased to hundreds of volts, more ions will be collected, until all of the primary ions produced by an alpha particle are detected and the output signal remains constant over a considerable voltage range. This is the region of saturation collection, and in this region the instrument can operate as an alpha detector.

As the voltage is further increased, gas amplification occurs as a result of the acquisition of sufficient energy by the primary electrons formed to cause further ionization.

A voltage difference is applied between anode and cathode through the resistance, R, shunted by the capacitor, C (Fig. 95.3).

In this range of applied voltage the multiplication varies exponentially with voltage, but each electron acts independently, giving rise to its own avalanche of secondary electrons within a small volume adjacent to the anode. This is the voltage region corresponding to the proportional counter, since each electron will produce a number of secondary electrons proportional to its initial energy. The pulse height is not of sufficient magnitude to be detected without considerable external amplification.

Further increase in applied potential above this region results in interaction of the individual avalanches and subsequent discharge along the entire anode. The charge collected is independent of the extent of initial ionization, that is, every ionizing event yields the same collected charge. In this "Geiger" region pulse heights are of the order of volts, requiring minimal additional amplification for detection.

The entire duration of the event, from initiation to discharge, is of the order of a fraction of a microsecond, in which time the electrons

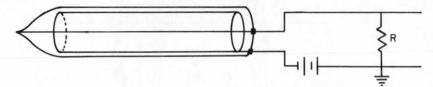

Fig. 95.4. A tube-type Geiger-Müller counter.

are collected. The positive ions, because of their much lower mobility, remain as an essentially stationary sheath, which terminates the discharge by the reduction in electric-field strength near the wire. The positive ions then migrate toward the cathode; when they reach the cathode, additional electrons may be released, giving rise to another discharge. Spurious counts from such an additional discharge can be avoided electronically, but this technique has the great disadvantage of long resolving times. Modern counters, as shown in Figs. 95.4 and 95.5, contain a small percentage of a quenching gas, which may be a halogen or organic gas, in addition to the counting gas. The quenching gas has a lower ionization potential than the counting gas, and upon collision the charge is transferred from the counting gas. At the cathode, the energy released produces dissociation of the quenching gas molecule rather than additional electrons.

When operated near the Geiger-Müller threshold, organic quenched tubes have lifetimes of the order of 10^{10} counts. Depletion of the organic quenching gas is indicated by an increase in the slope and a shortening of the plateau in the counting-rate versus applied-voltage curve. The diatomic halogen gas molecules used as quenchers have the advantage that they recombine after dissociation, thus extending the life of the

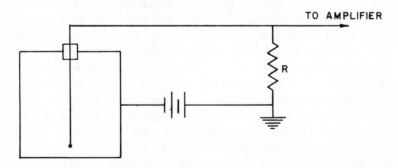

Fig. 95.5. A schematic diagram of an end-window Geiger-Müller counter.

tube and making it possible to operate the tube at higher voltages, yielding output pulses of 10 V or more.

Geiger tubes are subject to dead times, or resolving times, of 100–300 μsec, the duration being variable even for the same tube from pulse to pulse. The dead time is a consequence of the positive-ion sheath which remains after the discharge of an initial ionizing event and decreases, for a time, the electric field in the G-M tube. During this interval additional ionizing events are not recorded, and observed counting rates must be corrected for such coincidence losses.

The resolving time of the proportional counter is much shorter as a result of the fact that the positive-ion sheath is localized, and a second ionizing event can be detected if the discharge occurs at another position on the anode. However, as the voltage is increased, thus raising the gas multiplication factor, the area covered by the positive ion sheath increases, with a resulting rise in dead time. It is often the case in proportional counting that the dead time is determined by the electronics rather than by the counter, and may be less than microseconds.

C. PLATEAU

It is quite obvious that a plateau occurs in the curve of voltage versus counting rate for G-M counting, since in this region of applied voltage all entering ionizing events produce output pulses of the same magnitude. The reason for a plateau in the proportional region is less obvious, but it is easily understood if we refer to a typical bell-shaped beta spectrum and consider the fraction of the emitted particles detected as a function of applied voltage, Fig. 95.2.

As the positive voltage applied to the anode is increased to the beginning of the proportional counting region, V_1, only the most energetic of the particles in the beta spectrum are detected. As the applied voltage is increased further, greater and greater fractions of the total beta spectrum are detected. Additional increase in voltage above V_2 results in very small increments in the counting rate, thus giving rise to a plateau in the curve of counting rate versus applied voltage.

D. SCINTILLATION DETECTORS

Scintillation detectors have their origin in the observation by Sir William Crookes that a screen of zinc sulfide appeared luminous when exposed to the radiations from radium and could be viewed with a low-power microscope. He also observed that the luminosity was discontinuous when viewed in a darkened room with a radioactive source of only moderate intensity. Each scintillation which results from the impact of a single alpha particle lasts for the order of 25 μsec, and consequently its momentary brilliance is very intense.

Modern application of scintillation detectors began in 1947 with the use of a photomultiplier tube coupled to a scintillator. The scintillator can be any of a number of phosphors, depending on the particular application, which can be excited by radiation and subsequently emit a fraction of the absorbed energy as blue visible light. The number of photons produced is proportional to the energy of the absorbed radiation; for example, if a gamma ray produces a 1-MeV electron in the scintillator, and if this electron in turn excites the molecules in the scintillator, using up 10 eV in each excitation process, then 100,000 such excitations are possible. Furthermore, if only 5% of the excited states result in the emission of light, 5000 photons are produced by the interaction of a single gamma ray in the scintillator.

In a typical gamma counter, the scintillator is optically coupled to the photosensitive face of a photomultiplier tube surrounded by reflector. The light incident on the photocathode yields of the order of 1 photoelectron per 10 photons, and the electrons are accelerated by a potential of about 100 V to the first electrode (dynode), where each produces several secondary electrons. A photomultiplier tube with 10–14 dynodes can multiply the charge of the original photoelectron in the order of 10^5–10^6. The burst of electrons arriving at the anode flows through a resistor, giving a voltage drop which is coupled to the counter through a blocking capacitor; this causes a negative output pulse which will have an amplitude of a few millivolts to a few volts.

The most familiar scintillation detector used for gamma-ray detection and energy determination is the thallium-activated sodium iodide crystal NaI(Tl). Sodium iodide has a high density and a moderately high effective atomic number, and is highly transparent to its own fluorescent light. Pure NaI is, however, a low-efficiency emitter at room temperature, and the addition of approximately 0.1% thallium results in high efficiency of emission at room temperature and at a wavelength (4300A°) compatible with many types of photomultiplier tubes.

Organic scintillators, available in liquid, gel, and solid form, are widely used for beta counting, and it has become common practice to detect low-energy beta emitters by internal counting. The beta emitter is incorporated into the organic scintillator; this affords the advantage of eliminating any losses due to self-absorption, to absorption in a counter window or air, and to scattering. Another advantage is that the detector is a 4π solid-angle device.

A typical liquid scintillator is p-terphenyl or diphenyl oxazole solute dissolved in benzene, toluene, or xylene. Alcohol or dioxane can be added to the liquid scintillator to increase the solubility of an aqueous solution, thus making it possible to count a small amount of aqueous solution by incorporating it directly into the organic scintillator. When the wavelength of the light emitted by the phospher is not compatible with the sensitivity of the photomultiplier tube, it is often possible to introduce a "shifter molecule," which absorbs the original phosphorence and re-emits at an appropriate wavelength. The process may thus consist of (1) radiation excitation of solvent materials, (2) transfer of energy to the scintillator molecules (concentration approximately 0.5%), (3) energy transfer to the shifter (concentration approximately 0.001%), and (4) light emission.

E. SOLID-STATE DETECTORS

When charged particles traverse any matter, energy is transferred to the atomic electrons of the absorber. Most of the electrons which are ejected are released with an energy less than the ionization potential; some few, however, may be ejected with higher energies, up to the maximum:

$$W_{max} \cong \frac{4mE_p}{M}$$

where M is the mass of the incident particle with energy E_p, and m is the mass of the electron. The equation derives from classical kinetics, assuming $M \gg m$, as is the case for an alpha particle, for example. The probability for formation of these energetic electrons, delta rays, is small, and the energy dissipated in the absorber is divided between excitation of atoms without ionization and emission of electrons with some average energy.

The energy absorbed per electron released is nearly the same, about 30 eV, regardless of the particle type or energy in a given counting gas, and differs in different counting gases by only a factor of less than 2. For solid-state detectors, however, the energy absorbed is an order of magnitude less: 3.5 eV for silicon and 2.8 eV for germanium. The advantage of the solid-state detector is inherent in the lower energy absorption per electron produced. The energy of the incident radiation is determined by the ionization produced in the detector, and the accuracy of the measurement by the fluctuation in the number of ionization events produced. Thus the greater number of events produced in the solid-state detector affords a means for high-resolution energy determination; resolution of 3 keV (FWHM) can be achieved with solid-state detectors.

The lower energy for the excitation of an electron in a solid arises from the fact that the electrons exist in bands of many close levels rather than in levels characteristic of the individual atoms, as in a gas. The positions of the bands are a property of the crystal, and in the ionization process electrons are raised from one band to unoccupied levels in another band. In a collection of individual atoms such as a crystal, the individual atomic energy levels are changed to an extent determined by the proximity of the atoms and the periodicity of the electric field due to the nuclei. At absolute zero, the available electrons fill the lowest available energy levels, with one or more energy bands completely filled (valence bands) and separated from the next highest band (conductance band) by an energy interval, E_g, in which there are no allowed levels. Therefore, at absolute zero, there can be no conduction because the conduction band is empty; and since the total momentum in a filled valence band is always zero, there can be no change in the momentum of an electron without an equal and opposite change in another electron. Thermal excitation occurs at higher temperatures, with electrons being raised from the valence band to the conduction band. Under equilibrium conditions, the probability of an electron being at an energy level E is:

$$f(E) = \frac{1}{1 + \exp\ (E - \zeta)/kT} \tag{30}$$

where k is Boltzmann's constant, T is absolute temperature, and ζ is the Fermi energy, that is, the energy where the occupation probability of an allowed state is $\frac{1}{2}$. For a pure semiconductor is in the center of the energy gap $(E_g - \zeta = E_g/2$; and since it is usually the case that $E_g \gg kT$, the Fermi function becomes $\exp(-E_g/2kT)$. The number of electrons in the conduction band, n_i, and concomitantly the number missing from the valence band, p_i, is given by the product of the probability of an electron being at a level E and the number of allowed energy states in the energy interval:

$$n_i = N(E) \exp\ (-E_g/2kT) = p_i \tag{31}$$

Under normal conditions both the conduction band and the valence band are partially filled, and upon application of an electric field energy and momentum are imparted to the electrons in each band. There are usually relatively few electrons in the conduction band, and they may be considered as acting independently of each other. Conversely, the valence band is almost full, and the motion of electrons depends on the available vacant levels into which the electrons can be accelerated by the applied electric field. Under these conditions the electron has a quantum-mechanical effective mass, m^*, relating its energy and momentum, which is given by the relation

$$\frac{1}{m^*} = \frac{\partial^2 E}{\partial k^2} \tag{32}$$

The effective mass of an electron near the top of the valence band is negative; and rather than consider the collective motion of electrons of negative effective mass in an almost full valence band, it is conventional to consider the vacant states as having positive charge and mass. The valence band current can then be calculated by considering the motion of the vacant states as independent entities, holes. Applying a field, E, to a semiconductor then gives a total current density

$$J = Ee(n\mu_n + p\mu_p) \tag{33}$$

where μ_n and μ_p are the mobilities of the electrons in the conduction band and the holes in the valence band, respectively; n and p are their respective numbers; and e is the charge of the electron.

In a pure, or intrinsic, semiconductor, the number of holes and electrons are equal, as seen by equation (31). However, the introduction of impurities or distortions in the lattice structure can create localized energy levels in the forbidden energy gap. Ionization may be induced at these sites, with an electron going to the conduction band (donors) or being accepted from the valence band (acceptors) with a required energy for these processes less than that corresponding to the energy gap. When either of these processes occurs, only one free charge is created in one of the bands and there is no longer an equality between the numbers of holes and electrons. Such a semiconductor is called extrinsic, and the number of free charges at any given temperature is a function of the number of impurities and their activation energy. However, it has been shown for extrinsic semiconductors that $np = n_i^2$, where n_i is the corresponding intrinsic carrier density.

Localized distortions, which can be due to impurity atoms introduced substitutionally or interstitially into the lattice or to distortions in the lattice, can also have a detrimental effect. They can act as traps for holes or electrons and as sites where holes and electrons can recombine and be annihilated.

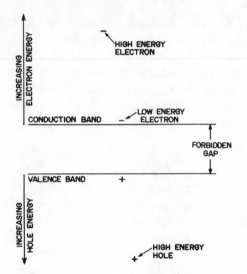

Fig. 95.6.   The band structure diagram.

The following series of diagrams represents the physical processes occurring in a semiconductor detector. The band structure diagram, Fig. 95.6, shows the electrons which take part in the valence bonding process in the lattice to be in the valence band; electrons which have absorbed sufficient energy to break the bond are free to move and are said to be in the conduction band; the in-between region, the forbidden gap or band gap, represents by its height the amount of work necessary to move an electron from the valence band into the conduction band. The more energetic electrons are shown further up from the conduction band edge, and the more energetic holes further down from the valance band edge.

If we now consider a solid-state detector consisting of a crystal with two conducting electrodes and an electric field across it, as in Fig. 95.7, an ionizing particle traversing the lattice breaks bonds and leaves a trail of electron–hole pairs. These carriers are swept out by the electric field and collected, giving rise to a measurable quantity of charge which is proportional, with high accuracy, to the energy of the incident radiation.

In order to achieve such a detector, the crystal has to meet certain stringent requirements. It must have the property that not much energy is required to create a hole–electron pair and that both the hole and the electron are free to move all the way to the conducting electrodes. The motion of the carriers, that is, the holes and electrons, depends on their respective mobilities, which are generally different, and the velocity of a carrier is the mobility multiplied by the electric field. The lattice is constantly vibrating; hence the motion of the carriers is essentially a drift superimposed on a random motion due to collisions with the lattice. The lifetime of a carrier, which is also generally different for holes and electrons, is of the order of 10^{-4} to 10^{-9} sec before it is trapped. The distance the carrier can travel is the product of its velocity and its lifetime; thus a good solid-state detector must also have a good lifetime–mobility product. In addition, the material must have the properties of an insulator, that is, an absence of free carriers, in order that a significant current over the background due to incident radiation can be detected.

Two materials currently available with good lifetime–mobility products are silicon and germanium, neither of which is ideal because of insufficient resistivity. However, it is possible to improve the insulating properties by the expedient of introducing more electrons into the conduction band and, since np is constant, decreasing the hole population. If, for example, phosphorus with five valence electrons is introduced

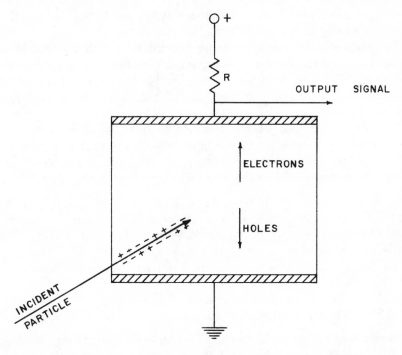

Fig. 95.7. Production of electron-hole pairs in a solid-state detector.

substitutionally into a silicon lattice, the extra electron will be only weakly bound to the phosphorus atom. The lattice is always vibrating at room temperature, and the corresponding phonons have energies of some tens of millivolts with occasional phonons of energies greater than a few electron volts, sufficient to break a covalent bond. When one of these phonons is incident on the site of the phosphorus atom, the additional electron is always raised to the conduction band. Introducing such a donor atom into the silicon lattice, a process called doping, yields an n-type silicon, so-called because it contains a large number of negative charge carriers, electrons, and very few holes. Similarly, a p-type silicon is obtained by doping with boron, which has one less valence electron than silicon and results in a material that has very few electrons and conducts primarily by the motion of holes.

An effective detector is made by creating a p-n junction in one piece of silicon, for example, when opposite faces of a silicon crystal are doped with phosphorus and boron, respectively. Applying an electric field to such a doped crystal will result in the migration of electrons and holes to the respective opposite poles, leaving between them a region depleted in carriers. An ionizing particle incident on this depletion layer creates a trail of hole–electron pairs which are swept out by the electric field and collected by the external circuit.

In practice, a detector is made by forming a very thin n-type layer, about 0.1 μ, on the surface of a p-type silicon. Imposing a reverse bias makes it possible to maintain the required electric field without a large leakage current, and this in turn creates a depletion layer very close to the surface.

The incident particle penetrates the n-type layer, depositing only insignificant energy because the n-type layer is so thin and deposits the major fraction of its energy in the depletion region as shown in Fig. 95.8.

The characteristics of silicon detectors can be summarized as follows:

1. Since only 3.5 eV is required to form an electron–hole pair, the detector yields good statistics and, concommitantly, high resolution.

2. The energy required to produce an electron–hole pair is independent of the type of incident charged particle.

3. Discrimination between particles of different ranges can be achieved by changing bias voltage.

4. It is possible to achieve a negligible dead layer.

5. The detector is insensitive to gamma rays and to neutrons.

Because of the much higher atomic number of germanium, lithium-drifted germanium detectors made by analogous techniques are effective

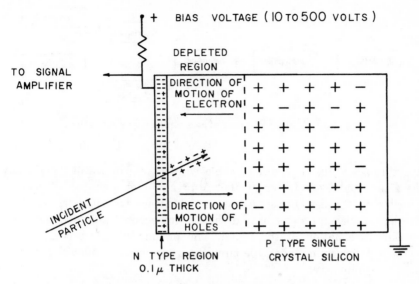

Fig. 95.8. Schematic diagram of a solid-state detector.

for the detection and characterization of gamma rays. It is necessary to maintain them at temperatures below 90°K in order to reduce current noise, but the resolution attainable, 3 keV (FWHM), has led to their widespread application.

IV. INSTRUMENTATION FOR RADIATION DETECTION

The small output signals from the various types of detectors, ranging from 10^{-15} to 10^{-10} coulomb, require amplification and, often, pulse shaping and discrimination to screen out noise and to sort pulses according to their amplitude, before they can be recorded. A schematic diagram for a complete counting system with provision for either integral counting or differential pulse-height analysis is shown in Fig. 95.9.

A. PREAMPLIFIER

In cases where the detector can be located close to the amplifier, the preamplifier can be eliminated. Usually, however, the detector and amplifier are connected by a long length of low-impedance cable. The capacitance of the detector may be about 10–25 pF, and the signal travel-

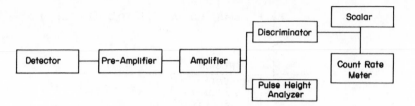

Fig. 95.9. A block diagram of a complete counting system.

ing through a 100-ft length of cable with a capacitance of 25 pF may be attenuated by a factor of 100. A preamplifier mounted as near as possible to the detector serves as an impedence transformer, matching the high-impedence detector and the low-impedence cable and thus preventing attenuation of the signal. The preamplifier may consist of a cathode follower for tube-type circuitry or an emitter follower for transistorized circuitry.

B. AMPLIFIER

As we saw in Table 95.II, the output pulse from the various detectors may range from a low of microvolts from an ionization chamber to 10 V from a Geiger-Müller counter. Consequently, before the detected events can be registered, further amplification, usually over a range of factors from 10 to 1000, is necessary. The optimum characteristics for such an amplifier are (a) stability, that is, constant gain; (b) low noise; (c) good overload characteristics, particularly when energy discrimination is desired; and (d) linearity, that is, the output amplitude is accurately proportional to the input amplitude.

C. PULSE SHAPING

The shape of the pulse delivered at the detector terminals is deliberately changed in the various stages of a counting system in order (1) to prevent overlap of the individual events in the detector and (2) to improve the signal-to-noise ratio. Pulse shaping provides a modified signal of individual pulses the result of which is a signal whose amplitude corresponds to the energy of the radiation and which can be accurately interpreted by a pulse-height analyzer.

There is an inherent conflict at high counting rates, that is, at rates greater than 100 counts/sec, in the choice of the pulse shaping method for optimizing the signal-to-noise ratio and minimizing overlap. Conse-

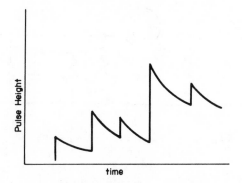

Fig. 95.10. Preamplifier output signals.

quently it is a great advantage to design experiments to avoid, in so far as possible, sacrifice of one for the other.

The output signal from the commonly used preamplifier is a step function with a rise time determined by the circuitry. This rise time may be about 0.1 μsec. The decay of the signal is exponential and may have a time constant of 50 μsec. Even at ordinary counting rates, pulses will be superimposed on the tail of preceding pulses; and since radioactive decay itself is random and, also, the amplitude of pulses will vary, the preamplifier output is irregular, as shown in Fig. 95.10.

When the energy is high enough and the counting rate is also high, pileup will occur, and the output signal from the preamplifier will deviate from linearity. This is usually ascertained by superimposing the known energy output signal from a pulse generator and verifying its position on an energy scale, for example, with a pulse-height analyzer.

The information of value in the preamplifier output is contained in the rise portion, only, of each pulse. The pulse-shaping circuits, usually incorporated in the main amplifier, have time constants much shorter than those corresponding to the decay of the preamplifier signal. Consequently the flow component of the preamplifier signal, the decay tail, is eliminated, and the signals produced are individual pulses whose amplitudes correspond to the energy of the radiation. Optimum time constants for the different detectors vary from 2–10 μsec for gas counters to 0.25-2 μsec for semiconductor detectors.

D. DISCRIMINATOR

The discriminator circuit screens out pulses with amplitude less than a given value and thus may serve to eliminate electronic noise. An exam-

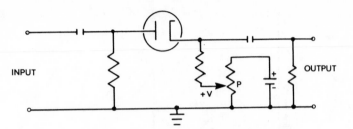

Fig. 95.11. Biased diode discriminator.

ple of a simple type of discriminator is the biased diode shown in Fig. 95.11.

Under normal conditions current will flow from the anode to the cathode. When the cathode is back biased, however, that is, when a positive voltage, V, is applied to the cathode, only pulses of magnitude greater than V will give a current. The bias voltage, V, can be varied by means of the potentiometer, P, thus excluding unwanted signals of lower energy.

E. SCALARS

The various radiochemical techniques of interest to the analytical chemist may involve measurements of counting rates that vary from less than 1 count per minute to greater than thousands of counts per second. Pulses from the detector, after suitable amplification and shape modification, can be recorded on a scalar and the count rate determined with an appropriate timer.

A major contribution to the development of scalars was the "flip-flop" circuit, which employs electronic tubes in either an "on" or an "off" position. This is a binary circuit with the first pulse transferring the tube from the "off" state to the "on" state, and the second pulse restoring the tube to the "off" state and also generating an output signal. Thus one signal is generated for every two input pulses; with six such stages a scale of 64 is achieved. With each stage connected to an indicator lamp, the first input pulse lights the lamp corresponding to the first stage. The second input pulse extinguishes this lamp, resetting the tube to the "off" state, but lighting the lamp corresponding to the second stage since the second pulse has also generated an output signal from the first stage into the second stage. The process is repeated until 63 events are recorded, at which time all indicator lamps are lit. The sixty-fourth input pulse causes all the indicator lamps to be extinguished and delivers an output pulse from the sixth stage to a mechanical register.

Such scalars with scaling factors of some power of 2 (64, 128, 256, etc.) are relatively inconvenient to use and have in recent years been supplanted by various decade scalars. One of the earlier decade scalars consists of four binaries and a "feedback" circuit to convert from a scale of 16 to a scale of 10. When the last binary indicates the eighth input pulse, the circuit feeds a signal into the second and third binaries, adding in effect 6 counts, indicating 2 in the second binary and 4 in the third binary. The ninth input pulse is added normally, and upon the tenth input pulse an output signal is generated. Thus the normal scale of 16 has been converted to give an output signal for every 10 input pulses.

The newer decade scalars may use glow transfer tubes and/or beam switching tubes, both of which are true decade scaling devices having 10 stable positions. The glow tubes have the advantage of being lower in price but are significantly slower than either the binary circuits or the beam switching tubes. Some scalars incorporate a binary circuit for the initial decade, following this with glow transfer tubes for higher scaling factors. Others incorporate the fast beam switching tube as the initial high-speed scaling stage and follow it with glow transfer tubes.

F. COUNT-RATE METERS

It is often convenient to measure the counting rate directly and continuously as, for example, when monitoring radiation levels. This is accomplished by feeding an input pulse of constant amplitude into a storage capacitor which is shunted by a resistor. Each input pulse delivers a fixed amount of charge to the storage capacitor. When the charge being fed to the capacitor equals the charge leaking off through the resistor, an equilibrium voltage is maintained across the capacitor. This voltage is measured by a linear or logarithm vacuum tube voltmeter, and its magnitude is displayed on a panel meter. Frequently, an audio circuit and speaker are included to indicate the radiation level by audible signal.

G. PULSE-HEIGHT ANALYSIS

All the detectors studied, with the exception of the Geiger-Müller counter, produce a pulse whose height is proportional to the energy of the incident radiation. Thus the possibility for sorting pulses according

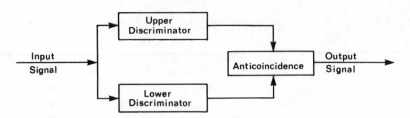

Fig. 95.12. Diagrammatic representation of the single-channel analyzer.

to their height, and determining energy spectra, is implicit in the characteristics of these detectors. Pulse-height analysis was first accomplished by means of the single-channel analyzer, which is represented by two discriminator circuits and an anticoincidence circuit, as shown in Fig. 95.12.

The lower discriminator is biased to pass only pulses of magnitude greater than V, and the upper discriminator is biased to pass pulses of magnitude greater than $V + \Delta V$. The anticoincident circuit will, however, permit an output signal when only the lower discriminator is triggered. Both the base line, V, and the window width, ΔV, can be varied manually, and the latter can be fixed to correspond to a constant energy interval. The energy spectrum is obtained by determining the counting rate as a function of the base line setting. The single-channel analyzer suffers from two serious defects: (1) the acquisition of data is very time consuming, making it difficult to determine energy spectra of short-lived nuclides; and (2) instrument drifts can affect succeeding counting intervals in a very different way.

Early attempts to resolve these difficulties led to construction of multichannel analyzers consisting of a stacked array of single-channel analyzers whose bias levels are progressively greater. This became very expensive as the number of channels was increased, and it also proved difficult to maintain constant and stable window widths. Contemporary multichannel analyzers with as many as 4000 channels are based on techniques developed for digital computers, which also permit large storage capacities per channel. The method depends upon "analog-to-digital conversion," in which an input pulse to the analyzer charges a condenser to a voltage proportional to the input pulse height. The condenser discharges at constant current; thus the time to discharge back to zero voltage, which is measured digitally, is also proportional to the input pulse height. The output from the analog-to-digital converter (ADC) counted by the address scalar determines the channel number in which the pulse is to be stored in a ferrite core memory unit. The memory cycle consists of selection of channel number by the address scalar, recording the number of counts already stored in that channel in a scalar, adding 1 to the number, and recording the new number back into the memory. The net result is considerable dead time during which the analyzer is not capable of recording any pulses. However, the spectrum is not distorted under these conditions; and by incorporating an electronic clock which is stopped while the analyzer is processing a count, it is possible to count for a known amount of "live time."

V. ANALYTICAL APPLICATIONS OF RADIOCHEMICAL AND NUCLEAR TECHNIQUES

The very nature of nuclear radiation leads to the general classifications of (*1*) methods dependent on tracer characteristics and (*2*) the use of isotopes as a source of ionizing radiation.

Tracer methods afford extremely sensitive techniques for the investigation of a large number of phenomena in any physical or chemical state. The fact that a given nuclide may be radioactive does not in any way affect its chemical properties, and because the tracer atoms are detected by virtue of their radioactive decay, they behave normally up to the moment of detection; thereafter, their fate is no longer of any consequence. The sensitivity of tracer techniques is inherent in the fact that as little as 10^{-16} to 10^{-19} g of a radionuclide (with a half-life ranging from hours to years) may be present in the sample.

Radioactive tracers have proved extremely useful in a great variety of migration problems, particularly in studies of self-diffusion. The erosion and corrosion of a surface can be measured with great sensitivity if the surface to be tested can be made intensely radioactive. Radioactive gases or vapors can be detected in small concentrations, and the leakage, flow, and diffusion rates of gases can therefore be studied by the tracer method.

Many exchange systems have been studied by radioactive techniques which have yielded valuable information to the analytical chemist with regard to exchange between different oxidation states and between complexed and uncomplexed species.

In the field of reaction kinetics and mechanisms, tracer techniques yield information that can be obtained in no other way. For example, let us consider the reversible reaction

$$HAsO_2 + I_3^- + 2H_2O \rightleftharpoons H_3AsO_4 + 3I^- + 2H^+$$

The equlibrium constant, K, equals k_f/k_r, where k_f and k_r are the specific rate constants of the forward and reverse reactions, respectively. Application of more conventional techniques makes it possible to measure K at equilibrium and k_f or k_r far from equilibrium. This system has been studied with radioactive arsenic, and the rate law and the specific rate constant at equilibrium have been determined.

More directly of interest to the analytical chemist is the application of radioactive tracers to follow the progress and test the completeness of chemical separation procedures. The solubilities of quite insoluble precipitates are conveniently determined by tracer methods; the approach to equilibrium between solid and solution phases can be followed by simply counting the activity of the supernatant solution. A somewhat analogous method is applicable to the determination of small vapor pressures. Solvent extraction procedures in particular lend themselves to tracer methods for the determination of distribution coefficients and interferences.

Analytical procedures by tracer methods for species which themselves are not radioactive have been introduced and given the name radiometric analysis. An example of such a procedure is the determination of silver by precipitation of AgI with the radioactive isotope ^{131}I ($t_{1/2} =$ 8 days).

Activation analysis, which depends on nuclear reactions as well as the tracer characteristics of the radionuclide produced, can also be extended to take advantage of the prompt reaction by detection of the characteristic gamma rays or charged particles from the various reactions possible, for example, η, γ; n, α; and n, p.

Isotope dilution, which depends on determination of the change in specific activity after equilibration of the radioactive-tagged species with the identical nontagged species, has found widespread application in analysis. A modification of this technique, substochiometric isotope dilution, affords a great improvement in convenience when the exchange between the radioactive species and the nonactive species is complete and sufficiently fast.

The use of isotopes as a source of ionizing radiation has perhaps its most obvious application in radiation thickness gages, which are now familiar in industrial practice. They have their basis in the fact that nuclear radiations are attenuated to a predictable amount by a certain mass of material. It is perhaps worthy of note that current estimates of industrial applications indicate a bare 10% saturation of the market.

Beta gages use ^{90}(Sr-Y), a strong beta-emitting material, for many applications, and ^{14}C for gaging exceptionally thin materials.

The absorption of beta radiation by matter has long been applied to determine thickness. A more recent application is the Cenco beta-ray C/H meter, which takes advantage of the fact that the ratio of atomic electrons to atomic weight is approximately twice as great for hydrogen as for the other light elements.

Gamma gages are useful for measuring thicknesses of more dense material; an example is the pipewall gage. This instrument takes advantage of the fact that a gamma source at a point of contact and a detector in a tangent plane determine a chord which contains a constant amount of material for a given wall thickness, irrespective of the radius of curvature of the pipe.

The principle of gamma attenuation has been used to determine the concentration of uranium in solution with an apparatus consisting of an ^{241}Am source (60-keV gamma ray) and a $\frac{1}{2}$-mm-thick NaI(Tl) scintillator to minimize contribution from higher-energy gamma rays. A fission product concentration of less than 200 μC/liter for uranium concentrations of 50–100 g/liter yields an error of only about 1%.

Neutron attenuation has also been applied for the analysis of materials with high capture cross sections. Elements such as boron and cadmium, whose isotopes with extremely high capture cross sections do not yield reaction products with useful radiation characteristics, can be determined by this principle.

X-ray flourescence by means of beta-ray excitation also affords a technique which should be capable of extensive application. Standards can be made by placing a target of a single element over a pure beta emitter, such as ^{147}Pm or ^{90}Sr. Characteristic X-rays of the target element are generated, usually the $K\alpha$ radiation, although L X-rays are predominant for very thin targets. For energy calibrations greater than 17–20 keV it is best to use thin metallic foil targets or pressed oxide disks. Standards for energies less than 17 keV are best prepared by mixing a solution of the beta emitter with a few milligrams of a salt of the target element, and evaporating to dryness on a micro cover glass. The lowest-energy $K\alpha$ X-ray clearly observed by this technique was that of vanadium at 4.96 KeV, using a $1\frac{1}{2} \times \frac{1}{2}$ in. thallium-activated sodium iodide crystal equipped with a 0.005-in. beryllium window. The use of these standards affords a simple and convenient technique for checking nonlinear response in the low-energy region for scintillation detector systems, measuring the noise level of multiplier phototubes, and identifying radioactive elements which emit converted gamma rays.

The thickness of metallic deposits has been measured by beta excitation of the characteristic X-ray spectrum and determination of intensities. Under optimum conditions, it is possible to take advantage of the fact that the characteristic X-rays of the deposit increase with thickness and simultaneously the characteristic X-rays of the backing decrease. The source, ^{90}(Sr-Y), is placed in an aluminum block, a magnetic field is imposed to deflect scattered β^- radiation, and the X-rays from the samples are determined with a proportional counter (beryllium window).

Scattering phenomena, a characteristic of both neutrons and beta rays, have also proved a useful means of analysis. Neutron scattering is by now well known in oil-well logging and the determination of moisture in a variety of substances. The basis for this technique is the relatively high total cross section of hydrogen. Other elements which have similarly high scattering cross sections, such as boron, may also be determined in an appropriate matrix. The usual apparatus consists of a neutron source (Ra-Be) located inside a block of moderator (paraffin) and cadmium filters located between source and sample and sample and detector.

The back scattering of beta rays can be described by a simple linear function within periods of elements corresponding to the noble-gas configurations. This technique has been applied to the analysis of hydrocarbons.

REFERENCES

1. Bethe, H. A., and P. Morrison, *Elementary Nuclear Theory*, 2nd Ed., Wiley, New York, 1956.
2. Blatt, J. M., and V. F. Weisskopf, *Theoretical Nuclear Physics*, Wiley, New York, 1952.
3. Brown, W. L., and S. Wagner, eds., "Semiconductor Materials for Gamma Ray," Report of the Meeting Held in New York, June 24, 1966, National Research Council Committee on Semiconductor Particle Detectors.
4. Choppin, G. R., *Experimental Nuclear Chemistry*, Prentice-Hall, Englewood Cliffs, N.J., 1961.
5. Friedlander, G., J. W. Kennedy, and J. M. Miller, *Nuclear and Radiochemistry*, Wiley, New York, 1964.
6. Halliday, D., *Introductory Nuclear Physics*, 2nd Ed., Wiley, New York, 1955.
7. Harvey, B. G., *Introduction to Nuclear Physics and Chemistry*, 2nd Ed., Prentice-Hall, Englewood Cliffs, N.J., 1968.
8. Kaplan, Irving, *Nuclear Physics*, 2nd Ed., Addison-Wesley, Cambridge, Mass., 1962.
9. Kramer, A. W., "Nuclear Energy—What It Is and How It Acts," *Atomics*, Reprints Vols. I, II, III, and IV (1961).
10. Mann, Wilfred B., and S. B. Garfinkel, *Radioactivity and Its Measurement*, Van Nostrand Momentum Book No. 10, published for The Commission on College Physics, Van Nostrand, Princeton, N.J., 1966.
11. Miller, G. L., "The Physics of Semiconductor Radiation Detectors," *Brookhaven Lecture Ser.* No. 9 (1961).

12. O'Kelley, G. D., "Detection and Measurement of Nuclear Radiation," Subcommittee on Radiochemistry, *Nat. Acad. Sci.-Nat. Res. Council, Nucl. Sci. Ser., Radiochemical Tech.*, 1962.
13. Ortec Instrument Co., Catalog 1001, "Instrumentation for Research," Oak Ridge, Tenn., 1969.
14. Overman, R. T., and H. M. Clark, *Radioisotope Techniques*, McGraw-Hill, New York, 1960.
15. Rutherford, E., J. Chadwick, and C. D. Ellis, *Radiations from Radioactive Substances*, Cambridge University Press, 1930.

Chapter 96

RADIOCHEMICAL SEPARATIONS

By C. E. CROUTHAMEL AND R. R. HEINRICH, *Chemical Engineering Division, Argonne National Laboratory, Argonne, Illinois*

Contents

I. INTRODUCTION

The discovery of radioactivity was made in 1896 by Becquerel (10). The first chemical separations to isolate and identify the elements responsible for the observed radiation were made by the Curies (44,45) and by Rutherford and Soddy (196). During the 15 years following, the techniques of radiochemical separations developed rapidly into a fairly distinct branch of chemistry.

Early books written in this new field (24,93,152,194,223) dealt with the application of inorganic chemistry to the isolation and identification of the products of the radioactive decay processes. The single unifying aspect of the field of radiochemistry is the unique mode of detection of the atoms by their radiation. This permits the measurement of single atomic events. The success or failure of a radiochemical separation, therefore, may depend on the separation of extremely small quantities of interfering radioactive elements, for example, approximately 10^{-12} g for atoms with half-lives measured in days, and 10^{-6} g for atoms with half-lives measured in thousands of years. Paneth has defined radiochemistry simply as the chemistry of bodies which are detected through their nuclear radiations.

After Becquerel's historical find, interest in radioactivity was brought to its present level by a series of important discoveries. The earliest of these was the first artificially induced nuclear reaction in 1919 by Rutherford (195), who bombarded ^{14}N with 7.7-MeV alpha particles from RaC (^{214}Po), and produced ^{17}O and protons. From 1919 to 1932, the only known sources of particles which would induce nuclear reactions were the natural alpha emitters, and the only reaction known was the (alpha, proton) reaction.

Two events in the year 1932—Chadwick's discovery of the neutron (31) and Feather's noting of the neutron's efficiency in producing transmutations (55)—were followed in rapid succession by many additional reactions and improvements in the production of artificially induced radioactivity, discovered and reported in 1934 by Curie and Joliot (43,114), and also by Amaldi, Fermi, et al. (3). The reasoning of Fermi was important in slowing down fast neutrons in multiple collisions with hydrogen nuclei (contained in paraffin), so that with an energy distribution over the thermal range the neutrons would exhibit the expected larger capture cross sections.

In 1939, Hahn and Strassmann (84) observed and reported the fission of ^{235}U by neutrons. The year 1942 saw the first sustained nuclear chain reaction at Chicago under the direction of Fermi. The nuclear reactor remains the most important single source of radioactive material.

The most comprehensive and best known works which had been written before the exploitation of the fission process were those of Paneth (186), Hahn (82), and Hevesy and Paneth (94). After 1939, the very large effort to exploit the fission process necessitated an extremely rapid development of the techniques of radiochemical separation. During the years of World War II, practically no information was released, but the record of the radiochemical work accomplished in these years was published from 1949 to 1951 in five volumes of the *National Nuclear Energy Series*. Three of these were edited by Coryell and Sugarman (39), and two by Seaborg, Katz, and Manning (215). A later volume edited by Seaborg and Katz (215) was a new edition issued in 1954.

Today, more than ever before, radiochemical separations are defined by a particular objective and the conditions associated with an experiment. The type of radiation, the type of detector, the matrix, and other experimental conditions determine the success or failure of a separation.

Two working assumptions must generally be applied in making radiochemical separations. The first is the independence of radioactivity from the chemical and physical states of the elements. Although it is not rigorously true that all modes of radioactive decay are completely independent of the chemical or physical state, the exceptions in the laboratory are extremely minor. Thus, radioactive decay modes which involve the interaction of an orbital electron and the nucleus of the same atom (e.g., isomeric transition or the electron capture process) may be affected very slightly, in certain special instances (8,22,131), by the chemical or physical state of the element. This effect may have more practical implications in dealing with cosmic nuclei. Consequently, ^{55}Fe or ^{7}Be and certain other K-capture and isomeric states may decay at very slow rates when the orbital electrons are stripped from the nuclei as in primary cosmic rays.

The second working assumption which is generally made is that in chemical behavior the various nuclear species or isotopes of an element are identical to each other. Many attempts were made, in the 15 years following the discovery of radioactivity, to separate chemically ^{230}Th and ^{232}Th and also ^{228}Ra and ^{226}Ra. The chemical differences between the heavy isotopes are extremely small, and even today, with the very sensitive ion-exchange and paper chromatographic methods of separation, the isotopes can be fractionated to detectable levels only with the light elements, and then in very carefully controlled experiments. The differences in the rates of reaction of isotopes (23,164), however, are more likely to affect fractionating of the isotopes of the lighter elements, under ordinary conditions. Thus, when a reaction is not allowed to go to completion or equilibrium before effecting the separation, the assumption of identical reaction rates of isotopic species may lead to error. The probability that this may occur to an appreciable extent increases with the elements of lower atomic number.

The problems of radiochemical separations are similar to those of ordinary chemical separations except that the amounts of material involved in the former may be extremely small (e.g., 10^{-6} to 10^{-12} g). These minute amounts of substances may be present as either the species whose isolation is sought, or as the contaminants from which isolation is to be made, or as both. The small amounts of materials involved in radiochemical separations result directly from the extreme sensitivity of the methods of detection. The scintillation, gas proportional, solid-state, or even the old Geiger-Müller counters can detect single atomic events with up to 100% efficiency of the radiation incident on the sensitive volume of the detectors.

The uniqueness of radiochemical separations, then, is this extreme sensitivity in detection, and the analyst must be cognizant of the principles and methods used in dealing successfully with these extremely low concentrations of atoms.

We have classified radiochemical separations into five general modes of operations, as follows:

1. Separations in which the total atom concentrations are in the macro region. It should be pointed out that the lower boundary which defines the macro concentration region is somewhat arbitrary. What is meant is that the total atom concentrations are high enough so that various special phenomena observed at very low concentrations of atoms do not affect the usual chemical behavior of the radioactive isotope, as viewed by a sensitive nuclear detector. The concentration levels required to minimize such effects as, for example, those of radiocolloids, surface adsorption, coprecipitation, and reactions with very small amounts of impurities will vary with both the element and the experimental conditions. In many instances, it is possible to add such a large excess of a stable carrier isotope that this question is avoided. In the other cases, actual experience with the procedure under identical experimental conditions must be obtained.

2. Separations in which no additional carrier isotopes are added purposely to the radioactive product formed, and in which the radioactive products are different in atomic number with the target matrix. These are generally referred to as carrier-free separations. The term *carrier-force* is used rather loosely in radiochemistry to refer to tracer concentrations which are composed of nearly a single isotopic species, or possibly a tracer of relatively high specific activity level. In many (n, γ) reactions, for example, where both the interaction cross section and the thermal neutron flux are high and the exposure approaches saturation, the resulting specific activity (disintegrations per second per gram) may be so high that the total atom concentrations employed in an experiment are in the submicrogram concentration region. In this case, the problems of handling the small concentrations of undiluted tracer are identical to those of handling a carrier-free solution.

3. Separations which are based on the response of radiation in a detector as in proportional gas ionization, scintillation, and solid-state pulse-height analyzers; or possibly the response of radiation in the special particle detectors, such as alpha counting in the presence of a high beta-gamma background, fission fragment detection, neutron detectors, or Čerenkov radiation detectors. In addition, separations based on the physical characteristics or interactions of radiation with matter, such as half-life, coincidence, differences in absorption in matter, and annihilation in matter (e.g., positron annihilation followed by detection of the characteristic-MeV photons).

4. Separations and counting of low-level activity, which require unique skills and knowledge. This a specialized branch of radiochemical separations and detection has developed in relatively recent years. It would be difficult, however, to characterize low-level methods with any specific example because they now range over a very wide area of interests. Low-level radioactivity can be defined in an arbitrary way as a determination involving small absolute amounts (~ 10 disintegrations/min), as well as a sample-to-background counting-rate ratio of approximately unity. Libby (146), a pioneer in low-level radiochemistry who received the Nobel award for his method of dating ancient objects by radiocarbon, says, "It is something like the discipline of surgery—cleanliness, care, seriousness, and practice."

The use of low-level methods may then apply to almost any field utilizing radioactivity; for example, (a) studies of naturally occurring activity; (b) determination of contamination of the earth's environment, living plants, and animals; (c) cosmic ray-induced activity in extraterrestrial and atmospheric matter; (d) low-level analysis of the products of nuclear reactions, as in the determination of nuclear-reaction cross sections where the beam intensity is low, where the cross section is very low, or where the element being activated in the target is present at very low concentration.

5. Separations of activation products which are based on recoil of the atomic nucleus. Chemical bond energies range from approximately 0.5 to 5 eV (10–100 kcal/mole). Recoil energies which are imparted to nuclei by gamma-ray emission are in this range when the gamma-ray energy is several million electron volts or more. The recoil energy imparted to the nucleus also decreases with increasing mass. The classical work was done in 1934 (237), and the names of the authors, Szilard and Chalmers, are used in referring to these reactions.

More recently, the methods of nuclear recoil chemistry, or "hot atom chemistry," have been expanded from applications involving carrier-free separations of isotopes to the labeling of compounds. This is especially true of organic compounds which can be labeled, for example, by ^{14}C or tritium recoil processes. Reactor-irradiated organic substances may produce fragmentation products, both produced and labeled as the result of nuclear recoil. These products are usually isolated by adding carrier compounds. Certain very sensitive techniques, however, such as paper chromatography or gas–liquid chromatography may be applied to the isolation of nearly carrier-free species. A relatively large and complex literature has developed in the last 10 years on both recoil and knock-on displacements in materials. These techniques are being applied for labeling, isolation of carrier-free isotopes, and formation of synthesis or fragmentation labeled products. The radiochemical separation of recoil or knock-on products requires the consideration of all possible products or impurities in a particular matrix, removal of gross impurities, and removal of radiochemical impurities. A detailed and specific consideration of the various problems involved in this type of chemical separation will not be given in this chapter.

It should now be apparent that a successful radiochemical separation is defined principally in terms of a particular experiment. Also, in most instances, a successful radiochemical separation will utilize more than one of the five modes of operations just defined. For example, the analyst may couple chemical separations with some special detector discrimination. The unique requirements of low-level radiochemical analysis usually dictate the use of both special chemical separations and counting techniques. In an appreciable number of instances, the chemical separations can be eliminated completely and the separations made by electronic pulse-height analysis of the detector pulse spectrum. Proportional ionization gas counters and scintillation and solid-state spectrometers are all available commercially as detectors. Multichannel pulse-height analyzers make the analysis and display of the spectrum relatively easy. In each nonroutine radiochemical separation, the analyst must select the separation steps, decide on his order of application, and verify the success of the separation by devising appropriate tests. In addition, for quantitative radiochemical analysis, isotopic exchange of the carrier with the radiochemical species must be achieved and a weighable stoichiometric compound of the carrier and radioactive isotopes synthesized. An alternative quantitative method is the use of a known quantity of a second tracer isotope which can be determined independently by mass analysis or by pulse-height analysis and other special counting techniques.

The technical information applied to radiochemical separations ranges over a major part of the chemical principles which comprise the contents of Volumes 1 and 2 of this Treatise. Additional knowledge of nuclear physics will be needed to understand the interactions of nuclear particles with matter, the principles of nuclear decay, and nuclear detectors. Much of the latter information is contained in the chapters on radioactive methods in this volume. Many separations must be planned and conducted with this knowledge in mind. A few examples may serve to illustrate what is meant. The interaction of alpha radiation with matter requires the mounting of a sample with very little inert material. If the alpha spectrum is to be observed, rather than just the gross alpha counting rate, there should be less than 190 γ of total sample mass per square centimeter of sample surface. To achieve this, the separation methods must be planned carefully so that inert noncombustible deposits do not appear on the final sample plate. In other separations, the rate of decay or the rate of growth of a daughter will dictate that the separation must be accomplished in a short time. In low-level radiochemical separations, a great deal of care must be taken to prevent the contamination of samples from both natural and artificial sources. It has been pointed out (234), for example, that drawing a few liters of air through a filter will deposit several counts per minute of beta activity in the sample. The source of this contamination is air-borne radon. The effect may increase at night or with high humidity. In low-level separations this and many similar sources of contamination must be avoided by proper planning of the radiochemical separation procedures.

II. ISOTOPIC EXCHANGE

In most radiochemical separations, the addition of a carrier of natural isotopic composition, a stable isotope tracer, or a second radioactive tracer will be made. If quantitative yields are to be determined in the course of the separation, there must be an initial complete interchange of the isotopes with all the chemical species present.

Fortunately, there are many reactions which have high exchange rates. This applies even to many heterogeneous systems, as in the heterogeneous catalysis of certain electron transfer reactions. In 1920, Hevesy, using ThB (^{212}Pb), demonstrated the rapid exchange between active lead nitrate and inactive lead chloride by the recrystallization of lead chloride from the homogeneously mixed salts. The ionization of these salts leads to the chemically identical lead ions, and a rapid isotopic exchange is expected. Similar reversible reactions account for the majority of the rapid exchange reactions observed at ordinary temperatures. Whenever possible, the analyst should conduct the isotopic exchange reaction through a known reversible reaction in a homogeneous system. The true homogeneity of a system is not always obvious, particularly when dealing with the very low concentrations of the carrier-free isotopes. Even the usually well-behaved alkali-metal ions in carrier-free solutions will adsorb on the surfaces of their containment vessels or on colloidal and insoluble material in the solution. This is true especially of the heavier alkali metals, rubidium and cesium. Cesium ions in aqueous solution have been observed to adsorb appreciably on the walls of glass vessels when the concentrations were below 10^{-6} g/ml. This can be observed by assaying a solution over a period of several days to several months, and by noting a continual drop in the assay of the long-lived isotope ^{137}Cs. The decrease may exceed 50% of the original content. The addition of 100 γ of carrier with stirring for several hours will usually return the solution to its expected assay value. The transition metals tend to form radiocolloids in solution, and in these heterogeneous systems the isotopic exchange reaction between a radiocolloid and inactive carrier added to the solution is sometimes slow and, more often, incomplete. Elements which show a strong tendency to form radiocolloids, even in macro concentrations and acid solutions, are titanium, zirconium, hafnium, niobium, tantalum, thorium, and protactinium, and, to a lesser degree, the rare earths. Other metals also may form radiocolloids, but generally offer a wider choice of valence states which may be stabilized in aqueous solutions.

Harris (91), Buton et al. (17), Myers and Prestwood (176), Norris (179), and Friedlander and Kennedy (63) have discussed the kinetics of the isotope exchange. The simple homogeneous exchange reaction occurring at equilibrium can be expressed as

$$AX + BX^* \rightarrow AX^* + BX$$

The rate of appearance of AX^* in the initially untagged AX reactant will follow a simple exponential law:

$$Rt = -\frac{[AX]\,[BX]}{[AX] + [BX]}\ 2.3 \log (1 - F)$$

In this equation, [AX] and [BX] are the total molar or atom concentrations in the homogeneous medium, and F is the fraction of exchange equilibrium of radioactivity attained in time t. The rate, R, is then the total number of exchanges per unit time in the concentration units employed. The adherence to the exponential law will be observed regardless of the reaction mechanism, the formulas of the chemical species, or the equilibrium concentrations of the reactants. Especially useful derivations of the exchange law are given in references 176 and 63. In practical radiochemical separations, the difficult problem usually involves the analyst's recognition of the state of a particular system. For example, it is hard to determine whether one is dealing with a simple or complex homogeneous system, or with a heterogeneous system. The exchange of lead in lead nitrate and lead chloride is a simple homogeneous system in aqueous solution because the salts ionize to produce chemically equivalent lead ions in solution:

$$Pb(NO_3)_2 + Pb^*Cl_2 \rightleftharpoons Pb^*(NO_3)_2 + PbCl_2$$

The system, however, may suddenly change to a heterogeneous one if lead ions precipitate as an insoluble compound or if they are adsorbed on a surface. This situation becomes more probable at very low lead ion concentrations.

An example of a complex homogeneous system is the easily demonstrated nonequivalence of sulfur atoms in thiosulfate. In the following sequence of reactions starting with tagged sulfur

$$S^* + Na_2SO_3 \rightleftharpoons Na_2S^*SO_3$$
$$2Ag^+ + S^*SO_3{}^{2-} + H_2O \rightarrow Ag_2S^* + SO_4{}^{2-} + 2H^+$$

the final silver sulfide precipitate will contain all the original activity, with none appearing in the sulfate. This indicates that the sulfur atoms of thiosulfate are not chemically equivalent and do not exchange as a result of the sequence of chemical operations. In the reaction

$$S^*O_3{}^{2-} + SSO_3{}^{2-} \rightleftharpoons SS^*O_3{}^{2-} + SO_3{}^{2-}$$

the silver sulfide precipitated by reaction with thiosulfate will be inactive. The radioactive sulfur in sulfite exchanges at a measurable rate with only one atom of sulfur in thiosulfate.

Information on isotopic exchange rates has been compiled by Myers and Prestwood (176) for both homogeneous and heterogeneous systems. Also, Edwards (52) gives a summary of exchange rates and mechanisms of exchange. Burgus (18) has tabulated a number of nonexchangeable species.

Probably the best way to give the reader a feeling for the ways in which isotopic exchange is achieved in practice is to note some specific examples from radiochemical procedures. The elements which show strong tendencies to form radiocolloids in many instances may be stabilized almost quantitatively as a particular complex species and an exchange effected. Zirconium, for example, is usually exchanged in strong nitric acid–hydrofluoric acid solution. In this medium, virtually all the zirconium forms a ZrF_6^{2-} complex. Niobium exchange is usually made in an oxalate or fluoride acid medium. The exchange of ruthenium is accomplished through its maximum oxidation state, Ru(VIII), which can be stabilized in a homogeneous solution and distilled as RuO_4. Exchange may also be achieved by cycling the carrier through oxidation and reduction steps in the presence of the radioactive isotope. An iodine carrier with possible valence states from -1 to $+7$ is usually cycled through its full oxidation–reduction range to ensure complete exchange. In a large number of cases, isotopic exchange is not a difficult problem; however, the analyst cannot afford to relax his attention to this important step. He must consider in each analysis the possibility of both the slow exchange of certain chemical species in homogeneous solution and the possible very slow exchange in heterogeneous systems. In the latter case, this may consist simply of examining the solutions for insoluble matter and taking the necessary steps to either dissolve or filter it and to assay for possible radioactive content.

III. CARRIER-FREE SEPARATIONS

A radioactive carrier-free isotope, or tracer, as it is commonly called, can be defined as a radioisotope which does not contain weighable amounts of stable isotopes of the given radioelement. The chemical properties of a carrier-free radioisotope are identical with those of macroscopic amounts of the same element; however, because such minute quantities of the element are involved, special precautions must be taken to ensure a clean and efficient recovery. Trace impurities which could normally be neglected in the macro chemical procedure now become quite significant, and care must be taken to prevent radiocolloid formation (107,209,229) and hydrolysis (115), or to prevent adsorption of the radioisotope on the walls of containers. The same methods that are used for macro radiochemical separations are generally used in isolating carrier-free activities. The most versatile, of course, are the methods which isolate the tracer directly, namely, solvent extraction, ion exchange, electrodeposition, and volatilization. The other methods of chromatography, radiocolloid formation, and coprecipitation are also used, but to a lesser degree. Each of the aforementioned methods will be discussed in detail and can be considered applicable to carrier-free separations, even if the latter are not specifically mentioned. References for specific carrier-free separations are given in the Appendix.

A. COPRECIPITATION

Coprecipitation generally means the carrying of a radioactive tracer by a precipitate which is formed in the presence of the radioactive tracer, and no particular mechanism by which the radioactive tracer is carried is implied. The two general types of carrying processes are the incorporation of the radioactive tracer in the crystal lattice of the precipitate, and the adsorption of the tracer on the surface of the precipitate either after or during its formation.

Hahn (82) has further classified the carrying processes as the following:

1. Isomorphous replacement. The radioactive tracer is incorporated into the crystal lattice of the precipitate, presumably isomorphously, since macro amounts of tracer and precipitate form isomorphous mixed crystals.

2. Adsorption. The radioactive tracer simply adsorbs on the precipitate.

3. Anomalous mixed crystal formation. The radioactive tracer also is incorporated into the crystal lattice of the precipitate, but macro amounts of tracer and precipitate do not form isomorphous mixed crystals.

4. Internal adsorption. The radioactive tracer is adsorbed on the surface of growing carrier crystals, where it is trapped, resulting in a non-homogeneous incorporation of the tracer in the precipitate.

Of the four processes listed, the first two are by far the most clearly understood. The third and the fourth are mentioned because it is impossible to assign every carrying process to either *1* or *2*.

The separation of minute (submicrogram) quantities of radioactive isotopes from solution by means of nonisotopic precipitates more often than not involves a subsequent separation of the desired radioactive isotope from the carrier precipitate. It is more convenient, therefore, to choose coprecipitating agents which can be separated with the least amount of chemical or physical manipulation. For this reason, isomorphous replacement and mixed crystal formation, although extremely reproducible and independent of precipitating conditions, are not of so great a practical importance as other processes in the isolation of carrier-free radioisotopes. The inherent difficulties encountered in the subsequent separation of the radioisotope from the precipitant at times considerably outweigh the desirable features of this process.

Generally, the separation of carrier-free activities employs a "scavenging" type of precipitation. In this case, the radioactive tracer is carried as a result of the adsorption phenomena. Typical scavenge precipitates are ferric hydroxide precipitated from hydrochloric acid, and manganese dioxide precipitated from nitric acid. Unfortunately, this type of precipitation reaction is quite sensitive to changes in experimental conditions and is not so specific as the amorphous and mixed crystal process. It does have the important advantage, however, that the scavenging precipitate can be chosen so as to permit a relatively easy and rapid subsequent separation of the carrier-free radioisotope. One major disadvantage of the coprecipitation processes in general is the fact that there is involved the addition of macro amounts of carrier material, which may contain impurities isotopic in character with the desired radioisotope. Also, if a certain precipitation of a carrier is intended to separate from a specific radioimpurity, it may be ineffective because of the peculiar absorptive properties of trace concentrations of the impurity. [Under certain conditions, some elements carry as traces but do not carry when present in macro amounts (69–71).] This problem can be eliminated to a great extent, however, by the addition of a small amount of holdback carrier before precipitation. The holdback carrier, since it is inactive material isotopic with the impurity, and since it is in great excess of the impurity, reduces the concentration of the impurity on the precipitate.

A few examples of coprecipitation are illustrated by the classical separations of neptunium (56), plutonium (92), americium, and curium (241) by lanthanum fluoride precipitation; strontium (68) and barium (85) precipitated as the carbonate; and bismuth (167) and niobium (85) scavenged by ferric hydroxide. Plutonium has recently been separated from its different oxidation states by the coprecipitation of mandelic and p-bromomandelic acid (166). Zirconium (161) has also been separated using the same reagent. Strontium (248) has been separated from the other alkaline earths with the reagent potassium rhodizonate.

B. SOLVENT EXTRACTION

The distribution of a substance between two immiscible solvents is a phenomenon which is of great practical use to the radiochemist. The extraction of a radioactive tracer as a nonpolar compound or complex from an aqueous solution by an immiscible organic solvent is a very selective and effective method, and perhaps the one most widely used for separating tracers from nonisotopic materials, whether they be present in micro or macro amounts. This method can also be used to remove macro amounts of nonisotopic materials from aqueous solutions containing the carrier-free tracer (78,79). This is important, since a precipitation separation of the undesirable macro constituent could result in a loss of the carrier-free tracer by coprecipitation.

The distribution of the material between the two immiscible solvents follows the Berthelot-Nernst distribution law, which states that the ratio of the equilibrium concentrations of a substance in two phases is a constant at a given temperature. The distribution of the material at macro concentrations, therefore, should be the same as the distribution at micro or tracer concentrations, providing that the activity coefficient of the material does not change on dilution. Generally, the distribution ratio of an extractable material is independent of its initial concentration; stant at a given temperature. The distribution of the material at macro or micro concentrations, this does not necessarily mean that submicrogram amounts of the material will be extractable. For example, the distribution ratio of ferric chloride at high concentrations is large enough to warrant use of diethyl ether extraction as a quantitative separation process; however, ferric iron in very low (carrier-free) concentrations cannot easily be extracted quantitatively into diethyl ether at the same acid concentration (60). More efficient extractions at low concentrations of ferric iron are achieved by using higher acid concentrations and other organic reagents, such as isopropyl ether or β,β'-dichloroethyl ether. Many other carrier-free radioactive species at concentrations of $10^{-15}M$ have extraction coefficients which are of the same order of magnitude as those obtained at macro concentrations.

Solvent extraction can be highly selective, for example, in the extraction of radioiodine and radiobromine (1) with CCl_4 from an aqueous solution of the fission products of uranium. Since some of the iodine is present as iodate and periodate and the bromine is present as bromate, a reduction of these to the iodide and bromide must be done first, and then the halide oxidized to the free halogen before extraction. Other examples of specificity are the extraction of uranyl nitrate from $1M$ HNO_3–$2M$ $Ca(NO_3)_2$ solution into diethyl ether (66), and the extraction of tantalum into diisopropyl ketone from sulfuric acid solution containing a small amount of fluoride (230). It is also possible to change the conditions of the solvent extraction so that the desired nuclide extracts back into the aqueous phase. This back extraction can further increase the specificity of the separation and can also return the nuclide to a more convenient medium for subsequent purification.

Another type of solvent extraction, chelation, is also used extensively in carrier-free radiochemistry work and is very selective. Some of the more common reagents used in forming the metal complex are cupferron, 8-quinolinols, acetylacetone, and thenoyltrifluoroacetone (TTA). Ethylenediaminetetraacetic acid (EDTA) also serves as a complexing agent, but its use is primarily for masking. For example, in the extraction and separation of ^{233}U from thorium and all other alpha emitters, the uranium is complexed as the 8-quinolinol chelate in methyl isobutyl ketone, and EDTA is used to mask thorium and other metals (35). Similarly, EDTA serves to mask the extraction of copper, bismuth, and lead in the selective extraction of uranium with acetylacetone (132). TTA extractions of zirconium from nitric acid solutions (170), and the extraction of plutonium (174), neptunium (172), cerium (221), and protactinium (14) from strong mineral acid media, are other selective examples of the chelation type of solvent extraction.

General usage of the solvent extraction technique in radiochemical separations has been thoroughly reported by Freiser and Morrison (61). Especially useful are the tables containing lists of elements which are extractable by forming ion-association or chelation complexes with various reagents. Moore (172) has written a monograph on the liquid–liquid extraction of elements using high-molecular-weight amines. White and Ross (250) have written a similar article on solvent extraction separations in which tri-n-octyl phosphine oxide was employed. Some of the more recent liquid–liquid extraction systems considered worthy of mention are the tributyl phosphate (TBP) system for the extraction of zirconium, thorium, cerium, promethium, and yttrium (220) from $HClO_4$ media, and the extraction of the actinides plutonium, americium, curium, berkelium, californium, and einsteinium from nitrate (13) and chloride (187) media. The diisobutyl carbinol system has been used for the extraction of protactinium from irradiated thorium (180) and also for the extraction of antimony (149). Extraction of the actinide elements, thorium through americium, using tri-n-octyl amine (TOA), has been reported by several authors (116,120,169,217). Other extractions employing amines are the separation of uranium and plutonium from thorium and fission products using tri(isooctyl) amine (171), and the separation of technetium, neptunium, and uranium using tertiary amines (36). Quaternary amines have been employed by Maeck et al. (151) for the separation of several elements. The use of methyl-substituted pyridine derivatives for the extraction of technetium and rhenium from alkali media (190) has also been demonstrated. In an even more versatile technique for technetium separation described by Salaria and his associates (201), the pertechnetate ion is extracted into a chloroform solution of methyltricapryl–ammonium chloride from either high acid or basic media. The determination of rare earths has been greatly facilitated by the use of solvent extraction, and techniques for determining individual rare earths (21,51,153,188) and total rare earths (159) are given by various authors.

Two of the difficulties which may be encountered when employing the solvent extraction technique are emulsion formation and the slow attainment of equilibrium. An example of the latter is the extraction of zirconium into TTA in benzene, the rate of which is dependent on the rate of formation of the complex at the interface (38). Even though extraction under equilibrium conditions is desirable, it is sometimes possible to use a slow extraction rate of a particular component to improve selectivity of the process (175). Another difficulty encountered is the extraction of reagents from the aqueous phase into the organic phase, and thus alteration of the original extraction conditions. This can be remedied somewhat in the case of a hydrochloric acid and diethyl ether medium by equilibrating the ether phase with the acid before extraction of the radioelement. The chelation-type extractions require a controlled pH if complete extraction of the radioelement is desired. Also, in some extractions, radiocolloids may appear at the interface between phases, and these in themselves are not extractable (209). Finally, as mentioned earlier in regards to iron (48), substances which normally extract in macro amounts may not extract as readily when present in trace quantities.

The innumerable papers pertaining to solvent extraction published in the last 3 years indicate the popularity of the method as a rapid, simple, and very specific technique. Moreover, it should be noted that the references listed represent only a minute portion of the work that has been done in this field.

C. ION EXCHANGE AND PAPER CHROMATOGRAPHY

In the past decade the method of selective elution of adsorbed ions from ion-exchange resins has become extremely useful to the radiochemist; in fact, it is the only method which is applicable to both carrier-free tracers and macro amounts of material. Presently, the ion-exchange resins used most extensively are of the synthetic organic polymer type. The exchange properties of the type of resin being used are the result of attached groups which make hydrogen or hydroxyl ions readily available for exchange. Usually the cation exchanger consists of an insoluble resin to which sulfonic acid, carboxylic acid, or phenolic acid groups are firmly attached. The anion-exchange resins are commonly insoluble resins containing basic groups, for example, amino, substituted amino, or quaternary ammonium groups. The difference in distribution of two ions with the resin and surrounding solution determines how well the ions can be separated. Usually the ions are introduced and adsorbed at the top of the column resin in a narrow band, and then an eluting solution is passed through the column, equilibrating the resin and ionic species. At equilibrium, the differences in the distribution coefficients of the ions to be separated will be largest, and the ion with the lower distribution coefficient will move more rapidly down the column than the other ion. Thus different ions will have different flow rates and hence will appear in the eluate at different times. In general, the distribution coefficient of ions increases with increasing charge and decreases with increasing ionic radius.

The problems of decreasing separation time and control of elution order have been somewhat overcome by the technique of complexing the aqueous ions with a buffered complexing agent. For example, citrate solutions complex Pr(III) more strongly than Ce(III); therefore, the Pr(III) will be removed from a cation resin more rapidly than the Ce(III). Selection of the type of resin used, either cation or anion, can also control elution order and can even determine which ions will be adsorbed and which ones will not. The ion that is complexed most strongly will be adsorbed least by cation resins, and the reverse is usually true for anion resins. Cation resins were used in developing the separation of the rare earths in the Manhattan Project (224,244,245), and have also been employed in the separation of alkaline earths (243) and the hafnium-zirconium pair (232). Numerous reagents have been used in improving the separation of alkali metals (19,254), the rare earths and trivalent actinides (15,34,50,74,97,158,225,231), and the alkaline earths (99,144). Reagents which are the most widely employed presently for the separation of these ions are ethylenediaminetetraacetic acid (EDTA) (225), lactate (15), and α-hydroxyisobutyrate (34,225).

The principle of organic complexing of ions can also be applied to anion exchangers. The technique is somewhat varied in that the complexing agent is adsorbed on the exchanger and is not added to the eluting solution. This technique, using exchangers in acetate, EDTA, and citrate forms, has been successful in separating alkali metals and alkaline earths from other elements (202–206). Lithium was separated from sodium (178) by anion exchange, as were the rare earths (103). Also, oxalates have been used in anion-exchange separations of tin, antimony, and tellurium (222); molybdenum and tungsten (165); and niobium and zirconium (247).

Most of the emphasis in anion exchange has been with chloride solutions. Many metals form chloride complexes when dissolved in HCl, and these complexes are strongly adsorbed by anion exchangers, making separations quite effective. Separations for a number of elements have been summarized in a review by Kraus and Nelson (126), and early work with uranium and transuranic elements has been reviewed by Hyde (108). At moderate HCl concentrations, uranium and the transuranic elements when present in the 6+ or 4+ oxidation state form strong chloride complexes which are highly adsorbed, whereas little or no adsorption is shown by the 3+ oxidation state. The difference in oxidation state of these elements enables efficient separation of uranium from plutonium, and each of these can be separated from the other transplutonic elements (106,242) and from thorium (12) and the rare earths (110,129,130,167) by the same principle.

Some elements which do not form chloride complexes, or which form complexes at very high HCl concentration, can be separated by cationic exchange. Separation is based on specificity of the resin for the element and on the decrease of adsorbability with increased eluting ion concentration. Also, elements which form extremely strong chloride complexes, such as the platinum group, are difficult to remove from anion resins and can be separated by cation exchange. The removal of mercury from gold (20) and the removal of base metals from platinum and rhodium (150) were accomplished in this way.

Elements which have a tendency to hydrolyze or to precipitate even at high acid concentrations are complexed by HF or HCl–HF mixtures. Elements of the third, fourth, and fifth group, and a few in group six, fall in this category and have been effectively separated by anion exchange (59,81,102,124,207). Other elements of groups 3–6, including Fe(III), have been studied in $1M$ HF over various HCl concentrations, and the data have been summarized by Kraus and Nelson (128). Cation exchange can also be utilized for fluoride complexes, as is illustrated by the separation of aluminum from zirconium (62).

Another inorganic complexing agent that has been used successfully in anion-exchange separations is thiocyanate in the separation of iron from aluminum (238), from gallium (123), and from chromium (246). The actinides and lanthanides in the +3 oxidation state are also adsorbed on strongly basic anion columns from thiocyanate solutions (37,236). Cyanide solutions have been used also in the separation of iron from titanium (257). Adsorption from nitric acid solution has found numerous applications in radiochemical separations (16,54,127,128). Some specific examples are the separation of thorium from rare earths

(33,122) and uranium (29), and the purification of plutonium (198,199) and neptunium (200). Also, nitric acid has been used as an eluent in the separation of rhenium and technetium from tungsten and molybdenum, respectively (104), and dilute nitric acid–alcohol medium has been employed in the separation of yttrium, scandium (53), magnesium, and calcium (65). Phosphate (98) and pyrophosphate (197) media have also been used for both anion- and cation-exchange separations.

The widespread use of ion exchange can be attributed primarily to its simplicity of operation and to its insensitivity to concentration. The method is fairly rapid and efficient; the efficiency is dependent, however, on the following factors:

1. The type of resin used. Some ions will be adsorbed more strongly on anion resins than on cation resins. The choice is dependent on the ion to be separated. Amberlite IR-1 and Dowex-50 have seen extensive use as cation exchangers and anion exchangers.

2. The eluting solution. The type of complexing agent is also dependent on the ion to be separated. Concentration and pH are equally important, the latter factor being more sensitive in anion exchange than in cation exchange.

3. Flow rate and grain size of resin. Equilibrium is enhanced by a slow flow rate and small grain size.

4. Column size. Increased length of the column increases the efficiency of separation. Large diameters may be needed when substantial amounts of the cation or anion must be separated.

Paper chromatographic techniques have also been used successfully for both tracer and macro amounts of material, since each has the same R_f values* in numerous solvents containing strong acids. This technique has been used to identify tracer amounts of contaminants in pile-produced isotopes (64), for the study of spallation products of target material (27), for the study and separation of fission products (41,77,155,217), and for the separation of the actinide elements (117). Preparation of carrier-free isotopes is simple by this method (72,119,139–141); however, it is somewhat restricted in its applicability by the capacity of the paper for the carrier. Usually, carrier-free separations are preceded by the removal of bulk target material. At separation, the zone of activity is usually measured, cut out, and either ashed or eluted; the final tracer

---

* R_f is defined as the ratio of the distances of the solute frontal boundary and the solvent boundary, both as referred to the original sample position.

then contains only minute mineral constituents of the paper as impurities.

The technique of paper electromigration, or paper electrophoresis, as it is commonly called, is based in principle on the ability of a substance to move under the influence of an electric field. This principle differs fundamentally from that of paper chromatography, in which the separation of a substance is dependent solely on its distribution coefficient between the stationary (paper) phase and the mobile (solvent) phase. In paper electrophoresis, the paper or supporting medium serves to prevent mixing due to convection, while the liquid electrolyte serves to keep the migrating substance mobile and suitably charged. An interesting application of this technique demonstrated by Schumacher (210) is formulated on the principle that, when an electric potential is applied to a complex-forming system, there will exist in the paper a gradient of migrating complex anions and cations. Somewhere along this gradient there will be a point at which the migrating ions behave as neutral particles, and at this point they become concentrated. This "focusing" point, as it is called, will vary with the stability of the ion complex, and since the number of adaptable complexing agents is large, the technique can be utilized for the separation of many inorganic cations (211). Separation can be extremely rapid at high potentials (200 V) and is very selective since the focusing points are extremely narrow bands, usually just a few millimeters in width. This focusing technique has been applied to radiochemistry for the separation of ^{90}Y (212,213), ^{140}La, and ^{144}Pr (213) from their radioactive parents. The separations of rare earths, fission products, and ^{137}Cs have also been reported (157,214,219).

Since the field of inorganic ion-exchange separations is extremely large, it is impossible to cover the subject adequately in this section. Several books and review articles (7,108,125,126,134,142,143,177,181,240) are available which should enable the reader to become fully acquainted with this technique.

D. VOLATILIZATION

Volatility separations have an important advantage in that the desired activity can be isolated from large amounts of nonisotopic target material quite rapidly. One example is the volatilization of astatine from a mixture of astatine and polonium, the entire separation taking less than 0.5 sec (226). The use of an inert carrier gas is often helpful in transferring the isolated activity to a more suitable medium and in increasing the speed of separation and the yield.

Volatilization of carrier-free tracers from solutions and melts proceeds in the same manner as for macro amounts of material; however, volatilization of carrier-free tracers from solid materials depends on several factors. The first is the nature of the tracer to be isolated; the second, the nature of the surface material from which the tracer is volatilized, and the third, the atmosphere in which the tracer is volatilized. The last of these, depending on the concentration of impurities it contains, can alter the nature of both the tracer and surface material.

Volatilization separations of noble-gas fission products have been reported by many authors (2,6,138,148,189,193,233), as have distillations of numerous compounds of other elements. These are Ru, Ge, As, Se, Re, Fr, Tc, Sb, Os, At, Po (73,75,76,109,112,216,239,251–253). The halogens and Hg have been separated as the elements.

More recently, the applicability of the vacuum distillation of radioactive metals has been evaluated by DeVoe and Meinke (47)· The distillation apparatus used in these studies is shown in Fig. 96.1. The vapor pressures of the metals, oxides, and halides vary over a relatively wide range, and volatilization or sublimation should be capable of very good separations. The pure metal, metal oxide, and metal halide vapor pressures are not generally known with accuracy; however, some of the current data for metal and metal chlorides is summarized in Tables 96.I and 96.II. This will give the reader an approximate basis for designing or evaluating the potential of distillation separations. The expected

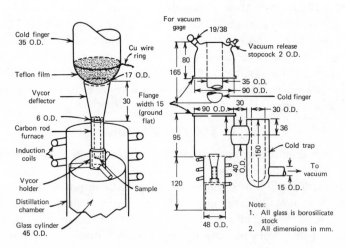

Fig. 96.1. Distillation apparatus. From reference **47**, with permission of *Analytical Chemistry*.

TABLE 96.I
Melting and Boiling Points of Metal Chlorides

Chloride	M.p., °C	B.p., °C	Chloride	M.p., °C	B.p., °C	Chloride	M.p., °C	B.p., °C
$LaCl_3$	852	1747	$LuCl_3$	892	1480	$MoCl_5$	194	268
$CeCl_3$	802	1727	Alkali chlorides	615–800	1300–1500	$NbCl_5$	212	243
$PrCl_3$	776	1707	Alkali earth chlorides	715–1000	1420–2025	$PbCl_2$	498	954
$NdCl_3$	760	1687	$AlCl_3$	193	447	$PbCl_4$	−15	140
$PmCl_3$	737	1667	Al_2Cl_6	193	180	$ScCl_3$	960	967
$SmCl_2$	740	2027	$CdCl_2$	568	960	$SiCl_4$	−67	57
$SmCl_3$	678	Decomp.	$CoCl_2$	Subl.	1049	Si_2Cl_6	−1	139
$EuCl_2$	727	2027	$CrCl_2$	815	1302	$SnCl_2$	246	623
$EuCl_3$	623	Decomp.	$CrCl_3$	1152	947	$SnCl_4$	−33	114
$GdCl_3$	609	1580	$CrCl_4$	−28	160	$TaCl_5$	207	234
$TbCl_3$	588	1550	$CuCl_2$	498	Decomp.	$TiCl_4$	−23	136
$DyCl_3$	654	1530	Cu_2Cl_2	422	1366	UCl_4	590	792
$HoCl_3$	718	1410	$FeCl_2$	677	1026	VCl_4	−26	164
$ErCl_3$	774	1400	Fe_2Cl_6	304	319	YCl_3	700	1507
$TmCl_3$	821	1390	$HfCl_4$	432	317	$ZnCl_2$	262	732
$YbCl_2$	727	1930	$MoCl_4$	310	320	$ZrCl_4$	437	331
$YbCl_3$	854	Decomp.						

separations from mixed metal solutions by distillation are subject to very wide deviations from those of ideal solutions because of the frequent and strong interactions of mixed metal atoms. The interaction of mixed halides and oxides, on the other hand, is usually smaller. In general, the method is fastest with high-vacuum distillation; for example, the apparatus of Frauenfelder (59a) is capable of measurements on isotopes with half-lives of a few seconds. The separations can be made carrier free and are comparable to ion-exchange and liquid–liquid extraction in decontamination in a single step. In cases where many impurities are also volatile, preliminary separations can be used. Auto-reduction of mercury on copper followed by volatilization of the mercury is an example.

IV. ELECTROCHEMICAL METHODS

The electrochemical methods used in the separation of carrier-free tracers are electrolytic deposition and reduction by metals. The former

TABLE 96.II

Vapor pressure of Metals in the Range 100–1100°C

Metal	Temp., °C	Vapor pressure, mm Hg	Metal	Temp., °C	Vapor pressure, mm Hg	Metal	Temp., °C	Vapor pressure, mm Hg
Ag	700	1.9×10^{-6}	Ga	700	1.5×10^{-6}	Pd	1100	4.9
	800	4.6×10^{-5}		800	3.2×10^{-5}	Pt	1000	2.0×10^{-7}
	900	6.4×10^{-4}		900	4.1×10^{-4}		1100	2.7×10^{-6}
	950	2.0×10^{-3}		1000	3.5×10^{-3}	Rb	1100	1.6×10^{-11}
Al	1000	1.8×10^{-4}		1100	2.2×10^{-2}		100	1.1×10^{-4}
	1050	5.4×10^{-4}	Ge	1000	3.8×10^{-6}		200	3.1×10^{-2}
	1100	1.5×10^{-3}		1100	4.1×10^{-5}		300	1.1
As$_4$	350	0.86	Hf	1100	2.9×10^{-13}		400	13.3
	400	4.7	Hg	100	0.282		500	82.1
	450	20		200	17.4		600	326
	500	71		300	248.0	Sb$_4$	400	1.9×10^{-5}
	550	220	In	700	2.5×10^{-5}		500	1.9×10^{-3}
	600	580		800	4.0×10^{-4}		600	6.4×10^{-2}
Au	900	6.2×10^{-8}		900	4.0×10^{-3}	Sb$_2$	700	1.1×10^{-2}
	1000	1.2×10^{-6}		1000	2.7×10^{-2}		800	9.5×10^{-2}
	1050	4.6×10^{-6}		1100	0.14		900	0.56
Ba	500	1.8×10^{-5}	K	100	1.9×10^{-5}		1000	2.5
	550	1.0×10^{-4}		200	6.8×10^{-3}		1100	8.5
	600	4.9×10^{-4}		300	0.30	Sc	1000	6.5×10^{-6}
	650	2.0×10^{-3}		400	4.16		1100	6.9×10^{-5}
	750	2.0×10^{-2}		500	28	Se$_4$	250	3.6×10^{-2}
	800	5.2×10^{-2}		600	122		300	0.24
	900	2.9×10^{-1}	Li	300	8.0×10^{-7}		400	4.76
Be	850	1.0×10^{-6}		400	1.0×10^{-4}		500	43.2
	950	1.6×10^{-5}		500	3.7×10^{-3}		600	240
	1050	1.7×10^{-4}		600	5.8×10^{-2}	Sn	800	1.1×10^{-6}
	1100	4.8×10^{-4}		700	0.51		900	1.7×10^{-5}

Bi (total) (Bi + Bi₂):

Element	T	Value
Bi (total) (Bi + Bi₂)	500	3.2×10^{-5}
	600	8.5×10^{-4}
	700	1.1×10^{-2}
	800	8.9×10^{-2}
	900	0.48
	1000	2.0
	1100	6.5
Ca	450	5.0×10^{-5}
	550	2.1×10^{-3}
	650	3.9×10^{-2}
	750	0.40
	850	2.7
	950	13.0
	1050	49.0
	1100	88.0
Cd	350	0.27
	450	4.4
	550	36.0
	650	180
	750	655
Co	1000	1.2×10^{-7}
	1100	2.2×10^{-6}
Cr	1000	1.8×10^{-6}
	1100	2.6×10^{-5}
Cs	100	5.4×10^{-4}
	200	7.7×10^{-2}
	300	1.9
	400	17.0
	500	83.5
	600	281
	700	724
Cu	1000	4.6×10^{-5}
	1100	3.5×10^{-4}
Fe	1000	4.9×10^{-7}
	1100	6.7×10^{-6}

Element	T	Value
	800	2.9
	900	12.5
	1000	42.2
	1100	119.0
Mg	300	3.0×10^{-5}
	400	2.7×10^{-3}
	500	7.6×10^{-2}
	600	0.97
	700	6.6
	800	30.0
	900	106
	1000	303
	1100	737
Mn	800	2.5×10^{-4}
	900	3.1×10^{-3}
	1000	2.5×10^{-2}
	1100	0.14
Mo	1100	1.5×10^{-14}
Na	225	6.4×10^{-4}
	325	4.0×10^{-2}
	425	0.70
	525	6.5
	625	36.0
	725	137
	825	433
	883	760
Nb	1100	2.6×10^{-18}
Ni	1000	1.7×10^{-7}
	1100	3.0×10^{-6}
Pb	500	1.6×10^{-5}
	600	4.6×10^{-4}
	700	6.4×10^{-3}
	800	5.4×10^{-2}
	900	0.32
	1000	1.4

Element	T	Value
	1000	1.7×10^{-4}
	1100	1.2×10^{-3}
	550	6.0×10^{-3}
Sr	600	2.5×10^{-2}
	700	0.29
	800	1.88
	900	8.7
	1000	31
	1100	92
Ta	1100	3.5×10^{-20}
Te₂	300	1.2×10^{-4}
	400	2.1×10^{-2}
	500	0.61
	600	5.44
	700	29.6
	800	113
	900	330
Ti	1100	4.63×10^{-8}
Tl	500	3.1×10^{-4}
	600	6.7×10^{-3}
	700	7.6×10^{-2}
	800	0.54
	900	2.7
	1000	10.6
	1100	33.8
V	1100	1.8×10^{-9}
Zn	200	3.8×10^{-6}
	300	1.2×10^{-3}
	400	6.7×10^{-2}
	500	1.3
	600	11.2
	700	60.4
Zr	1100	1.8×10^{-13}

method is utilized more than the latter because of its applicability to other fields. For example, the deposited material can be extremely thin and uniform, making it ideal for nuclear physics experiments (alpha-particle range determinations, fission counter sources, etc.) and counting samples for absolute radiochemical work.

In choosing this type of separation, the first factor which must be considered is the possible oxidation state of the tracer. For many tracers this will be determined by their position in the periodic table; however, for the heavier elements, such as the actinides, where several stable oxidation states can exist, accurate predictions of oxidation state may become questionable. The second factor is the potential of the element to be separated with respect to its target material. This difference in potential is actually the basis for separation and is dependent on the oxidation state and the concentration of the tracer, the entire dependence being expressed by the Nernst equation:

$$E = E_0 - \frac{RT}{nF} \ln Q$$

where E = potential for the reaction under experimental conditions;
E_0 = potential for the reaction when the thermodynamic activities of the substances are unity;
R = gas constant;
T = absolute temperature;
n = number of faradays of electricity in the reaction;
F = value of the faraday;
Q = the product of the activities of the resulting substances divided by the product of the activities of the reacting substances, each activity raised to that power whose exponent is the coefficient of the substance in the chemical equation.

The evaluation of Q is difficult, since concentrations are ordinarily substituted for activities, and these concentrations usually are calculated from the radioactivity of the tracer. Early work by Hevesy and Paneth (96) and also by Haissinsky (90) first indicated that electrodeposition of bismuth on gold and silver at concentrations of 10^{-4} to 10^{-12} M obeyed the Nernst equation. Work by Rogers et al. (191) indicated that the Nernst equation was apparently valid only for concentrations up to 10^{-5} M. In all of the above work, the thermodynamic activity of the deposited metal was assumed to be unity. For concentrations less than 10^{-5} M, the Nernst relation does not necessarily apply, apparently because the thermodynamic activity of the deposited metal may no longer be unity. The deviation depends on the nature of the electrodes and the electrolyte and is in the direction of increased nobility or greater ease of deposition. A modified Nernst equation to correct for this deviation was proposed by Rogers and Stehney (192). Numerous radionuclides have been separated by this technique, for example, iron (87), copper (228), cadmium (256), technetium (57), and indium (111). This method, since it is ideal for the preparation of alpha sources, has been described in numerous publications on the electrodeposition of the heavy elements. Procedures for polonium, thorium, uranium, neptunium, and plutonium have been published in records of the Manhattan Project (30,49,105), and an article giving procedures for all the actinide elements through curium has been written by Ko (121).

A. ISOTOPIC EXCHANGE SEPARATIONS

Isotopic exchange has been successfully utilized as a method in radiochemical separations. Studies by Meinke et al. have demonstrated procedures which are rapid and which have generally higher decontamination factors than other methods in a single step. Isotopic separations are based on the rapid exchange of many ions in solution at tracer concentrations with either a solid or liquid phase, for example, silver ions in solution with silver chloride solid supported on a platinum surface or with metal ions in solution with an amalgam of the metal. In each case, if at zero time there are many more inactive atoms of the element in the solid or the amalgam, a rapid exchange results with virtually all the radioactive atoms being incorporated in the solid or amalgam phase. The increased decontamination obtained by isotopic separations is due to the inherent specificity of the isotopic exchange reaction itself, and also to the fact that a large mass transfer is not involved in the separation. The techniques of precipitation, electrodeposition, distillation, extraction, and ion exchange as generally applied involve a relatively large mass transfer which is then limited in effectiveness by occlusion, coprecipitation, and entrainment. Isotopic exchange has the minimum transfer of material and thus the best possibility for high decontamination in a single step. In the procedures which have been developed, it is not unusual to find isotopic exchange decontamination factors at least 10 better than those of the next best separation technique for a particular element. The references to detailed procedures for radiochemical separations employing isotopic exchange have been summarized in the Appendix.

TABLE 96.III
Radiochemical Determination of Silver[a]

Ag, Ce, Co, Cr, Cs, I, Ir, Ru, Sb, Se, Sn, Sr, Ta, Zr

1. Plate 10 mg of Ag on platinum gauze.	Decontamination factors:
2. Electrolyze as anode in HCl to change to AgCl.	10^3 Ir, Ru, Se, Sn
3. Bring in contact with silver tracer solution for 15 min at room temperature, stirring during contact.	10^4 Cr, I, Sb, Ta, Zr
	10^5 Co
4. Wash gauze with 8M HNO$_3$ for 1 min; rinse with water.	10^6 Ce, Cs, Sr

Isotopic Exchange with Supported AgCl   Yield 100%

5. Dissolve silver chloride in 3M NaCN–0.5M NaOH plating solution.	Decontamination factors:	
	10^4	Sn
6. Electrolyze at 4 V for 15 min.	10^5	Ir
7. Wash electrode with water and acetone, and dry.	2×10^5	Sb, Ta
	5×10^5	Cr
	10^6	Ru, Zr
	2×10^6	Se
	4×10^6	Ce
	2×10^7	I
	4×10^7	Sr
	2×10^8	Co
	10^{10}	Cs

Electrodeposition of Silver   Yield 100%

8. Weigh gauze to determine yield.
9. Prepare for counting.

[a] From D. N. Sunderman and W. W. Meinke, *Anal. Chem.*, **29**, 1585 (1957).

Table 96.III illustrates a typical isotopic exchange procedure for the separation of radioactive silver. The isotopic exchange separation step may be represented as:

$$Ag^{*+} + AgCl(s) \rightleftharpoons Ag^*Cl(s) + Ag^+$$

where the asterisk denotes the radioactive species and the silver chloride solid is supported in a thin layer over a platinum gauze.

In the isotopic exchange separation of cadmium, the reaction can be represented by:

$$Cd(Hg) + Cd^{*2+} \rightleftharpoons Cd^*(Hg) + Cd^{2+}$$

The procedure is merely the equilibration by shaking of inactive cadmium amalgam with an aqueous solution containing tracer concentrations of radioactive cadmium.

Table 96.IV illustrates the percentage of contaminant tracers separated in an isotopic exchange step and an elution step with thallous ion and the cadmium amalgam. One of the limitations of the procedure is the reduction of certain ions in the aqueous solution by the cadmium amalgam. An eluent such as an aqueous solution of thallous ion can then be employed to give additional selectivity through the preferential oxidation of cadmium in the amalgam, leaving the contaminating elements in the mercury.

B. FOAM SEPARATION

Although foam separation of radioactive ions from solution is not a generally recognized analytical technique in radiochemistry, it is considered promising and potentially important. Foam separation appears to be particularly suited to very low concentrations and to be capable of separations on samples of very large volume.

The agent used to foam metal ions from a solution must have two characteristics: first, there must be preferential attraction for the ions to be removed, and, second, the resulting interaction molecular species must be surface active. Actually, many chemicals with combined surface-active and chelating groups incorporated in a single molecule are available. Apparently there are two reaction mechanisms which will enrich the metal-ion content of the foam. One is the formation with the surface-active agent of a metal chelate, and the other is an electrostatic attraction by an oppositely charged surface-active agent.

The method has been used to foam out of underground tanks ^{90}Sr and ^{137}Cs concentrate (32). Kevorkian and Gaden (58,118) have made basic studies.

V. RADIOCOLLOIDS

Appreciation of the low concentration of radioactive tracers, 10^{-12} to 10^{-22} gram atom, is necessary to achieve efficient recovery of the radiosubstance in question. Trace impurities which are normally neglected in the ordinary chemical practice become very significant in the radiochemical procedure. A special problem which occurs when dealing with these submicro concentrations is the formation of radiocolloids. The classic example of the existence of such colloidal phenomena was first observed by Paneth (183,184) in 1913, and even today complete explanation of radiocolloid behavior is still in dispute. Some radiocolloids appear to be composed primarily of inactive colloidal material (258) onto which the radioactive tracer has adsorbed, while other radiocolloids may appear to be composed of pure radioactive tracer (182,185,227). The latter ex-

TABLE 96.IV
Separation of Cadmium and Contaminants (Amalgam Exchange Procedure)[a,b]

Tracer[c]	Solution for exchange (no carrier added)[d]	Reduction potential[e]	% Separated Exchange step	% Separated Elution step	% Separated Total separation
^{131}I	I$^-$, 0.2N HNO$_3$ (C.F.)		0.005	100	0.005
^{137}Cs	0.1N HCl (20 γ)	-2.92	<0.01	f	<0.01
^{140}Ba–^{140}La	0.5N HCl (C.F.)	$-2.90, -2.52$	<0.002	f	<0.002
^{90}Sr–^{90}Y	H$_2$O (C.F.)	$-2.89, -2.37$	<0.0001	f	<0.001
^{144}Ce–^{144}Pr	1.5N HNO$_3$ (C.F.)	$-2.48, -2.47$	<0.01	f	<0.01
^{95}Zr–^{95}Nb	0.5N H$_2$C$_2$O$_4$ (C.F.)	$-1.53, -1.1$	<0.001	f	<0.001
^{65}Zn	1.0N HCl (0.8 mg)	-0.76	<0.001	f	<0.001
^{51}Cr	0.2N HCl (0.1 γ)	-0.74	<0.001	f	<0.001
^{115m}Cd	(18 γ)	-0.40	78 ± 3	98 ± 2	76 ± 3.3
^{114m}In	0.5N HNO$_3$ (50 γ)	-0.34	8.3	0.20	1.6
^{204}Tl	0.5N HNO$_3$ (0.83 mg)	-0.34	100	5	5
^{60}Co	0.5N HNO$_3$ (7 γ)	-0.28	1	<0.01	<0.01
^{113}Sn	0.9N HCl (1.1 mg)	-0.14	20	0.1	0.02
^{124}Sb	SbO$^+$, 0.5N HNO$_3$ (10 γ)	$+0.21$	100	0.1	0.1
^{104}Ru	RuCl^{2-} 0.5N HNO$_3$, 0.7N HCl (6 γ)	$+0.60$	17	<0.001	<0.0005
^{75}Se	H$_2$SeO$_3$ 0.1N HCl (0.12 mg)	$+0.74$	50	33	17
^{192}Ir	IrCl^{-3} 0.1N HCl (0.2 γ)	$+0.77$	33	0.001	0.0003
^{203}Hg	0.5N HNO$_3$ (0.2 mg)	$+0.79$	100	0.05	0.05
^{110m}Ag	0.8N HNO (20 γ)	$+0.80$	100	0.1	0.1

[a] J. R. DeVoe, H. W. Nass, and W. W. Meinke, *Anal. Chem.*, **33**, 1714 (1961).

[b] Average of duplicate runs.

[c] Elements listed in order of reduction potentials.

[d] Weight of inactive element before separation indicated in parentheses. C.F. = carrier free.

[e] Standard reduction potential of form $M^{a+} + ne^- \rightarrow$ M, where M^{n+} is the lowest stable oxidation state of the element (unless otherwise stated in previous column). Data taken from Latimer (5).

[f] Not detectable, since no measurable radioactivity in amalgam.

[g] Cadmium yields are average of separate runs made with each of listed contaminants (inactive this time) in medium specified. Error is "standard deviation."

planation, however, has met with numerous arguments (135,136,209), the strongest of which asserts that the solubility product of the colloidal substance is not exceeded and therefore it is not possible to form discrete particles.

One factor influencing the formation of radiocolloids is whether the radioactive substance has a tendency to hydrolyze in a solution, or whether it forms an insoluble compound with some component of the solution. Hahn and Werner, in a series of experiments (86), confirmed the fact that the colloidal behavior of thorium C (^{212}Bi) is related to the hydrolyzability of the isotope and its tendency to form slightly soluble products of hydrolysis. For example, thorium B (^{212}Pb) and thorium C, whose hydroxides are slightly soluble in ammoniacal solution, definitely formed radiocolloids, but thorium X (^{224}Ra), whose hydroxide is a strong soluble base, did not. Addition of a complexing agent to radioactive solution [for instance, sodium citrate in the case of thorium B (^{212}Pb)] decreases the radiocolloid formation because of a lead(II) citrate complex. The addition of hydrochloric acid in the case of thorium C (^{212}Bi) (82) represses hydrolysis of the bismuth ion and consequently represses the formation of radiocolloids.

Another equally important factor in the formation of radiocolloids is adsorption of radioactive tracers on foreign particles, such as silica and dust, which are suspended in solution. Werner showed by conclusive experiments (249) that radiocolloid formation could be suppressed simply by filtering the water used in the dilution of a radioactive solution.

Another factor which influences the formation of radiocolloids in a solution is the action of electrolytes on either the tracer ions or the suspended foreign particles, or their action on both. The nature of the solvent used in the solution may or may not enhance radiocolloid formation, for example, lead and bismuth form radiocolloids in water, benzene, ether, methyl and ethyl alcohol, but not in acetone or dioxane (82,83). Time can also be a determining factor in the amount of radiocolloid formed, since the longer the solution stands the more radiocolloid formation is observed.

Methods for determining radiocolloids are generally the same as those used for nonradioactive colloidal dispersions. Dialysis and ultrafiltration are the most widely used methods.

A. LOW-LEVEL RADIOCHEMICAL SEPARATIONS

In general, radiochemical procedures used for low-level determinations are very similar to those employed in normal radiochemical work. In the low-level procedure, however, decontamination steps may be repeated several times to ensure high chemical purity, and more care is exercised to achieve maximum chemical yield. The method of separation is usually chosen for the particular nuclide in question. Those which have been found the most successful are ion exchange, solvent extraction, and volatilization. Carriers are always utilized in low-level work; in fact, the amounts of carrier added are usually larger than in the normal radiochemical procedure because of the tendency toward reduced yields resulting from additional decontamination steps.

The basic criterion of low-level counting is simply to maximize the number of desired particle events and minimize all others. The former, the sample counting rate, is dependent on sample size, counting efficiency, and time; the latter is the background rate. In considering counting statistics, it has been established that the figure of merit for choosing between detector systems should be S^2/B, where S is the sample counting rate, and B the background rate. The system giving the largest figure of merit and likewise the greatest statistical accuracy, therefore, is the most sensitive for the measurement of the nuclide in question. Since the sample count rate appears in the numerator as the square and the background rate in the denominator as the first power, the reduction in background must be very large in order to be effective. Better methods for improving the sensitivity consist of increasing the counting rate by improvement of geometry, reduction of absorption, and increased sample size. Increased sample size, however, may be somewhat limited by the availability of the sample, and the cost and complexity of large counters. Attaining suitable backgrounds for low-level counting is primarily dependent on the type of detector used. Methods of background reduction have been discussed by Libby (145), Moljk et al. (168), and Arnold (5). Problems of low-level counting have been reviewed and discussed by numerous authors (9,60,67,101,113,133) and will not be enumerated here.

Equally as important as low counting backgrounds is the ability of the radiochemist to reduce the blank correction to a minimum. A radiochemical blank means that a certain portion of the observed counting rate of a sample is due to the radioactivity of reagents and other constituents used in a particular procedure. Different procedures will have different blanks, depending on the type of counting method used in analysis; therefore, decisions as to how the blank should be measured are difficult. The figure of merit mentioned previously should be an aid in choosing the proper counting method.

Sugihara (234), in a monograph which emphasizes the importance of negligible blanks, gives a number of steps which, if taken, will reduce the blank correction to a minimum. These are as follows:

1. Contamination of reagents should be guarded against. The Alkaline earths and rare earths are almost always contaminated with measurable amounts of thorium, thorium daughters, radium, and naturally occurring activities such as ^{40}K. Of course, potassium salts and other reagents which may contain naturally occurring activities should be avoided if the counting method chosen is sensitive to this type of activity. A complete review (46) of radioactive contamination of materials used in scientific research has been published and should prove invaluable to those attempting to detect minimal amounts of radioactivity.

2. Aqueous reagents should be prepared with deionized water or water that has been distilled from a glass or quartz still, in order to avoid any possible contamination of metal ions from the still itself. In evaluating ordinary distilled water, it is usually adjudged unsafe for use in low-level work if it has a measurable blank when titrated with EDTA or when used in trace element colorimetry.

3. Since a large amount of the blank activity is made in the last step of the procedure, for example, in precipitation, extreme care should be taken in choosing the best precipitant. In general, heavy metal-containing reagents are usually avoided. In the case of cesium, however, a chloroplatinate precipitate (28) has been found to give lower blanks than the perchlorate.

4. Suction filtration should be avoided whenever possible to eliminate the possibility of air contamination, which may or may not be appreciable, depending on the procedure (11,156). Any air that does pass over the sample should be prefiltered and trapped to remove such airborne activity.

5. Inert carriers for nuclides frequently contain active contaminants, and to separate the carrier from these contaminants so as to obtain a low-blank carrier can be difficult and tedious. For this reason, the use of a nonisotopic carrier of suitable chemical characteristics is much more desirable, providing the nonisotopic carrier is free or can be separated easily from any active contaminants. A convenient technique is the isolation of ^{90}Y from ^{90}Sr by coprecipitation of the yttrium on ferric hydroxide (235).

6. Strict segregation of the low-level laboratory and facilities, such as glassware, hardware, centrifuges, ovens, balances, counters, and absorbers, from laboratories used for ordinary radiochemical work is mandatory.

B. RAPID RADIOCHEMICAL SEPARATIONS

Radionuclides which decay with half-lives of 20 min or less require special techniques to facilitate separation. An excellent review of these techniques has been compiled by Kusaka and Meinke (137), and the reader is referred to this source for detailed examples.

In experiments in which a rapid radiochemical separation is to be used, the choice of sample or target material and the facilities for transferring the sample after irradiation are very important. The chemical and physical form of the sample depend more or less on the type of separation to be used. Most suitable, perhaps, is a material in salt form, since it can be dissolved easily in an appropriate solvent. Metallic samples should be as small or as thin as possible to enable rapid dissolution before separation. A "rabbit" facility is usually used for the speedy transference of irradiated samples to the area designated for separation. This type of facility is a pneumatic tube system in which a small sample-holding container called a "rabbit" is pushed to a desired position in a reactor or cyclotron, irradiated, and quickly returned for the chemical operations necessary for separation. In reactors, the pneumatic tube system is operated by vacuum, and the cyclotron system, by compressed air. Details on the operation of this type of facility are given by Meinke (162,163).

The general types of separation processes used for rapidly separating a short-lived radionuclide are the same as those encountered in normal radiochemical work, namely, solvent extraction, ion exchange, precipitation, etc. The detailed methods used are, of course, dependent on the sample to be isolated, and some of the techniques have been used in combination with one another. A few examples are given here to illustrate what can be done to facilitate rapid separation; if more specific examples are desired, they can be found in reference 137.

Solvent extraction is perhaps the best method for making a fast radiochemical separation whether the nuclide be present in macro or micro amounts. The method is simple, convenient, and clean, and separation can be carried out with the least amount of equipment; usually a separatory funnel is all that is required. Short-lived iron isotopes and iron reaction products have been separated in this manner. For example, 8.9-min ^{53}Fe produced by the reaction ^{50}Cr$(\alpha, n)^{53}$Fe (147) and iron reaction products of high-energy deuteron-bombarded arsenic (100) have been extracted into diethyl ether as the chloride from 6N HCl. Also, 2.6-min ^{64}Ga produced by the reaction ^{64}Zn$(p, n)^{64}$Ga (40) has been extracted in the same manner.

The ion-exchange technique, although generally classified as a slow method, can be utilized effectively for the separation of short-lived daughter activities from relatively long-lived parent activities. Some examples are the separation of 24-sec ^{90m}Nb from ^{90}Mo, using an anion-exchange column and $6N$ HCl as eluent (154); the separation of 59-sec ^{109m}Ag from ^{109}Cd, using the same type of resin and $10M$ HCl as eluent (208); and the separation of 4.9-sec ^{191m}Ir from ^{191}Os, using anion-exchange resin and $6N$ HCl as eluent (25).

An example of the volatilization method is the distillation separation of 0.53-sec ^{73m}Ce from a hydrochloric acid solution of ^{73}As (26) and the separation of 9-min ^{79}As and arsine from a mixture of fission products (42).

REFERENCES

1. Abelson, P., *Phys. Rev.*, **56**, 1 (1939).
2. Adams, R. M., and H. Finston, *Natl. Nucl. Energy Ser., Plutonium Project Record*, Vol. 9, Paper 313, McGraw-Hill, New York, 1951, p. 1791.
3. Amaldi, A., O. D'Agostino, E. Fermi, B., Pontecorvo, F. Rasetti, and E. Segré, *Proc. Roy. Soc. (London)*, **A149**, 522 (1935).
4. Anderson, E. C., and F. N. Hayes, *Ann. Rev. Nucl. Sci.*, **6**, 303 (1956).
5. Arnold, J. R., in W. C. Peacock, ed., "Conference on Measurements and Standards of Radioactivity," *Natl. Acad. Sci.-Natl. Res. Council Publ.* No. 573, 1958.
6. Ayres, J. A., and I. B. Johns, *Natl. Nucl. Energy Ser., Plutonium Project Record*, Vol. 9, Paper 311, McGraw-Hill, New York, 1951, p. 1763.
7. Bailey, R. A., *Natl. Acad. Sci.-Natl. Res. Council, Nucl. Sci. Ser. Radiochem. Tech.*, NAS-NS-3106, June 1962.
8. Bainbridge, K. T., M. Goldhaber, and E. Wilson, *Phys. Rev.* **90**, 430 (1953).
9. Barendsen, G. W., *Proceedings of the Second International Conference on the Peaceful Uses of Atomic Energy, Geneva, 1958*, Vol. 3, Paper 554.
10. Becquerel, H., *Compt. Rend.*, **122**, 423, 501, 559, 762, 1086 (1896); **128**, 771 (1899).
11. Belousov, A. Z., Yu, V. Novikov, V. F. Oreshko, and B. I. Polivoda, *Gigiena i Sanit.*, **23**, 17 (1958); *Nucl. Sci. Abstr.*, **13**, 2693 (1959).
12. Berman, S. S., L. E. McKinney, and M. E. Bednoc, *Talanta*, **4**, 153 (1960).
13. Best, G. F., E. Hesford, and H. A. C. McKay, *J. Inorg. Nucl. Chem.*, **12**, 136 (1959).
14. Bouissieras, G., and J. Vernois, *Compt. Rend.*, **244**, 2508 (1957).
15. Bunney, L. R., E. C. Freiling, L. D. McIsaac, and E. M. Scadden, *Nucleonics*, **15**, No. 2, 81 (1957).
16. Bunney, L. R., N. E. Ballou, J. Pascual, and S. Foti, *Anal. Chem.*, **31**, 324 (1959).
17. Bunton, C. A., D. P. Craig, and E. A. Halevi, *Trans. Faraday Soc.*, **51**, 196 (1955).
18. Burgus, W. A., *Radioactivity Applied to Chemistry*, Wiley, New York, 1951, p. 466, Table 8A.

19. Buser, W., *Helv. Chim. Acta,* **34,** 1635 (1951).
20. Butler, C. K., *Ind. Eng. Chem.,* **48,** 711 (1956).
21. Butler, T. A., and E. E. Ketchen, *Ind. Eng. Chem.,* **53,** 651 (1961).
22. Byers, D. H., and R. Stump, *Phys. Rev.,* **112,** 77 (1958).
23. Cahill, A. E., and H. Taube, *J. Am. Chem. Soc.,* **74,** 2312 (1952).
24. Cameron, A. T., *Radiochemistry,* London, 1910.
25. Campbell, E. C., and F. Nelson, *J. Inorg. Nucl. Chem.,* **3,** 233 (1956).
26. Campbell, E. C., and F. Nelson, *Phys. Rev.,* **107,** 502 (1957).
27. Carleson, G., *Acta Chem. Scand.,* **8,** 1697 (1954).
28. Caron, N. L., and T. T. Sugihara, "Annual Progress Report Contract AT-(30-1)-1930," *U.S. At. Energy Comm. Rep.* NYO-7759, 1959.
29. Carswell, D. J., *J. Inorg. Nucl. Chem.,* **3,** 384 (1957).
30. Casto, C. C., and C. J. Rodden, N. R. Furman, E. H. Huffman, L. L. Quill, T. D. Price, and J. I. Watters, eds., *Analytical Chemistry of the Manhattan Project, Natl. Nucl. Energy Ser.,* Div. VIII-1, McGraw-Hill, New York, 1950, Chap. 23.
31. Chadwick, J., *Proc. Roy. Soc. (London),* **A136,** 692 (1932).
32. *Chemical Engineering,* "Foam Separation," April 1961, p. 100.
33. Chen, Y., C. Lin, and T. Chen, *J. Chinese Chem. Soc.* [2], **2,** 111 (1955); *Chem. Abstr.,* **50,** 9220 (1956).
34. Choppin, G. R., B. G. Harvey, and S. G. Thompson, *J. Inorg. Nucl. Chem.,* **2,** 66 (1956).
35. Clayton, R. F., W. H. Hardwick, M. Moreton-Smith, and R. Todd, *Analyst,* **83,** 13 (1958).
36. Coleman, C. F., F. A. Kappelmann, and B. Weaver, *Nucl. Sci. Eng.,* **8,** 507 (1960).
37. Coleman, J. S., et al., *J. Inorg. Nucl. Chem.,* **3,** 327 (1956).
38. Connick, R., and W. H. McVey, *J. Am. Chem. Soc.,* **71,** 3182 (1949).
39. Coryell, C. D., and N. Sugarman, eds., *Natl. Nucl. Energy Ser.,* IV-9, Books 1, 2, 3, McGraw-Hill, New York, 1951.
40. Crawemann, B., *Phys. Rev.,* **90,** 995 (1953).
41. Crouthamel, C. E., and A. J. Fudge, *J. Inorg. Nucl. Chem.,* **5,** 240 (1958).
42. Cuninghame, J. G., *Phil. Mag.,* **44,** 900 (1953).
43. Curie, I., and F. Joliot, *Compt. Rend.,* **198,** 559 (1934).
44. Curie, P., and M. Curie, *Compt. Rend.,* **127,** 175 (1898).
45. Curie, P., M. Curie, and G. Bemont, *Compt. Rend.,* **127,** 1215 (1898).
46. DeVoe, J. R., "Radioactive Contamination of Materials Used in Scientific Research," *Natl. Acad. Sci.-Natl. Res. Council Publ. No. 895,* 1961.
47. DeVoe, J. R., and W. W. Meinke, *Anal. Chem.,* **35,** 2 (1963).
48. Dodson, R. W., G. J. Forney, and E. H. Swift, *J. Am. Chem. Soc.,* **58,** 2573 (1936).
49. Dodson, R. W., A. C. Graves, L. Helmholz, D. L. Hufford, R. M. Potter, and J. G. Povelites, in A. C. Graves, and D. K. Froman, eds., *Miscellaneous Physical and Chemical Techniques of the Los Alamos Project, Natl. Nucl. Energy Ser.,* McGraw-Hill, New York, 1951, Chap. 1.
50. Dybczynski, *Chem. Anal. (Warsaw),* **4,** 531 (1959); *Anal. Abstr.,* **7,** 2652 (1960).
51. Dyrssen, D., and L. D. Hay, *Acta Chem. Scand.,* **14,** 1100 (1960).
52. Edwards, R. R., "Isotopic Tracers in Chemical Systems," *Ann. Rev. Nucl. Sci.,* **1,** 301 (1952).
53. Faris, J. P., and J. W. Warton, *Anal. Chem.,* **34,** 1077 (1962).

54. Faris, J. P., and R. F. Buchanan, *U.S. At. Energy Comm. Rept.* TID-7606, 1960.
55. Feather, N., *Proc. Roy. Soc. (London)*, **A136**, 709 (1932).
56. Fields, P., in *Natl. Nucl. Energy Ser.,* Div. IV, Vol. 14B, G. T. Seaborg, J. J. Katz, and W. M. Manning, eds., McGraw-Hill, New York, 1949, p. 1072.
57. Flagg, J. F., and W. E. Bleidner, *J. Chem. Phys.,* **13**, 269 (1945).
58. "Foam Separation," final report, Radiation Applications, Inc., Long Island City, N.Y., June 1960.
59. Forsling, W., *Arkiv Kemi,* **5**, 489, 503 (1953).
59a. Frauenfelder, H., *Helv. Phys. Acta,* **23**, 347 (1950).
60. Freedman, A. J., and E. C. Anderson, *Nucleonics,* **10**, No. 8, 57 (1952).
61. Freiser, H., and G. H. Morrison, *Ann. Rev. Nucl. Sci.,* **9**, 221 (1959).
62. Freund, H., and F. J. Miner, *Anal. Chem.,* **25**, 564 (1953).
63. Friedlander, G., and J. W. Kennedy, *Nuclear and Radiochemistry,* Wiley, New York, 1955, p. 315.
64. Frierson, W. J., and J. W. Jones, *Anal. Chem.,* **23**, 1447 (1951).
65. Fritz, J. S., and H. Waki, *Anal. Chem.,* **35**, 1079 (1963).
66. Furman, N. H., R. J. Mundy, and G. H. Harrison, *U.S. At. Energy Comm. Doc.* AECD-2938, 1950.
67. Geiss, J., and C. Gfeller, *Proceedings of the Second International Conference on the Peaceful Uses of Atomic Energy, Geneva, 1958,* Vol. 3, Paper 236.
68. Gerhardt, L., *Naturwissenchaften,* **33**, 24 (1946).
69. Gest, H., N. E. Ballou, B. M. Abraham, and C. D. Coryell, *Natl. Nucl. Energy Ser. Plutonium Project Record,* Vol. 9, Paper 12, McGraw-Hill, New York, 1951, p. 145.
70. Gest, H., W. H. Burgus, and T. H. Davies, *Natl. Nucl. Energy Ser., Plutonium Project Record,* Vol. 9, Paper 13, p. 157.
71. Gest, H., and L. E. Glendenin, *Natl. Nucl. Energy Ser. Plutonium Project Record,* Vol. 9, Paper 14, p. 170.
72. Ghosh-Mazumdar, A. S., and M. Lederer, *J. Inorg. Nucl. Chem.,* **3**, 379 (1957).
73. Gile, J. D., W. M. Garrison, and J. G. Hamilton, *J. Chem. Phys.,* **18**, 1419 (1950).
74. Glass, R., *J. Am. Chem. Soc.,* **77**, 807 (1955).
75. Glendenin, L. E., *Natl. Nucl. Energy Ser., Plutonium Project Record,* Vol. 9, Paper 260, McGraw-Hill, New York, 1951, p. 1549.
76. Glendenin, L. E., *Natl. Nucl. Energy Ser., Plutonium Project Record,* Paper 259, p. 1545.
77. Gotte, H., and D. Patze, *Z. Elektrochem.,* **58**, 636 (1954).
78. Grahme, D., and G. Seaborg, *J. Am. Chem. Soc.,* **60**, 2524 (1938).
79. Grahme, D., and G. Seaborg, *Phys. Rev.,* **54**, 240 (1938).
80. Haenny, C., and P. Mivelaz, *Helv. Chim. Acta,* **31**, 633 (1948).
81. Hague, J. L., E. D. Brown, and H. A. Bright, *J. Res. Natl. Bur. Std.,* **53**, 261 (1954).
82. Hahn, O., *Applied Radiochemistry,* Cornell University Press, Ithaca, N.Y., 1936.
83. Hahn, O., *Z. Elektrochem.,* **38**, 511 (1932).
84. Hahn, O., and F. Strassmann, *Naturwissenschaften,* **27**, 11 (1939).
85. Hahn, O., F. Strassmann, and W. Seelman-Eggebert, *Z. Naturforsch.,* **1**, 545 (1946).

86. Hahn, O., and O. Werner, *Naturwissenschaften,* **17,** 45 (1945).
87. Hahn, P. F., *Ind. Eng. Chem., Anal. Ed.,* **17,** 45 (1945).
88. Haissinsky, M., *Compt. Rend.,* **199,** 1397 (1934).
89. Haissinsky, M., *J. Phys. Radium,* **10,** 312 (1949).
90. Haissinsky, M., *J. Chim. Phys.,* **32,** 116 (1935).
91. Harris, G. M., *Trans. Faraday Soc.,* **47,** 716 (1951).
92. Harvey, B., H. Heal, A. Maddock, and E. Rowley, *J. Chem. Soc.,* **1947,** 1010.
93. Henrich, F. A. K., *Chemie und chemische Technologie radioaktiver Stoffe,* Berlin, 1918.
94. Hevesy, G., and F. A. Paneth, *A Manual of Radiochemistry,* 2nd Ed., Oxford University Press, New York, 1938.
95. Hevesy, G., and F. A. Paneth, *Sber. Akad. Wien.,* **122,** 1937 (1913), **123,** 1619 (1914).
96. Hevesy, G., and F. Paneth, *Sitzber. Akad. Wiss. Wien, Math.-naturw. Klasse,* **Abt. IIa, 123,** 1909 (1914); *Monatsh.,* **36,** 75 (1915), *Physik Z,* **15,** 797 (1914).
97. Holleck, L., and L. Hartinger, *Angew. Chem.,* **66,** 586 (1954).
99. Honda, M., *Japan Analyst,* **3,** 132 (1953); *Chem. Abstr.,* **48,** 9868 (1954).
100. Hopkins, H., Jr., Procedure in: Meinke, W. W., *U.S. At. Energy Comm. Rept.* AECD-2738, 40, 1949.
101. Hosain, F., *Naturwissenschaften,* **45,** 107 (1958).
102. Huffman, E. H., and R. C. Lilly, *J. Am. Chem. Soc.,* **73,** 2402 (1951).
103. Huffman, E. H., and R. L. Oswalt, *J. Am. Chem. Soc.,* **72,** 3323 (1950).
104. Huffman, E. H., R. L. Oswalt, and L. A. Williams, *J. Inorg. Nucl. Chem.,* **3,** 49 (1956).
105. Hufford, D. L., and B. F. Scott, in G. T. Seaborg, J. J. Katz, and W. M. Manning, eds., *The Transuranium Elements, Natl. Nucl. Energy Ser.,* Div. IV, Vol. XIV-B, McGraw-Hill, New York, 1949, p. 1149.
106. Hulet, E. K., R. G. Gutmacher, and M. S. Coops, *J. Inorg. Nucl. Chem.,* **17,** 350 (1961).
107. Hume, D. N., *Radiochemical Studies: The Fission Products,* Vol. 3, McGraw-Hill, New York, 1951, p. 1495.
108. Hyde, E. K., *Proceedings of the International Conference on the Peaceful Uses of Atomic Energy, Geneva, 1956,* Vol. 7, p. 281.
109. Hyde, E. K., A. Ghiorso, and G. T. Seaborg, *Phys. Rev.,* **77,** 769 (1950).
110. Ishimori, T., and H. Okuno, *Bull. Chem. Soc. Japan,* **29,** 78 (1956).
111. Jacobi, E., *Helv. Phys. Acta,* **22,** 66 (1949).
112. Johnson, G. L., R. F. Leininger, and E. Segré, *J. Chem. Phys.,* **17,** 1 (1944).
113. Johnston, W. H., *Science,* **124,** 801 (1956).
114. Joliot, F., and I. Curie, *Radioactivité artificielle,* Hermann, Paris, 1935.
115. Katz, J., and G. T. Seaborg, *The Chemistry of the Actinide Elements,* Wiley, New York, 1957, pp. 225, 297.
116. Keder, W. E., J. C. Sheppard, and A. S. Wilson, *J. Inorg. Nucl. Chem.,* **12,** 327 (1963).
117. Keler, C. *J. Chromatog.,* **7,** 535 (1962).
118. Kevorkian, V., and E. L. Gaden, Jr., *A.I.Ch.E.J.,* **3,** No. 2, 183 (1957).
119. Kiha, E., S. Ohashi, and S. Tado, *Bull. Chem. Soc. Japan,* **29,** 745 (1956).
120. Knoch, W., *Z. Naturforsch.,* **16a,** 525 (1961).
121. Ko, R., *Nucleonics,* **15,** No. 1, 72 (1957).
122. Korkisch, J., and F. Tera, *Anal. Chem.,* **33,** 1264 (1961).

123. Korkisch, J., and F. Hecht, *Mikrochim. Acta,* **1956,** 1230.
124. Kraus, K. A., and G. E. Moore, *J. Am. Chem. Soc.,* **73,** 9, 13, 2900 (1951).
125. Kraus, K. A., and F. Nelson, *Ann. Rev. Nucl. Sci.,* **7,** 31 (1957).
126. Kraus, K. A., and F. Nelson, *Proceedings of the International Conference on the Peaceful Uses of Atomic Energy, Geneva, 1956,* Vol. 7, pp. 113, 131.
127. Kraus, K. A., and F. Nelson, *J. Am. Chem. Soc.,* **76,** 5916 (1954).
128. Kraus, K. A., and F. Nelson, "Symposium on Ion Exchange and Chromatography in Analytical Chemistry," *ASTM Spec. Bull. No.* 1ᶠ . 27, 1956.
129. Kraus, K. A., G. E. Moore, and F. Nelson, *J. Am. Chem. Soc.,* **78,** 2692 (1956).
130. Kraus, K. A., F. Nelson, and G. E. Moore, *J. Am. Chem. Soc.,* **77,** 3972 (1955).
131. Kraushaar, J. J., E. D. Wilson, and K. T. Bainbridge, *Phys. Rev.,* **90,** 610 (1953).
132. Krishen, A., and H. Freiser, *Anal. Chem.,* **29,** 288 (1957).
133. Kulp, J. L., in W. C. Peacock, ed., "Conference on Measurements and Standards of Radioactivity," *Natl. Acad. Sci.-Natl. Res. Council Publ.* No. 573, 1958.
134. Kunin, R., et al., Annual Reviews, *Anal. Chem.,* **27,** 1191 (1955); **28,** 729 (1956); *Ind. Eng. Chem.,* **47,** 565 (1955); **48,** 540 (1956); **49,** 507 (1957).
135. Kurbatov, J. D., and M. N. Kurbatov, *J. Chem. Phys.,* **13,** 208 (1945).
136. Kurbatov, J. D., and M. H. Kurbatov, *J. Phys. Chem.,* **46,** 441 (1942).
137. Kusaka, Y., and W. W. Meinke, "Rapid Radiochemical Separations," *Nucl. Sci. Ser. Rept.,* NAS-ND 3104, 1961.
138. Lad, R. A., and T. F. Young, *Natl. Nucl. Energy Ser., Plutonium Project Record,* Vol. 9, Paper 317, McGraw-Hill, New York, 1951, p. 1833.
139. Lederer, M., *Anal. Chim. Acta,* **8,** 134 (1953).
140. Lederer, M., *Anal. Chim. Acta,* **11,** 528 (1954).
141. Lederer, M., *Anal. Chim. Acta,* **12,** 14 (1955).
142. Lederer, T., and M. Lederer, *Chromatography,* Elsevier, 2nd English Ed., Amsterdam, 1957.
143. Lederer, M., *Paper Electrophoresis,* Elsevier, Amsterdam, 1955.
144. Lerner, M., and W. Rieman, *Anal. Chem.,* **26,** 610 (1954).
145. Libby, W. F., *Radiocarbon Dating,* 2nd Ed., University of Chicago Press, Chicago, Ill., 1955.
146. Libby, W. F., "Radiocarbon Dating," *Science,* **133,** 621 (1961).
147. Livingood, J. J., and G. T. Seaborg, *Phys. Rev.,* **54,** 51 (1938).
148. Livingston, R., E. Shapiro, and N. Elliot, *Natl. Nucl. Energy Ser., Plutonium Project Record,* Vol. 9, Paper 316, McGraw-Hill, New York, 1951, p. 1823.
149. Lowe, R. W., S H. Prestwood, R. R. Rickard, and E. I. Wyatt, *Anal. Chem.,* **33,** 874 (1961).
150. MacNevin, W. M., and I. L. Lee, *Anal. Chim. Acta,* **12,** 544 (1965).
151. Maeck, W. J., G. L. Booman, M. E. Kussy, and J. E. Rein, *Anal. Chem.,* **33,** 1775 (1961).
152. Makover, W., *The Radioactive Substances,* K. Paul, French, Frübner and Company, London, 1908.
153. Marsh, S. F., W. J. Maeck, G. L. Booman, and J. E. Rein, *Anal. Chem.,* **34,** 1406 (1962).
154. Mathur, H. B., and E. K. Hyde, *Phys. Rev.,* **98,** 79 (1955).
155. Matsuura, J., *Japan Analyst,* **4,** 242 (1955).

156. Matusek, J. M., and T. T. Sugihara, *Anal. Chem.* **33**, 35 (1961).
157. Maydan, D., J. Toicher, and F. Feidenberg, *Israel At. Energy Rept.*, No. IA 619, July 1961.
158. Mayer, S. W., and E. C. Freiling, *J. Am. Chem. Soc.*, **75**, 5647 (1953).
159. McCown, J. J., and R. P. Larson, *Anal. Chem.*, **33**, 1003 (1961).
160. McHutchison, J., *Proc. Roy. Soc. (London)*, **A111**, 134 (1946).
161. Meadows, J. W. T., and G. M. Matlack, *Anal. Chem.*, **32**, 1607 (1960).
162. Meinke, W. W., *Nucleonics,* **17**, No. 9, 86 (1959).
163. Meinke, W. W., A. Chiorso, and G. T. Seaborg, *Phys. Rev.*, **81**, 782 (1951).
164. Melander, Lars, *Isotope Effects on Reaction Rates,* Ronald, New York, 1960.
165. Meloche, V. W., and A. F. Preuss, *Anal. Chem.*, **26**, 1911 (1954).
166. Merz, E., *Z. Anal. Chem.*, **166**, 417 (1959).
167. Misumi, S., and T. Taketatsu, *J. Inorg. Nucl. Chem.*, **20**, 127 (1961).
168. Moljk, A., R. W. P. Drever, and S. C. Curran, *Proc. Roy. Soc. (London)*, **A239**, 433 (1957).
169. Moore, F. L., *Anal. Chem.*, **32**, 1075 (1960).
170. Moore, F. L., *Anal. Chem.*, **28**, 997 (1956).
171. Moore, F. L., *Anal. Chem.*, **30**, 908 (1958).
172. Moore, F. L., *Anal. Chem.*, **29**, 941 (1957).
173. Moore, F. L., *Natl. Acad. Sci.-Natl. Res. Council, Nucl. Sci. Ser. Radiochem. Tech.*, NAS-NS-3101, December 1960.
174. Moore, F. L., and J. E. Hudgens, *Anal. Chem.*, **29**, 1767 (1957).
175. Morrison, G. H., and H. Freiser, *Anal. Chem.*, **30**, 632 (1958).
176. Myers, O. E., and R. J. Prestwood, in A. C. Wahl and N. A. Bonner, eds., *Radioactivity Applied to Chemistry,* Wiley, New York, 1951, Chap. 1, and Part 2, p. 323, Table 1A, 1B.
177. Nachod, F. G., and J. Schubert, eds., *Ion Exchange Technology,* Academic, New York, 1956.
178. Nelson, F., *J. Am. Chem. Soc.*, **77**, 813 (1955).
179. Norris, T. H., *J. Phys. Colloid Chem.*, **54**, 777 (1950).
180. Oliver, J. R., J. R. Meriwether, and R. H. Rainey, ORNL-2668, April 1959.
181. Osborn, G. N., *Synthetic Ion Exchangers,* Chapman and Hall, London, 1955.
182. Paneth, F., *Ber.*, **51**, 1704 (1918).
183. Paneth, F., *Kolloid-Z,* **13**, 1 (1913).
184. Paneth, F., *Kolloid-Z,* **13**, 297 (1913).
185. Paneth, F., *Z. Elektrochem.*, **24**, 298 (1918).
186. Paneth, F., *Radioelements as Indicators,* McGraw-Hill, New York, 1938.
187. Peppard, D. F., G. W. Mason, W. J. Driscoll, and S. McCarty, *J. Inorg. Nucl. Chem.*, **12**, 141 (1959).
188. Peppard, D. F., G. W. Mason, and I. Hucher, *J. Inorg. Nucl. Chem.*, **18**, 245 (1961).
189. Raynor, J., *Natl. Nucl. Energy Ser., Plutonium Project Record,* Vol. 9, Paper 312, McGraw-Hill, New York, 1951, p. 1775.
190. Rimshaw, S. J., and G. F. Malling, *Anal. Chem.*, **33**, 751 (1961).
191. Rogers, L. B., D. P. Krause, J. C. Greiss, Jr., and D. B. Ehrlinger, *J. Electrochem. Soc.*, **95**, 33 (1949).
192. Rogers, L. B., and A. F. Stehney, *J. Electrochem. Soc.*, **95**, 25 (1949).
193. Rubinson, W., *Natl. Nucl. Energy Ser., Plutonium Project Record,* Vol. 9, Paper 314, McGraw-Hill, New York, 1951, p. 1833.

194. Russell, A. S., *Introduction to the Chemistry of the Radioactive Substances*, J. Murray, London, 1922.
195. Rutherford, E., *Phil. Mag.*, **37**, 537 (1919).
196. Rutherford, E., and F. Soddy, *Phil. Mag.*, **4**, 569, 580 (1898); **5**, 441, 561 (1903).
197. Ryabchikov, D. I., and V. E. Osipova, *J. Anal. Chem. USSR* (English transl.) **11**, No. 3, 285 (1956).
198. Ryan, J. L., and E. J. Wheelwright, *Ind. Eng. Chem.*, **51**, 60 (1959).
199. Ryan, J. L., and E. J. Wheelwright, *U.S. At. Energy Comm. Rept.* HW-55893, 1959.
200. Ryan, J. L., *U.S. At. Energy Comm. Rept.* HW-59193 (REV), 1959.
201. Salaria, G. B. S., C. L. Rulfs, and P. J. Elving, *Anal. Chem.*, **35**, 983 (1963).
202. Samuelson, O., *Chim. Anal.*, **37**, 191 (1955).
203. Samuelson, O., L. Lunden, and K. Schramm, *Z. Anal. Chem.*, **140**, 330 (1953).
204. Samuelson, O., and B. Sjöberg, *Anal. Chim. Acta*, **14**, 121 (1956).
205. Samuelson, O., and E. Sjöström, *Anal. Chem.*, **26**, 1908 (1954).
206. Samuelson, O., E. Sjöström, and S. Forsblom, *Z. Anal. Chem.*, **144**, 323 (1955).
207. Schindewolf, U. L., and J. W. Irvine, Jr., *Anal. Chem.*, **30**, 906 (1958).
208. Schindewolf, U. L., J. W. Wonchester, and C. D. Coryell, *Phys. Rev.*, **105**, 1763 (1957).
209. Schubert, J., and E. E. Conn, *Nucleonics*, **4**, No. 6, 2 (1949).
210. Schumacher, E., *Helv. Chim. Acta*, **40**, 221 (1957).
211. Schumacher, E., and H. J. Streiff, *Helv. Chim. Acta*, **40**, 228 (1957).
212. Schumacher, E., and H. J. Streiff, *Helv. Chim. Acta*, **40**, 234 (1957).
213. Schumacher, E., and W. Friedli, *Helv. Chim. Acta*, **43**, 1706 (1960).
214. Schumacher, E., and H. J. Streiff, *Helv. Chim. Acta*, **41**, 824 (1958).
215. Seaborg, G. T., J. J. Katz, and W. M. Manning, eds., *Natl. Nucl. Energy Ser.*, Div. IV, Vol. 14B, Parts I and II, McGraw-Hill, New York, 1949; G. T. Seaborg, and J. J. Katz, eds., *Natl, Nucl. Energy Ser.*, Div. IV, Vol. 14A, McGraw-Hill, New York, 1954.
216. Seiler, J. A., *Natl. Nucl. Energy Ser., Plutonium Project Record*, Vol. 9, Paper 270, McGraw-Hill, New York, 1951, p. 1589.
217. Shevekenko, V. B., V. S. Shmidt, and E. A. Mezov, *Zh. Neorg. Khim.*, **5**, 1911 (1960).
218. Shukla, S. K., and M. Lederer, *J. Chromatog.*, **9**, 255 (1962).
219. Shvedov, V. P., T. Ten, and A. V. Stepanov, *Zh. Anal. Khim.*, **15**, 16 (1960).
220. Siekierski, S., *J. Inorg. Nucl. Chem.*, **12**, 129 (1959).
221. Smith, G. W., and F. L. Moore, *Anal. Chem.*, **29**, 448 (1957).
222. Smith, G. W., and S. A. Reynolds, *Anal. Chim. Acta*, **12**, 151 (1955).
223. Soddy, F., *Chemistry of the Radioelements*, London, 1911 and 1914.
224. Spedding, F. H. et al., *J. Am. Chem. Soc.*, **69**, 2777, 2786, 2812 (1947).
225. Spedding, F. H., J. P. Powell, and E. E. Wheelwright, *J. Am. Chem. Soc.*, **76**, 612 (1954).
226. Spiess, F. N., *Phys. Rev.*, **94**, 1292 (1954).
227. Starick, I. E., *Z. Physik. Chem.* (*Leipzig*), **A157**, 269 (1931).
228. Steigman, J., *Phys. Rev.*, **53**, 771 (1938).
229. Steinberg, E. P., *Radiochemical Studies: The Fission Products*, Vol. 3: McGraw-Hill, New York, 1951, p. 1495.

230. Stevenson, P. G., and H. G. Hicks, *U.S. At. Energy Comm. Doc.* UCRL-2009, 1952.
231. Stewart, D. C., and J. P. Faris, *J. Inorg. Nucl. Chem.*, **3**, 64 (1956).
232. Street, K., Jr., and G. T. Seaborg, *J Am. Chem. Soc.*, **70**, 4268 (1948).
233. Sugarman, N., *Natl. Nucl. Energy Ser., Plutonium Project Record*, Vol. 9, Paper 315, McGraw-Hill, New York, 1951, p. 1808.
234. Sugihara, T. T., in C. E. Crouthamel, ed., *Analytical Chemistry*, Vol. III, (*Progr. in Nucl. Energy Ser.* IX), Pergamon, 1961, Chap. 1.
235. Sugihara, T. T., H. I. James, E. J. Troianello, and V. T. Bowen, *Anal. Chem.*, **31**, 44 (1959).
236. Surl, J. P., Jr., and G. R. Choppin, *J. Inorg. Nucl. Chem.*, **4**, 62 (1957).
237. Szilard, L., and T. A. Chalmers, *Nature*, **134**, 462 (1934).
238. Teicher, H., and L. Gordon, *Anal. Chem.*, **23**, 930 (1951).
239. Templeton, D. H., J. J. Howland, and I. Perlman, *Phys. Rev.*, **72**, 758 (1947).
240. Thomas, H. C., and G. R. Frysinger, *Ann. Rev. Phys. Chem.*, **7**, 137 (1956).
241. Thompson, S. G., L. Morgan, R. James, and I. Perlman, *Natl. Nucl. Energy Ser.*, Div. IV, Vol. 14B, G. T. Seaborg, J. J. Katz, and W. M. Manning, eds., McGraw-Hill, New York, 1949.
242. Thompson, S. G., B. G. Harvey, G. R. Choppin, and G. T. Seaborg, *J. Am. Chem. Soc.*, **76**, 6229 (1954).
243. Tompkins, E. R., *J. Am. Chem. Soc.*, **70**, 3520 (1948).
244. Tompkins, E. R., J. X. Khym, and W. E. Cohn, *J. Am. Chem. Soc.*, **69**, 2769 (1947).
245. Tompkins, E. R., and S. W. Mayer, *J. Am. Chem. Soc.*, **69**, 2859, 2866, (1947).
246. Venturello, G., and C. Gualandi, *Ann. Chim.*, **46**, 229 (1956); *Chem. Abstr.*, **50**, 12739° (1956).
247. Wacker, R. E., and W. H. Baldwin, *U.S. At. Energy Comm. Rept.* ORNL-637, 1950; *Nucl. Sci. Abstr.*, **4**, 469 (1950); *Brit. Abstr.*, **1951**, C244.
248. Weiss, H. V., and W. H. Shipman, *Anal. Chem.*, **29**, 1764 (1957).
249. Werner, O., *Z. Physik Chem.* (*Leipzig*), **A156**, 89 (1931).
250. White, J. C., and W. J. Ross, *Natl. Acad. Sci.-Natl. Res. Council. Nucl, Sci. Ser. Radiochem. Tech.*, NAS-NS-3102; February 1961.
251. Wilkinson, G., and H. G. Hicks, *Phys. Rev.*, **77**, 314 (1950).
252. Winsberg, L., *Natl. Nucl. Energy Ser., Plutonium Project Record*, Vol. 9, Paper 228, McGraw-Hill, New York, 1951, p. 1440.
253. Winsberg, L., and L. E. Glendenin, *Natl. Nucl. Energy Ser., Plutonium Project Record*, Vol. 9, Paper 229, p. 1443.
254. Wish, L., *Anal. Chem.*, **33**, 53 (1961).
255. Wish, L., Anal. Chem., **33**, 1002 (1961).
256. Wright, B. T., *Phys., Rev.*, **71**, 839 (1947).
257. Yoshino, Y., and M. Kojima, *Bull. Chem. Soc. Japan*, **23**, 46 (1960).
258. Zsigmondy, *Kolloid-Z*, **13**, 105 (1913).

APPENDIX: RADIOCHEMICAL PROCEDURES

Radiochemistry has come to be known as the study and chemical manipulation of radioactive sources, at either the macro or the tracer concentration level. This field of study has no journal or other publication outlet of its own, and material must therefore be published (usually in abbreviated and summarized form) in analytical chemistry and nuclear science journals. The one exception to this is the Russian journal, *Radiochemistry*, which first appeared in 1959. Partial translations of Volume 1, Numbers 1 and 2, and abstracts of articles in Numbers 4, 5, and 6, edited by Dr. A. G. Maddock, were published in 1960 by Pergamon Press Limited, Oxford, England. Subsequent editions have not been published for reasons unknown. Lack of a specialized publication outlet for radiochemistry and its applicability to many branches of chemistry are the reasons for diversified sources for radiochemical separation procedures.

The Subcommittee on Radiochemistry of the U.S. National Research Council has realized the need for current compilations of radiochemical information and procedures, and has sponsored the writing of a series of monographs on the radiochemistry of the elements. Each monograph includes information required for radiochemical work with a particular element or group of closely related elements, for example, nuclear and chemical features pertaining to radiochemistry, counting techniques, and a very important collection of radiochemical procedures as found in the literature. At this time, 56 of the monographs have been published and are available at a nominal cost (about one cent per page) from the Office of Technical Services, Department of Commerce, Washington, D.C. See Appendix Table I. The sequence numbering is NAS-NS-3001

APPENDIX TABLE I
National Academy of Sciences Nuclear Science Series on Radiochemistry[a]

NAS-NS Number	Radiochemistry of—	Authors
3001	Cd	James R. DeVoe
3002	As	Harold C. Beard
3003	Fr	Earl K. Hyde
3004	Th	Earl K. Hyde
3005	F, Cl, Br, I	Jacob Kleinberg; G. A. Cowan
3006	Am, Cm	R. A. Pennemen; T. K. Keenan
3007	Cr	J. Pijck
3008	Rh	G. R. Choppin
3009	Mo	E. M. Scadden; N. E. Ballou
3010	Ba, Ca, Sr	D. N. Sunderman; C. W. Townley
3011	Zr, Hf	Ellis P. Steinberg
3012	At	Evan H. Appelman
3013	Be	A. W. Fairhall
3014	In	D. N. Sunderman; C. W. Townley

NAS-NS Number	Radiochemistry of—	Authors
3015	Zn	Harry G. Hicks
3016	Pa	H. W. Kirby
3017	Fe	J. M. Nielsen
3018	Mn	G. W. Leddicotte
3019	C, O, N	J. Hodis
3020	Rare earths, Sc, Y, Ac	P. C. Stevenson; W. E. Nervik
3021	Tc	Edward Anders
3022	V	J. L. Brownlee, Jr.
3023	Sn	W. E. Nervik
3024	Mg	A. W. Fairhall
3025	Rare gases	F. J. Momyer, Jr.
3026	Hg	J. Roesmer; P. Kruger
3027	Cu	F. F. Dryer; G. W. Leddicotte
3028	Re	G. W. Leddicotte
3029	Ru	E. I. Wyatt; R. R. Richard
3030	Se	G. W. Leddicotte
3031	Transcurium	G. H. Higgins
3032	Al, Ga	J. E. Lewis
3033	Sb	W. J. Maeck
3034	Ti	Chong Kuk Kim
3035	Cs	H. L. Finston; M. T. Kinsley
3036	Ag	J. F. Emery; G. W. Leddicotte
3037	Po	P. E. Figgins
3038	Te	G. W. Leddicotte
3039	Nb, Ta	E. P. Steinberg
3040	Pb	W. M. Gibson
3041	Co	L. C. Bate; G. W. Leddicotte
3042	W	W. T. Mullins; G. W. Leddicotte
3043	Ge	J. A. Marinsky
3044	Pt	G. W. Leddicotte
3045	Ir	G. W. Leddicotte
3046	Os	G. W. Leddicotte
3047	Ag	D. N. Sunderman; C. W. Townley
3048	K	W. T. Mullins; G. W. Leddicotte
3049	Si	W. T. Mullins; G. W. Leddicotte
3050	U	James E. Gindler
3051	Ni	L. J. Kirby
3052	Pd	Ove T. Høgdahl
3053	Rb	G. W. Leddicotte
3054	S	G. W. Leddicotte
3055	Na	W. T. Mullins; G. W. Leddicotte
3056	P	W. T. Mullins; G. W. Leddicotte

a These reports are available from the Office of Technical Services, Department of Commerce, Washington, D.C.

through NAS-NS-3056. The reader is referred to these publications as an excellent source of radiochemical material for a particular element. These monographs are being revised (under the same numbers) as the situation warrants and will presumably be available for many years to come. The references for radiochemical procedures are divided into six classifications: general, carrier-free, rapid, low-level, isotopic exchange, and volatilization. The general category includes detailed as well as general procedures found in radiochemistry. The carrier-free classification includes procedures used specifically in carrier-free work. It should be mentioned that numerous procedures found in the general classification can also be applied to carrier-free separations, as for example, reference I-7. For rapid and low-level classifications the reader is referred to the most recent review articles.

REFERENCES FOR RADIOCHEMICAL PROCEDURES

General

I-1. Beaufit, L. J., and H. R. Lukens, Jr., "Handbook of Radiochemical Analysis," Vol. I: "Radiochemical Techniques"; Vol. II: "Radiochemical Procedures," *U.S. At. Energy Comm., Rept.* PB-121690 and PB-121689, 1953–52.

I-2. Finston, H. L., and J. Miskel, Jr., "Radiochemical Separation Techniques," *Ann. Rev. Nucl. Sci.,* **5,** 269 (1955).

I-3. Freiser, H., and G. H. Morrison, "Solvent Extraction in Radiochemical Separations," *Ann. Rev. Nucl. Sci.,* **9,** 221 (1959).

I-4. Harley, J. O., and I. B. Whitney, "Manual of Standard Procedures; Health and Safety Laboratory, U.S. AEC New York Operations Office," *U.S. At. Energy Comm. Rept.* NYO-4700, 1960.

I-5. Hyde, E. K., "Radiochemical Separation Methods for the Actinide Elements," *Proceedings of the International Conference on The Peaceful Uses of Atomic Energy, Geneva, 1955,* Vol. 7, Paper 728, p. 281.

I-6. Kahn, B., E. R. Eastwood, and W. J. Lacy, "Use of Ion Exchange Resins to Concentrate Radionuclides for Subsequent Analysis," *U.S. At. Energy Comm. Rept.* ORNL-2321, 1957.

I-7. Kleinberg. J., et al., "Collected Radiochemical Procedures," *U.S. At. Energy Comm. Los Alamos Rept.* LA-1721, 2nd Rev., 1958.

I-8. Kraus, K. A., and F. Nelson, "Anion Exchange Studies of the Fission Products," *Proceedings of The International Conference on The Peaceful Uses of Atomic Energy, Geneva, 1955* Vol. 7, Paper 837, p. 113.

I-9. Kraus, K. A., and F. Nelson, "Radiochemical Separation by Ion Exchange," *Ann. Rev. Nucl. Sci.,* **7,** 31 (1957).

I-10. Lindner, M., "Radiochemical Procedures in Use at University of California Radiation Laboratory (Livermore)," *U.S. At. Energy Comm. Rept.* UCRL-4377, 1954.

I-11. Meinke, W. W., "Chemical Procedures Used in Bombardment Work at Berkeley," *U.S. At. Energy Comm. Rept.* AECD-2738, 1949; Addendum No. 1, AECD-2750, 1949; Addendum No. 2, AECD-3084, 1951.

I-12. Oak Ridge National Laboratory, "Radiochemical Methods (Oak Ridge)," *Master Analytical Manual*, Section 2, TID-7015, 1957.

I-13. Oak Ridge National Laboratory, "Nuclear Analyses Methods (Oak Ridge)," *Master Analytical Manual*, Sections 3 and 5, TID-7015, 1957.

I-14. Rodden, C. J., "Analytical Chemistry of the Manhattan Project," *Natl. Nucl. Energy Ser. VIII*, Vol. 1, McGraw-Hill, New York, 1950.

I-15. Stevenson, P. C., and H. Hicks, "Separation Techniques Used in Radiochemistry," *Ann. Rev. Nucl. Sci.*, **3**, 221 (1953).

I-16. Stewart, D. C., "Rare Earth and Trans-plutonium Element Separation by Ion Exchange Methods," *Proceedings of The International Conference on The Peaceful Uses of Atomic Energy, Geneva, 1955*, Vol. 7, Paper 729, p. 321, 1956.

I-17. Sunderman, D. N., and W. W. Meinke, "Evaluation of Radiochemical Separation Procedures," *Anal. Chem.*, **29**, 1578 (1957).

I-18. Wilkenson, C., and W. E. Grummitt, "Chemical Separation of Fission Products," *Nucleonics*, **9**, No. 3, 52 (1951).

I-19. World Health Organization, "Methods of Radiochemical Analysis," *Tech. Rept. Ser.* No. 173, also *FAO At. Energy Ser.* No. 1.

Carrier-Free

II-1. de Pasquali, G., E. von Goeler, and R. N. Peacock, "Separation of Carrier-free Indium-III from Silver by Ion Exchange, "*J. Inorg. Nucl. Chem.*, **11**, 257 (1959).

II-2. de Soete, D., J. Hoste, and G. Leliaert, "Preparation of Carrier-free Cr-51," *Intern. J. Appl. Radiation Isotopes*, **8**, 134 (1960).

II-3. Garrison, W. M., and J. C. Hamilton, "Production and Isolation of Carrier-free Radioisotopes," *Chem. Rev.* **49**, 237 (1951). Also as UCRL-1067, available from Publication Board Project, Library of Congress, Washington, D.C.

II-4. Gruverman, I. J., and P. Kruger, "Cyclotron-Produced Carrier-free Radioisotopes—Thick Target Yield Data and Carrier-free Separation Procedures," *Intern. J. Appl. Radiation Isotopes*, **5**, 21 (1959).

II-5. Kahn, M., and A. L. Langhorst, "A Novel Method for the Preparation of Carrier-free Lead-212," *J. Inorg. Nucl. Chem.*, **15**, 384 (1960).

II-6. Kahn, M., K. H. Jones, and K. Lawson, Bibliography (1950–57) of 1216 references to unclassified work on the preparation, separation, purification, etc., of carrier-free tracers, *Rept.* LA 2331 UNM, July 1959.

II-7. Mayer, W. J., and R. L. Anderson, "Radioactive Parent-Daughter Separations *(Indium-113)*," *Ind. Eng. Chem.*, **52**, 993 (1960).

II-8. McDonald, D., "Carrier-free Separation of Praseodymium-144 from Cerium-144 by Anion Exchange," *Anal. Chem.*, **33**, 1807 (1961).

II-9. Meadows, J. W. T., and G. M. Matlock, "Radiochemical Determination of Ruthenium by Solvent Extraction and Preparation of Carrier-free Ruthenium Activity," *Anal. Chem.*, **34**, 89 (1962).

II-10. Murin, A. N., V. D. Nefedov, and I. A. Yutlandov, "The Production and Separation of Carrier-free Radioactive Isotopes," *At. Energy Res. Establ. (Harwell) Rept.* AERE-Lib/Trans. 722, 1956.

II-11. Saito, K., S. Ikeda, and M. Saito, "Separation of Radioactive Arsenic from Germanium Irradiated with Protons," *Bull. Chem. Soc. Japan*, **33**, 884 (1960).

II-12. Taylor, D. M., "Preparation of Carrier-free Ag-111," *Intern. J. Appl. Radiation Isotopes*, **12**, 66 (1961).

II-13. Wahl, A. C., and N. A. Bonner, Radioactivity Applied to Chemistry, Wiley, New York, 1951, Table 6A-6F, pp. 393–460.

II-14. Wish, L., and E. D. Freiling, "Quantitative Radiochemical Carrier-free Procedures," *U.S. Naval Radiological Defense Lab. Rept.* USNRDL-464, July 1960.

Rapid

III-1. Greendale, A. E., and D. L. Love, "A Rapid Radiochemical Procedure for Tin," *U.S. Naval Radiological Defense Lab. Rept.* USNRDL-TR-632, February 1963.

III-2. Greendale, A. E., and D. L. Love, "Rapid Radiochemical Procedure for Antimony and Arsenic," *Anal. Chem.*, **35**, 632 (1963).

III-3. Kusaka, Y., and W. W. Meinke, "Rapid Radiochemical Separations," *Natl. Acad. Sci. Natl. Res. Council, Nucl. Sci. Ser. Radiochem. Tech.*, NAS-NS-3104, 1961.

III-4. Love, D. L., and D. Sam, "Radiochemical Determination of Sodium-24 and Sulfur-35 in Seawater," *U.S. Radiological Defense Lab. Rept.* USNRDL-TR-552, July 1961.

Low-Level

IV-1. Sugihara, T. T., in C. E. Crouthamel, ed., Analytical Chemistry, Vol. III, (*Progr. in Nucl. Energy Ser. IX*), Pergamon, 1961, Chap. 1.

Isotopic Exchange Separations

V-1. DeVoe, J. R., H. W. Nass, and W. W. Meinke, "Radiochemical Separation of Cadmium by Amalgam Exchange," *Anal. Chem.*, **33**, 1713 (1931).

V-2. Langer, A., *J. Chem. Phys.*, **10**, 321 (1942).

V-3. Qureshi, I. H., and W. W. Meinke, "Radiochemical Separation of Strontium by Amalgam Exchange," *Talanta*, **10**, 737 (1963).

V-4. Ruch, R. R., J. R. DeVoe, and W. W. Meinke, "Radiochemical Separation of Indium by Amalgam Exchange," TID-13224, 1961.

V-5. Sunderman, D. W., and W. W. Meinke, *Science*, **121**, 777 (1955).

V-6. Sunderman, D. W., and W. W. Meinke, "Evaluation of Radiochemical Separation Procedures (Silver by isotopic exchange)," *Anal. Chem.*, **29**, 1585 (1957).

Volatilization Separations

VI-1. DeVoe, J. R., and W. W. Meinke, "Some Applications on Vacuum Distillation of Metals to Radiochemical Separation," *Anal. Chem.*, **35**, 2 (1963).

VI-2. Frauenfelder, H., *Helv. Phys. Acta.*, **23**, 347 (1950).

VI-3. Kahn, M., and A. L. Langhorst, "Preparation of Carrier-free Lead-212," *J. Inorg. Nucl. Chem.*, **15**, 384 (1960).

VI-4. Zimmerman, J. B., and J. C. Ingles, "Isolation of the Rare Earth Elements," *Anal. Chem.*, **32**, 241 (1960).

Part I
Section D-6

Chapter 97

TRACER TECHNIQUES

By William Seaman, *American Cyanamid Company, Bound Brook, New Jersey* (retired)

Contents

I. INTRODUCTION

A. THE TRACER PRINCIPLE

A tracer, in general terms, is something which enables one to follow the location of something else by virtue of its being attached to the latter, thus allowing the combination to be recognized among other things not so identified. In chemical terms, this concept requires labeling or tagging the chemical entities—molecules or atoms—which one wishes to trace (or, less frequently, other molecules or atoms which remain associated with them) so that they may be located, and under some conditions their amounts determined, in the presence of foreign atoms or molecules. Tracing can be accomplished by altering the normal isotopic composition of the atomic species of the substance being tagged. This principle, together with instruments and techniques for detecting and measuring the abnormal composition, is the basis of all chemical tracer methods.

B. HISTORICAL DEVELOPMENT AND TYPES OF TRACER APPLICATIONS

Very soon after Soddy announced the principle of isotopy in 1913, F. Paneth and G. von Hevesy (1) first applied tracers to chemical analysis. They utilized the fact that they had not been able to separate radium D (^{210}Pb) from ordinary lead as a method for determining the solubility of lead sulfide and lead chromate in water by using this isotope as a tracer. Later Hevesy (2) used another lead isotope, thorium B (^{212}Pb), to determine the exchange of lead between a solution of lead nitrate and metallic lead or solid lead compounds.

These applications exemplify two classes into which tracer methods may be divided. The first has been called "analytical" and the second "kinetic," although both may be applied to analysis. Problems which are solved by "analytical" tracer methods could conceivably also be solved by other available chemical or physical procedures. Problems of exchange and transport of atoms or molecules in the presence of the identical atoms or molecules can be solved only by tracer methods. Both types assume identical behavior for all isotopes of an element (exceptions to this rule will be discussed later); however, in the "analytical" method the tracer maintains constant specific activity in all phases and chemical forms of the system, whereas in the "kinetic" method the specific activity may change with time and may differ among phases or chemical forms.

The use of radiotracers remained limited in scope during the period when only natural radioisotopes were available. The discovery of artificial radioisotopes by Joliot and Curie in 1934 led to some increase in use. Not until 1946, however, when radioisotopes from nuclear reactors started to become generally available, was a real impetus given to such applications. These years also saw a rise in the availability of many stable isotopes, which led to their more widespread use.

C. CAPABILITIES AND LIMITATIONS OF TRACER METHODS

For the many applications for which only tracer methods are suitable a discussion of their advantages and disadvantages would be academic; for applications where other methods might be used one should weigh relative merits and disadvantages. The most striking feature of tracer methods is their extreme sensitivity. This will vary greatly with different conditions, but with radioactive tracers the level of detection may be 10^{-14} g or even lower, with a precision around 1% or better. With stable isotopes, accuracies of $\pm 1\%$ may be attained, with high sensitivities. Another advantage of these methods is their specificity, which is unmatched by most other procedures. Tracer analyses may also frequently be carried out with little or no previous chemical or physical separations.

On the darker side of the picture is the need for rather expensive equipment. This is particularly true for the mass spectrometric methods used for most work with stable isotopes, although, for deuterium, methods based on infrared spectra and on the determination of the density of the water obtained by combustion of the sample have been employed. For radioactive tracers much work can be carried out with considerably less expensive equipment.

Although one must carefully consider the problem of safety, this is minimal for the small amounts of activity which are needed for most tracer uses. In fact, the U.S. Atomic Energy Commission permits the purchase of small amounts of many radioactive isotopes without special license. Nevertheless the problem must be recognized and met. Related to it is the need for meticulous work to prevent contamination of the laboratory and of counting equipment.

An inconvenience is the frequent need to synthesize the requisite tagged substances; this remains true in spite of the ever-widening list of commercially available compounds. Furthermore, there is the necessity for purification, not merely to a normal state of chemical purity, but to a state of radiochemical purity as well. This requires eliminating even minute amounts of impurities with high specific activities.

II. REQUIREMENTS FOR TRACERS AND LABELED COMPOUNDS AND THEIR PREPARATION

A. FACTORS AFFECTING CHOICE OF TRACER

1. Availability and Cost

Since chemical tracer methods depend on the use of an element which has had its isotopic composition altered from its normal state, either stable or radioactive isotopes may be employed. There is no difference in principle between the use of one and that of the other, but techniques, instruments, adaptability for particular analytical needs, cost, and other factors are not the same for the two types.

A great variety of isotopes is available. *International Directory of Isotopes* (3) lists the radioisotopes and stable isotopes available from suppliers throughout the world in various physical and chemical forms, as well as other related information. Stable isotopes may be purchased or, in some cases, rented, either in elemental form or incorporated into tagged compounds, with various enrichment factors, in some cases close to 100%. The Oak Ridge National Laboratory issues a catalog listing a number of enriched stable isotopes. They can also be obtained from several commercial suppliers, as well as from governmental and institutional sources outside of the United States (4).

Stable isotopes may be acquired without a license from the U.S. Atomic Energy Commission. Artificial radioisotopes, as well as so-called source materials (uranium and thorium and their ores) and special nuclear material, however, require a license for their purchase. Such a license may be obtained from the Atomic Energy Commission by anyone who satisfies certain requirements which demonstrate his ability to use these materials safely. Certain specified small quantities, which may be purchased and used without a license, are often sufficient for a given tracer experiment. Generally, though, it is advisable for anyone contemplating more than an occasional, sporadic experiment to acquire a license.

The cost of isotopes and tagged compounds varies considerably. When they can be used in the chemical and physical forms in which they are available from Oak Ridge or similar installations, their cost is not, as a rule, excessively higher than that of many reagent grade chemicals, particularly when one considers that very little activity is usually required for a tracer experiment. When tagged compounds must be synthe-sized, however, they can be rather expensive, either in terms of money if purchased, or in terms of working time if prepared in one's own laboratory.

2. Health Protection

No problem of health protection exists in regard to stable isotopes except the considerations normal to all chemical work. In using radio-isotopes, however, health protection must be considered, but for tracer work the protection problem is usually fairly simple, since the amounts of activity used are generally at the microcurie level. For gamma emit-ters or high-energy beta emitters some simple shielding may on occasion be necessary, but often can be avoided by utilizing distance. For moder-ately energetic and weak beta emitters the usual glass vessels provide sufficient shielding. It is important to carry out experiments with scrupu-lous cleanliness to avoid possible ingestion, absorption, or inhalation of radioactive material. Aside from any considerations of health, one must avoid contamination of laboratory surfaces and equipment so as not to raise the background radiation or otherwise endanger the accuracy of further work.

Table 97.I classifies a number of radioisotopes into three groups ac-cording to relative hazard on absorption into the body, and defines ac-

TABLE 97.I

Relative Internal Hazard of Various β- and γ-Emitting Isotopes[a]

Radiotoxicity group	Activity limits		
	Low	Intermediate	High
Slight hazard Na^{24}, K^{42}, Cu^{64}, Mn^{52}, As^{76}, As^{77}, Kr^{85}, Hg^{197}	Up to 1 mc	1–10 mc	Above 10 mc
Moderately dangerous H^3, C^{14}, Na^{22}, S^{35}, Cl^{36}, Mn^{54}, Fe^{59}, Co^{60}, Sr^{89}, Nb^{95}, $Ru^{103,106}$, P^{32}, Te^{127}, Te^{129}, I^{131}, Cs^{137}, Ba^{140}, La^{140}, Ce^{141}, Pr^{143}, Nd^{147}, Au^{198}, Au^{199}, $Hg^{203,205}$	Up to 0.1 mc	0.1–1 mc	Above 1 mc
Very dangerous Ca^{45}, Fe^{55}, Sr^{90}, Y^{91}, Zr^{95}, Ce^{144}, Pm^{147}, Bi^{210}	Up to 0.01 mc	0.01–0.1 mc	Above 0.1 mc

[a] Compiled from "*Safe Handling of Radioactive Isotopes*," National Bureau of Standards, Handbook 42, U.S. Dept. of Commerce, 1949, page 4.

tivity levels to be handled in the laboratory in relation to their intake hazards.

3. Specific Activity or Enrichment

The specific activity or enrichment required for a given tracer problem will depend on the dilution to which the tracer will be subjected, considered in conjunction with the efficiency of the method of detection. In many biochemical problems high dilutions are necessary, and introduction of substances of low specific activity could require undesirably large amounts. Tracers of low specific activities can often be used with an efficient detection procedure, but high specific activities are required with a less efficient detection system. With organic compounds tagged with stable isotopes such as ^2H, ^{13}C, ^{14}N, and ^{18}O it is often possible to analyze to better than 1% with the mass spectrograph. Radioactivity detection efficiencies may vary considerably, depending on the technique used. Organic compounds labeled with ^{14}C are generally available commercially with specific activities of 1 or more millicuries per millimole, but for preparation by ordinary organic techniques, rather than by vacuum line and microtechniques, lower specific activities such as millicuries per mole are preferable. For high specific activities, organic compounds tagged with tritium can be made.

A more subtle, but sometimes important, influence of specific activity in organic systems, particularly for soft beta emitters such as ^{14}C, ^{35}S, and tritium, is that due to radiation decomposition caused by absorption of the emitted radioactivity within the system. With high specific activities excessively high levels of impurities due to decomposition by irradiation may occur. For example, it has been calculated (5) that at a certain level of specific activity under a certain set of conditions all of the tagged compound containing ^{35}S would be destroyed, thus necessitating the use of much lower specific activities to avoid such an effect.

4. Half-life

To ensure efficient detection, the half-life of a tracer atom must be long enough so that its activity will not decline excessively during the course of an experiment. With elements such as oxygen and nitrogen the only existing radioisotopes have half-lives of just seconds or minutes, so that tracer work with such elements is better carried out with stable isotopes. Except for such extremes, however, the use of radioisotopes of brief half-lives may be advantageous in that these usually accompany high specific activities. On the other hand, there is an advantage in using long-lived radioisotopes when tagged substances must be synthesized, since the tagged substances then have a much longer shelf life.

5. Types and Energy of Radiations

Beta radiation is often preferable in tracer work to gamma or alpha. The low penetration of alpha particles is inconvenient; the high penetration of gamma rays sometimes poses shielding problems, although some of these are minimized by the use of moderately energetic gamma radiation. Efficient counting of gamma emitters may also require scintillation counting, with the attendant more expensive equipment.

For beta radiation, on the other hand, even fairly energetic particles are usually sufficiently shielded by the containers used in ordinary analytical methods. Efficiency of counting is high with Geiger tubes, and counting may usually be carried out in the Geiger region with less expensive equipment. Weak beta emitters such as ^{14}C and ^{35}S may require more complicated methods such as liquid scintillation counting, but even with these isotopes, at a sufficiently high level of activity, end-window counting is often feasible. Tritium, a very weak beta emitter, is preferably measured by liquid scintillation or gas counting, although some problems may be handled with a windowless gas-flow counter.

6. Isotope Effect

The basic assumption underlying the use of tracers is that all isotopes of an element behave identically. That this is not rigorously true is indicated by the fact that isotopes of light elements, such as carbon and nitrogen, are separated commercially by multistage exchange reactions. Diffusion methods are used to separate uranium isotopes in gaseous compounds by porous membranes, since the lighter isotopes diffuse more rapidly. Thermal diffusion, high-speed centrifugation and other methods are also used.

Rates of reaction may be affected by isotopic tagging because of differences in bond energies. Thus a C–H bond has more vibrational energy and is less stable than a C–D bond, so that the rates of reactions involving such bonds will depend on the isotopic composition. Chemical equilibrium constants are also affected, but to a smaller degree than reaction rates, because the equilibrium constant involves both reactants and products, with some cancellation of effects.

The extent of an isotope effect is governed by the percentage difference between the masses of the isotopes concerned. It may accordingly be

TABLE 97.II
Isotope Effect on Rate Constants[a]

Isotopes compared	k/k'
$^1H-^2H$	31
$^{18}F-^{19}F$	1.09
$^{35}Cl-^{36}Cl$	1.03
$^{35}Cl-^{38}Cl$	1.11
$^{79}Br-^{80}Br$	1.01
$^{79}Br-^{82}Br$	1.03
$^{127}I-^{129}I$	1.01
$^{127}I-^{131}I$	1.02

[a] Reprinted from *J. Chem. Educ.*, **35**, 625. Copyright 1958, by Division of Chemical Education, American Chemical Society.

more serious for the lighter elements than for the heavier, although even for the former it may be negligible for applications for which isotopic mass differences are without influence. Table 97.II gives the order of magnitude of the isotope effect for a number of comparisons of different pairs of isotopes, expressed as the ratio of rate constants k/k'. For most tracer work it is necessary to consider the possibility of an isotope effect only for the lighter elements, for which there is a high percentage difference between the masses of the isotopes. For light isotopes such as ^1H, ^2H, and ^3H, the possibility of an effect should always be considered. However, for such elements the effect can be turned to advantage in studying reaction mechanisms. For example, if a proposed mechanism involves the breaking of a particular C–H bond, H may be replaced by ^2H or ^3H and the reaction rate studied to see whether a change occurs which is consistent with the expected influence of such a replacement.

B. PREPARATION AND PURIFICATION OF LABELED SUBSTANCES

1. Chemical Synthesis

Some tracer experiments may be carried out with elemental tracers or tracers combined as chlorides, nitrates, oxides, and similar simple inorganic compounds. Many of these can be purchased in the required form from installations connected with the U.S. Atomic Energy Commission or commercial facilities. If not available in the needed form, they can be converted by simple reactions, often quantitative in nature.

Different from such tagged substances are the many compounds, chiefly organic, which must be synthesized to incorporate tagged atoms in specific positions or groups. There are no special restrictions concerning radioisotopes for such syntheses; among those used are ^{14}C, ^3H (tritium), ^{35}S, ^{131}I, and ^{32}P. Among stable isotopes ^2H (deuterium), ^{13}C, ^{15}N, and ^{18}O may be mentioned. One could start a desired synthesis with the simplest (also cheapest) available compound, such as $BaCO_3$ for ^{14}C or H_2SO_4 or BaS for ^{35}S. But generally it is more practical to purchase a tagged intermediate, of which a large variety is available from a number of sources (3,4).

Methods of synthesis for preparations of low specific activity or enrichment may usually follow conventional procedures, preferably reduced to a semimicro scale, and suitably modified so as to maximize the yield based on the isotope (7,8). In a synthesis consisting of several steps, it is desirable to introduce the tagged intermediate as close to the last step as possible in order to reduce losses of isotope and avoid accumulation of impurities of high specific activity or enrichment. For tagged preparations of high specific activity or enrichment, semimicro techniques are essential. Often vacuum trains are needed in which small quantities of intermediates can be transferred, reacted, and stored without loss (7–10).

Among the various isotopes used in synthesizing tagged compounds, ^{14}C is by far the most common for organic and biochemical problems. On account of the rather low specific activities of ^{14}C compounds, however, there has been an increasing tendency to use tritium as a tag. Very high activities are available at a cost considerably less than that for ^{14}C. The main interest in tagging with tritium is in connection with the exchange reaction, to be discussed next. Other techniques, when applicable, may, however, be more advantageous, in enabling one to obtain much higher specific activities with tagging at definite sites rather than at more or less random locations. These techniques may also help to avoid the main difficulties of radiation decomposition which arise with the exchange methods because of the high activities used. Reduction of a double bond, a carbonyl group, or other reducible group with tritium gas makes possible the introduction of as much as 58 curies/millimole. Sodium borotritide and lithium aluminum tritide (12) may be used as reducing agents by conventional procedures to introduce tritium. Hydrolysis of the Grignard reagent with tritiated water may also be utilized (13). A method for replacing sulfur or bromine in steroid derivatives with deuterium in the presence of "deuterized" Raney nickel (11) should also be applicable with tritium.

2. Exchange Reactions

Exchange of normal atoms with isotopically enriched atoms is used for the preparation of various types of tagged compounds. Certain organic halogen compounds have been labeled by exchange with active aluminum halide (20,21) lithium halide (22), or alkyl halide (23). Cystein has been tagged with ^{35}S by contact with $H_2^{35}S$ in the presence of enzymes (24).

Exchange reactions are widely used for tagging with deuterium or tritium. Under normal conditions hydrogen atoms in organic compounds differ in their stability to exchange. They range from stable hydrogen atoms, such as those of paraffin hydrocarbons, through the semilabile, such as the enolizable hydrogen of a ketone (which exchanges with water under the influence of alkali), to the labile, such as carboxyl hydrogen (which will exchange with water quite easily). However, even stable hydrogen atoms may be made to exchange under sufficiently vigorous conditions when one wishes to prepare deuteriated or tritiated compounds. Concentrated sulfuric acid, such as T_2SO_4 or D_2SO_4 (14), catalyzes the exchange of hydrogen in aromatic systems. To use this for the preparation of tagged compounds one must minimize the concurrent sulfonation reaction. Exchange under the influence of a previously reduced platinum catalyst is frequently carried out in the presence of labeled water or acetic acid at an elevated temperature (15,16).

The most useful and widespread method for tritium labeling by exchange is Wilzbach's gas-exposure technique (17). In this procedure it is merely necessary to expose the compound to tritium gas, usually for a period of days. Introduction of tritium into the compound by exchange for hydrogen takes place under the influence of the absorbed energy released by the disintegration of the tritium. Considerable tritium activity may be purchased at a cost small in comparison, for example, to that of ^{14}C, and labeled compounds of high specific activity can be prepared. There are, however, two disadvantages. First, labeling may occur at all possible positions, but not uniformly, so that if a tag in a specific position is desired, other methods may be preferable. Second, the considerable quantity of radiation energy absorbed leads to much radiation damage (see Section II.B.4) with the formation of impurities of high specific activity. Nevertheless many compounds can be prepared with specific activities of 10^{-3} to 10^{-1} curie/g without excessive radiation damage (18,19). Before a compound which has been tagged in this way is further purified, it must be treated with large quantities of a

TABLE 97.III

Tritium Tracers Produced with $TH_2PO_4.BF_3$ at 23°[a]

Labeled compound	Specific activity before purification, $\mu C/g$	Specific activity, purified tracer, $\mu C/g$
Benzene	106.6	108.2
Toluene	550.0	554.0
Tetralin	536.0	531.0
Naphthalene	1422.0	1440.0
Anisole	146.0	146.0
Methylcyclohexane	9.25	9.36
Decalin	32.2	31.4
Cyclohexane (99.6%)	0.96	0
n-Octane (98.6%)	16.3	0

[a] From Yavorsky and Gorin, *J. Am. Chem. Soc.*, **84**, 1072 (1962).

hydroxylic solvent such as water or alcohol in order to remove labile tritium.

A method has been reported (19A,19B) for tagging with tritium by exchange with tritiated $H_3PO_4 \cdot BF_3$. Although there are some advantages over the Wilzbach method, it is not so universally applicable. It has been applied to hydrocarbons by shaking them with the viscous liquid reagent. Tagged materials are obtained which are apparently radiochemically pure. (See Table 97.III.) The specific activities vary considerably. Hydrogen on aromatic or tertiary carbon atoms exchanges easily; hydrogen on unbranched aliphatic carbon atoms undergoes little or no exchange.

3. Other Methods of Preparation of Labeled Compounds

a. BIOSYNTHESIS

Many labeled compounds of biological interest cannot be synthesized by the usual methods of organic chemistry. Some may be tagged by exchange, for example, with tritium. Numerous methods of biosynthesis have been described (9,25,26). Some use living organisms cultivated in the presence of radioactive nutrients. An interesting example is the biosynthesis of vitamin B_{12} tagged with ^{60}Co by adding the isotope to a suitable fermentation broth (27). The purified compound is used in the assay of the vitamin. As another example, penicillin has been prepared biosynthetically with a ^{35}S label (28,29).

b. Neutron Activation

In some cases it has been possible to label a compound by irradiating it with thermal neutrons in order to convert a component atom into a radioactive isotope. This method has been utilized for tagging cystine with ^{35}S (30) and vitamin B_{12} with ^{60}Co (31,32). The efficacy of this approach depends not only on the presence of an atom with a suitable cross section for neutron activation but also on the maintenance of the activated atom, at least in part, in its original state of combination. The occurrence of the Szilard–Chalmers process (see Section II.B.3.c) could, depending on its extent, either negate such an approach or make special purification necessary.

c. Szilard–Chalmers and Other Recoil Processes

When the transformation of an atom to a radioactive isotope takes place as a result of irradiation with thermal neutrons, the recoil energy due to the emitted gamma ray may be sufficient to disrupt the bond holding the transformed atom to the rest of the molecule. The atom may then fly off and combine with molecular fragments from the original molecule. This is known as the Szilard–Chalmers process. If there is no exchange with inactive atoms in other molecules of the target, and if the chemical nature of the tagged compound permits its separation from the target compound, the Szilard–Chalmers process may be used to prepare tagged compounds.

Most work with Szilard–Chalmers reactions has been done with halogen compounds (33). By solvent extraction, organic halogen derivatives may be separated from the inorganic halides which are formed.

Another process involving recoil may occur, as illustrated by the formation of ^{14}C-labeled compounds by irradiation of nitrogenous materials with slow neutrons. With acridine, for example, conversion of the nitrogen atom to a ^{14}C atom results in the formation of ^{14}C-tagged anthracene (39). It is thought that this occurs by virtue of the fact that some of the ^{14}C formed by the $^{14}N(n, p)^{14}C$ reaction, after losing most of its kinetic energy in other collisions, replaces nitrogen atoms in acridine to form anthracene. Carbon-14-tagged acridine is also obtained.

4. Decomposition of Labeled Compounds by Self-Irradiation

In addition to whatever normal chemical decomposition can be expected of substances containing only stable atoms, radioactive tagged substances undergo two special types of decomposition. The first is the disintegration of the tracer atoms. If the daughter atoms are stable, their quantity is usually too small to be of importance; but if they too are radioactive, their presence may have to be considered.

What may be a more serious decomposition is due to the effects of the absorbed radiation energy in causing chemical changes to occur in a tagged molecule either during its preparation or in storage after purification. The mechanisms are too complex (34,35) to discuss here; they involve a number of variables, including, for example, the intensity, penetration, and energy of the radiation, and the physical condition of the substance. The chemical nature of the compound and its molecular weight may be of particular importance. With the absorption of a minimum of radiation energy, some compounds, such as some unsaturated hydrocarbons and some quaternary ammonium salts, may undergo extensive decomposition by means of a chain reaction, whereas others, such as aromatic hydrocarbons, are highly resistant to decomposition. So-called G values can be found in the literature (36,37). These indicate the number of molecules of a compound produced or destroyed by a given reaction with the absorption of 100 eV of energy. References 36 and 38 discuss this subject in some detail (see Table 97.IV).

It is wise to take precautions to minimize radiation self-decomposition when tagged preparations are to be stored. Storage at low temperature, dilution in an appropriate nonreactive solvent, the exclusion of air, and

TABLE 97.IV

Radiation Destruction Coefficients and Radiation Yields[a]

Compounds	Destruction coefficients, molecules/100 eV	Radiation yields, molecules/100 eV	Radiation-induced reactions
Saturated hydrocarbons	6–9	H_2, 1–7	Fragmentation, dimerization
Unsaturated hydrocarbons	14–2,000	H_2, 1	Polymerization
Aromatic hydrocarbons	1	H_2, 0.3	Polymerization
Alcohols	4–12	H_2, 1–3	Fragmentation, dimerization
Ethers	7[b]	H_2, 2–3	Fragmentation, dimerization
Ketones	7[b]	H_2, 1–5; CO, 1–3	Fragmentation at CO
Carboxylic acids	5[b]	H_2, 1; CO_2, 1–4; CO, 0.1–0.5	
Esters	4[b]	H_2, 2–3	Fragmentation at ester group
Organic halides	4	HX or $\frac{1}{2}X_2$, 2–5	
Quaternary ammonium salts	5–2,000	Amines and aldehydes	
Amino acids	2–30	Ammonia, amines, CO_2	

[a] From Tolbert, *Nucleonics*, **18**, No. 8, 74 (1960).
[b] Estimated values.

spreading out the substance in a thin layer to minimize self-absorption of radiation are some of the methods used to minimize decomposition.

5. Radioactive and Radiochemical Purity

a. SIGNIFICANCE IN TRACER APPLICATIONS

Assurance of the purity of a tagged compound by the usual chemical and physical criteria is vital, but even more important is assurance of radioactive and radiochemical purity, or at least a knowledge of the nature and behavior of the impurities in the traced process under study. "Radioactive purity" refers to the absence of foreign radioisotopes, and "radiochemical purity" to the absence of radioactivity in any chemical species other than that in which it is desired.

The amounts of any of these types of impurities which can be tolerated will depend on the particular application. In some cases chemical impurities may not be as serious as radioactive or radiochemical impurities; in other applications, often in biological work, they may be quite serious. If a given percentage of a chemical impurity separated from the tagged compound during the course of a tracer experiment, the effect on the specific activities would be proportionate to the concentration of the impurity. With radioactive or radiochemical impurities of high specific activity, however, even a slight increase or decrease in the degree of contamination of the tagged compound by the impurity during the course of the experiment could cause a disproportionately large change in the specific activity and lead to large errors.

b. METHODS OF PURIFICATION AND CRITERIA OF PURITY

The usual chemical methods of purification of a tagged substance, such as recrystallization, fractional distillation, or chromatographic separation, should be carried out in a preparation until at least the normal requirements for chemical purity are met. The next step is to determine specific activities on fractions thus purified. Lack of constancy of specific activity is evidence of radioactive or radiochemical impurities, and further purification is then required.

Testing for the absence of radioactive impurities—foreign radioisotopes—may be carried out by the usual radiometric methods such as determination of decay curves and absorption curves or by pulse-height analysis. Radioactive impurities can be removed by precipitating or otherwise separating in the presence of added inactive carrier, as often as needed to achieve the desired purification. Chromatographic procedures, particularly paper or gas chromatography, are potent tools for determining whether radioactivity is present in more than one constituent and sometimes for effecting desired purifications.

III. INDICATOR ANALYSIS

A. DEFINITION AND COMMON FEATURES

Indicator analysis, as a branch of tracer techniques, is distinguished in its most quantitative aspect by the following features. The entire quantity, of known mass, of the substance to be followed in a system is treated with the tracer, and the latter is present as a known amount of activity or, for stable isotopes, as a known concentration of tracer isotope. The tracer must be uniformly distributed throughout the entire mass of the traced material at the beginning of the experiment, regardless of whether the system is homogeneous or heterogeneous. With these conditions fulfilled, the measurement of activity or isotopic ratio on samples from any part of the system provides a measurement of the mass of the traced substance.

A corollary of these conditions is that attainment of the maximum information in indicator analysis is possible only for systems which are under the control of the experimenter before the tracer is added. Consider, for example, a system in which it is known that the tracer has been uniformly distributed, but the mass of traced material is unknown. For such a system one could obtain some limited information with respect to the qualitative distribution of tagged material but could not determine from the activity or isotope ratio the masses involved in the distribution. Consider also, for example, the injection of a calcium tracer into an animal. Uniform distribution cannot be attained because of the slow exchange of calcium in the bones, so that the specific activity of the calcium in the blood will be greater than in the bones. Therefore, in such an experiment one could not determine the distribution of the calcium in the body or its total mass.

One of the advantages of indicator analysis is its extreme sensitivity, for both stable and radioactive isotopes, but particularly for the latter. The sensitivity depends on a number of factors, including the type of activity, its half-life, and the type of detector, but can be as low as 10^{-18} mole for some conditions. Another advantage is the fact that often no chemical separations are necessary before a measurement is made, since inactive impurities do not interfere with the measurement if the absorption of radiation by the impurity can be neglected or corrected for. Sometimes measurements can be made without destroying the sample, for example, by measuring activity on a chromatographic sample or even directly in solution. Finally, it is sometimes even possible to make measurements of systems in inaccessible places by proper location of detectors on the surface of containers.

As previously noted, the use of indicator techniques for the determination of a substance which is present in an unknown quantity in a system by the addition of an isotopic tracer is impossible because the specific activity of the substance must be not only uniform but also known. (One exception is mentioned in Section III.B.2.) Isotope dilution techniques (see Section IV) must be used for this purpose. An important role of indicator methods in chemical analysis is to provide information by which to test or improve analytical methods.

B. TYPES OF INDICATOR APPLICATIONS

1. Physical

Under physical methods there will be considered some applications which have no very clear chemical implications, even though the borderline between physical and physicochemical methods may not be clearly defined. Physical applications, in general, include the use of tracers for following and measuring the movement of molecules and other entities in gases, liquids, and solids, in both homogeneous and multiphase systems. One class of such applications involves the movement of one substance in the presence of others. To choose a tracer method to follow such a substance rather than to use some other type of method may be a matter of convenience, sensitivity, or accuracy, rather than of necessity. For systems in which the movement of a substance within a matrix of the same substance is to be followed (self-diffusion), however, no other type of method will serve.

Much of the use of radioisotopes in industrial processes involves the application of radioactive sources for such purposes as radiography, level measurement and control, and density measurements. There are other process applications, however, which may best be classified with physical indicator techniques, even though some of these applications may not conform to the requirements for the most quantitative aspects of indicator measurements necessary for the determination of masses in a system. Two examples might be the measurement of residence times in a process plant and the detection of leaks in a distillation plant (40a). As an example of the measurement of residence time, a salt of ^{24}Na (half-life 15 hr) was added to an aqueous system at the inlet and the time required for it to appear at the outlet was measured. In a two-phase system it may be necessary to measure the residence time in both phases. In such investigations a soluble inorganic salt (e.g., ^{24}NaCl or NH$_4$^{82}Br) has been used to label the aqueous phase, and a stable organic compound (e.g., C$_6$H$_4$^{84}Br$_2$) to label the organic phase.

In using radioisotopic tracers in plant processes it is essential to avoid contamination of products with radioactivity. One method of accomplishing this is to use isotopes of short half-lives so that the activity decays before the product must be handled. It has been proposed to produce such isotopes at the site of application by irradiation with radioisotopic neutron sources such as polonium–beryllium or americium–beryllium (40b).

In testing for leaks in a distillation plant, ^{24}Na or ^{82}Br has been injected into water-cooled units and sought at sites where leaks are suspected. In testing steam-heated units, ^{133}Xe has been injected into the steam supply.

Another useful method in industrial plant studies is the measurement of large flow rates in plant equipment containing channels of nonuniform section. The procedure involves the measurement of isotope dilution and hence might perhaps have been included in the discussion of isotope dilution methods in Section IV, but for convenience it will be considered here. In this method (40a) a small quantity of radioactive solution with specific counting rate A_1 is injected into the stream at an accurately measured rate, V_1. The counting rates, A_2, of samples drawn far enough downstream to ensure thorough mixing are determined. The required flow rate, V_2, is given by the expression:

$$V_2 = \frac{A_1 V_1}{A_2} \text{ (provided } V_2 \gg V_1)$$

In one effluent system, rates up to several million gallons per hour were measured, using ^{24}Na. The flow rates of various gases, up to rates of 10^5 m^3/hr, were measured with ^{133}Xe.

The determination of vapor pressures may be made quite conveniently. For example, ^{32}P has been used to determine the vapor pressure of white phosphorus (40) by introducing the vapor into a container surrounding a counter. The vapor pressure of pure metals such as silver (41) and gold (42) and the partial vapor pressure of gold in a gold–copper alloy (42) have also been determined.

The measurement of self-diffusion in gases with tracers is carried out in an apparatus wherein two gas chambers united by a narrow tube either consist of ionization chambers or are arranged to contain scintillation crystals for counting. The self-diffusion of carbon dioxide (43) and of argon (44) has been determined in this manner. In a similar manner self-diffusion in liquids has been measured (45,46,47). Many other tracer methods for diffusion in liquids have been reported, references to several of which are given (48,49,50). Diffusion in solids has been studied by a number of investigators. Their methods differ in the manner in which the active and inactive solids are brought into contact with each other and in the manner of measuring the transfer of activity. Three review articles are available (51,52,53). Of particular technical significance is the application of tracers to study wear by friction in machine parts, such as piston rings, by using parts which have been activated by neutron irradiation (54).

2. Physicochemical

Physicochemical applications, like other types of indicator applications, are numerous and limited only by the ingenuity of the investigator. Studies have been reported on the adsorption of ions from solution onto glass surfaces (55) and on their influence on change in potential (56). The increase in concentration of surface-active substances at liquid surfaces has been measured by using a low-energy beta emitter, with a short range, and determining the difference in count between the surface of a solution containing the tagged surface-active material and that of a solution containing the same isotope not combined as a surface-active material (57).

Solubility determinations of slightly soluble substances are particularly suitable for study by means of radioactive tracers because of the high sensitivity possible. Many determinations of the solubility of difficultly soluble salts have been made by measuring the activity of the saturated solution derived from the tagged solid (58). Thus the solubility of water in hydrocarbons has been determined with tritiated water (59).

Partition between liquid phases can be determined easily and quickly, and many such studies have been carried out. Solvent extraction studies have been facilitated by tracer methods (60).

One method has been reported (61) in which it was possible to determine, by distribution data, the absolute mass of a substance being traced without knowing the specific activity. The distribution of barium between a solution and a precipitate of iron hydroxide, under definite conditions of volume, ammonium chloride concentration, and pH, is a function of the concentration of barium ion. The addition of a negligibly small mass of radioactive barium tracer to the solution, and the determination of its distribution between the iron hydroxide and the solution, permit measurement of the barium concentration within well defined limits. Other possibilities for such applications may exist.

3. Chemical

a. ANALYTICAL METHODS

Isotopic indicators are used in chemical analysis for the purpose of studying analytical methods, either to gain a better understanding of procedures already in use or to help in devising new ones. This statement refers to purely indicator methods; isotope dilution and radiometric analysis are discussed in Sections IV and V. Attention should be called, however, to "radiometric correction" or "radioactive yield" determinations," whereby a tracer enables one to correct for inefficient recovery in analytical methods not otherwise based on the use of isotopes. This special case of isotope dilution will be covered in Section IV.

Mention should also be made of the role of natural radioactivity in analysis. One can use natural radioisotopes as tracer materials in the same manner as artificial radioisotopes. This requires no special consideration. However, some elements naturally contain a fixed concentration of at least one radioactive isotope, thus, in effect, fulfilling the condition set down for indicator analysis—that the specific activity must be known and uniform in the system. The elements potassium, rubidium, samarium, lutetium, rhenium, francium, and uranium have specific activities which do not vary with the age and source of the samples. Therefore they may, in principle, at least, be determined by measuring their radioactivity. Other naturally radioactive elements vary in isotopic composition with source and age and have mixtures of long-lived and short-lived isotopes.. Although procedures have also been worked out for analyzing some of them by their natural radioactivity, these methods are not simple (6).

Potassium, an example of an element with unvarying specific activity, is determined by means of its natural radioactivity due to ^{40}K (62). The radioactivity of a solution of the potassium salt, which emits negative beta particles, positrons, X-rays, and gamma rays, is measured with a Geiger counter. The detection apparatus is first calibrated against known concentrations of potassium in order to determine constants for an equation relating counts and concentration with the self-absorption of the solution and the efficiency of the counter. An accuracy of several per cent and a limit of detection of 0.4 g potassium can be attained.

It should, perhaps, be pointed out here that quantitative analysis of mixtures of nonradioactive compounds, without first changing the natural isotopic ratios by adding stable isotopic tracers, is not analogous to determinations by natural radioactivity. The mass spectrometric methods used for that purpose, which depend on spectroscopic "finger-prints" for each compound, are quite analogous to other spectrometric procedures used for similar analyses.

The investigation of analytical procedures is sometimes approached quantitatively, sometimes qualitatively. In quantitative studies one works with systems of known composition, at least with respect to the constituent or constituents to be determined. Addition of known amounts of tracers makes it possible to obtain quantitative data regarding various possible sources of errors, such as coprecipitation, incomplete precipitation, losses by volatilization, and mutual contamination by several constituents, which could lead to compensating errors. Wherever possible, the experiments should be arranged to permit direct counting of that part of the sample which constitutes the loss or contaminant, rather than of the main sample before and after loss. In this way the effect of normal radioactive counting errors on the results is minimized. The addition of tracers to systems of unknown quantitative composition leads only to qualitative information.

The literature abounds in reports of indicator studies of analytical methods using radioactive tracers; stable tracers have been employed more for isotope dilution and mechanism studies. A few examples will be mentioned. A method for the precipitation of ammonium phospho-molybdate, which was shown to have errors of as much as 1.5%, was improved as a result of tracer studies (63). In the precipitation of ortho-phosphate in the presence of pyrophosphate it was shown that coprecipitation of pyrophosphate was reduced by the use of zinc or cadmium instead of magnesium (64). With the aid of antimony and tin tracers a rapid method was developed for the separation and determination of bismuth, antimony, and tin by controlled cathode electrolysis (65). A colorimetric method for the determination of niobium and tantalum in steels was developed with the aid of tracers to determine the completeness of precipitations (66). (See Table 97.V). Considerable work has been carried out on the separation of rare earths (67). The possibility of losses by volatilization during an analytical procedure has been tested in connection with the determination of ruthenium by cupellation (68) and the ashing of petroleum for the determination of sodium, calcium, and iron (69). The time required for volatilization of indium in emission spectrography of samples containing low levels of this element could be determined by following the activity of added tracer in the sample (70).

TABLE 97.V
Effect of Concentration of Niobium and Tantalum on
Their Precipitation at pH of 0.8[a]

Precipitation of Niobium

| Sample | Niobium present, Mg. | Counts/min due to betas | | Niobium lost, %[b] |
		A, ignited precipitate	B, filtrate	
1	1.0	14743	4169	22.0
2	2.0	14606	1267	8.0
3	3.9	14227	587	4.0
4	5.9	13713	284	2.0
5	7.8	13178	277	2.1

Precipitation of tantalum

| Sample | Tantalum present, Mg. | Counts/min due to betas | | Tantalum lost, %[b] |
		A, ignited precipitate	B, filtrate	
1	1	2060	16008	88.6
		1402	15780	91.8
2	4	11588	10424	47.4
		10386	10631	50.6
3	10	14111	5292	27.3
		13117	5013	27.7
4	16	17119	1013	5.6
		16655	1155	6.5
5	20	14972	580	3.7
		14314	452	3.1

[a] From Boyd and Galan, *Anal. Chem.*, **25,** 1570 (1953).

[b] $\dfrac{B}{A + B} \times 100.$

Compensating errors may be quite troublesome in analysis, and tracer methods are most useful in elucidating them. The determination of sulfate by precipitation as the 4-chloro-4'-aminodiphenyl sulfate gives satisfactory results under well defined conditions. By means of a radioactive label it was shown, however, that, on the one hand, reagent in excess over the stoichiometric requirement was used, and, on the other hand, part of the sulfate was not precipitated and part was lost upon washing the precipitate (71).

An interesting example of the application of radioactive tracers to the development of new analytical methods is their use in developing organic coprecipitants. These are organic analytical reagents which are used for the selective concentration of elements out of very dilute solutions for the purpose of subsequent examination by spectrographic, colorimetric, or other analytical methods. Not only has the use of radioactive isotopes facilitated the development of such reagents, but also without these tracers the work would not have been possible (72).

b. REACTION MECHANISMS

There are two broad approaches to the use of isotopes, radioactive or stable, in studying reaction mechanisms. In one approach isotopes are chosen for the study of suitable reactions for which there is reason to believe that tagging a molecule will not result in any essential difference in its reactions and those of the untagged molecule. As has already been noted, this is likely to apply to the heavier elements and may be less true for the lighter ones. Whether or not an isotopic mass effect occurs will also depend on whether the reactions studied involve breaking bonds associated with the isotopes. In the other approach, light isotopes such as deuterium or tritium are purposely chosen in order to utilize the mass isotopic effect. From an examination of the speeds and products of reaction it is often possible in such cases to obtain evidence leading to confirmation or rejection of mechanisms involving bonds linking such isotopes to other atoms.

The contribution of tracer studies to the fields of chemical mechanisms and kinetics cannot be overemphasized, and the basis of much of our modern understanding of these fields has been laid by tracer uses, particularly, but not exclusively, in the field of organic chemistry. As a result, in great part, of such tracer studies, it is now evident that very few reactions occur by simple one-stage mechanisms. Hence there is a need for extreme caution in devising a tracer approach to the study of reaction mechanisms. Competition between fast and slow reactions and the possibility of fast or slow exchange of isotopes and of different elements, as well as many other complications, exist. The problems are essentially physicochemical and organic in nature rather than analytical. Their consideration is beyond the scope of this chapter except for a few references to typical problems in this field that have been handled by tracers.

Reference 72 gives a number of papers which are more or less typical. They include "Investigations on the Mechanism of Catalysis Based on Isotopic Data," p. 1; "Elucidation of the Mechanism of the Michael

Reaction with Carbon-14 as Tracer," p. 21; and "A Study of the Hofmann Degradation of Quaternary Ammonium Bases Using Tritium-Labeled Compounds," p. 31. Reference 72a is a thorough treatment of the theoretical principles involved in using isotopes for studying chemical reactions.

Average molecular weights for γ-polyoxymethylenes have been determined by preparing these compounds from a mixture of paraformaldehyde and radioactive methanol of known specific activity (73). Reaction between a solid phase and a monomolecular layer has been followed by spreading layers of ^{14}C-tagged stearic acid on various surfaces, removing the stearic acid with solvents, and measuring the residual activity, which is explained as being due to salt formation (74).

In the study of reaction mechanisms, stable isotopes are frequently used, such as deuterium, ^{13}C, ^{14}N, ^{15}N, and ^{18}O. The principles are the same as for radioactive isotopes, but measurements are usually made with a mass spectrometer. The latter is essential in the analysis of ^{14}N and ^{13}C. For deuterium, good methods of measurement exist which depend on determining the density of the water produced by combustion; instead, the water may be decomposed to elementary hydrogen and the latter examined mass spectrometrically.

4. Biological

The multistage character of even apparently simple organic reactions seems uncomplicated by comparison with biochemical processes. In chemical and physical studies it is usually possible to isolate and study independently the reactions of a limited number of constituents. In biological studies, on the other hand, every reaction takes place in, and may be influenced by interaction with, the overall biological medium. In addition, heterogeneous systems abound in which problems of transport are added to an already complex situation. To make the matter even more involved, the biochemist is concerned not only with the reactions of given substances but also with their past history in the systems.

Even this brief and very general statement of the nature of biochemical problems suggests the importance of tracer methods. Their value was recognized by biochemists even before the advent of isotopic tracers, and they carried out other types of labeling. For example, studies with fatty acids in organisms were made with fatty acid derivatives containing at one end a stable group, such as a benzene ring. Metabolic fragments and reaction products could thus be traced during biological oxidation processes. Another device was to introduce some easily detected element such as chlorine into the molecule under study. With such schemes the possibility always existed, however, that the labeling would affect the course of reaction.

It is small wonder, then, that isotopic tracer methods, both radioactive and stable, have achieved greater prominence in biology than in any other branch of science. One need only mention the tracer work which has been done on the chemistry of alcoholic fermentation, on the synthesis of urea, and on photosynthesis to suggest the great dependence of modern biochemistry on isotopic tracers. Such studies have also moved logically into medical diagnosis.

The methods used in biology and biochemistry differ more in application than in essence from other tracer methods discussed in this chapter. Nevertheless, because it is a highly specialized and very widespread area, any detailed consideration must be omitted here. References 75–79 may be consulted for more information.

5. Chromatography and Related Separation Procedures

The great power of chromatographic methods in effecting separations and making possible identifications may often be increased by using radioisotopic tracers. They add extraordinary sensitivity and sharpness of separation to these methods. This applies to the various types of chromatography, such as column, paper, and gas, as well as to similar or related procedures, such as ion exchange and electrophoresis. Like other indicator tracer methods, the use of radiotracers may serve either as an integral part of a procedure or as a means for developing chromatographic procedures which will not themselves employ radioisotopes.

A number of variations in approach are possible. Thus one may subject to chromatography radioactive samples which were first tagged and then used in a tracer experiment. One may, alternatively, take untagged samples and add radiotracers before chromatographing. This may be done by adding radioactive tracers, by reacting the samples with tagged reagent, by activating them with neutrons, or by subjecting the samples to exchange with radioisotopes. The first two methods are the most common.

The separated fractions may be detected and measured by the most appropriate of a number of procedures. With paper chromatography one may merely scan the paper with a suitable detector, such as a Geiger counter, or cut out the active areas and measure them directly or after elution with a suitable solvent. Column chromatography may be followed either by elution of active zones or, for gamma emitters, by activity

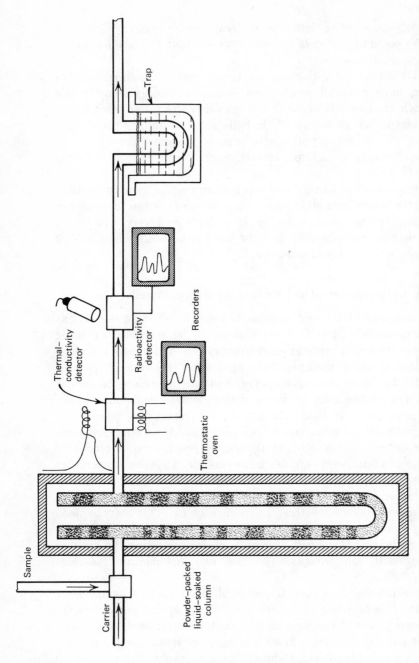

Fig. 97.1. Gas chromatography apparatus separates components of gas mixture by allowing carrier gas to carry them through liquid-bearing column of powder. As components emerge, thermal-conductivity cell and radioactivity detector indicate arrivals. Traps can be used to isolate different components. From Cacace and Fulvio, *Nucleonics*, **19**, No. 5 (1961).

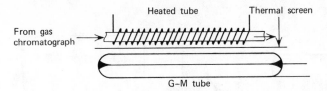

Fig. 97.2. G-M detector counts gammas as effluent passes through heated tube. From Cacace and Fulvio, *Nucleonics*, **19**, No. 5 (1961).

measurements at the surface of the column without elution. In gas chromatography gaseous fractions may be measured in an ionization chamber or other detector (89) (Figs. 97.1–97.3). One may also use auto-radiographic procedures to detect and to measure the activity on paper chromatograms.

An important function of these methods is their use as means of identification or as proof of radiochemical purity. If a sample is tagged, either as an active preparation for tracer work or as part of an analytical study, proof of its identity may be obtained by mixing it with inactive material with which it is thought to be identical. Concentration of all of the radioactivity in the same location as the added inactive material is proof of identity, particularly if this can be achieved under diverse conditions, including different types of adsorbents, solvents, and eluents. If, in addition, fractional separation of the tagged material and determination of the specific activity of the separated fractions lead to identical values, one may conclude that no trace amounts of impurities of high specific activity are present. By using double tagging with ^{131}I and ^{35}S, an identification of γ-aminobutyric acid in mouse brain was

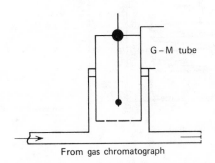

Fig. 97.3. Special mounting for soft-beta emitters. Special mounting puts soft emitters near thin-window G.M. tube. From Cacace and Fulvio, *Nucleonics*, **19**, No. 5 (1961).

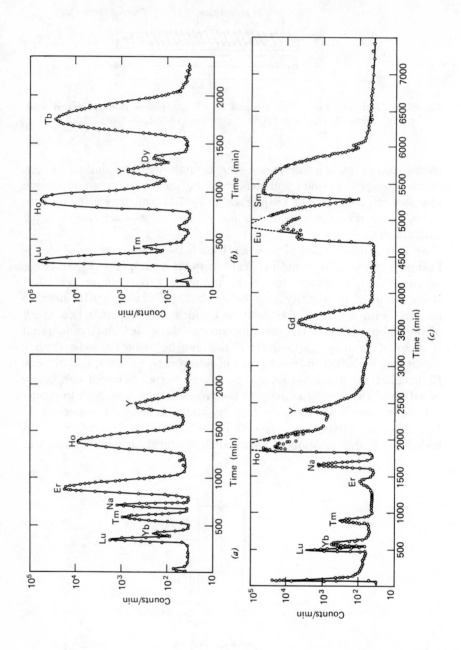

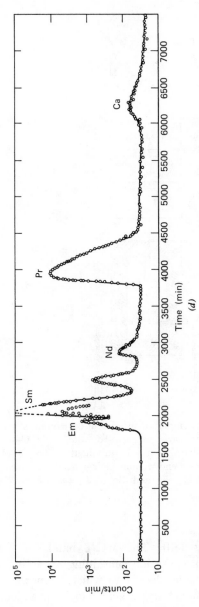

Fig. 97.4 Demonstrations of rare earth separations effected with a 270/325 mesh Dowex-50 column at 100°: bed dimensions, 97 cm by 0.26 cm²; flow rate, 1.0 ml/cm²/min except in a where 2.0 ml/cm²/min was used: (a) (pH 3.20); (b) fractionation of activities produced by neutron irradiation, 0.8 mg spectrographic grade Er_2O_2 (Hilger) (pH 3.20); (b) fractionation of heavy rare earth mixture consisting of 0.1 mg each of Lu_2O_3, Yb_2O_3, Ho_2O_3, and Tb_2O_3 [Tm, (Er) Y, and Dy present as impurities] (pH 3.20); (c) fractionation of intermediate rare earth mixture consisting of 0.1 mg Ho_2O_3 and 1.0 mg each of Dy_2O_3, Gd_2O_3, Eu_2O_3 and Sm_2O_2 (Cl, Lu, Yb, Tm, Er, and Na present as impurities, pH 3.25 for 4550 min, then pH 3.33); (d) fractionation of light rare earth mixture consisting of 0.1 mg each of Sm_2O_3 and Nd_2O_3 plus 0.01 mg each of Pr_2O_3, Ce_2O_3, and La_2O_3 (Eu present as impurity 61 produced by 1.7 h $Nd^{149} \rightarrow 47$ h 61^{149}, pH 3.33 for 1610 min, then pH 3.40). From Ketelle and Boyd, $J.$ $Am.$ $Chem.$ $Soc.$, 69, 2808 (1947).

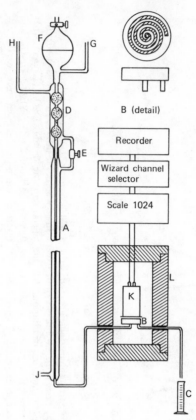

Fig. 97.5. Experimental arrangement employed in adsorption column separations: A—adsorbent bed, Amberlite IR-1 or Dowes-50; B—counting cell; C—receiver; D—Allihn condenser; E—throttle valve; F—gas entrainment bulb; G—elutriant inlet; H—thermostat fluid inlet; K—mica end-window Geiger-Müller counting tube; L—2-in. lead radiation shield. From Ketelle and Boyd, *J. Am. Chem. Soc.*, **69**, 2804 (1947).

made by paper chromatography (80). The active bands of a paper chromatogram, obtained from a preparation containing an active amino acid derivative made with pipsylchloride–^{131}I to which a known pipsyl-^{35}S-amino acid derivative of γ-aminobutyric acid had been added, were each cut into several transverse segments, and the ratio of ^{131}I to ^{35}S in each segment was determined radiometrically. That a particular band was due to γ-aminobutyric acid was established by the constancy of the ratio of ^{35}S to ^{131}I throughout the band.

In column chromatography, as well as in gas chromatography, it is possible to measure the activity of an eluate continuously. Continuous measurements of this kind offer obvious advantages in separation procedures. A good example of this technique is the development of separation methods for rare earths after the addition of tracers (81) (Figs. 97.4 and 97.5).

Much tracer work has been done on ion-exchange methods, paper electrophoresis, and electrochromatography. The General References may be consulted for sources of information.

6. Spatial Distribution

The determination of the spatial distribution of tagged substances is an obvious application of the indicator method, and has been an implied part of several of the applications mentioned in other connections in this section. Diffusion studies, biological transport mechanisms, and the detection of activity in chromatograms are examples. The techniques of autoradiography (82) are very important. By exposing tagged radioactive specimens to a photographic film, a record may be obtained of the location of the radioactive particles. These methods are much more prominent in biological and medical investigations than in chemistry. Another rapidly growing body of techniques in medical diagnosis and physiological studies involves the three-dimensional location of radioactive tracers in human beings and animals by various systems of measurement with more than one counter external to the organism. Such applications need not be restricted, however, to higher animal organisms; for example, the locations of ^{60}Co-tagged insects or worms in the ground have been followed by counters above ground (83).

IV. ISOTOPE DILUTION

A. DEFINITION AND AREAS OF APPLICATION

Indicator tracer methods are most useful for analytical determinations on systems in which the substances of interest are present in known amounts and, at the time of addition of the tracer, in equal distribution throughout; in other words, where the problem is to determine, not the total amount of a substance in a system, but rather the redistribution of a known amount. Such a system must, for the achievement of optimum results, be under the control of the experimenter. Problems of perhaps more widespread interest in analytical chemistry are those requiring the determination of the total amount of a substance in a system which has not been under the control of the experimenter. This is the area of application of the isotope dilution method, with the use of isotopic tracers, stable or radioactive.

Isotope dilution methods merely offer another choice among the numerous possibilities in the analytical spectrum. There is no area, such as exists among indicator methods in the determination of the movement of a substance in a matrix of the same substance, where any but isotope dilution methods are theoretically impossible. The problem, then, is to define the areas of application for which isotope dilution methods either are much better than other alternatives or perhaps constitute the only feasible approach.

One of the most important of these areas is the determination of a substance which cannot be quantitatively isolated in pure form for weighing or for determination by other methods. A second involves systems, many in biological studies, in which exceedingly small amounts of substances are to be determined in the presence of difficultly separable substances which will interfere with other analytical methods. A third is the determination of substances for which other analytical methods are available but for which the extent of the system is unknown or poorly defined. For such a system the usual chemical analysis will give values of concentration but not of total mass, except under the (usually impractical) condition that the entire system serve as the sample for analysis.

A basic assumption in the isotope dilution method is that, once the tagged substance has been mixed uniformly in the system, it will not be separated even partially from the substance being determined by any subsequent process used to isolate a chemically pure specimen of that material. This means that ordinarily the tagged substance should be chemically identical with the substance being determined. In this connection, in addition, radiation self-decomposition (see Section II.B.4) before use should be considered. Furthermore, for tagged molecules, the tracer atom should be one which is not foreign to the substance being determined, nor should it undergo exchange, reactions, or isotopic fractionation processes which will alter the isotopic composition of the tagged molecule. When a foreign atom is used as a label, as, for example, when ^{131}I associated with insulin serves as a tag for the latter (84,85), it is essential to demonstrate complete inseparability of the foreign tag from the molecular species being determined.

Another basic assumption is uniform mixing. Although in most cases this requires little thought, occasionally it poses a problem. For example, in determining vitamin B_{12} (86) it is necessary to break up any combination of the vitamin with protein; otherwise that portion of the vitamin will not be available for mixing with the tagged vitamin.

Finally, it should be pointed out that isolation of a pure specimen is essential only if its mass is to be determined by weighing. If an analytical method is available which will determine the substance in the presence of such a quantity and type of impurities as may accompany it when isolated, it may be possible to use isotope dilution without the isolation of a pure specimen. In isotope dilution with stable isotopes the samples measured need be free only of contamination with the element entering into the isotope-ratio determination; other impurities introduce no error.

Several different variations of the basic principle of isotope dilution analysis will be developed in this section. At this point the general principles will be discussed, and the individual variations will then be taken up in turn.

In general, in isotope dilution a known mass of an element or compound, of known specific activity or isotopic content, is added to a system containing an unknown mass of the same element or compound with a different specific activity or isotopic content. Depending on the particular variation of the technique, the original or added material may be of either normal or artificial isotopic content, radioactive or stable. The added material is then dispersed uniformly throughout the system. A sample is withdrawn, and a portion of the substance to be determined is isolated in pure condition. The ratio of the specific activity or isotopic content of the pure sample to that of the added material is used to calculate the total mass of the substance being determined.

The generalized calculations for isotope dilution (87,75), regardless of type, are as follows. (For a less general derivation for radioisotopes see reference 88.) For stable or radioactive isotopic dilution, let $X =$ total mass of the substance to be determined in the system; $C_x =$ atom-percentage of (stable or radoactive) indicator isotope in X; W and $C_w =$ corresponding values for the added substance; $C_f =$ atom-percentage of indicator isotope in the pure material isolated after isotopic dilution; and M_x and $M_w =$ average molecular weghts of the originally present and added substances, respectively. M_x and M_w each equal $\Sigma p_i M_i$, where p_i is the fractional abundance of the species with molecular weight M_i.

Then

$$C_f = \frac{(C_x X/M_x) + (C_w W/M_w)}{(X/M_x) + (W/M_w)} \tag{1}$$

A discussion of the accuracy of the isotope dilution method is given separately for each of its several modifications in reference 88a.

B. DIRECT ISOTOPE DILUTION

1. Derivation

In the direct isotope dilution method the sample being analyzed contains substances of natural isotopic concentration. There is added tracer of unnatural isotopic composition (radioactive or stable). Equation (1) may then be modified to avoid the necessity of introducing atom-percentages by using instead atom-percentages in excess of the natural content. Let C_0 be the natural atom-percentage and Δ_f, Δ_x, and Δ_w be the differences between the natural atom-percentage and the actual atom-percentage corresponding to the isolated material, the material to be determined, and the added material, respectively. Then

$$\Delta_x = C_x - C_0, \quad \Delta_w = C_w - C_0, \quad \text{and} \quad \Delta_f = C_f - C_0$$

Since $C_x = C_0$, $\Delta_x = 0$. The following equation may then be derived:

$$X = \left(\frac{\Delta_w}{\Delta_f} - 1\right) W \left(\frac{M_x}{M_w}\right) \tag{2}$$

The ratio M_x/M_w may be neglected if the molecular weights of the original and added substances are not significantly different. This will be the case except when the molecular weights are small and the atom-percentage of isotopic content is high. For radioactive isotopes the atom-percentage of enrichment is practically always negligibly small, so that $M_x = M_w$ and the factor M_x/M_w may be eliminated. Furthermore, in Equation (1) the M_x and M_w terms drop out; and if, in addition, specific activities S_f, S_x, and S_w are substituted for C_f, C_x, and C_w, Equation (1) becomes

$$S_f = \frac{S_x X + S_w W}{X + W} \tag{3}$$

In direct isotope dilution $S_x = 0$; therefore

$$X = \left(\frac{S_w}{S_f} - 1\right) W \tag{4}$$

In equation (4) specific activities are used only as ratios. It is unnecessary, therefore, to make any absolute determinations of activity, but measurements must be made under comparable and reproducible conditions.

2. Radiometric Correction or Radioactive Yield Determination

Equation (4) may be written as

$$X = W\left(\frac{S_w}{S_f}\right) - W \tag{5}$$

When W is very much smaller than X, it may be neglected as a correction term and equation (5) then becomes

$$X = W\left(\frac{S_w}{S_f}\right) \tag{6}$$

Equation (6) applies when the added tracer is of very high specific activity or is carrier-free. If in equation (6) we substitute for S_w the expression A_w/W and for S_f the expression A_f/Z, where A_w is the counting rate for the (negligible and undetermined) weight W of added material and where A_f is the counting rate for the weight Z of pure material isolated after isotopic dilution, we get

$$X = W\left(\frac{A_w/W}{A_f/Z}\right) = Z\left(\frac{A_w}{A_f}\right) \tag{7}$$

This variation of isotope dilution permits the calculation of the weight of the substance being determined from the total counting rate of the added and the recovered materials and the weight of the latter, without the necessity of knowing the specific activity of the added material. It is sometimes referred to as the "radiometric correction" or "radioactive yield" determination because the method can be used to correct for losses in analytical or other chemical process. But in this respect it differs in no manner but convenience from the straight isotope dilution method.

3. Precision and Accuracy

If in equation (4) W is much greater than X because of the low specific activity of the tagged material, the term $(S_w/S_f) - 1$ becomes very small and S_w/S_f approaches unity. The normal variations in counting then lead to large errors. Of course, the usual considerations in counting radioactive samples, such as reproducibility of geometry, self-absorption, physical characteristics of the sample, and statistics of counting must also be considered. Sometimes it is possible to avoid some of the experimental difficulties in controlling some of these factors by using experimentally determined corrections (90).

Radioisotopes may be diluted with normal material to a considerably greater degree than stable isotopes and still be measured with good precision. However, considerable work in biochemistry and organic chemistry can be carried out satisfactorily with stable isotopes even with this limitation. For small differences between the isotope ratios of the samples to be measured, a mass spectrometric apparatus has been used (91) which permits rapid alternate measurement of sample and standard and double collection so as to record a direct ratio. Relative isotope ratios can thus be measured with an accuracy of 0.01%. With an overall attainable accuracy of 0.25–0.50% in the ratio determination, pure ^{13}C would have an uncertainty in its final determined concentration of 100% if a 20,000-fold dilution were made, and could be diluted only 2000 times for a required 10% accuracy. Less pure ^{13}C would be capable of even less dilution (92). Table 97.VI (93) gives the percentage error in X [equation (2)] caused by a 1% error in Δ_w for various ratios of W/X. Similar relationships would hold for equation (4). However, a weighing error in W would not be multiplied in calculating X. For favorable W/X ratios radioactivity measurements could give an overall precision of about 2%; for mass spectrometric measurements with a 0.5% error an overall precision of about 0.5% could be attained.

The frequent necessity for using high stable-isotope concentrations may result in a significant change in the molecular weights which enter into equation (1). Therefore the molecular weights would have to be

TABLE 97.VI
Isotope Dilution Errors[a]

Percentage error in X for 1% error in Δ_w [equation (2)]	Ratio W/X [equation (2)]
∞	∞
10	9
3	2
1.5	0.5
1.1	0.1
1.01	0.01

[a] Reprinted from *Nucleonics*, **1**, No. 2, 51. Copyright 1947, by McGraw-Hill Publishing Company, Inc.

used in the calculations to avoid an error which does not normally occur with radioisotopes. For example, deuterium and ^{15}N are often used in concentrations of 99 and 32 atom-per cent, respectively.

C. REVERSE ISOTOPE DILUTION

1. Derivation

In reverse isotope dilution there is added to a sample containing an unknown weight of tagged material of known specific activity a known weight of untagged material. Thereafter the method is the same as for direct isotope dilution. In addition to determining an originally labeled material, the method can also be used for an unlabeled material if it is converted to a labeled derivative (see Section IV.D).

By substituting $\Delta_w = 0$ in equation (1) for C_w, Δ_x for C_x, and Δ_f for C_f and transposing, equation (1) becomes

$$X = \frac{1}{(\Delta_x/\Delta_f) - 1} \cdot W \left(\frac{M_x}{M_w}\right) \tag{8}$$

and by substituting $S_w = 0$ in equation (3) the latter becomes

$$X = \frac{W}{(S_x/S_f) - 1} \tag{9}$$

These equations may then be used for reverse isotope dilution in the same manner as similar equations were used for direct isotope dilution.

Reverse isotope dilution possesses the advantage over the direct method that it can be applied for the determination of traces of labeled materials. The biochemist studies metabolic pathways by adding labeled compounds and recovering the original substance and labeled metabolites. When the specific activities of the metabolite and precursor are identical but quantitative recoveries are impossible, both metabolite and precursor may be assayed by reverse isotope dilution. Reverse isotope dilution is also used in determining the recoveries of activated elements in activation analysis.

2. Precision and Accuracy

Whereas, in direct isotope dilution, the determination of too low a quantity of a substance may lead to large errors because it is necessary to use a high ratio of added labeled material to unlabeled material, in the reverse isotope dilution method this problem never arises. The method is always useful for trace quantities. In equation (9) X is always much smaller than W and therefore $(S_x/S_f) - 1$ is always very large, so that it becomes practically equal to S_x/S_f. Therefore the errors in determining X may depend only on those in the radioactivity measurements.

Establishment of radiochemical purity is extremely important with this method, since even traces of impurities of high specific activity may cause large errors. This is true of the direct method also, but there the addition of radiochemical impurities can sometimes be avoided more easily because one adds a previously purified preparation. In the reverse method, however, the adventitious occurrence of such difficultly removable impurities may not be avoidable. It is, therefore, important to test for radiochemical purity by the methods previously discussed.

D. ISOTOPIC DERIVATIVE DILUTION

1. General

In the isotopic dilution methods discussed thus far, labeled substances chemically identical with those being determined are needed. This is sometimes inconvenient, especially when many substances are to be determined. The isotopic derivative dilution method (94,95) simplifies this problem by using a labeled reagent which forms a derivative with the substance being determined. A single labeled reagent can thus serve to determine many chemically related substances.

The reagent must be of known specific activity, thus fixing the specific activity of the derivative. The reaction should be stoichiometric, and the derivative should be capable of being freed from excess reagent and separated in a pure state (but not necessarily quantitatively) from other reaction products. Although the method may be made quite sensitive by using an appropriate reagent, it may suffer from lack of specificity.

Equations (8) and (9) for reverse isotope dilution are applicable to derivative dilution if moles instead of weights, and activity or isotope content per mole or fraction of a mole rather than per unit weight, are used.

Under some circumstances the compound being determined may be reacted quantitatively with a labeled reagent and recovered quantitatively. This will be discussed in Section V.

2. Unlabeled Carrier

In a complex mixture, such as that of amino acids in a protein hydrolyzate, or other biological samples, quantitative isolation of a derivative of a single substance is often not feasible, although quantitative reaction with the reagent occurs. Reaction of the mixture of chemically related substances with the labeled reagent may then be followed by addition of a considerable quantity of the unlabeled derivative which it is desired to determine. This derivative may then be isolated and purified without the necessity for quantitative recovery.

In equation (9), let X = number of moles of the derivative of the substance being determined; W = number of moles of unlabeled carrier derivative added; S_x = specific activity (counting rate per mole) of the pure derivative prepared with the same batch of reagent (or its equivalent) used for the determination; and S_f = specific activity (counting rate per mole) of the recovered, pure derivative obtained after quantitative reaction of the substance being determined and addition of W. From the known composition of the derivative, the weight corresponding to X moles may be calculated.

Equation (9) may be transposed to the form $XS_x = XS_f + WS_f$. If X is very small compared to W, the term XS_f may be neglected and the equation reduces to

$$X = \frac{WS_f}{S_x} \tag{10}$$

3. Two Isotopic Labels

In using an unlabeled carrier, as in Section IV.D.2, there is an implicit assumption not only that it is possible to form the derivative quantitatively but also that, after the addition of carrier, the derivative can be isolated in a quantity sufficient for weighing or for determination by some other method which will allow S_f to be calculated. Sometimes, however, this cannot be done, because, for example, the isolation involves merely recovery of a spot on a paper chromatogram. A second tracer may then be used to obtain information concerning the recovery. For example, a derivative of an amino acid, which is to be determined in a protein hydrolyzate, is made with ^{131}I-iodophenylsulfonyl chloride (pipsyl chloride). After formation of the derivative, but before isolation, a definite amount of ^{35}S-labeled pipsyl derivative of known specific activity is added to the mixture. After chromatography the activity due to ^{131}I and that due to ^{35}S in the spot are both determined by an appropriate radiometric method, such as counting with, and without, an aluminum absorber. The ^{35}S activity serves as a measure of the fraction of the derivative present in the chromatographic spot. This value may be used to correct the ^{131}I count to one which can serve for the calculation of the amino acid content according to Section IV.D.2.

As has already been noted, it is assumed, in using the method of Section IV.D.2, that all of the substance being determined reacts with the radioactive reagent to form the desired derivative. If this is not achieved, it is still possible to use derivative dilution by adding to the sample, before reacting it with the tagged reagent, a definite quantity of known specific activity of the substance to be determined, which has been tagged with an isotope other than that used for the reagent, and which is capable of being counted separately. (The quantity added should not be greater than the quantity already present.) After reaction, unlabeled carrier derivative is added and a pure specimen is isolated in the usual manner. The specific activities of both tracer isotopes are determined. From these values one can calculate both the total yield (product of the yield of formation of derivative and of recovery of derivative formed) and the amount of reagent in the recovered pure derivative. From the specific activity of the reagent, the amount of substance of interest may be calculated. The amount of inactive carrier added does not enter into the calculation. This method may be used for the determination of very small quantities.

4. Unlabeled Derivative with Small Addition of Labeled Derivative

Another variation of the derivative dilution method is an adaptation of direct isotope dilution. Instead of adding a tagged material to a sample in which the untagged substance is to be determined, one first quantitatively converts the inactive substance to an inactive derivative and then adds a small known amount of this derivative of a known specific activity. This will have been separately prepared with a tagged reagent and purified. After proper mixing, a sample of the derivative is isolated, purified, and counted in the same manner as for the direct isotope dilution method (Section IV.B).

This method has the advantage over the ordinary direct isotope dilution method that a labeled compound need not be prepared. One tagged reagent can serve to determine a number of different compounds; and in preparing a derivative, excess of inactive standard rather than excess of labeled reagent may be used. Furthermore, in preparing the unlabeled derivative, it may be possible to achieve quantitative conversion by means of large excesses or high concentrations of reagent or by the use of vigorous conditions possibly not acceptable with tagged reagents because of considerations of safety.

E. DOUBLE ISOTOPE DILUTION

When it is necessary to determine an isotopically tagged material under conditions where the specific activity or isotope ratio is unknown, the reverse isotope dilution method as previously described cannot be applied. Only a qualitative indication of the presence of the isotopic label would be obtained. Such a situation sometimes occurs in biological metabolism studies, as well as in analyses of naturally radioactive materials, for example, in the determination of lead in a uranium mineral containing both active and inactive lead.

In such a situation one can sometimes use a method involving two reverse isotope dilutions on separate aliquots of the same sample, to each of which different amounts of inactive carrier are added (96). A quantity of inactive carrier W, equation (9), is added to an aliquot of the sample and a different quantity, W', to an equal aliquot, and the corresponding specific activities, S_f and $S_{f'}$ of the isolated pure materials are determined. Two solutions are thus obtained for equation (9):

$$X = \frac{W}{(S_x/S_f) - 1} \quad \text{and} \quad X = \frac{W'}{(S_x/S_{f'}) - 1}$$

S_x, the specific activity of the substance being determined, which is unknown, may be eliminated and the following equation obtained:

$$X = \frac{S_{f'}W' - S_f W}{S_f - S_{f'}} \tag{11}$$

Alternatively, X may be eliminated and the following equation obtained:

$$S_x = \frac{S_f S_{f'}(W - W')}{S_f W - S_{f'} W'} \tag{12}$$

Similar equations may be derived for an analysis in which the same amount of carrier is added to different sizes of aliquots. Another variation (97) is to treat the same sample successively with two different weights of carrier.

Although these methods are suitable for determining very small amounts, the errors of determination may be extremely high. For high dilution with carrier the $S_f W$ and $S_{f'}W'$ terms in equation (11) approach each other in magnitude, so that the small difference between them suffers from large relative errors. It has been calculated (96) that, for an error of 1% in counting and ten- to twenty-fold dilutions with carrier, the probable error in X would be 20%. For metabolic experiments with inherently high biological variations, errors of such magnitude may be acceptable.

Still another variation of the double dilution method (98) is applied to the reverse dilution technique, whereby labeled carrier is added to a mixture being analyzed. The method has been employed for isotope dilution with stable isotopes in order to reduce errors of mass spectrographic measurement which result from the use of isotopic tracers of low enrichment. Equations similar to (11) and (12) may be derived which are based on equation (1), the form required for stable isotope dilution. There is no advantage in using this method for radioisotopes, since radioactivity measurements may generally be reduced to the desired statistical error by counting for a sufficient length of time and are therefore not as severely limited as are mass spectrographic measurements by the degree of isotopic enrichment.

F. SUBSTOICHIOMETRIC ISOTOPE DILUTION

In the foregoing isotope dilution methods using radioisotopes, it is necessary to determine specific activities at two points in the analysis, in accordance with equations (4), (9), and (11) or variations of these, for the determination of X. As has been pointed out previously, the determination of specific activity means isolating a sufficiently large sample for weighing or for an analytical determination, in addition to radioactive counting. This requirement imposes severe limitations on the use of isotope dilution for trace analysis.

In equation (4), and similarly in the other equations which are solved for X, one may substitute, for S_w and S_f, A_w/W and A_f/Z, respectively. If experimental conditions are then so arranged that $W = Z$, the weights cancel out and counting rates may be used instead of specific activities. A body of methods, referred to as *substoichiometry*, has been devised to utilize this principle in trace analysis (98a). The use of substoichiometry in isotope dilution methods enhances the sensitivity of determination of some elements so that under favorable circumstances even lower quantities may be determined than is possible by activation analysis.

The isolation of equal and very small amounts of tagged substances in isotope dilution with high precision and reproducibility may be achieved by various methods, including extraction by complexing agents, ion exchange, and electrolysis. The term substoichiometry applied to these methods refers to the fact that the element of interest is isolated by using for the samples, before and after dilution, under appropriate analytical conditions, identical quantities of a given reagent which will react completely to isolate less than the total amount of the element present. The counting rate of the samples so isolated may then be used, instead of their specific activities, in the calculations. Low activities may be handled, but for good accuracy the blank value of the reagents for the element being determined must be no higher than about one-tenth of that being determined. This method is therefore most suitable for easily soluble samples which require for solution a minimum quantity of reagent which could contribute to the blank (98a).

Reference 98b gives an example of an electrolytic deposition method of substoichiometry; references 98a, 98c, 98d, 98e, and 98h provide examples of solvent extraction methods; and 98f and 98g illustrate ion-exchange methods.

G. APPLICATIONS

The applications of isotope dilution techniques have been extensive and varied. No attempt will be made here to do more than to mention briefly examples of some of the different types. The bibliography includes references to sources from which many more examples can be gathered.

1. Inorganic and Elemental Analysis

The isotope dilution method was first used for the microdetermination of lead (99). Radioactive lead (radium D) was used to determine the loss in anodic deposition as lead dioxide. Instead of the previously assumed 100% deposition, this method showed that only a few per cent was actually deposited. Zinc-65 (100) measured the radioactive correction in the electrogravimetric determination of small quantities of zinc in aluminum alloys.

Potassium-42 chloride was added to mixtures of alkali nitrates to determine potassium after precipitation as perchlorate (101,102).

Cerium was determined by precipitation, followed by solution in sulfuric acid and spectrophotometric measurement. Losses incurred during the separation from other constituents were followed by isotope dilution with ^{144}Ce (103) (Table 97.VII).

Mass spectrometric methods have been used for isotope dilution measurements of stable isotopes in the determination of small quantities of elements (104). The rubidium (105) content of samples of sea water containing about 0.01 γ of Rb_2SO_4 was determined mass spectrometri-

TABLE 97.VII
Determination of Cerium[a]

Run	Cerium taken Mg	Cerium found (spectro.), Mg	Std. dev., %	Recov. tracer (radiom.), %	Std. dev., %	Corrected cerium, Mg	Std., dev., %	Exptl. error, %
I A	8.026[b]	7.44 ± 0.08	1.1	93.1 ± 0.3	0.3	7.99 ± 0.09	1.1	−0.5
		7.30 ± 0.08	1.1	90.9 ± 0.2	0.2	8.03 ± 0.09	1.1	0.0
I B	16.05[b]	15.11 ± 0.17	1.1	94.7 ± 0.1	0.1	15.96 ± 0.18	1.1	−0.6
		14.81 ± 0.09	0.6	93.3 ± 0.2	0.2	15.87 ± 0.09	0.9	−1.1
II	0.1892[c]	0.00974 ± 0.00056	5.7	5.11 ± 0.04	0.7	0.191 ± 0.011	5.7	+0.8
		0.03432 ± 0.00058	1.7	18.08 ± 0.07	0.4	0.1897 ± 0.0034	1.8	+0.3
		0.0963 ± 0.0006	0.6	50.13 ± 0.04	0.7	0.1921 ± 0.0017	0.9	+1.7
		0.0981 ± 0.0006	0.6	52.13 ± 0.02	0.4	0.1884 ± 0.0013	0.7	−0.4
III	1.605[d]	0.753 ± 0.015	2.0	46.82 ± 0.02	0.5	1.613 ± 0.034	2.1	+0.5
		1.046 ± 0.015	1.4	65.99 ± 0.03	0.5	1.588 ± 0.024	1.5	−1.1
		0.996 ± 0.014	1.4	61.84 ± 0.02	0.4	1.614 ± 0.024	1.5	+0.6
		0.924 ± 0.029	3.1	57.97 ± 0.02	0.4	1.598 ± 0.050	3.1	−0.4
IV	1.605[e]	0.775 ± 0.016	2.0	48.42 ± 0.04	0.8	1.601 ± 0.035	2.2	−0.3
		0.814 ± 0.014	1.7	50.53 ± 0.04	0.7	1.610 ± 0.029	1.8	+0.3
		0.858 ± 0.015	1.7	53.49 ± 0.03	0.5	1.604 ± 0.029	1.8	−0.1

Errors in columns 3, 5, and 7 represent standard deviations of mean. Columns 4, 6, and 8 show corresponding percentage values. Column 9 represents percentage differences between known mg. of cerium in solutions and experimental values.

[a] From Freedman and Hume, Anal. Chem., **33**, 933 (1950).

[b] Soln. contained 60 mg La as nitrate.

[c] Soln. contained 0.6 mg La as nitrate.

[d] Soln. contained 20 mg Nd, 18 mg Sm, 20 mg Th as chlorides, plus 30 mg La as nitrate.

[e] Soln. contained 20 mg Fe as nitrate, 20 mg V as NH_4VO_3, and 20 mg Cr as $K_2Cr_2O_7$.

cally with stable rubidium isotopes, and lithium (107a) and rubidium (107b) were determined similarly in sea water and graphite.

The double dilution technique was applied to the determination of antimony in lead at about the $10^{-5}\%$ level by the use of radioactive isotopes (106). Difficultly separable elements, such as pairs of Nb–Ta, Nb–Zr, and Ta–Zr, have been analyzed with ^{95}Nb, ^{182}Ta, and ^{95}Zr by isotope dilution (90).

Zinc has been determined by substoichiometric isotope dilution (98d) after a single extraction with dithizone, in amounts of 0.3–33 μg, with an average precision of about ±1%, and in amounts of 0.03 μg with an average precision of ±15%. Mercury was determined similarly (98h). Copper (98e) was also determined with dithizone in amounts of 0.002–1.3 μg in the presence of large excesses of other extractable metals, and iron has been determined by extraction with cupferron (98a,98i). Ion exchange has been applied to the substoichiometric determination of iron in amounts of 0.002–0.1 μg by using ethylenediaminetetraacetic acid (EDTA) and cation exchanger Dowex-50 (98f). Similarly, but with a preliminary cupferron extraction, indium has been determined in amounts of 0.00006–0.2 μg (98g). By electrodeposition a substoichiometric isotopic dilution determination of ^{110}Ag was carried out (98b).

Elemental analyses of organic compounds have been carried out for oxygen (108,108a,108b), carbon (109,110), nitrogen (111), and sulfur (112) by means of stable isotopes of those elements. A measured amount of stable isotope in a convenient form such as $^{18}O_2$ or $^{15}NH_3$ is added to the sample, which is then burned or pyrolyzed to achieve uniform isotopic distribution. Mass spectrometric measurement follows. Since mass spectrometry does not require the removal of nonisotopic impurities, it has been suggested (92, p. 3496) that a universal reagent containing deuterium, ^{13}C, ^{15}N, ^{18}O, and ^{34}S could be used to determine carbon, hydrogen, nitrogen, oxygen, and sulfur simultaneously.

2. Organic and Biochemical Analysis

The first applications of isotope dilution to organic analysis were carried out with stable isotopes. For example, an application involving the determination of amino acids by means of a deuterium label was reported in 1939 (113). In the same period work was also reported with deuterium-labeled palmitic acid on the determination of the palmitic acid content of animal fat (114). The composition of tissue phospholipids was studied with ^{15}N (115).

Amino acid analysis with ^{15}N was carried out (116), with the separation of optical isomers and calculations for the degree of racemization.

The reverse isotope dilution procedure with derivative formation was carried out (94) by reacting amino acids with ^{131}I-labeled p-iodophenylsulfonyl chloride (pipsyl chloride) and adding inactive derivative.

Direct isotope dilution was used to determine naphthalene in coal tar with ^{14}C-labeled naphthalene (117), vitamin B$_{12}$ with ^{60}Co-tagged vitamin (118), and γ-hexachlorobenzene in a mixture of isomers by means of ^{36}Cl (119) or ^{14}C (120).

In physiological investigations there have been a number of applications of isotope dilution for the purpose of determining a constituent which is of indefinite extent but in which it is possible to distribute labeled material uniformly; for example, the determination of water in the human body by intravenously injected radioactive water (121) or deuterium-labeled water (122). Reference 78 has a discussion of such methods. Analytical applications of these methods to closed systems, especially in chemical engineering problems, should be feasible.

TABLE 97.VIII
Analyses of Known Mixtures[a]

Compound analyzed, Mg	Other compounds added,[b] Mg		Found, Mg	Deviation, %
2,4-D				
202.0	2,4-D homolog mixt.	40	204	+1.0
251.8		40	250	−0.7
301.4		40	303	+0.5
503.2	2,4,5-T	500	503	0
505.0	2,4-D homolog mixt.	80	498	−1.4
	2,4,5-T	500		
	2,4,5-T homolog mixt.	60		
2,4,5-T				
225.8	2,4,5-T homolog mixt.	30	225	−0.3
301.4		30	299	−0.8
222.2	2,4-D	200	221	−0.5
221.6	2,4,5-T homolog mixt.	30	222	+0.2
	2,4-D	200		
	2,4-D homolog mixt.	40		
MCP				
400.7	MCP homolog mixt.	200	398	−0.7
292.4		200	294	+0.5
256.2		100	256	−0.5
4-Chlorophenoxy-acetic acid				
204.4	4-chlorophenoxyacetic	30	206	+0.8
253.2	acid homolog mixt.	30	254	+0.3
303.0		30	307	+1.3

[a] From Sorensen, *Anal. Chem.*, **26**, 1585 (1954).

[b] 2,4-D homolog mixture. Equal amounts of 2-chlorophenoxyacetic acid, 4-chlorophenoxyacetic acid, 2,6-dichlorophenoxyacetic acid, and 2,4,6 trichlorophenoxyacetic acid.

2,4,5-T homolog mixture. Equal amounts of 2,5-dichlorophenoxyacetic acid, 2,3,6-trichlorophenoxyacetic acid, and 2,3,4,6-tetrachlorophenoxyacetic acid.

MCP homolog mixture. 20% 2-methylphenoxyacetic acid, 60% 2-methyl-6-chlorophenoxyacetic acid, and 20% 2-methyl-4,6-dichlorophenoxyacetic acid.

4-Chlorophenoxyacetic acid homolog mixture. Equal amounts of phenoxyacetic acid, 2-chlorophenoxyacetic acid, and 2,4-D.

The determination of organic acids, alcohols, and amines by conversion to chlorinated derivatives, followed by isotope dilution with ^{36}Cl-labeled compounds, has been discussed (123) (Table 97.VIII).

Finally, separation of sulfonamide mixtures was carried out by ion-exchange chromatography and the constituents were determined by isotope dilution (124).

V. ANALYSIS WITH RADIOACTIVE REAGENTS (RADIOMETRIC ANALYSIS)

A. DEFINITION AND AREAS OF APPLICATION

The use of radioactive reagents for the determination of inactive substances by isotope dilution has been discussed in Section IV. Those methods could have been included here but were placed in Section IV for convenience. In this section various other methods will be discussed in which radioactive reagents are used for the determination of inactive substances.

The original stimulus for the use of these methods arose before the availability of artificial radioisotopes. The paucity of natural radioisotopes restricted applications requiring labeled substances. Therefore the use of these methods was extended by means of applications whereby natural radioisotopes could be employed for determining inactive substances. Radioactive lead was used, for example, to determine chromate by precipitating the latter with an excess of a solution of known lead content and activity, followed by a measurement of the activity of the filtrate.

The advantages of these methods led to considerable extension of such applications when artificial radioisotopes became available. The term "radiometric analysis" used for these methods has become too broad in view of subsequent developments and should perhaps be replaced by "analysis with radioactive reagents."

Reports in the literature on the use of radioactive reagents for applications in which no particular advantages were thereby gained should not obscure the fact that, for certain applicaaions, this type of analysis offers real advantages over conventional methods.

The determination of inactive substances with radioactive reagents may be carried out according to a number of principles, as will be seen in the rest of this section, so that it is not possible to lay down general rules. Some applications require a quantitative and stoichiometric reaction of reagent and substance; some do not require quantitative reaction; others may be satisfied by an empirically constant relationship, even though it is not stoichiometric. For some applications it is not necessary to know the specific activity of the reagent. When this information is essential, it is sometimes possible to calibrate one batch of reagent to serve for more than one specific determination.

In addition to other applications, radioactive reagents offer particular advantages for the determination of functional groups, either in substances where such groups are few in number, as in the case of end groups in polymers of high molecular weights, or in microdeterminations. Substituting activity determinations for weighing, or similar measurements, may add much to the sensitivity of a determination. Furthermore, doubts concerning the identity of reaction products may sometimes be resolved by adding inactive material of known identity and testing for the constancy of specific activity during a series of purifications.

B. STOICHIOMETRIC PRECIPITATION METHODS

A radioactive reagent may be used to precipitate and determine an inorganic or organic substance by quantitatively forming a precipitate of stoichiometric composition. If the specific activity of the reagent has been previously determined, one may calculate the amount of substance precipitated by (1) measuring the activity of the precipitate; or (2) measuring the activity of the excess reagent in the filtrate which remains unprecipitated; or (3) determining the radiometric end point in a titration of the substance with the reagent or vice versa.

Using a radioactive reagent, instead of weighing a precipitate, may be advantageous when the amounts of precipitate are too small to weigh accurately, the precipitates offer difficulties in drying to constant weight, or precipitation is accompanied by impurities which do not, however, interfere with counting. It is possible also to add inactive carrier to achieve complete precipitation without vitiating the analysis, provided absorption of radiation is negligible or a correction for this can be applied.

1. Measurement of Activity of Precipitate

With the variety of radioisotopes available, it is not difficult to procure tagged reagents with suitable beta or gamma radiations to permit direct measurement of the activity in a precipitate. For low-energy beta particles, particularly, such as those of ^{14}C, tritium, or ^{35}S, weighing the precipitate may be necessary in order to apply a self-absorption correction.

Thallium has been determined by precipitating with radioactive iodide and centrifuging in a special tube, which can be taken apart to serve as a planchet for counting the precipitate. The thallium content is calculated from a calibration curve prepared by counting precipitates of a number of solutions of known thallium content precipitated with the same tagged iodide solution (125). In still another method for thallium, ^{60}Co-tagged hexamminecobaltic trichloride was used to precipitate the compound $Tl[Co(NH_3)_6]Cl_6$. The precipitate was counted after redissolving it in hydrogen peroxide and acetic acid and drying (126). The method is capable of determining 0.5 μg of thallium.

Organic substances have also been analyzed. Thus 10 μg or less of protein has been determined by precipitating with radioactive tungstic acid and counting the precipitate (127). A method has been reported for determining minute quantities of phenobarbital by precipitating it with ^{203}Hg perchlorate ($^{203}Hg(ClO_4)_2$) (128). A drop of the radioactive reagent, applied to filter paper, is treated with a drop of the sample solution. The precipitate is washed free of excess reagent with water on the paper, and the precipitated phenobarbital containing ^{203}Hg is counted under a Geiger tube. As little as 1 γ may be determined. A similar method has also been reported for determining fatty acids on paper by precipitating with ^{60}Co acetate, counting, and comparing with standards (129).

2. Measurement of Activity of Excess Reagent

Frequently it is more convenient to arrive at the amount of radioactive reagent used for precipitating a substance by adding a measured amount of the reagent in excess over the amount required for precipitation, withdrawing an aliquot of the solution after filtering or centrifuging, and measuring the activity of the aliquot. The volume (or weight) of the solution before the addition of reagent, as well as of the added reagent, must be known in order to calculate the total excess from the activity of the aliquot.

The activity of the aliquot includes not only that due to excess reagent but also that due to any dissolved precipitate (as well as the normal background). For extremely insoluble precipitates, such as lead chromate, this may be small enough to be disregarded, but for somewhat less insoluble precipitates, for example, lead sulfate, it cannot be ignored. It must be determined separately and subtracted from the activity of the aliquot. In order to make such a correction, the fact must be borne in mind that increasing the excess concentration of the precipitating ion will decrease the solubility of the precipitate. The amount of excess should thus be kept within a range which will ensure that the correction used actually corresponds to the conditions of the determination.

It is also important, from considerations of the statistical variations of counting radioactivity, to keep the excess as small as possible. The activity determined for the aliquot by counting will have a random error depending on the number of counts. However, the effect of this error on the final answer depends on the magnitude of the difference between the activity of the reagent added and that of the excess. If this difference is large (corresponding to a small excess), the error of counting the excess will result in a smaller relative error in the difference than if the difference is small (corresponding to a large excess).

Finally, it should be noted that determination of a solubility correction may be avoided, for precipitates for which it would otherwise be required, if a calibration curve is set up by measuring the activity of the filtrates of a series of different concentrations of the substance to be determined when a given amount of reagent is added. The answer can be read off the curve, and it is also unnecessary to know the activity of the reagent, provided the same reagent is used throughout. Such a method is convenient for repetitive analyses.

Measurement of excess reagent was used as far back as 1925 for the determination of chromate with radioactive lead (130). Before the advent of artificial radioisotopes, Ehrenberg carried out a number of ingenious determinations with active lead as the reagent in the last of a series of steps in indirect analyses for which no radioactive reagent was available for direct application. For example, potassium was precipitated as inactive hexanitritocobaltate(III); the precipitate was oxidized with excess permanganate; the excess permanganate was reduced with oxalate; and the excess oxalate was precipitated with radioactive lead (131).

Thorium was determined by precipitation with an excess of active pyrophosphate solution, and the activity of the filtrate was determined by direct measurement (132). To avoid differences in absorption of the ^{32}P beta rays, the densities of the solutions were adjusted by adding sodium nitrate.

3. Titration Methods

a. GENERAL CONSIDERATIONS

When determining inactive elements or compounds that can be precipitated by an active reagent, or active substances that can be precipitated by either an inactive or active reagent, titration methods may be used, rather than a determination of the amount of substance precipitated by means of measuring the activity of the precipitate or filtrate. In titration methods the appearance or disappearance of activity is used to determine an equivalence point. Schematic curves are given in Figs. 97.6, 97.7, and 97.8 for three general possibilities. Figure 97.6 represents a titration in which an inactive substance is being titrated with an active precipitant. As the latter is added, the original background activ-

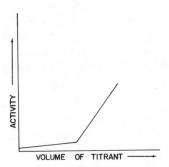

Fig. 97.6. Titration of inactive sample with radioactive titrant.

ity is supplemented by activity caused by the solubility of the precipitate, with its extent dependent on that solubility. It tends to rise as the titration proceeds because of the decrease in concentration of the excess component. After the end point the activity rises as a result of excess radioactive reagent. The two lines may be plotted and the equivalence point obtained graphically from their intersection, or it may be calculated from two points on the rising part of the curve. An actual curve may be rounded near the intersection because of increased solubility due to negligible excess of precipitating ions. Figure 97.7 represents a titration of an active substance with an inactive reagent, and Figure 97.8 a titration in which both constituents are active. The resemblance of the titration curves to conductometric titration curves is obvious.

The experimental approach may be varied according to conditions. The simplest (and most cumbersome) method is to withdraw and count aliquots of known volume or weight and to relate these volumes or weights to the total volume or weight. A more rapid procedure makes

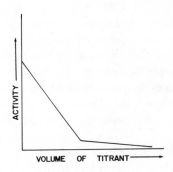

Fig. 97.7. Titration of radioactive sample with inactive titrant.

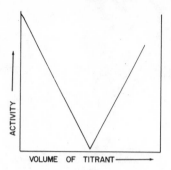

Fig. 97.8.   Titration of radioactive sample with radioactive titrant.

use of an apparatus (133) (Fig. 97.9) in which the titration mixture, after each addition of reagent, is sucked up through a sintered glass disk into a container surrounding a Geiger tube, where the activity is measured. The solution is then returned to the main reaction mixture, and the titration continued. To prevent adsorption of radioactive substances from solution it has been recommended that all the glass surfaces, except the sintered glass filter disk, be treated with a methylchlorosilane hydrophobing agent (134).

Several points should be kept in mind in designing precipitation titrations. Since the precision obtainable is limited by the counting statistics, activities high enough to obtain good statistics without the need for awkwardly long counting periods are desirable. Furthermore, substances with a very low solubility product are preferable for two reasons. First, there is a smaller contribution of dissolved reaction product to the total count, with correspondingly less inaccuracy caused by variations in this factor. Second, and more significant, is the fact that only when the solubility product is negligibly low (perhaps $<10^{-6}$) can the titration curves be truly linear. (For more highly soluble products linear curves may be obtained by using high concentrations and small, precisely measured volumes, but at the expense of sensitivity.) Ordinarily, for more highly soluble products the titration curves will not be truly linear, but may be close enough to true linearity for confusion to result and for apparent end points to be obtained which are incorrect. Duncan and Thomas have discussed the theory of this subject thoroughly (135).

Radiometric titrations are useful for determinations for which ordinary titration methods cannot be used: for turbid or colored solutions, for which visual indicators are inadequate; for corrosive or reactive solutions, which might destroy visual indicators or prevent the use of electrometric or similar methods; for extremely dilute solutions; and

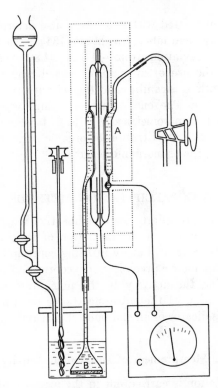

Fig. 97.9. Experimental arrangement. A—tube counter for liquids shielded with lead; B—fritted-glass immersion filter; C—counting meter. From Langer, *J. Phys. Chem.*, **45**, 641 (1941).

for systems for which no other end point indication is available. Finally, these methods are not restricted to inorganic systems or to aqueous solvents.

b. APPLICATIONS

Titrations have been carried out with radioactive phosphate ($Na_2H^{32}PO_4$) for the determination of magnesium, uranyl ion, silver, lead, and thorium (133). Chloride and bromide have been titrated with radioactive silver (136).

An illustration of the titration of a radioactive substance with an inactive reagent is the addition of ^{204}Tl tracer to a thallium sample and titration with $0.1M$ potassium iodide solution. Milligram samples of thallium have thus been determined with deviations from the true values ranging from -2.6 to $+3.2\%$ of the total (134a). Milligram samples of thallium treated with ^{204}Tl were also determined by titration with phosphotungstic acid labeled with ^{32}P (134a). The curve was similar to that in Fig. 97.8. Deviations from -1.9 to $+4.8\%$ were obtained.

An example of the use of an organic tagged reagent is the titration of copper (137) with anthranilic acid tagged with tritium by the Wilzbach method (17–19). The end point determined was almost identical with that obtained amperometrically (137). It should be possible to extend such methods to the titration of organic substances by precipitation with tagged organic or inorganic reagents.

C. COPRECIPITATION METHODS

In some instances precipitation by radioactive reagents may be used for analysis, not by virtue of the formation of stoichiometric insoluble compounds with the reagent, but by utilizing coprecipitation phenomena. As is true for stoichiometric precipitations, coprecipitation can be utilized either by measuring the activity in the precipitate or in the residual solution for calculation of the substance to be determined, or by using the activity as an indicator of an end point in a titration.

1. Measurement of Distributed Activity

When a precipitation is carried out in the presence of a coprecipitable foreign ion, the latter is partitioned between the precipitate and the solution in a ratio which depends on a number of conditions, including the nature of the substances and their concentrations. If the foreign ion is radioactive, a measurement of the activity of the precipitate, or of the residual activity of the solution left after removing the precipitate, will serve to determine the amount of precipitate. This principle may be used to calculate the quantity of precipitate if the partition coefficient under the conditions of the analysis is known with sufficient precision. Usually, however, this is not the case, so that the principle is actually useful only if it is applied by means of a calibration curve of activity versus amount of determined substance, which has been set up under well defined conditions. This requirement presupposes that the conditions for reproducibility are well enough understood to make the construction and use of such a curve feasible.

The determination of either calcium or oxalate has been carried out by this method (138,139). For the calcium determination, there are added first a $(10^{-6}\ N)$ acid solution of lead nitrate tagged with thorium B (^{212}Pb) and then an excess of ammoniacal, saturated ammonium oxalate solution. The calcium oxalate, with coprecipitated lead, is removed by centrifugation and the activity of the solution is measured. For an oxalate determination, excess calcium chloride solution which has been tagged with thorium B is added to the oxalate solution.

2. Titration Methods

During the course of precipitation of an anion with a cation, or vice versa, there is a sudden reversal of the polarity of the precipitate at the equivalence point, where an excess of the precipitating ion, of opposite charge to the ion which has been precipitated, is suddenly present. At low concentrations, an ion which is adsorbed by, but not isomorphous with, the precipitate will undergo a sudden change in distribution between precipitate and solution. If the adsorbable ion is radioactive, the sudden change in activity of the solution will indicate the occurrence of the equivalence point of the precipitation reaction.

This principle has been used for the titration of sulfate with barium (140). A little radioactive phosphate is added to the barium solution. Before the barium sulfate equivalence point the anionic phosphate is hardly adsorbed at all on the negatively charged barium sulfate, but at the equivalence point it is strongly adsorbed. The drop in activity of the solution serves to indicate the end point.

D. MISCELLANEOUS METHODS

1. Radioactive Gas Evolution (Radiochemical Release)

Since radioactive gases can be counted either directly or after freezing out, dissolving, or otherwise trapping, it follows that a reaction which liberates (or consumes) a radioactive gas either quantitatively or, conceivably, reproducibly, although not quantitatively, could be considered for use as a possible analytical method. One example of such a method is the determination of moisture by using [36]Cl-tagged aluminum chloride (141). Reaction of this reagent with moisture quantitatively liberates tagged hydrogen chloride, which can be frozen out and counted.

A more systematic exploration of the analytical possibilities of liberation of radioactive gases has been carried out (142–144). To determine sulfur dioxide, the gas is allowed to pass over [36]Cl-tagged solid sodium chlorite in order to liberate chlorine dioxide ([36]ClO$_2$) in an amount proportional to the sulfur dioxide content. The activity can be counted as a measure of sulfur dioxide. It was also found that gases such as ozone liberate radioactive ^{85}Kr from a solid krypton–quinol clathrate. As little as parts per billion of ozone may be determined in this way. Since the chlorine dioxide liberated by reaction with sulfur dioxide may in turn be made to liberate krypton from the clathrate, the possibility for amplification of effects should permit attainment of sensitivities down to about 10^{-13} to 10^{-14} g. A method is also described for determining chemically active hydrogen, in any of its varied forms, by reacting it with lithium aluminum tritide and measuring the radioactivity of the tritium evolved (142a). Our experience with this method indicates the need for caution if a calibration curve based on one substance containing chemically active hydrogen is to be used for the analysis of other substances (142b,142c).

A novel tracer method which may be of interest in a number of applications—analytical and nonanalytical—involves the use of so-called solid kryptonates (144a,144b,144c,144d). These are solids in which ^{85}Kr has been incorporated either by ionizing it and impelling it into the solid by a potential drop or by perfusing it into the solid at high temperature and pressure. Evidently any solid substance may be kryptonated, and there is little leakage of krypton from the solid at temperatures below that at which it was introduced. Any process, chemical or physical, which disturbs the solid will cause the liberation of krypton, so that the activity of the released gas or the decrease in activity of the solid source may serve as a measure of the extent of reaction. For example, kryptonated metal foils release krypton when oxides form at the surface and may be used to detect or measure oxygen. Kryptonated calcium carbide may be employed to detect water either as a gas or in solution, and kryptonated magnesium or zinc may be used to analyze for acids in aqueous solution. Many other potential applications exist in chemical kinetics studies, surface temperature measurements, and friction and wear studies.

In addition, kryptonated solids may be utilized as flow tracers and for similar applications. The biological and chemical inertness of ^{85}Kr, its long half-life, and the absence of any considerable gamma radiation are advantageous; and for applications where it could be used, there would be no need to synthesize compounds containing special isotopes.

2. Paper Chromatography

In Section III.B.5 the use of radioisotopic tracers in paper chromatography was discussed mainly from the point of view of the use of tagged materials. However, untagged materials may be separated by paper chromatography and treated on the paper with a radioactive reagent. The amount of constituent is determined from the radioactivity of the spot and the known specific activity of the reagent, together with the stoichiometry of the reaction, or by comparison against standards similarly treated.

Radioactive iodine bromide has been used for the determination of double bonds (158), radioactive cobalt acetate for fatty acids (129), and active hydrogen sulfide for traces of metals (146). It has also been proposed to treat amino acids, separated on a paper chromatogram, with radioactive methyl iodide (147) (Fig. 97.10). Fluorine has been determined by paper chromatography with the aid of ^{32}P-tagged phosphoric acid (145). The latter is used to form active $(ZrO)_3(PO_4)_2$ on paper,

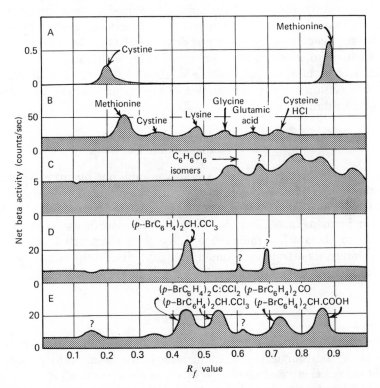

Fig. 97.10. Paper chromatograms. A—chromatogram of protein hydrolysate from wheat grown on S^{35}; B—strip spotted with known amino acids and exposed to I^{131}-labeled CH_3I; C—neutron-activated chromatogram of $C_6H_6Cl_6$ isomers; D—neutron-activated chromatogram of a bromine analog of DDT; E.—neutron-activated chromatogram of bromine analogs of DDT derivatives. From Winteringham et al., *Nucleonics*, **10**, 53 (1952).

which liberates active H_3PO_4 by reaction with fluorine. The activity of the H_3PO_4 spot serves as a measure of the fluorine present.

3. Solid and Surface Constituents

The sensitivity of radioactivity methods may be utilized in determining traces of substances on surfaces by reacting them with radioactive reagents. Radioactive iodine in an amyl iodide solution is taken up by metals such as silver, chromium, lead, and copper. One method (148) utilizes this effect, as an example, to determine the amount of copper deposited on a steel plate by a copper point drawn over it. The plate is immersed in the radioisotopic iodine solution, and the amount of copper deposited by abrasion of the point determined autoradiographically.

In an example of the determination of functional groups in solids, the carboxyl content of cellulose was determined (149) by allowing cellulose samples to attain adsorption equilibrium with radioactive cerium by immersion in a cerium acetate solution, followed by counting the adsorbed cerium activity. Evaluation of the cerium content was carried out with calibration curves.

4. Oxidation–Reduction

Oxidation–reduction reactions are also suitable for quantitative determinations with radioactive reagents. An example is the determination of iodine (150). The iodine is converted to iodate by reaction with bromine or permanganate, and the iodate is treated with excess iodide-131 solution to liberate free radioactive iodine. This is extracted with carbon tetrachloride, its activity is counted, and the iodine in the original sample is calculated from the known specific activity of the tagged iodide.

5. Complex Formation

The formation of complexes may be used with either the complexed element or the complexing agent in radioactive form. Since it is easier to obtain radioactive elements, however, radiometric methods have emphasized determinations with complexed radioactive elements.

As with precipitation, complexation may be applied to a radiometric determination by one of several general procedures. One method involves segregation of the reaction product, measurement of its activity, and calculation of the amount of sample from the activity and the known specific activity (or from calibration curves). Another method involves treatment of the radioactive element with successive measured volumes of standard complexing reagent, extraction of the complex, and measurement of the residual activity after each addition. A curve similar to that of Fig. 97.7 is obtained, from which an equivalence point may be calculated. A third method depends on titration of an inactive element with an inactive complexing agent in the presence of an insoluble compound containing a radioactive element that does not react as easily with the complexing agent as does the element being determined. As soon as excess reagent is present, some of the radioactive element is solubilized. A curve such as that of Fig. 97.6 is obtained, from which an equivalence point may be determined.

By means of the first approach, the solubility of dithizone in water was determined (151). An excess of radioactive silver was added to the aqueous dithizone solution, and the silver complex was extracted with chloroform. From a measurement of the activity, the amount of extracted silver was determined; from this could be calculated the amount of dithizone in solution.

A method for determining amino acids is based on the use of ^{64}Cu-tagged copper phosphate (152). This finely divided solid reagent is mixed with the amino acid solution, and after standing, the mixture is filtered. The activity of the solution, which is caused by the formation of a soluble copper complex with amino acids, is measured, and the amino acid content calculated from a calibration curve. The method is so sensitive that it may be used to determine amino acids in single paper chromatographic spots.

Another example of a complexation titration method is used for cations of the transition elements (153). Some of the radioactive form of the cation is added to the solution of the cation being determined. Measured volumes of standard dithizone solution in carbon tetrachloride are mixed successively with the sample, in order to transfer some of the cation to the organic phase. The activity of the aqueous solution is measured after each addition. From the titration curve, equivalence points are determined. Sensitivities of 10^{-7} to 10^{-5} g of cation are obtained.

A method with end-point indication provided by solubilization of a radioactive compound has been reported for the determination of copper (154). A chelating material, "complexon III," was used. Silver-110 iodate served as the insoluble tagged indicator. The standardized complexon solution was titrated into the copper solution containing the suspended indicator, and the activity of the solution was measured. From a plot

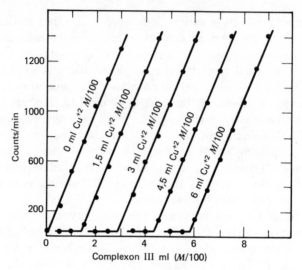

Fig. 97.11. Ag^{110}IO$_3$ indicator for titration with complexon III. From Braun et al., *Nature,* **182,** 936 (1958).

of volume of complexon added versus activity the equivalence point was obtained (Fig. 97.11).

6. Covalent Reactions

The previous examples of analysis with radioactive reagents have involved mainly ionic reactions. Covalent reactions also are being utilized in ever-increasing numbers.

This approach is especially important for determining functional groups in high-molecular-weight compounds for the purpose of assay, to establish a structure or, if the structure is known, to determine the average molecular weight. One example is the determination of free amino groups in egg albumin by means of ^{131}I-labeled p-iodophenylsulfonyl chloride (pipsyl chloride) (155,156). After treatment of the albumin with the tagged pipsyl chloride, the sample is hydrolyzed to form a mixture containing the pipsyl amino acids, which are extracted with ether. The activity serves as a measure of the total number of amino groups in the albumin. Identification and determination of the individual amino acids may be achieved by paper chromatographic separation of the pipsyl derivatives and application of isotope dilution procedures.

Aldehyde or ketone carbonyl groups in polysaccharides have been determined by reaction with ^{14}C-tagged cyanide to form cyanohydrins

(157). Hydrolysis to acids and determination of the ^{14}C activity lead to values for the carbonyl content of the polysaccharide.

The double-bond unsaturation in fats has been determined with tagged iodine bromide on small samples (158). Some of the fat solution is dropped onto paper, and the solvent evaporated. The spot is treated with the tagged reagent, and excess reagent is washed out with ethanol and water. After the activity on the paper has been measured, the number of double bonds is calculated by comparison with standards. The method depends on excess reagent being washed out. If the derivative is soluble in alcohol too, it can first be converted to an insoluble compound (in the case of acids, for example, to a heavy metal salt).

7. Determination of Free Radicals

Radioisotopes have been used to determine free radicals either by reaction of a gas stream with a tagged metallic mirror or by reaction of a gas or liquid with a labeled nonmetallic reagent.

In an example of the first method a metallic mirror of radioactive lead is allowed to react with a gas stream of tetraphenyl- or tetramethyl-lead (159). If either of these compounds is heated or subjected to ultraviolet radiation before it reaches the metallic mirror, free radicals are formed which react with the mirror. The active lead compounds may be condensed and counted. A similar method was used to study the formation of free radicals by pyrolysis of acetaldehyde (160).

Iodine is a so-called scavenger for free radicals, reacting with them to form iodides, which can be identified as an indication of the nature of the free radicals. The use of radioactive iodine makes a sensitive determination possible. The first experiments were made with gaseous streams (161). Vaporized radioactive iodine was injected near the point of pyrolysis or photolysis of the sample. The alkyl iodides were frozen out, carriers added, and the mixtures separated by fractional distillation. The activity of the fractions was then measured. It was found in this way, for example, that propyl radicals could be decomposed to ethylene and a methyl radical. Later studies (162) subjected vaporized mixtures of acetone or methyl ethyl ketone and radioactive iodine to ultraviolet radiation. The excess iodine was removed with finely divided silver, and the alkyl iodides were fractionated after addition of carriers.

Irradiation with gamma rays of radioactive iodine dissolved in liquid *n*-pentane led to a mixture of products in which methyl, ethyl, propyl, butyl, and amyl iodides could be identified (163).

8. Biochemical Processes

Although most biological work with tracers has involved the direct use of isotopes as indicators, it is possible even in biological systems to use radioactive reagents. One example (164) is the determination of thiouracil in tissue. The parts of animals in which thiouracil is to be determined are fed to rats, which are later injected with radioactive iodine and, after a further period of time, are killed. The activity of the thyroid gland is measured, the uptake of iodine by this gland being a function of the thiouracil content. Reference is then made to standard curves drawn up from experiments with animals injected with known amounts of thiouracil for a determination of the thiouracil values in the animal parts of interest. A similar method for the determination of desoxycorticosterone depends on its effect on the excretion of ^{42}K-labeled potassium (165).

GENERAL REFERENCES

"Source Material for Radiochemistry," *Nucl. Sci. Ser., Rept.* No. 27 (rev. 1), Division of Physical Sciences, National Research Council-National Academy of Sciences, 2101 Constitution Ave., Washington, D.C., Publication 825, July 1960. References to books and other material on various aspects of radiochemistry, with a brief description of each reference.

"Nucleonics. Review of Fundamental Developments in Analysis," Meinke, W. W., *Anal. Chem.,* **28,** 736 (1956); **30,** 686 (1958); **32,** 104R (1960). Leddicotte, G. W., *Anal. Chem.,* **34,** 143R (1962); **36,** 419R (1964). Lyon, W. S., E. Ricci, and H. H. Ross, *Anal. Chem.,* **38,** 251R (1966); **40,** 168R (1968). Biennial reviews of current references with extensive coverage of tracer applications. (See also *Analytical Chemistry* reviews for previous years by other authors.)

Isotopes—A Five-Year Summary of U.S. Distribution, August 1951. Available from Superintendent of Documents, U.S. Government Printing Office, Washington, D.C. A detailed account of all isotope investigations initiated from the beginning of the distribution program in August 1946 through June 30, 1951. Contains a bibliography of 1400 references.

Isotopes—An Eight-Year Summary of U.S. Distribution and Utilization, March 1955. Available from Superintendent of Documents, U.S. Government Printing Office, Washington, D.C. Resumé of isotope distribution during the first 8½ years of the U.S. Atomic Energy Commission distribution program. Includes a narrative account of isotope uses in medicine, biology, agriculture, industry, and research and a title listing of 7200 references appearing in report and journal literature since 1951.

Isotopes—A Bibliography of United States Research and Application, 1955–1957, TID 3076, May 1958. Available from Office of Technical Services, U.S. Department of Commerce, Washington, D.C. Titles listed for about 6000 references, arranged by various categories of application.

Hevesy, G., *Radioactive Indicators*, Interscience, New York, 1948. A thorough review of the application of radioactive indicator techniques to the fields of animal physiology, pathology, and biochemistry.

Kamen, M. D., *Isotopic Tracers in Biology*, 3rd Ed., Academic, New York, 1957. Tracer methodology in biological research for biochemists, physiologists, and others with little previous contact with the field.

Broda, E., and T. Schönfeld; see Specific Reference 139. Broad treatment (in German) of tracer techniques, with copious literature references; not restricted to microchemistry (in spite of the title).

Radioisotopes in World Industry, TID 6613 (Suppl. 4), U.S. Atomic Energy Commission, Division of Technical Information, July 1963. Bibliography on the use of radioisotopes in world industry.

Wahl, A. C., and N. A. Bonner, *Radioactivity Applied to Chemistry*, Wiley, New York, 1951. Summary of work up to 1950 in many areas of chemistry, including such topics as isotope exchange, kinetics, structural chemistry, self-diffusion, analysis, carrier-free tracers, and surface chemistry. Many references tabulated by subject field.

Monograph Series on the Radiochemistry of the Elements. Separate volumes from NAS-NS 3001, January 1960, to NAS-NS 3014, May 1960. Available from Office of Technical Services, U.S. Department of Commerce, Washington, D.C. Monographs on individual elements or groups of elements which provide information required for radiochemical work with the elements. (See "Source Material for Radiochemistry," pp. 16–19, for titles of individual monographs.)

Reynolds, S. A., and G. W. Leddicotte, "Radioactive Tracers in Analytical Chemistry," *Nucleonics*, **21**, No. 8, 128 (August 1963). General discussion of tracer techniques with literature references through 1961, but with scattered coverage in 1962, of inorganic and typical organic applications.

SPECIFIC REFERENCES

1. Paneth, F., and G. von Hevesy, *Z. Anorg. Chem.*, **82**, 323 (1913).
2. Hevesy, G. von, *Physik. Z.*, **16**, 52 (1915).
3. *International Directory of Isotopes*, 3rd Ed., International Atomic Energy Agency, Kärntner Ring, Vienna I, Austria, 1964 (Unipub, Inc., P.O. Box 433, New York, 10016).
4. *The Isotope Index*, Scientific Equipment Co., P.O. Box 5686, Indianapolis, Ind., issued annually.
5. Hentz, R. R., *J. Chem. Educ.*, **35**, 625 (1958).
6. Finney, G. D., and R. D. Evans, *Phys. Rev.*, **48**, 503 (1935).
7. Murray, A., III, and D. L. Williams, *Organic Syntheses with Isotopes*, Part I: *Compounds of Isotopic Carbon*; Part II: *Organic Compounds Labeled with Isotopes of the Halogens, Hydrogen, Nitrogen, Oxygen, Phosphorus and Sulfur*; Interscience, New York, 1958.
8. Weygand, F., and H. Simon, "Herstellung isotopenhaltiger organischen Verbindungen," in *Methoden der Organischen Chemie*, 4th Ed., Vol. IV, Georg Thieme Verlag, Stuttgart, 1955, Part 2, p. 539; through ref. 139, p. 272.
9. Calvin, M., C. Heidelberger, J. C. Reid, B. M. Tolbert, and P. E. Yankwich, *Isotopic Carbon*, Wiley, New York, 1949.

10. Cheronis, N. D., A. R. Ronzio, and T. S. Ma, "Micro and Semimicro Methods," in A. Weissberger, ed., *Technique of Organic Chemistry*, Vol. VI, Interscience, New York, 1954.
11. Fukushima, D. K., S. Lieberman, and B. Praetz, *J. Am. Chem. Soc.*, **72**, 5205 (1950).
12. Smith, N. H., K. E. Wilzbach, and W. G. Brown, *J. Am. Chem. Soc.*, **77**, 1033 (1955).
13. Wagner, C. D., and D. P. Stevenson, *J. Am. Chem. Soc.*, **72**, 5785 (1950).
14. Ingold, C. K., C. G. Raisin, and C. L. Wilson, *J. Chem. Soc.*, **1936**, 915.
15. Van Heyningen, W. E., D. Rittenberg, and R. Schoenheimer, *J. Biol. Chem.*, **125**, 495 (1938).
16. Eidinoff, M. L., and J. E. Knoll, *J. Am. Chem. Soc.*, **75**, 1992 (1953).
17. Wilzbach, K. E., *J. Am. Chem. Soc.*, **79**, 1013 (1957).
18. Wilzbach, K. E., "The Gas Exposure Technique for Tritium Labeling," in *Proceedings of the Symposium on Tritium in Tracer Applications*, sponsored by New England Nuclear Corporation, 575 Albany Street, Boston, Mass., Atomic Associates, Inc., and Packard Instrument Company, Inc., New York City, Nov. 22, 1957.
19. Wilzbach, K. E., *Atomlight*, December 1960, No. 15, 1, New England Nuclear Corporation, 575 Albany Street, Boston, Mass.
19a. Yavorsky, P. M., and E. Gorin, *U.S. At. Energy Comm.*, N.Y. 0-9143, 1961, 56 pp.; through *Chem. Abstr.*, **56**, 5030 (1962).
19b. Yavorsky, P. M., and E. Gorin, *J. Am. Chem. Soc.*, **84**, 1071 (1962).
20. Wallace, C. H., and J. E. Willard, *J. Am. Chem. Soc.*, **72**, 5275 (1950).
21. Blau, M., and J. E. Willard, *J. Am. Chem. Soc.*, **73**, 442 (1951).
22. Kieffer, F., and P. Rumpf, *Bull. Soc. Chim. France*, 5° Serie, **18**, 584 (1951M).
23. Chatterjee, S. D., and D. K. Banerjee, *J. Indian Chem. Soc.*, **17**, 712 (1940).
24. Smythe, C. V., and D. Halliday, *J. Biol. Chem.*, **144**, 237 (1942).
25. "Abstracts of a Conference on the Use of Isotopes in Plant and Animal Research," *Kansas Agr. Exp. Stat. Rept.* No. 4, Kansas State College, June 1952 (U.S. Government Printing Office, Washington, D.C., TID-5098, April 1953).
26. *Selected Reference Material on Atomic Energy*, Vol. VII; *Eight-Year Isotope Summary*, International Conference on Peaceful Uses of Atomic Energy, Geneva, August 1955. (Superintendent of Documents, U.S. Government Printing Office, Washington, D.C.), through ref. 7, p. 21 (bibliography with 345 references).
27. Chaiet, L., C. Rosenblum, and D. T. Woodbury, *Science*, **111**, 601 (1950).
28. Rowlands, S., D. Rowley, and E. L. Smith, *J. Chem. Soc.*, **1949**, S405.
29. Rowley, D., P. O. Cooper, P. W. Roberts, and E. L. Smith, *Biochem. J.*, **46**, 157 (1950).
30. Ball, E. G., A. K. Solomon, and O. Cooper, *J. Biol. Chem.*, **177**, 81 (1949).
31. Anderson, R. C., and Y. Delabarre, *J. Am. Chem. Soc.*, **73**, 4051 (1951).
32. Smith, E. L., *Biochem. J.*, **52**, 384 (1952).
33. Willard, J. E., *Ann. Rev. Nucl. Sci.*, **3**, 193 (1953); *Ann. Rev. Phys. Chem.*, **6**, 141 (1955).
34. Burton, M., *J. Chem. Educ.*, **28**, 404 (1951).
35. Hart, E. J., *J. Chem. Educ.*, **34**, 586 (1957).
36. Tolbert, B. M., *Nucleonics*, **18**, No. 8, 74 (1960).

37. Tolbert, B. M., and R. M. Lemmon, *Radiation Res.*, **3**, 52 (1955).

38. Bayly, R. J., and H. Weigel, *Nature*, **188**, 384 (1960).

39. Wolf, A. P., and R. C. Anderson, *J. Am. Chem. Soc.*, **77**, 1608 (1955).

40. Dainton, F. S., and H. M. Kimberley, *Trans. Faraday Soc.*, **46**, 912 (1950).

40a. Johnson, P., R. M. Bullock, and J. Whiston, *Chem. & Ind.*, **750** (May 11, 1963).

40b. *Studies of On-Stream Production of Short-Lived Intrinsic Radiotracers for Industrial Process Control*, BMI-1606, U.S. Atomic Energy Commission, Division of Isotopes Development, Dec. 18, 1962.

41. Schadel, H. M., Jr., and C. E. Birchenall, *J. Metals,* **188,** *Trans.* 1134 (1950); through *Chem. Abstr.*, **44**, 9756 (1950).

42. Hall, L. D., *J. Am. Chem. Soc.*, **73**, 757 (1951).

43. Timmerhaus, K. D., and H. G. Drickamer, *J. Chem. Phys.*, **19**, 1242 (1951); **20**, 981 (1952).

44. Hutchinson, F., *J. Chem. Phys.*, **17**, 1081 (1949).

45. Cuddeback, R. B., and H. G. Drickamer, *J. Chem. Phys.*, **21**, 597 (1953).

46. Cuddeback, R. B., R. C. Koeller, and H. G. Drickamer, *J. Chem. Phys.*, **21**, 589 (1953).

47. Koeller, R. C., and H. G. Drickamer, *J. Chem. Phys.*, **21**, 267, 575 (1953).

48. Wang, J. H., and J. W. Kennedy, *J. Am. Chem. Soc.*, **72**, 2080 (1950).

49. Wang, J. H., *J. Am. Chem. Soc.*, **75**, 2777 (1953).

50. Burkell, J. E., and J. W. T. Spinks, *Can. J. Chem.*, **30**, 311 (1952).

51. Harwood, J. J., *Nucleonics*, 2 (1), 57 (1948).

52. Hutter, J. C., *Bull. Soc. Chim. France*, 5° Serie, **18**, 45D (1951).

53. Lindner, R., and G. Johansson, *Acta Chem. Scand.*, **4**, 307 (1950) (in German); through *Chem. Abstr.*, **44**, 9760 (1950).

54. Burwell, J. T., and S. F. Murray, *Nucleonics*, 6 (1), 34 (1950).

55. Wittand, J. E., *J. Phys. Chem.*, **57**, 129 (1953).

56. Palacios, J., and A. Baptista, *Nature*, **170**, 665 (1952).

57. Judson, C. M., A. A. Lerew, J. K. Dixon, and D. J. Salley, *J. Chem. Phys.*, **20**, 519 (1952).

58. Cooley, R. A., and H. O. Banks, *J. Am. Chem. Soc.*, **73**, 4022 (1951).

59. Black, C., G. G. Joris, and H. S. Taylor, *J. Chem. Phys.*, **16**, 537 (1948).

60. Morrison, G. H., and H. Freiser, *Solvent Extraction in Analytical Chemistry*, 1st Ed., Wiley, New York, 1957.

61. Kurbatov, M. H., and J. D. Kurbatov, *J. Am. Chem. Soc.*, **69**, 438 (1947).

62. Barnes, R. B., and D. J. Salley, *Ind. Eng. Chem., Anal. Ed.*, **15**, 4 (1943).

63. Ferla, F., *Ann. Chim. Applicata*, **28**, 331 (1938); through *Chem. Abstr.*, **33**, 2434 (1939).

64. Straaten, H. van der, and A. H. W. Aten, *Rec. Trav. Chim.*, **69**, 561 (1950).

65. Dean, J. A., and S. A. Reynolds, *Anal. Chim. Acta*, **11**, 390 (1954).

66. Boyd, T. F., and M. Galan, *Anal. Chem.*, **25**, 1568 (1953).

67. Boyd, G. E., and D. N. Hume, *Natl. Nucl. Energy Ser.*, Div. VIII, 1, *Anal. Chem. Manhattan Project*, 662 (1950); *Chem. Abstr.*, **45**, 1899 (1951).

68. Thiers, R., W. Graydon, and F. E. Beamish, *Anal. Chem.*, **20**, 831 (1948).

69. Morgan, L. D., and S. E. Turner, *Anal. Chem.*, **23**, 978 (1951).

70. Kimura, K., *Proceedings of the International Conference on Peaceful Uses of Atomic Energy, Geneva, 1955*, Vol. 15, p. 220 (pub. 1956); *Chem. Abstr.*, **50**, 13620 (1956).

71. Geilmann, W., and W. Gebauhr, *Angew. Chem.*, **66**, 453 (1954).
72. R. C. Extermann, ed., *Radioisotopes in Scientific Research-Proceedings of the First (UNESCO) International Conference*, Vol. II, Pergamon, New York, 1958.
72a. S. Z. Roginsky, *Theoretical Principles of Isotope Methods for Investigating Chemical Reactions*, translated by Consultants Bureau, Inc., U.S. AEC-tr-2873, available from the Office of Technical Services, U.S. Department of Commerce, Washington, D.C., $2.00.
73. Croatto, U., and G. Giacomello, *Gazz. Chim. Ital.*, **82**, 712 (1952); through *Chem. Abstr.*, **47**, 7292 (1953).
74. Beischer, D. E., *J. Phys. Chem.*, **57**, 134 (1953).
75. Kamen, M. D., *Radioactive Tracers in Biology*, 2nd Ed., Academic, New York, 1951.
76. Broda, E., *Radioactive Isotopes in Biochemistry*, Franz Deuticke, Vienna, 1958; through *Nucl. Sci. Abstr.*, **13**, 2694 (1959).
77. McCormick, J. A., *U.S. At. Energy Comm. Rept.* TID-3512, February 1958, pp. 1–44; through *Nucl. Sci. Abstr.*, **12**, 10346 (1958) (bibliography).
78. Hevesy, G., *Radioactive Indicators*, Interscience, New York, 1948.
79. Weygand, F., *Angew. Chem.*, **61**, 285 (1949).
80. Udenfriend, S., *J. Biol. Chem.*, **187**, 65 (1950).
81. Ketelle, B. H., and G. E. Boyd, *J. Am. Chem. Soc.*, **69**, 2800 (1947).
82. Boyd, G. A., *Autoradiography in Biology and Medicine*, Academic, New York, 1955.
83. Arnason, A. P., R. A. Fuller, and J. W. T. Spinks, *Science*, **111**, 5 (1950).
84. Kallee, E., *Z. Naturforsch.*, **7b**, 661 (1952); through *Chem. Abstr.*, **47**, 5979 (1953).
85. Kallee, E., *Klin. Wochschr.*, **32**, 508 (1954); through *Chem. Abstr.*, **48**, 9440 (1954).
86. Rosenblum, C., *Anal. Chem.*, **29**, 1740 (1957).
87. Gest, H., M. D. Kamen, and J. M. Reiner, *Arch. Biochem.*, **12**, 273 (1947); *Chem. Abstr.*, **41**, 4375 (1947).
88. Rosenblum, C., *Nucleonics*, **14**, No. 5, 58 (1956).
88a. Weiler, H., *Intern. J. Appl. Radiation Isotopes*, **12**, 49 (1961).
89. Cacace, F., *Nucleonics*, **19**, No. 5, 45 (1961).
90. Alimarin, I. P., and G. N. Bilimovitch, *Intern. J. Appl. Radiation Isotopes*, **7**, 169 (1960).
91. McKinney, C. R., J. M. McCrea, S. Epstein, H. A. Allen, and H. C. Urey, *Rev. Sci. Instr.*, **21**, 724 (1950).
92. Stewart, D. W., "Mass Spectrometry," in A. Weissberger, ed., *Technique of Organic Chemistry*, Vol. I, 3rd ed., Interscience, New York, 1960, Part IV, p. 3499.
93. Radin, N. S., *Nucleonics*, **1**, 48 (October 1947).
94. Keston, A. S., S. Udenfriend, and R. K. Cannan, *J. Am. Chem. Soc.*, **68**, 1390 (1946); **71**, 249 (1949).
95. Keston, A. S., S. Udenfriend, and M. Levy, *J. Am. Chem. Soc.*, **69**, 3151 (1947), **72**, 748 (1950).
96. Bloch, K., and H. S. Anker, *Science*, **107**, 228 (1948).
97. Mayor, R. H., and C. J. Collins, *J. Am. Chem. Soc.*, **73**, 471 (1951).
98. Berenbom, M., H. A. Sober, and J. White, *Arch. Biochem.*, **29**, 369 (1950); *Chem. Abstr.*, **46**, 5111 (1952).

98a. Růžička, J., A. Zeman, and J. Starý, *Substoichiometry in Activation Analysis and Isotope Dilution and its Use for Determination of Traces of Iron*, Symposium on Radiochemical Methods of Analysis, I.A.E.A., Salzburg, Austria, October 19–23, 1964, Paper SM-55/85.

98b. Růžička, J., *Coll. Czech. Chem. Comm.*, **25**, 199 (1958).

98c. Růžička, J. and J. Starý, *Talanta*, **8**, 228 (1961).

98d. Starý, J., and J. Růžička, *Talanta*, **8**, 296 (1961).

98e. Růžička, J., and J. Starý, *Talanta*, **9**, 617 (1962).

98f. Starý, J., and J. Růžička, *Talanta*, **8**, 775 (1961).

98g. Růžička, J., and J. Starý, *Talanta*, **11**, 691 (1964).

98h. Růžička, J., and J. Starý, *Talanta*, **8**, 535 (1961).

98i. Starý, J., J. Růžička, and M. Salamon, *Talanta*, **10**, 375 (1963).

99. Hevesy, G., and R. Hobbie, *Z. Anal. Chem.*, **88**, 1 (1932).

100. Theurer, K., and T. R. Sweet, *Anal. Chem.*, **25**, 119 (1953).

101. Sue, P., *Nature*, **157**, 622 (1946).

102. Sue, P., *Bull. Soc. Chim. France*, 5° Serie, **14**, M405 (1947).

103. Freedman, A. J., and D. N. Hume, *Anal. Chem.*, **22**, 932 (1950).

104. Inghram, M. G., *Ann. Rev. Nucl. Sci.*, **4**, 81 (1954).

105. Smales, A. A., and R. K. Webster, *Geochim. Cosmochim. Acta*, **11**, 139 (1957); through *Chem. Abstr.*, **51**, 7939 (1957).

106. Zimakov, I. E., and G. S. Rozhavsky, *Zavodskaya Lab.*, **24**, 922 (1958); through ref. 90.

107. Smales, A. A., and R. K. Webster, (a) *Anal. Chim. Acta*, **18**, 587 (1958); (b) **18**, 582 (1958).

108. Grosse, A. V., and A. D. Kirshenbaum, *Anal. Chem.*, **24**, 584 (1952).

108a. Kirshenbaum, A. D., A. G. Streng, and A. V. Grosse, *Anal. Chem.*, **24**, 1361 (1952).

108b. Kirshenbaum, A. D., and A. V. Grosse, *Anal. Chim. Acta*, **16**, 225 (1957).

109. Boos, R. N., S. L. Jones, and N. R. Trenner, *Anal. Chem.*, **28**, 390 (1956).

110. Grosse, A. V., S. G. Hindin, and A. D. Kirshenbaum, *Anal. Chem.*, **21**, 386 (1949).

111. Jones, S. L., and N. R. Trenner, *Anal. Chem.*, **28**, 387 (1956).

112. Kirshenbaum, A. D., and A. V. Grosse, *Anal. Chem.*, **22**, 613 (1950).

113. Ussing, H. H., *Nature*, **144**, 977 (1939).

114. Rittenberg, D., and G. L. Foster, *J. Biol. Chem.*, **133**, 737 (1940).

115. Chargaff, E., M. Ziff, and D. Rittenberg, *J. Biol. Chem.*, **138**, 439 (1941); **144**, 343 (1942).

116. Shemin, D., and Foster, G. L., *Ann. N.Y. Acad. Sci.*, **47**, 119 (1946); *Chem. Abstr.*, **40**, 7257 (1946).

117. MacDonald, W. S., and H. S. Turner, *Chem. Ind.*, **1952**, 1001.

118. Bacher, F. A., A. E. Boley, and C. E. Shonk, *Anal. Chem.*, **26**, 1146 (1954).

119. Craig, J. T., P. F. Tryon, and W. G. Brown, *Anal. Chem.*, **25**, 1661 (1953).

120. Palin, D. E., *Anal. Chem.*, **26**, 1856 (1954).

121. Pace, N., L. Kline, H. K. Schachman, and M. Harfenist, *J. Biol. Chem.*, **168**, 459 (1947).

122. Edelman, I. S., J. M. Olney, A. H. James, L. Brooks, and F. D. Moore, *Science*, **115**, 447 (1952).

123. Sorenson, P., *Anal. Chem.*, **26**, 1581 (1954); **27**, 388 (1955); **28**, 1318 (1956).

124. Hutchins, H. H., and J. E. Christian, *J. Am. Pharm. Assoc.*, **42**, 310 (1953); through *Chem. Abstr.*, **47**, 7163 (1953).

125. Moureu, H., P. Chovin, and R. Daudel, *Compt. Rend.*, **219**, 127 (1944).
126. Ishimori, T., *Bull. Chem. Soc. Japan*, **26**, 336 (1953); through *Chem. Abstr.*, **48**, 4362 (1954).
127. Shefner, A. M., R. Ehrlich, and H. C. Ehrmantraut, *Intern. J. Appl. Radiation Isotopes*, **2**, 91 (1957).
128. Paikoff, M., and J. E. Christian, *J. Am. Pharm. Assoc. ,Sci. Ed.*, **45**, 623 (1956); through *Chem. Abstr.*, **50**, 17337 (1956).
129. Kaufmann, H. P., and J. Budwig, *Fette Seifen*, **53**, 69 (1951); through *Chem. Abstr.*, **45**, 9893 (1951).
130. Ehrenberg, R., *Biochem. Z.*, **164**, 183 (1925); through *Chem. Abstr.*, **20**, 1188 (1926).
131. Ehrenberg, R., *Biochem. Z.*, **197**, 467 (1928); through *Chem. Abstr.*, **23**, 51 (1929).
132. Moeller, T., and G. K. Schweitzer, *Anal. Chem.*, **20**, 1201 (1948).
133. A. Langer, *J. Phys. Chem.*, **45**, 639 (1941).
134. Alimarin, I. P., and M. N. Petrikova, *Zh. Analit. Khim.*, **10**, 4, 251 (1955); through ref. 134a.
134a. Alimarin, I. P., I. M. Gibalo, and I. A. Sirotina, *Intern. J. Appl. Radiation Isotopes*, **2**, 117 (1957).
135. Duncan, J. F., and F. G. Thomas, *Australian Atomic Energy Symposium—Radio-isotopes–Physical Sciences*, 637 (1958), Land Newspaper Ltd., 57 Regent Street, Sydney, Australia.
136. Langer, A., *Anal. Chem.*, **22**, 1288 (1950).
137. Aylward, G. H., J. L. Garnett, J. W. Hayes, and S. W. Law, *J. Inorg. Nucl. Chem.*, **350** (1961).
138. Ehrenberg, R., in W. Böttger, ed., *Physikalische Methoden der analytischen Chemie*, Vol. 1, Akad. Verlagsges., Leipzig, 1933, p. 333; through ref. 139.
139. Broda, E., and T. Schönfeld, "Radiochemische Methoden der Mikrochemie" in F. Hecht and M. K. Zacherl, eds., *Handbuch der Mikrochemischen Methoden*, Vol. II: *Verwendung der Radioaktivität in der Mikrochemie*, Springer-Verlag, Vienna, 1955, p. 136.
140. Escue, R. B., and N. P. Bulloch, *Anal. Chem.*, **25**, 1932 (1953).
141. Wallace, C. H., and J. E. Willard, *J. Am. Chem. Soc.*, **72**, 5275 (1950).
142. Chleck, D. J., F. J. Brousaides, C. Hommel, W. Sullivan, L. Fishbein, and C. Ziegler, *U.S. At. Energy Comm. Rept.* AECU-4493, 1, July 1959.
142a. Chleck, D. J., F. J. Brousaides, W. Sullivan, and C. A. Ziegler, *Intern. J. Appl. Radiation Isotopes*, **7**, 182 (1960).
142b. Seaman, W., and D. Stewart, Jr., *Intern. J. Appl. Radiation Isotopes*, **15**, 565 (1964).
142c. Chleck, D. J., F. J. Brousaides, and C. A. Ziegler, *Intern. J. Appl. Radiation Isotopes*, **15**, 627 (1964).
143. Chleck, D. J., and C. A. Ziegler, *Nucleonics*, **17**, No. 9, 130 (1959).
144. Chleck, D. J., and C. A. Ziegler, *Chem. Process. Eng.*, **40**, 287 (1959).
144a. Chleck, D. J., R. Maehl, and O. Cucchiara, *Nucleonics*, **21**, No. 7, 53 (1963).
144b. Chleck, D. J., R. Maehl, O. Cucchiara, and E. Carnevale, *Intern. J. Appl. Radiation Isotopes*, **14**, 581 (1963).
144c. Chleck, D. J., and R. Maehl, *Intern. J. Appl. Radiation Isotopes*, **14**, 593 (1963).
144d. Chleck, D. J., and O. Cucchiara, *Intern. J. Appl. Radiation Isotopes*, **14**, 599 (1963).

145. Peixoto-Cabral, J. M., and H. Götte, *Z. Naturforsch.*, **10b**, 440 (1955); through *Chem. Abstr.*, **50**, 13656 (1956).
146. Erkelens, P. C. van, *Nature*, **172**, 357 (1953).
147. Winteringham, F. P. W., A. Harrison, and R. G. Bridges, *Nucleonics*, **10** (3), 52 (1952).
148. Rabinowicz, E., *Nature*, **170**, 1029 (1952).
149. Valls, P., A. M. Venet, and J. Pouradier, *Bull. Soc. Chim. France*, 5° Serie, **20**, C 106 (1953).
150. Raben, M. S., *Anal. Chem.*, **22**, 480 (1950).
151. Dyrssen, D., and B. Hök, *Svensk Kem. Tid.*, **64**, 80 (1952); through *Chem. Abstr.*, **46**, 5939 (1952).
152. Blackburn, S., and A. Robson, *Chem. Ind.*, **1950**, 614; *Boichem. J.*, **53**, 295 (1953).
153. Duncan, J. F., and F. G. Thomas, *J. Inorg. Nucl. Chem.*, **4**, 376 (1957).
154. Braun, T., I. Maxim, and I. Galateanu, *Nature*, **182**, 936 (1958).
155. Udenfriend, S., and S. F. Velick, *J. Biol. Chem.*, **190**, 733 (1951).
156. Velick, S. F., and S. Udenfriend, *J. Biol. Chem.*, **191**, 233 (1951).
157. Isbell, H. S., *Science*, **113**, 532 (1951).
158. Kaufmann, H. P., and J. Budwig, *Fette Seifen*, **53**, 253 (1951); through *Chem. Abstr.*, **45**, 7801 (1951).
159. Leighton, P. A., and R. A. Mortensen, *J. Am. Chem. Soc.*, **58**, 448 (1936).
160. Burton, M. J., E. Ricci, and T. W. Davis, *J. Am. Chem. Soc.*, **62**, 265 (1940).
161. Durham, R. W., G. R. Martin, and H. C. Sutton, *Nature*, **164**, 1052 (1949)
162. Martin, G. R., and H. C. Sutton, *Trans. Faraday Soc.*, **48**, 823 (1952).
163. Williams, R. R., Jr., and W. H. Hamill, *J. Am. Chem. Soc.*, **72**, 1857 (1950).
164. Salley, D. J., *Conference on Atomic Energy in Industry*, Nat. Ind. Conf. Board, New York, 1952; through ref. 139, p. 140.
165. Dorfman, R. I., *Proc. Soc. Exptl. Biol. Med.*, **70**, 732 (1949); through *Chem. Abstr.*, **43**, 5816 (1949).

Part I
Section D-6

Chapter 98

ACTIVATION ANALYSIS

By Vincent P. Guinn, *Department of Chemistry,*
University of California, Irvine, California

Contents

Activation analysis is a nuclear method of quantitative elemental analysis. It determines elements regardless of their chemical or physical state but gives no information about the chemical forms or valence states of the elements detected. Although its present form is quite sophisticated and highly instrumented, and hence often regarded as a very new analytical method, its history dates back to the mid-1930s. In that period, the first few activation analysis measurements were carried out, employing a cyclotron as the source of bombarding charged particles. With the advent of the nuclear reactor during the 1940s, neutrons—produced in copious quantities by the reactor—were soon shown to be a more generally useful bombarding particle, and neutron activation analysis (NAA) was born. During the next decade this form of the method was extensively developed, and it continues to be extended and refined. It is the most widely used form of the activation analysis method. During the 1950s it was also shown that relatively modest electron and charged-particle accelerators could provide more modest but still useful fluxes of slow and fast neutrons for NAA work. The availability of such moderate-cost accelerator neutron sources has greatly increased the use of NAA in industrial analytical laboratories. More recently, it has been shown that photonuclear activation analysis (PNAA), involving the activation of various elements by bombardment with high-energy gamma rays, can also be quite useful. The usual photon source for such work is the electron linear accelerator.

When the purely instrumental (nondestructive) form of the activation analysis method, based on gamma-ray spectrometry, can be employed, it can be a very fast method. In many cases samples can then be analyzed in a matter of minutes per sample. Also, if the highest of the available fluxes of neutrons (or of photons or charged particles) are used, the method is extremely sensitive for a large number of elements. In fact, in this situation, the method is the most sensitive one known at present for the quantitative determination of the majority of the elements of the periodic system. This is not true, of course, if relatively low fluxes are used. The dynamic range of the method is large, extending rigorously from the limit of detection up to the pure element itself (100% concentration).

Because of the high penetrability of neutrons and gamma rays, in most matrices, the range of practical sample sizes for NAAA and PNAA work is very large, extending from microscopic samples up to even 100 g (in some cases). Since the absolute limits of detection of the various elements, under prescribed irradiation and counting conditions, worsen only moderately with increasing sample size, throughout this range of sample sizes, the method is especially powerful for the detection of low concentrations. Thus, for a typical element, detectable down to 0.001 μg (by NAA in a nuclear reactor for 1 hr, at a thermal-neutron flux of 10^{13} n/cm²-sec), the concentration limit of detection is 0.1% in a 1-μg sample, 1 ppm in a 1-mg sample, 0.001 ppm in a 1-g sample, and 0.0001 ppm in a 10-g sample.

I. THEORY OF ACTIVATION ANALYSIS

For a particular stable nuclide (nucleus of a specific number of protons, Z, and of mass number, A) of an element that is capable of undergoing a particular nuclear reaction to form a radionuclide product, when bombarded with some specific nuclear particle, one can readily show that

$$\frac{dN^*}{dt} = N\phi\sigma \tag{1}$$

in which dN^*/dt is the rate of formation (in nuclei per second) of these radioactive product nuclei, N is the number of "target" stable nuclei of that type in the sample, ϕ is the average flux of bombarding particles (in particles per square centimeter per second) to which the sample is exposed, and σ is the nuclear reaction cross section (in square centimeters per nucleus) for that reaction with that particular kind of particle, at the particle energy employed. [Note: tables of nuclear reaction cross sections usually list σ values in "barns" (b), where 1 b = 10^{-24} cm²/nucleus.]

The radioactive product nuclei formed decay according to the first-order radioactive decay equation, both during and after the irradiation:

$$\frac{-dN^*}{dt} = \lambda N^* = \frac{0.693N^*}{T} \tag{2}$$

where λ is the decay constant of that radionuclide (in reciprocal seconds) and is equal to 0.693/T, where T is the half-life of the radionuclide.

Thus, during a uniform-flux irradiation, the net rate of formation of product radioactive nuclei of a given type is simply:

$$\frac{dN^*}{dt} = N\phi\sigma - \lambda N^* \tag{3}$$

When this is integrated to obtain an expression for N^* present after some irradiation time, t_i, one obtains the expression:

$$N^* = \frac{N\phi\sigma S}{\lambda} \tag{4}$$

where S, the so-called saturation term, is simply:

$$S = 1 - e^{-0.693 t_i/T} \tag{5}$$

Since at the end of the irradiation period λN^* is equal to $-dN^*/dt$ [see equation (2)], one can finally write the standard activation analysis equation:

$$A_0 = \left(\frac{-dN^*}{dt}\right)_0 = N\phi\sigma S \tag{6}$$

The quantity A_0, then, is the initial activity, in disintegrations per second (dps), of a particular radionuclide, formed by a specific nuclear reaction—just at the end of the irradiation.

A. IRRADIATION TIME

The saturation term, S, is a dimensionless quantity, ranging in value only from 0 (at $t_i = 0$) to 1 (at $t_i = \infty$, or $t_i/T \gg 1$). It asymptotically approaches a value of 1 with increasing t_i/T. Thus, at t_i/T values of 0, 1, 2, 3, 4, 5, . . . , S has values of 0, $\frac{1}{2}$, $\frac{3}{4}$, $\frac{7}{8}$, $\frac{15}{16}$, $\frac{31}{32}$, ... Each induced activity follows exactly the same curve of S versus t_i/T. Because of the rapid asymptotic approach of S to 1, in practice a sample is seldom irradiated for a period of time (t_i) longer than one or a few half-lives of the induced activity of interest. At very long irradiation times (relative to T), a steady state is approached (or a "saturation" condition), in which the rate of formation of new nuclei of that particular type, $dN^*/dt = N\phi\sigma$, is equal to the rate of decay of such nuclei already formed, $-dN^*/dt = \lambda N^*$. Thus, the activity level at saturation, $A_0(\text{sat'n})$, is equal to $N\phi\sigma$. Because of the saturation term, short-lived induced activities are enhanced, relative to longer-lived activities, by the use of a short irradiation period, followed quickly by counting. Similarly, longer-lived induced activities are enhanced by the use of a longer irradiation period, followed by an appreciable delay (for the decay of interfering short-lived activities), before counting. Optimum choices of irradiation and decay times are important in minimizing interferences.

B. RADIOACTIVE DECAY

Typically, the activated sample is counted at some decay time, t, after the end of the irradiation (t_0). Each induced activity decays according to equation (2), the basic first-order radioactive-decay equation. In integrated form, equation (2) becomes:

$$A_t = A_0 e^{-0.693t/T} \tag{7a}$$

or

$$\log A_t = \log A_0 - \frac{0.301t}{T} \tag{7b}$$

or

$$\log (A_t/A_0) = -\frac{0.301t}{T} \tag{7c}$$

Thus, for all radionuclides, the $\log (A_t/A_0)$ versus t/T relationship is the same straight line. The relationship between the activities at *any* two decay times (one not necessarily being t_0) is given by the same equations if t is replaced by Δt, the difference between the two decay times.

In practice, one seldom determines the actual disintegration rate of an induced activity, but rather measures its counting rate, under set counting conditions. The counting rate of a particular radionuclide, disintegrating at a given rate in a sample, depends on the type of counter used, the size of the counter, the decay scheme of the radionuclide, the sample-to-counter geometry (solid angle subtended by the counter), self-absorption or scattering within the sample, and, in some situations, other smaller factors as well. It is instructive to consider an illustrative case, for example, that of the ^{42}K gamma-ray photopeak (total-absorption peak) counting rate of a sample in which the ^{42}K disintegration rate, at the time of counting, is 1000 counts per second (cps). From the very useful "Table of Isotopes" by Lederer et al. (9), the decay scheme of 12.36-hr ^{42}K is seen to be a simple one: in 82% of its disintegrations, only a 3.52-MeV (E_{max})β^- particle is emitted; in the other 18%, a 2.00-MeV (E_{max})β^- particle is emitted, promptly followed by a 1.524-MeV gamma-ray photon. [Note: the common nuclear unit of energy is the electron volt (eV); multiples are 1 keV = 10^3 eV, 1

MeV $= 10^6$ eV.] If the sample is small (≤ 1 cm^3) and of low density (≤ 2 g/cm^3), absorption or Compton scattering of these relatively high-energy gamma-ray photons within the sample itself will be small or negligible. If the sample is in a small polyethylene vial, placed on top of a 1.5-cm-thick polystyrene β^--particle absorber, which in turn rests on top of an aluminum-encased thallium-activated sodium iodide scintillation crystal detector, the crystal being 3 in. in diameter and 3 in. high, the distance from the center of the sample to the top surface of the NaI(Tl) crystal will be about 2 cm. From Heath's excellent and highly useful *Scintillation Spectrometry Gamma-Ray Spectrum Catalogue* (7), one finds that the overall detection efficiency for such a crystal, for this sample-to-detector distance, for 1.524-MeV gamma rays, is 10.7% (26.5% geometry; 40.4% detection efficiency for those 1.524-MeV gamma rays that strike the crystal). Thus, the total ^{42}K gamma-ray counting rate of the sample, in this case, would be $1000 \times 0.18 \times 0.107$, or 19.3 cps, that is, 1160 counts per minute (cpm). However, in most instances one ignores the counts other than the characterizing photopeak counts. From Heath's *Catalogue* (7), one finds that the photofraction (peak-to-total ratio) for 1.524-MeV gamma rays, under these conditions of crystal size and geometry, is 32%. Thus, the ^{42}K photopeak counting rate, in the 1.524-MeV region of the observed pulse-height spectrum of the sample, would be 1160×0.32, or 370 cpm.

C. THE COMPARATOR TECHNIQUE

In regular activation analysis work, one very seldom uses equation (6) per se. There are several good reasons for not using it, as is, to compute the desired analytical value, N:

1. It is difficult, and often not very accurate, to convert an observed radionuclide counting rate to the corresponding disintegration rate at zero decay time, A_0 (because of often inadequate knowledge of the decay scheme and of the counting efficiency).

2. It is also difficult, and often not very accurate, to determine the true average value of the particle flux, ϕ, at the sample irradiation location.

3. Many of the reaction cross sections (σ) found in the literature are only approximate values.

One usually wishes to determine the amount of a particular element in the sample, rather than the amount of the particular stable nuclide of that element that could form the radionuclide detected. If so, one must also know the normal fractional isotopic abundance (a) of the particular stable nuclide that is activated (among the various stable nuclides of that element), and the chemical atomic weight (AW) of the element. The weight, w (in grams), of the element present is related to N and other quantities by the equation:

$$w = N \times \frac{AW}{N_A \times a} \tag{8}$$

where N_A is Avogadro's number (6.023×10^{23}).

In order to avoid dependence upon these various values, one usually employs a comparator technique, which is both simpler and more accurate than the "absolute" method. When a sample, or a series of samples, is to be analyzed for one (or several) elements of particular interest, one prepares standard samples of these elements, irradiates them at the same time as the unknowns and under exactly the same conditions, and counts them in exactly the same way that the unknowns are counted. Then, correcting all of the appropriate counting rates of a particular induced activity either to t_0 or to some specified decay time, t [by means of equation (7)], one can replace the disintegration rate values (A_0) of equation (6) by the corresponding counting-rate values (A_0') divided by ϵ, the fractional counting efficiency; replace the N of equation (6) by its equivalent from equation (8), $w \times N_A \times a/$AW; and divide the modified equation (6) for the unknown sample by that for the known standard of the element. Whatever their true, exact values may be, at least one knows that the values for ϵ, N_A, a, AW, ϕ, σ, and S are exactly the same for the unknown and the standard. Therefore, in dividing one equation by the other, these terms all cancel out, leaving the very simple final equation that is used in practice:

$$\frac{A_0'(\text{unk.})}{A_0'(\text{st'd})} = \frac{w(\text{unk.})}{w(\text{st'd})} \tag{9a}$$

or, for any specified decay time,

$$\frac{A_t'(\text{unk.})}{A_t'(\text{st'd})} = \frac{w(\text{unk.})}{w(\text{st'd})} \tag{9b}$$

Naturally, if it is not possible to irradiate the unknown sample and the standard at exactly the same flux, a correction factor, correcting both counting rates to what they would be at the same flux, must be applied before using equation (9). The relative counting rates of appropriate flux-monitor samples can be used to make such corrections. The same is true if it is not possible to count the unknown and the standard at exactly the same counting geometry. Such corrections, where needed, are obvious, and can usually be made quite accurately.

II. PARTICLE SOURCES AND TYPES OF NUCLEAR REACTIONS

Of the approximately 4000 papers published to date on the subject of activation analysis and its applications, the great majority have involved high-flux thermal-neutron activation of samples, in research-type nuclear reactors. A smaller number have involved fast-neutron activation, usually with accelerator-produced fast neutrons. Even smaller numbers have involved activation with high-energy photons or with energetic charged particles, both accelerator-produced. There are both historical and practical reasons for the popularity of reactor thermal-neutron activation analysis. However, the other forms of the method have some distinctive features and some particularly useful applications. All of these forms of activation analysis, as well as the particle sources they employ, are discussed briefly below.

A. THERMAL-NEUTRON AND FAST-NEUTRON ACTIVATION ANALYSIS WITH RESEARCH-TYPE NUCLEAR REACTORS

1. Research Reactors

The most prolific practical source of thermal neutrons (i.e., neutrons in kinetic-energy equilibrium with the surrounding temperature; at 18°C, this corresponds to a neutron kinetic energy of 0.025 eV) is the nuclear reactor. Modern research reactors especially suitable for NAA work are safe, relatively simple, and only moderately expensive ($150,000–$350,000, depending largely on the power level, and hence the neutron flux, required). They are all of the pool type, in which a lattice of enriched fuel elements, each containing some ^{235}U, is located in a deep pool of distilled, deionized water (20–25 ft deep). The water serves as the principal moderator and also as coolant and radiation shield. The reactor core contains slightly more than enough ^{235}U to make a critical mass under the conditions specified (i.e., geometry, reflector, moderator, etc.). When several boron-loaded control rods are inserted into the core, the thermal-neutron ^{235}U fission chain reaction is stopped by the appreciable absorption of slow neutrons by the boron, that is, the system is then subcritical, and the reactor is shut off. The thermal-neutron absorption cross section of ordinary boron is very high (759 barns) ; that of pure ^{10}B is higher yet (3837 barns).

To start the reactor, the control rods are withdrawn sufficiently to allow the system to become slightly supercritical. The ^{235}U fission chain reaction (initiated by a small isotopic or sealed-tube accelerator neutron source) then rapidly accelerates, and hence the power level and the neutron flux climb rapidly. When the power level (and hence the neutron flux) has reached the desired or prescribed level—typically, in a few minutes—the reactor operator reinserts the control rods just far enough to maintain criticality and steady operation at that level. Typically, modern pool-type reactors of various kinds and costs can be operated at steady power levels of from 10 to 1000 kW (1 MW), providing useful thermal-neutron fluxes, at locations in or near the core where samples can be placed for activation, in the range of 10^{11}–10^{13} n/cm^2 sec.

When a thermal neutron is captured by a ^{235}U nucleus, and fission of the $(^{236}$U$)^*$ excited compound nucleus results, two intermediate atomic-number radionuclides are formed, and two or three energetic neutrons are emitted. The most probable initial energy of these fission neutrons is about 1 MeV, with rapidly decreasing numbers of neutrons being emitted with higher energies, although a few have initial energies up to about 25 MeV. By collisions with moderator nuclei (e.g., hydrogen, beryllium, carbon) these fast neutrons are slowed down to thermal velocities (with some loss by escape or capture). Thus, especially within the core of a reactor, there are appreciable fluxes of fast neutrons, as well as high thermal-neutron fluxes. Roughly, the fast-neutron flux decreases by a factor of 10 for each 3-MeV increase in energy. Outside the core, the thermal/fast flux ratio is higher, but both fluxes are lower than within the core. For example, Table 98.I gives the thermal-neutron flux,

TABLE 98.I
Neutron Energy-Flux Distribution in the 250-kW TRIGA
Mark I Nuclear Reactor

Irradiation position	Neutron flux, n/cm^2-sec				
	Thermal	>10 keV	>1.35 MeV	>3.68 MeV	>6.1 MeV
D ring	4.9×10^{12}	9.0×10^{12}	2.3×10^{12}	4.1×10^{11}	6.2×10^{10}
F ring (rabbit terminus)	2.5×10^{12}	3.5×10^{12}	7.5×10^{11}	1.25×10^{11}	1.9×10^{10}
Rotary specimen rack	1.8×10^{12}	1.5×10^{12}	1.8×10^{11}	2.5×10^{10}	4.0×10^{9}

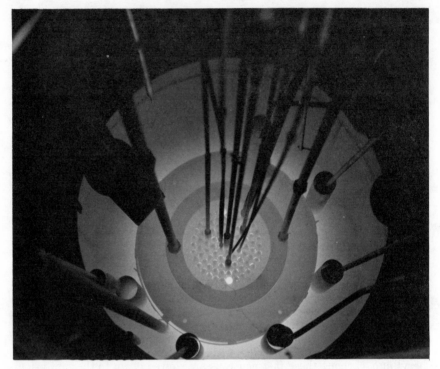

Fig. 98.1. TRIGA Mark I nuclear reactor.

and the fluxes of neutrons above various measured energies, at three different locations in one of the three TRIGA reactors at the author's former laboratory*—a 250-kW TRIGA Mark I reactor. A photograph of this reactor, looking down at the core through about 16 ft of water, is shown in Fig. 98.1. A cutaway scale diagram of the same reactor, showing the 40-tube rotary specimen rack in the annular graphite reflector, just outside the core, is shown in Fig. 98.2.

In the TRIGA type of reactor, the rotary specimen rack provided is especially helpful in activation analysis work, as up to 40 samples can be exposed simultaneously to exactly the same neutron flux. In practice, each sample (in a small polyvial) is placed inside a larger polystyrene or polyethylene tube and lowered into one of the 40 rotary-rack tubes, the rack is advanced one position, the next sample is lowered into place, and so on until all the samples are inserted. It takes only a matter of minutes to load the rack even to capacity. The rack motor

* Gulf General Atomic, San Diego, California.

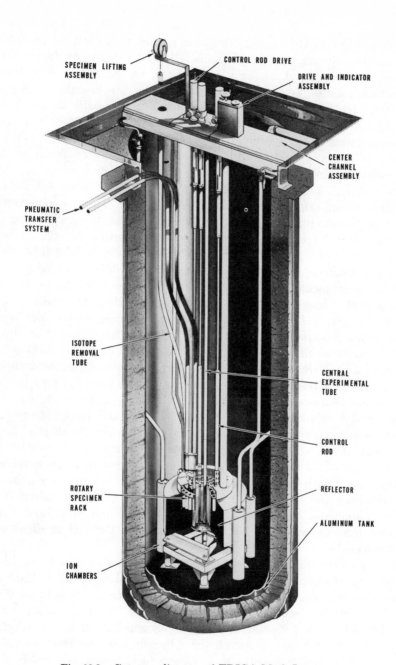

SPECIMEN LIFTING
ASSEMBLY

CONTROL ROD DRIVE

DRIVE AND INDICATOR
ASSEMBLY

CENTER
CHANNEL
ASSEMBLY

PNEUMATIC
TRANSFER
SYSTEM

ISOTOPE
REMOVAL
TUBE

CENTRAL
EXPERIMENTAL
TUBE

CONTROL
ROD

ROTARY
SPECIMEN
RACK

REFLECTOR

ALUMINUM TANK

ION
CHAMBERS

Fig. 98.2. Cutaway diagram of TRIGA Mark I reactor.

drive is then started (1 rpm), and the reactor is brought up to power. When the samples have been irradiated for the desired period of time (typically, 30 min, 1 hr, or a few hours), the reactor is shut down, the rack rotation is stopped, and the samples are brought up one at a time, each being briefly monitored, for safety, while still in the in-out tube, just as it arrives at the top of the pool. If the rack were not rotated during the irradiation, there would be small but significant differences (of the order of several per cent) in the neutron fluxes at the 40 different tube locations. These are averaged out, however, by the rack rotation. Thus one standard sample is all that is needed for up to 39 unknown samples—a considerable advantage. There is some vertical flux gradient in the sample tubes, but this can be measured to provide a calibration curve. Then, if desired, up to six small polyvials, stacked vertically, can be placed in each polystyrene or polyethylene tube, thus giving a total capacity of 240 samples (in half-dram polyvials).

The rotary rack is used whenever the main induced activity of interest is moderately long-lived, that is, when the half-life of the radionuclide is of the order of an hour, several hours, days, or longer. For such elements, a moderately long irradiation is helpful for increased sensitivity [because of the saturation term of equation (6)], and it is most economical to irradiate many samples at the same time. Because of the longer half-lives, the activated samples, after a suitable decay period (to allow shorter-lived interfering activities to decay out), can be counted one at a time without excessive loss of activity by decay. In the rotary-rack positions of the TRIGA Mark I reactor, the thermal-neutron flux (at 250-kW operation) is 1.8×10^{12} n/cm^2-sec. Especially in irradiations of 30 min or longer, one normally transfers the activated sample to a fresh polyvial before counting. This is done to eliminate contributions from activated small impurities in the polyethylene (sodium, aluminum, chlorine, titanium, vanadium) and also from ^{41}Ar formed from argon in the air that is present in the vial or dissolved in the sample.

For the determination of elements which form short-lived activities (half-lives of seconds or minutes), one uses instead a pneumatic transfer tube, and activates and counts one sample at a time. The sample vial is placed inside a sturdier polyethylene "rabbit" for irradiation. With the TRIGA Mark I reactor, the normal pneumatic tube has a transfer time of about 2.5 secs (between the adjacent counting room and the reactor core), and terminates in a vacant fuel-element position in the outermost ring (F ring) of fuel elements. In this position, the thermal-neutron flux (at 250-kW operation) is 2.5×10^{12} n/cm^2-sec. The desired irradiation time can be preset on a timer, and the irradiated sample then is returned to the counting room (where it was originally inserted) automatically and monitored automatically (for safety). The sample vial is then removed from the rabbit, ready for counting on the multi-channel gamma-ray spectrometer. A precise timer records the exact total irradiation time, and another timer is actuated as the rabbit leaves the reactor, giving a t_0 value.

The TRIGA reactor can also be pulsed to extremely high neutron fluxes (10^{16}–10^{17} n/cm²-sec). The pulse durations are in the range of 5–30 msec. When the TRIGA Mark I reactor is pulsed to a peak power level of about 1000 MW, it is found (18) that the amount of a given radionuclide activity generated in the pulse is $70/T$ times (where T is the half-life of the induced activity, in seconds) that which could be produced, even at saturation, with the reactor operated at its normal power level of 250 kW. This relationship applies to both thermal-neutron and fast-neutron reaction products. Thus, for products with half-lives of 0.1, 1, and 10 secs, the pulse enhancement values are about 700, 70, and 7, respectively. The pulsing technique is thus of real use for very short-lived induced activities. A special 0.5-sec transfer-time pneumatic tube, without rabbit, is used for pulsing NAA work.

2. Reactor Activation with Thermal Neutrons

Upon exposure to a flux of thermal neutrons, elements undergo the (n, γ) reaction, that is, when a nucleus captures a thermal neutron, the resulting excited compound nucleus promptly ($\sim 10^{-12}$ sec) drops down to the ground state of the product nucleus, emitting one or more "capture gamma-ray" photons to get rid of the excess excitation energy. The product nucleus is a nucleus of the same element, but now one atomic mass number higher (since the mass number of the neutron is 1). For example, if an ordinary stable ^{27}Al nucleus captures a thermal neutron, a nucleus of ^{28}Al is formed:

$$^{27}\text{Al} + {}^{1}\text{n} \rightarrow (^{28}\text{Al})^* \xrightarrow{\text{Prompt}} {}^{28}\text{Al} + \gamma$$

or, in simpler form, $^{27}\text{Al}(\text{n}, \gamma)^{28}\text{Al}$

If the product nucleus is a radioactive nucleus of that element, as in the case of ^{28}Al, its later decay can be detected and can thus be of use for NAA (^{28}Al decays with a half-life of 2.31 min, emitting a β^- particle and a 1.780-MeV gamma-ray photon in each disintegration). If the (n, γ) reaction only forms another stable nuclide of that element, the product is of no use in conventional NAA work (although prompt gamma rays can be utilized in a different form of activation analysis not discussed here). For example, the $^{16}O(n, \gamma)^{17}O$ reaction produces another stable nuclide of oxygen, ^{17}O. It should be remarked that, even with a fairly long irradiation at a thermal-neutron flux of 10^{13}, the elemental composition of a typical sample is not significantly altered. For an element with an (n,γ) cross section of 1 barn, for example, 1 hr at a 10^{13} flux results only in $4 \times 10^{-6}\%$ of the amount of that element present in the sample undergoing reaction.

3. Reactor Activation with Fast Neutrons

For a few elements, activation by reactor fast neutrons (fission-spectrum neutrons) is more sensitive than activation by reactor thermal neutrons, and/or forms a product that is more readily detected (perhaps a gamma emitter instead of a pure β^- emitter) or has a more convenient half-life relative to interfering activities or is more easily radiochemically separated. In such situations, one can employ one or more cadmium- or boron-encased sample vials for activation in the rotary rack, or, for short-lived activities, one can use a cadmium or boron liner in the pneumatic-tube terminus and employ a slightly smaller rabbit. With even a 1-mm annulus of cadmium or boron (ordinary or enriched in ^{10}B), the incident thermal neutrons are almost completely absorbed, whereas the incident reactor fast neutrons pass through, into the sample, undisturbed. The thermal-neutron absorption cross sections of cadmium, boron, and ^{10}B are, respectively, 2450, 759, and 3837 b. Epithermal neutrons are not greatly absorbed by the cadmium or boron, however, and still cause some (n, γ) activation of various elements in the sample. The ratio of the amount of (n,γ) activation of an element with no cadmium around it to the amount attained with a cadmium annulus is called the "cadmium ratio" at that reactor position. The cadmium ratio varies somewhat from element to element, because of the different shapes of the (n, γ) cross section versus neutron energy curves for the different elements. Their differences in the epithermal neutron and resonance-capture neutron regions, especially, affect the cadmium ratios. With a 1-mm annulus of cadmium, the attenuation of purely thermal neutrons is over 100,000-fold, but cadmium ratios in the range of 10–40, for example, are found experimentally in the pneumatic-tube (F-ring) position of the TRIGA Mark I reactor.

It is also possible to use an ordinary, or somewhat ^{235}U-enriched, annulus of uranium, instead of cadmium or boron. The ^{235}U also absorbs thermal neutrons quite well [$\sigma(n, \gamma) = 100$ b, $\sigma(f) = 580$ b] and, in addition, generates more fission-spectrum neutrons. Similarly, a lithium deuteride annulus (ordinary, or enriched in ^{6}Li) can be used. Lithium-6 is a strong absorber of thermal neutrons (σ for ordinary lithium = 70.7 b; for ^{6}Li = 953 b). When a ^{6}Li nucleus absorbs a thermal neutron, the ^{6}Li(n, t)^{4}He reaction results. The resulting energetic tritons, striking nearby deuterons, can cause the ^{2}H(t, n)^{4}He reaction, producing 14–15MeV neutrons.

Some elements that are sometimes determined with reactor fast neutrons, instead of with reactor thermal neutrons, are nitrogen, oxygen, fluorine, silicon, phosphorus, titanium, chromium, and iron. Knowledge of the possible fast-neutron products from samples is also important when one is conducting thermal-neutron activation analyses, since, as shown in Table 98.I, the samples are typically exposed to appreciable fast-neutron fluxes as well. For example, in analyzing samples for low levels of phosphorus [via the ^{31}P(n,γ)^{32}P reaction], one must ascertain that the samples do not contain sufficient sulfur and/or chlorine to produce relatively significant amounts of ^{32}P via the fast-neutron reactions ^{32}S(n, p)^{32}P and ^{35}Cl(n, α)^{32}P. This requires, in this case, activations with and without boron or cadmium around the samples, and a separate measurement of chlorine, via the ^{37}Cl(n, γ)^{38}Cl reaction. From the corresponding counting rates of ^{32}P and ^{38}Cl obtained from standard samples of phosphorus, sulfur, and chlorine, the amounts of ^{32}P generated in a sample and arising, respectively, from its contents of these three elements can readily be calculated. Fortunately, this multiplicity situation does not arise, or is not serious, for very many elements. Even when it occurs, it involves an appreciable correction only in situations where the desired (n, γ) reaction has a rather low cross section and/or involves a stable nuclide of rather low abundance (relative to the one or two possible fast-neutron interfering reactions from elements having atomic numbers one or two units larger), or where the ratio of interfering element(s) to the sought-for element is very high. A few other common difficult analyses of this type should, however, be mentioned:

1. Thermal-NAA of aluminum (or Al-rich matrices) for trace levels of sodium [since ^{24}Na is formed by both the ^{27}Al(n, α)^{24}Na and ^{23}Na(n, γ)^{24}Na reactions].

2. Thermal-NAA of silicon (or Si-rich matrices such as quartz, glass, or siliceous rocks) for trace levels of aluminum [since ^{28}Al is formed by both the ^{28}Si(n, p)^{28}Al and ^{27}Al(n, γ)^{28}Al reactions; high levels of phosphorus can also interfere, via the ^{31}P(n, α)^{28}Al reaction].

3. Thermal-NAA of chromium (or Cr-rich matrices) for trace levels of vanadium [since ^{52}V is formed by both the ^{52}Cr(n, p)^{52}V and ^{51}V(n, γ)^{52}V reactions].

Suitable measurements, analogous to those described above for the phosphorus, sulfur, and chlorine case, however, can provide accurate results, unless the interfering elements are present at relatively very high levels.

B. FAST-NEUTRON (AND THERMAL-NEUTRON) ACTIVATION ANALYSIS WITH ACCELERATOR-PRODUCED NEUTRONS

Alpha-neutron isotopic neutron sources, such as plutonium–beryllium and americium–beryllium, are not discussed here, since the maximum fluxes available with them are only around 10^5–10^6 n/cm²-sec—too low to be of much practical use. However, milligram amounts of the newly-available spontaneous-fission neutron source, ^{252}Cf, can provide thermal-neutron fluxes of $\sim 10^8$ n/cm²-sec, hence are as useful as some of the small-accelerator neutron sources discussed below.

1. Accelerator Sources of 14-MeV Neutrons

A commonly used, relatively low-cost, source of 14-MeV neutrons is a small Cockcroft-Walton deuteron accelerator. A typical commercial neutron generator of this type is shown in Fig. 98.3. These devices cost approximately \$20,000, and with a fresh tritium target can produce about 2×10^{11} neutrons per second (isotropic). At positions where one or a few samples of appreciable size (1–5 cm³) can be placed, this output rate corresponds to an average 14-MeV neutron flux within the sample of about 10^9 n/cm²-sec.

Typically, these machines operate as follows: (*1*) deuterium gas from a small cylinder is fed through a palladium leak into an ionizing chamber; (*2*) there it is ionized, forming D^+ and D_2^+ ions, either with a radio-frequency source or with a Penning source; (*3*) the ions emerge through a slit and are accelerated through a potential of 100–200 kV down an evacuated "drift tube"; (*4*) the collimated beam of accelerated D^+ and D_2^+ ions, usually defocused so that they cover an area of 1–10 cm² (to avoid melting a hole in the target), impinges upon a water-cooled metal-tritide target (typically, about 1 in. in diameter); and (*5*) $\sim$14-MeV neutrons are thus produced in the target, via the 17.6-MeV exoergic ^{3}H(d, n)^{4}He reaction, in which the neutron receives 14.0 MeV of the 17.6 MeV released. Because of the conservation of momentum, with 150-keV deuterons, the neutrons emitted in the forward direction largely

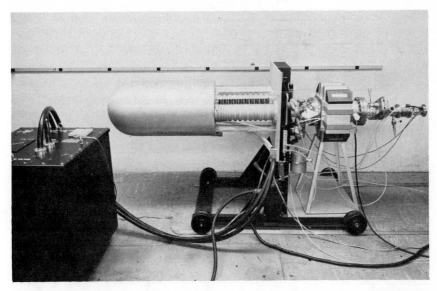

Fig. 98.3. Typical Cockcroft-Walton neutron generator. Courtesy of Texas Nuclear
Division, Nuclear-Chicago Corp.

range in energy (assuming a thick target) from 14.6 to 14.9 MeV; those
in the backward direction, from 13.7 to 13.4 MeV; those at right angles,
14.0 MeV.

The entire system is maintained at a reasonably good vacuum ($\sim 10^{-5}$
mm Hg) by means of a Vac-Ion or molecular pump (or, in some models,
by an oil- or mercury-diffusion pump), backed up by a mechanical
vacuum pump. For NAA work these accelerators are usually operated
steadily, but the neutron beam can also be pulsed, if desired, by either
pulsing the ion source, magnetically or electrostatically deflecting the
beam on or off the target, or both. Various pulse durations and repetition
rates can be selected. At maximum steady neutron output rate, such
a machine requires approximately 7 ft of concrete (or equivalent) shielding, in all directions, in order to reduce the fast-neutron flux outside
the shielding to a value less than the tolerance dose rate for fast neutrons
(~ 10 n/cm²-sec for 40-hr/week exposures). This also reduces the sub-
sidiary thermal-neutron and gamma-ray levels to less than tolerance.
Additional distance and/or additional thermal- and fast-neutron shielding (besides the usual gamma-ray shielding, lead) around the scintillation counter is also needed, to prevent increased counter background
and activation of the sodium and iodine in the NaI crystal.

The major limitation of such moderate-flux neutron sources is the target lifetime problem. The tritium targets used consist of a copper or other metal backing about 10 mils thick, on one face of which has been deposited a thin layer (1 or a few microns thick) of titanium, zirconium, or erbium, which has then been tritiated up to a level of 1–5 curies of tritium per square inch. Even when the impinging beam of 150-keV deuterons is spread over an area of several square centimeters, a 1–2 mA beam will result in a rapidly declining neutron output rate from the water-cooled target. Typically, at such full-power operation, the output rate will drop in half in an hour or less and will continue to decline with further use.

Various expedients can be employed to prolong the period before a target has ceased to generate enough neutrons per second to be of much practical use, and a fresh target must be installed. These expedients include the use of a larger-area, rotating target; use of thicker, higher curie-per-square inch targets; intermittent or continuous target replenishment with accelerated tritons; and use of refrigerated target coolants. Each of these helps, but each adds to the cost of the system and poses some complications. Target changing requires only a few minutes; but with the time required for subsequent pumping to a satisfactory vacuum, a total time of about 30–60 min is needed before operation with the new target can be started. Unless the very maximum neutron flux is needed, it is usually more desirable to start operation with a fresh target at a reduced deuteron beam current (perhaps 0.5 mA), and then gradually increase the beam current to maintain a constant neutron output rate. In this way an output rate of perhaps 5×10^{10} n/sec can be maintained for quite a few hours of on-target operation, until the beam current has been increased to its maximum attainable value with the accelerator being used (typically, about 2 mA). By using a neutron-flux monitor counter of almost any type, the gradual beam-current increase can be handled automatically, by a servo mechanism if desired. The target lifetime difficulty is largely circumvented by the use of a sealed-tube (rather than drift-tube) accelerator. These eliminate pumping and target changing. They contain a sealed-in source of deuterons and tritons, and can produce 10^{11} n/sec steadily for periods of 100 hours or more of on-target operation.

As with reactor installations, a pneumatic transfer tube is employed for work with short-lived induced activities (half-lives of seconds to a few minutes). Transfer times in the range of 1 or 2 sec are usually employed. Because of the difficulty in maintaining highly constant, reproducible, neutron fluxes with such generators (compared with a nuclear reactor), it is useful to employ a dual pneumatic-tube system—one for sample and the other for standard or flux monitor.

In the author's former laboratory, such a dual transfer system is used for routine oxygen determinations, with sample and standard both spinning during the irradiation period [oxygen is determined via the 7.14-sec half-life ^{16}N formed by the ^{16}O(n, p)^{16}N reaction, using 14-MeV neutrons]. The sample and a standard of known oxygen content (and of the same volume and shape), are simultaneously sent into position in front of the target, and the beam is turned back onto the target. After a preset irradiation period, the activated sample and the standard are sent back to their respective (cross-calibrated) counters, and counting is started simultaneously on both and continued for exactly the same preset length of time. Thus, sample and standard are processed identically, and hence, in this case, their respective oxygen contents are in exactly the same ratio as their net ^{16}N counts (corrected to the same counting efficiency). This statement presupposes that the neutron flux is the same for both sample and standard. For a perfectly symmetrical system, perfect centering of the beam, and a completely uniform target, this is true. In practice, the system is checked with each new target (or more frequently), using identical samples in the two tubes to provide a small correction factor if the two positions are exposed to slightly different neutron fluxes. For elements whose fast-neutron products are longer-lived, one can activate a number of samples simultaneously (usually for 5–10 min) in a spinning sample rack.

2. Accelerator Activation Analysis with 14-MeV Neutrons

With a source of 14-MeV neutrons, many elements can be suitably activated by means of (n, n′), (n, 2n), (n, p), or (n, α) reactions. The (n, n′) type of reaction is called a "fast-neutron inelastic scattering" reaction. If the product is a nuclear isomer (same Z and same A as the target nuclide) with an appreciable half-life, it can be of use for activation analysis. An example of this type of fast-neutron reaction is the ^{77}Se$(n, n′)$^{77m}Se reaction, forming 17.5-sec ^{77m}Se. The (n, n′) reactions are all endoergic, but usually only to the extent of 1 MeV or less. The (n, 2n) reactions result in a nuclide with the same Z, but one unit lower in A. An example of interest in activation analysis is the ^{14}N$(n, 2n)$^{13}N reaction, forming 9.96-min ^{13}N. All such reactions are also endoergic, usually to the extent of 6–12 MeV. The (n, p) reactions result in a product nuclide of the same A, but one unit lower in Z. An example of interest is the ^{28}Si(n, p)^{28}Al reaction, forming 2.31-

min ^{28}Al. Most (n, p) reactions are slightly endoergic (1 or a few MeV), and exhibit a Coulomb barrier. The (n, α) reactions form a product nuclide with an A three units lower than the target nuclide, and a Z two units lower. An important example of this type of fast-neutron reaction is the ^{27}Al(n, α)^{24}Na reaction, forming 14.96-hr ^{24}Na. Such reactions are often also slightly endoergic, and also exhibit a Coulomb barrier. All of these reactions can be produced, in many elements, not only by 14-MeV neutrons but also by reactor fission-spectrum neutrons of energies greater than the thresholds of the reactions involved.

3. Accelerator Activation Analysis with Thermal Neutrons

The same 14-MeV neutron generator can be used also to provide moderate thermal-neutron fluxes—by surrounding the target assembly with a tank of moderator (such as water, oil, or paraffin) into which sample tubes are inserted. At a position about 2 cm in front of the tritium target, the 14-MeV neutron flux, with a fresh, high-curie target, may be 1×10^9 n/cm^2-sec. With a moderator present, the 14-MeV neutron flux at this position is slightly lower, and the maximum thermal-neutron flux is a few centimeters further out and has a maximum value of perhaps 10^8 n/cm^2-sec. With full-power operation, the thermal-neutron flux of course declines rapidly also. However, a thermal-neutron flux of about 5×10^7 n/cm^2-sec can be maintained for quite a few hours of total on-target operation, by starting a fresh target with reduced beam current and gradually raising the beam current with continued usage, so as to maintain a constant neutron output rate. A moderator tank (cubic, cylindrical, or spherical) with maximum dimensions of about 2 ft is ample. A pneumatic tube can be inserted into the moderator tank for work with short-lived induced activities. For longer-lived activities, several (cross-calibrated) stationary tubes can be fixed in the moderator, usually in an arc, at a constant distance from the center of the target, or a spinning multisample rack can be installed in the moderator. Because of the target-deterioration problem, it is not very practical to employ irradiation periods longer than 5 or 10 min.

Less potent accelerator 14-MeV neutron sources, of both the continuously pumped and the sealed-tube types, can be obtained for about half the cost of the 2×10^{11} n/sec machines, but these give output rates about ten times lower, namely, around 1 or 2×10^{10} n/sec. The lower-output machines usually operate at 100 (instead of 150 or 200) kV and at a beam current of only 0.5–1 mA (instead of 2 mA)—hence the lower cost and lower output rate.

A somewhat more expensive accelerator—a 2-MeV Van de Graaff deuteron accelerator costing about \$40,000—can be used effectively to provide very constant thermal-neutron fluxes of about 5×10^8 n/cm²-sec (at a deuteron beam current of about 150 μA). Here the target employed is beryllium (water-cooled), and the nuclear reaction in the target is the ^{9}Be(d, n)^{10}B reaction. This target does not deteriorate appreciably with time and produces neutrons with energies in the 4–6 MeV range. These can be employed, unmoderated, to induce (n, p) and (n, α) reactions in many low-atomic-number elements, and (n, n′) reactions in some others. The energy is not high enough to induce (n, 2n) reactions in any element except beryllium (and in ^{2}H). With a moderator, as mentioned above, steady thermal-neutron fluxes of about 5×10^8 n/cm²-sec can be produced.

With much more expensive accelerators (higher-energy Cockcroft-Walton and Van de Graaff positive-ion accelerators, higher-energy Van de Graaff electron accelerators, electron linear accelerators, and cyclotrons), usable thermal- and fast-neutron fluxes in the range of about 10^8–10^{11} n/cm²-sec can be produced. By proper choice of particle, particle energy, and target, it is possible to obtain neutrons of various selected energy ranges. This flexibility can be advantageous for particular problems, often allowing one to use neutrons high enough in energy to activate the element of interest but below the threshold energy of one or more possibly interfering reactions of other elements present. However, the available neutron fluxes are still modest in comparison with those that can be produced in a modern research-type nuclear reactor, and the cost is as high as that of such a reactor, or even higher.

Sensitivities attainable for a number of elements, with 14-MeV neutron fluxes of 10^9 n/cm²-sec and thermal-neutron fluxes of 10^8 n/cm²-sec, are presented in a later section of this chapter.

C. Photonuclear Activation Analysis

Activation analysis can also be carried out by bombarding samples with high fluxes of energetic gamma rays (or bremsstrahlung). This form of the method is termed photonuclear activation analysis (PNAA). The types of nuclear reactions that can be induced in various elements by bombardment with energetic photons depend on the energy of the photons. Photoexcitation, (γ, γ') reactions can be produced in most elements with incident gamma-ray photons having energies of just several tenths of 1 MeV up to a few million electron volts. However, the photoexcited states of most nuclei promptly drop back to the ground state and hence are of no value from the activation analysis standpoint. Some, however, are metastable isomeric states which decay with quite appreciable half-lives and therefore are of interest.

Most nuclei will undergo the (γ, n) reaction with photons having energies in the range of 6–12 MeV, the normal range of neutron binding energies in most nuclei. Low-threshold exceptions are the 1.67-MeV endoergic $^9\text{Be}(\gamma, n)^8\text{Be}$ and 2.22-MeV endoergic $^2\text{H}(\gamma, n)^1\text{H}$ reactions, but these do not form a radioactive product (other than the neutron itself, which in the unbound state decays to a proton, β^-, and a neutrino, with a half-life of 11.7 min). A few important high-threshold exceptions are the 15.6-MeV endoergic $^{16}\text{O}(\gamma, n)^{15}\text{O}$ and 18.7-MeV endoergic $^{12}\text{C}(\gamma, n)^{11}\text{C}$ reactions. The products of (γ, n) reactions, being "neutron-deficient," often decay by positron (β^+) emission, as in the cases of 2.05-min ^{15}O and 20.34-min ^{11}C, cited above. Nitrogen-13 (half-life, 9.96 min), formed by the 10.5-MeV endoergic $^{14}\text{N}(\gamma, n)^{13}\text{N}$ reaction, is also a pure β^+ emitter.

At higher photon energies other reactions, such as the (γ, p), $(\gamma, 2n)$, (γ, np), and (γ, α) reactions, become possible or occur to an increasing extent. In addition to the basic threshold-energy requirement (Q value, related to the mass difference between reactants and products by the Einstein mass–energy relationship, $E = mc^2$), reactions in which a charged particle (p, d, t, or α; the abbreviations for proton, deuteron, triton, and alpha particle, respectively, i.e., ^1H, ^2H, ^3H, and ^4He) is produced have the Coulomb barrier energy requirement, for escape of the charged particle from the compound nucleus potential well. The magnitude of the Coulomb barrier increases with increasing atomic number of the target nucleus, thus tending to inhibit such reactions in high-Z elements unless correspondingly higher photon energies are employed.

The most prolific source of high-energy photons is the electron linear accelerator (Linac). For example, two such accelerators are used at the author's former laboratory—a 17-MeV (max) and a 45-MeV (max) machine. Each can be operated steadily at electron energies anywhere from around 2 or 3 MeV up to the maximum energy of the machine. At maximum power operation of the 45-MeV Linac (25-MeV, 0.5-mA average beam current), the beam power is 12.5 kW. Such accelerators cost approximately $500,000.

For the production of high-energy photons, the somewhat defocused electron beam is stopped in a thick water-cooled tungsten target (converter). Bremsstrahlung (continuous X-rays) ranging in energy from nearly zero up to the energy of the electrons are produced. Except for the absorption of the very low-energy photons ($<$ a few tenths of 1

MeV) in the tungsten, the photon flux produced at a particular electron energy decreases monotonically with increasing photon energy. For example, the calculated numbers of bremsstrahlung photons produced, per 10,000 25-MeV electrons hitting a thick tungsten converter, are approximately as follows:

~ 1 MeV	16,000	(± 0.5 MeV in each case)
~ 3	3,800	
~ 5	1,700	
~10	440	
~15	150	
~20	35	
~23	7	

The fraction of the total incident electron-beam energy (or power) that results in bremsstrahlung formation (the rest going to heat) increases with increasing target Z and increasing electron energy. For example, with a thick tungsten converter, the percentage of the electron-beam energy resulting in bremsstrahlung formation increases in the series 4%, 12%, 18%, 29%, 37%, 44%, 49%, and 53%, at electron energies of, respectively, 1, 3, 5, 10, 15, 20, 25, and 30 MeV. As the electron energy is increased, the fraction of the photons formed that are concentrated in the forward direction also increases. For example, at a distance of 6 mm in front of a tungsten converter, the percentage of all the photons produced that are contained within a cone making an angle of $\pm 30°$ with the forward direction line increases in the series 15%, 22%, 31%, 41%, 59%, and 74%, at electron energies of, respectively, 5, 10, 15, 20, 30, and 40 MeV. Such photon sources, in other words, are decidedly not isotropic—or monoenergetic.

By operating the 45-MeV Linac at 25 MeV and 0.5 mA, samples can be bombarded with appreciable photon fluxes in the pneumatic-tube position (center about 1.5 in. in front of the tungsten converter). Under these conditions, samples are exposed to photon fluxes of about:

7 $\times 10^{13}$/cm^2-sec	> 5 MeV
2 $\times 10^{13}$	>10
0.5 $\times 10^{13}$	>15
0.08 $\times 10^{13}$	>20

Most (γ, n) cross sections have maximum values (peak of the excitation curve, σ versus $E\gamma$) in the range of 1–1000 millibarns (mb), many of them averaging perhaps 10 mb (over the region from the threshold energy to 10 or 15 MeV above it). Thus, for the above-mentioned photon-flux energy distribution, an illustrative monoisotopic element of atomic weight 100, (γ, n) threshold of 8 MeV, forming a radionuclide product, would give an initial saturation activity of around 2×10^9 dps, or 1×10^{11} dpm, per gram of the element. If one assumes that the product emits a β^+ or a gamma in every disintegration, and that the photopeak counting efficiency is 10%, the initial disintegration rate corresponds to a photopeak counting rate of 1×10^{10} cpm. If a 100-cpm photopeak counting rate is arbitrarily defined as the lower limit of detection, the sensitivity of detection for such an element would then be 10^{-8} g (i.e., 0.01 μg or 10 ng). Many elements are found to give photonuclear activation analysis limits of detection, under the irradiation conditions defined above, in the range of 0.001–10μg, with a median-sensitivity element having a limit of around 0.1 μg.

Under such conditions, then, PNAA (with $\sim 10^{13}$/cm^2-sec fluxes of energetic photons), is, on the average, about 100 times less sensitive than thermal-NAA (at a thermal-neutron flux of 10^{13} n/cm^2-sec). However, since the thermal-NAA sensitivities vary widely from one element to another, and so do the PNAA limits, with no correlation between the two, certain elements are more sensitively detected by PNAA than by thermal-NAA (although the opposite is true for a much larger number of elements). For example, a few elements which are more sensitive to PNAA are carbon, nitrogen, oxygen, fluorine, chromium, and iron. For some other elements, the PNAA and thermal-NAA limits of detection (at 10^{13} fluxes) are comparable in magnitude with one another, but PNAA may sometimes be preferred because (*1*) the product is a β^+ or a gamma emitter, instead of a pure or almost pure β^- emitter, (*2*) the matrix interference with the detection of the induced activity of interest is less, (*3*) the half-life of the product is more convenient, or (*4*) extensive self-shielding, encountered with thermal neutrons in matrices rich in such elements as lithium, boron, cadmium or certain of the rare-earth elements, is eliminated. Thus, detection of the gamma emitter, ^{28}Al, produced from silicon by the ^{29}Si$(\gamma, \text{p})^{28}$Al reaction, may be preferred to the detection of the almost pure β^- emitter, ^{31}Si, formed from silicon by the thermal-neutron ^{30}Si$(\text{n}, \gamma)^{31}$Si reaction. Similarly, the (γ, n) product of phosphorus, ^{30}P (a β^+ emitter), is more readily detected than the (n, γ) product, ^{32}P (a pure β^- emitter). The large ^{24}Na interference produced by thermal-neutron activation of many sodium-rich matrices (such as sea water, blood, and urine) is eliminated if photon activation is employed, since sodium forms very little activity (2.62-year ^{22}Na) in a photon activation of 1 hr or less. The half-life of the (γ, n) product of fluorine (109.7-min, ^{18}F) may be more convenient than the half-life of its (n, γ) product (11.56 sec, ^{20}F), allowing one to process many samples at the same time, to wait for short-lived interferences to decay out, or to carry out radiochemical separations, if necessary.

Thus, whereas high-flux thermal-NAA is generally more sensitive than high-flux PNAA, the latter can still serve as a valuable supplement to the former. Activation analysis work to date has been carried out largely with neutrons, particularly thermal neutrons. Practical applications of PNAA have thus far been limited because of (1) the high cost of suitable electron linear accelerators, (2) the relative sparcity of information in the literature on photonuclear activation sensitivities, (3) the fact that reactor (n, γ) limits of detection, for most elements, are considerably better than photonuclear limits of detection, and (4) the fact that so many of the photonuclear products are pure β^+ emitters, posing a problem of unambiguous identification and accurate quantitation. Nonetheless, with the rapidly increasing interest in the determination of low levels of carbon, nitrogen, oxygen, and fluorine in many important high-purity materials; the increasing research and development on the PNAA method in the Saclay laboratories (in France), in the author's former laboratory, and in a few other laboratories; and the development of computer techniques for resolving β^+ activities into their components of different half-lives, there is no question but that PNAA will steadily attain increased importance. Like neutrons, high-energy photons are very penetrating in most matrices, and hence large samples can be utilized to attain better concentration sensitivities.

D. CHARGED-PARTICLE ACTIVATION ANALYSIS

Most elements can also be activated by bombardment with energetic charged particles, such as protons, deuterons, ^{3}He ions, or ^{4}He ions (α's). Many nuclear reactions are possible, depending on the element, the particle used, and its energy. At low bombarding energies, simple capture is possible—for example, the (p, γ) reaction, analogous to the (n, γ) reaction. At higher energies (several million electron volts and above), reactions such as the (p, n), (p, α), (d, n), (d, p), (^{3}He, n), (^{3}He, p), (^{3}He, α), (α, n), and the (α, p) can take place. At still higher energies ($\sim$15–20 MeV and above), even more complicated reactions occur.

Because of the complexity of products that can be formed by even a single stable nuclide of even a single element, at high bombarding energies, charged-particle activation analysis (CPAA) work is usually carried out at modest particle energies, typically in the range of 1–20 MeV. Because of the Coulomb barrier involved in charged-particle reactions, such low-energy work favors the activation of low-Z elements. It is quite useful, for example, for the determination of even very low levels of carbon, nitrogen, and oxygen in high-Z matrices, such as most of the metals.

The penetration of such moderate-energy charged particles in solid matrices is quite limited—typically in the range of 10 μ to a few hundred microns (depending on the type of particle, its energy, and the stopping power of the matrix for such particles). As the monoenergetic incident particles penetrate into the matrix, they rapidly slow down to a stop. Since each reaction cross section is markedly energy-dependent, matrix matching between sample and standard must be quite close if really quantitative results are to be obtained. Once the particle has slowed down to an energy less than the threshold (Q) value (if the reaction is endoergic), or to an energy appreciably less than the Coulomb barrier energy (even for exoergic reactions), the reaction ceases to be produced. Thus the sample depth within which a particular reaction can occur depends on the incident particle energy and the stopping power of the matrix. Even with a given matrix, a given particle, and a given incident particle energy, each induced reaction extends throughout a somewhat different sample thickness.

Not only does the small penetration of charged particles pose problems of matrix matching for quantitative work, but also it results in severe sample heating problems. For example, a 10-μA beam of 10-MeV particles might penetrate only 100 μ into a sample. Even with the beam spread over an area of 1 cm², 100 W of heat is released in a sample volume of only 0.01 cm³. Thus, in general, provisions for sample cooling must be very good, and samples must be quite thermally stable (although it is possible to employ even liquid samples if they are sealed into cooled sample cells with a thin foil window for entry of the charged-particle beam).

The most convenient source of 1–20 MeV charged particles is a small cyclotron. Fixed-energy machines of this type and energy are commercially available at a cost of about $250,000. With these, the fixed beam energy must be reduced down to any desired level by means of thin metallic foils of appropriate thickness (a variable-energy cyclotron would be appreciably more expensive). These cyclotrons are equipped with hydrogen, deuterium, and ³He sources. Van de Graaff positive-ion accelerators, of 1–2 MeV energy, cost much less but are of more limited usefulness for CPAA work. Single-stage Van de Graaffs up to 6 MeV, and compound ones up to 12–18 MeV, are more useful but become quite expensive. Cockcroft-Walton positive-ion accelerators of energies above even 1 MeV also are quite expensive.

At the present time, CPAA with a small cyclotron has limited use, although some of the applications can be rather important. It is very useful for the analysis of thin surface layers, such as oxide, sulfide, or corrosion layers on metals, and for the analysis of tiny particles. It is quite sensitive for the determination of even quite low concentrations (~ 0.001 ppm) of carbon, nitrogen, oxygen, and other low-Z elements. As mentioned earlier, the same machine can also be used to produce moderate fluxes of neutrons of various selected energies by the proper choice of particle, particle energy, and target.

III. FORMS OF THE ACTIVATION ANALYSIS METHOD

There are two forms of the activation analysis method: (1) the purely instrumental (nondestructive) form, and (2) the radiochemical-separation (destructive) form. (Note: in long irradiations at very high fluxes, radiolysis effects caused by the gamma radiation, fast neutrons, recoils, and radioactive-decay radiations can cause significant chemical and physical changes in samples, especially in organic and some inorganic materials; therefore, in such cases, the term "nondestructive" must be used with some reservations.)

A. THE INSTRUMENTAL FORM

As the name implies, this form of the method is entirely instrumental in nature, involving no chemical operations. If one has the necessary equipment to utilize this technique, it is the preferred form, wherever it is applicable, since it minimizes the use of a chemist's time. The instrumental technique involves only (1) activation of the sample, and (2) gamma-ray spectrometry of the activated sample.

1. Decay Schemes

The instrumental method requires that the induced activity (or activities) of interest emit gamma rays, characteristic X-rays, or positrons in an appreciable fraction of its disintegrations. Positron emitters are detected via the 0.511-MeV "annihilation" radiation produced when emitted positrons have slowed down (usually within the sample and its container, or in adjacent matter) to the point that they can be annihilated by reaction with electrons. The reaction is $\beta^+ + e^- \rightarrow 2\gamma$. From the masses of the electrons (both e^+ and e^- have the same mass), and the necessity of conservation of momentum, one should obtain from each annihilation event two 0.511-MeV gamma-ray photons, going in opposite directions, and such is found to be the case.

Characteristic X-rays can result from radioactive decay via "electron-capture" (EC) or "internal-conversion" (IC) mechanisms. Positron emission and/or electron capture are modes of radioactive decay that are common among "neutron-poor" (proton-rich) nuclei. If the decay energy (Q) available is less than 1.022 MeV (the energy equivalent of the rest mass of two electrons), only electron-capture decay can occur. If it is greater than 1.022 MeV, both EC and β^+ emission can occur; and there are many examples among the neutron-poor radionuclides in which a certain fraction of the disintegrations take place via EC and a different fraction via β^+ emission. The probability of EC decay rises with increasing atomic number. In both β^+ emission and EC decay, the product of the decay is a nucleus of the same mass number, A, but one atomic number unit lower (e.g., $^{13}_{7}\text{N} \rightarrow {}^{13}_{6}\text{C} + \beta^+$ and $^{55}_{26}\text{Fe} + \text{e}^- \rightarrow {}^{55}_{25}\text{Mn}$). Neutron-poor nuclides are often the products of (n, 2n) and (γ, n) reactions.

Electron-capture decay involves the conversion, within the nucleus, of a proton to a neutron by interaction with an orbital electron of that atom, usually a K-shell electron. This process thus leaves a vacancy in the K (or other) electron shell, which promptly results in an orbital-electron cascade to fill the vacancy, thereby producing characteristic X-rays of the daughter element. The K X-ray photon energies of practical interest range from about 10 keV (for zinc, $Z = 30$) to about 114 keV (for uranium, $Z = 92$), increasing with increasing Z.

In decay via IC, some fraction of the possible expected number of gamma-ray photons instead interact with an orbital electron of the atoms whose nuclei are disintegrating, ejecting the orbital electron, with the complete disappearance of the expected gamma-ray photon. This is then a "nuclear photoelectric-effect" process. If the gamma-ray energy available is greater than the binding energy of a K electron in the atom, ejection of a K electron is favored (otherwise, ejection of an L electron occurs). The energy of the ejected monoenergetic electron is equal to the gamma-ray energy minus the binding energy of the orbital electron. Again, as in EC decay, the filling of the resulting vacancy in the K (or L) shell by an orbital-electron cascade results in the emission of characteristic X-rays by the atom. The probability of internal conversion of gamma-ray photons increases with increasing Z and with decreasing gamma-ray energy. For any particular gamma-ray-emitting radionuclide, the percentage extent of IC has a definite value—anywhere from almost 0 to almost 100%.

Neutron-rich radionuclides [the type produced in most instances of thermal-neutron capture (n, γ) reactions and of fast-neutron (n, p) and (n, α) reactions] in most cases decay by β^- emission. The product of such decays is thus always a nucleus of the same mass number, but of one atomic number unit higher (e.g., $^{32}_{15}P \rightarrow ^{32}_{16}S + \beta^-$, and a neutrino).

Some radionuclides decay by isomeric transition (IT), in which a metastable nuclear isomer decays by simple gamma-ray emission. Here the product has the same A and Z as the nuclear isomer (e.g., $^{60m}Co \rightarrow ^{60}Co + \gamma$). In a number of instances, nuclear isomers are formed by (n, γ), (n, n'), and (γ, γ') reactions.

Whether the primary mode of decay is β^- emission, β^+ emission, or EC, the product may be formed in the ground state or in an excited state (depending on the decay scheme of the radionuclide), or some of each. If a certain percentage of the disintegrations result in the formation of an excited nucleus, each of these will promptly fall to the ground state, with the emission of one or more gamma-ray photons of characteristic discrete energies.

An excellent compilation of decay schemes is contained in the "Table of Isotopes" published by Lederer et al. (9).

2. The Scintillation Counter

The typical counting arrangement employed in gamma-ray spectrometry is described below. Usually, the detector is a solid cylindrical 3-in.-diameter $\times$ 3-in.-high thallium-activated sodium iodide single crystal, coupled to a photomultiplier tube (PMT), with the whole assembly made light-tight. The scintillation counter is placed inside a 4-in.-thick lead shield (to reduce the "background" counting rate) of reasonably large inside dimensions (perhaps 18 in. by 18 in. by 24 in. high). The reason for using such a large shield is to minimize the number of gamma-ray photons from the sample that are back-scattered from the shield to the NaI(Tl) crystal. Since such photons undergo approximately a 180° Compton scattering, their energies are related to the energies of the original gamma-ray photons by the equation:

$$E_{BS\gamma} = \frac{E_\gamma}{1 + 2E_\gamma/0.511} \text{ (in MeV)} \tag{10}$$

Thus, a sample emitting gamma-ray photons of just one particular energy will exhibit a pulse-height spectrum with an undesired small peak at the back-scattered energy, in addition to the desired total-absorption peak ("photopeak"). From equation (10), it can be calculated that the maximum back-scattered photon energy possible (i.e., from very high-energy gamma rays) is 0.255 MeV. For E_γ values of 0.1, 0.3, 0.5, 1.0, 2.0, and 3.0 MeV, the corresponding E_{BSY} values are 0.079, 0.138, 0.169, 0.204, 0.226, and 0.236 MeV, respectively. Samples emitting gamma rays of several energies thus can exhibit several back-scatter peaks, of various sizes and at various energies, or a broad composite back-scatter peak (due to overlapping of the various individual back-scatter peaks). Increasing the internal dimensions of the shield decreases the magnitude of the back-scatter peaks but does not eliminate them entirely. The lead shield is usually lined with a thin (0.030-in.) sheet of cadmium and a thin (0.015-in.) sheet of copper (in the order Pb–Cd–Cu), to absorb lead X-rays generated in the shield by interaction with sample gamma rays. The Pb X-rays are efficiently absorbed by the cadmium, and the resulting Cd X-rays by the copper. The resulting Cu X-rays are low in intensity and have an energy of only about 8 keV. Without these liners, an undesired photopeak would appear in the sample pulse-height spectrum at about 80 keV.

In gamma-ray spectrometry, it is desirable to eliminate the smearing of the pulse-height spectrum that can be caused by sample beta particles reaching the NaI(Tl) crystal. To accomplish this, one typically places a 1.5-cm-thick plastic (or other low-Z) absorber between the sample and the top of the detector crystal. The sample itself, if of appreciable size, will stop or slow down a large fraction of its beta particles. Then the sample vial, the plastic absorber, and the aluminum casing of the NaI(Tl) crystal combine to stop virtually all the remaining ones from reaching the crystal, without producing an appreciable number of bremsstrahlung photons.

When a gamma-ray photon strikes a NaI(Tl) crystal, four possible things can occur: (1) it can pass through the crystal with no interaction, (2) it can lose part of its energy to an electron by a Compton-scattering event, (3) it can lose all of its energy to an electron by a photoelectric absorption event, or (4) it can lose all of its energy by formation of an e^-/e^+ pair (a "pair-production" event). Pair production can occur only if the gamma-ray energy is greater than 1.022 MeV. In NaI(Tl), photoelectric absorption is the predominant mode of interaction for energies up to about 0.3 MeV, Compton scattering between 0.3 and about 6 MeV, and pair production above 6 MeV. For a given sample and a given counting geometry, both the counting rate and the fraction of the pulses in the photopeak increase with increasing crystal thickness. With the thicker crystals, many of the gamma-ray photons that first interact by Compton scattering are absorbed in a second (or third) interaction by photoelectric absorption. Since the two (or three) events occur within the resolving time of the detector, the output pulse size is the same as if the absorption had occurred in a single event. Thus, with larger crystals, the "peak-to-total" ratio (or "photofraction") of counts is increased, producing a more desirable pulse-height spectrum, inasmuch as the photopeak pulses serve to identify a particular radionuclide (via its characteristic gamma-ray energy or energies), whereas the smear of Compton pulses does not aid in the identification. With a 3 × 3-in. crystal the photofractions for gamma rays having energies of 0.1, 0.3, 0.5, 1, 2, and 4 MeV are about 0.99 (i.e., 99%), 0.84, 0.63, 0.41, 0.27, and 0.18, respectively. The corresponding overall detection efficiencies (geometry factor excluded) are approximately 100%, 72%, 59%, 46%, 37%, and 33%, respectively. At a 2-cm sample-to-detector distance, with a 3 × 3-in. detector, the geometry factor is 0.265 (i.e., 26.5%).

The maximum energy that a gamma-ray photon can lose by Compton interaction is given by the equation:

$$E_{CE} = \frac{E_\gamma}{1 + 0.511/2E_\gamma} \text{ (in MeV)} \tag{11}$$

This produces the "Compton edge" in the pulse-height spectrum.

As mentioned above, the most common NaI(Tl) detector is a solid cylindrical 3 × 3-in. crystal. However, for the detection and measurement of low-energy gamma rays (≤ 0.5 MeV) in the presence of considerable interference from higher-energy gamma rays, a thinner crystal is preferable. Conversely, for optimum detection of high-energy gamma rays (≥ 2 MeV), an even thicker crystal is desirable. One limitation on choice is that of cost. The approximate costs of canned 1×1, 2×2, 3×3, and 5×5-in. NaI(Tl) crystals are $50, $250, $750, and $2000, respectively. Also, larger crystals have higher background counting rates and usually somewhat poorer resolution.

Sodium iodide detectors have, in general, rather poor energy resolution; that is, the photopeaks obtained are quite broad, even though the result of interactions with monoenergetic gamma rays. For example, a 3 × 3-in. detector might show percentage resolutions (per cent full width at half-maximum of the photopeak) of about 15% (at 0.1 MeV), 11% (0.3 MeV), 9% (0.5 MeV), 7% (1 MeV), 5.5% (2 MeV), and 5% (3 MeV). The percentage resolution of a particular NaI(Tl) detector varies roughly according to the equation:

$$\% R = \frac{a}{b + E_\gamma^{1/2}}$$

where a and b are constants. This poor resolution accounts for the rounding of the pulse-height spectrum at the Compton edge and for the presence of pulses in the spectrum between the Compton edge and the photopeak.

When a gamma-ray photon interacts with an electron in the NaI(Tl) crystal, the struck electron slows to a stop in about a millimeter or less, giving up its kinetic energy to the crystal. About 10% of this energy appears as thallium fluorescence light, as a burst of ~3-eV photons, the other 90% going to heat. The photons produced are emitted isotropically; but, because of the efficiency of the diffuse white reflector coating on the outer surface of the crystal (except at the face adjacent to the photomultiplier tube), a high percentage of them strike the photocathode of the PMT essentially instantaneously. Since the photocathode has about a 10% photoelectric efficiency, about 10% of the incident 3-eV photons eject a photoelectron into the vacuum of the PMT. In the tube, the photoelectrons are accelerated by an electric potential and focused onto the first "dynode" (coated with a material having a high secondary-emission coefficient). The conditions typically employed are such that about three electrons are ejected from the dynode by each impinging electron. The process is repeated down the whole series of dynodes (usually 10), so that a large number of electrons appear as an output pulse, compared with the smaller initiating number that were produced at the photocathode. For example, the absorption of a single 1-MeV gamma-ray photon by the crystal produces some 33,000 light photons, which in turn generate about 3300 photoelectrons; these can result in an output pulse of the order of 10^8 electrons. It is important to note that the size of the output pulse is directly proportional to the amount of energy absorbed by the crystal in the interaction event.

3. Gamma-Ray Spectrometry

a. WITH NaI(Tl) DETECTORS

The complete gamma-ray spectrometer consists of (1) the scintillation detector, in its shield, (2) a pulse preamplifier and linear amplifier, (3) a pulse-height analyzer, and (4) one or more data-readout modes or displays. A typical modern transistorized multichannel pulse-height analyzer is shown in Fig. 98.4. Such analyzers operate on an amplitude-to-time conversion principle, sorting the incoming amplified electrical pulses into 100, 128, 200, 256, 400, 512, or a larger number of "channels" or pulse-height sizes. With NaI(Tl) detectors, a 400-channel spectrum is usually quite sufficient, so the observed pulse-height spectrum consists

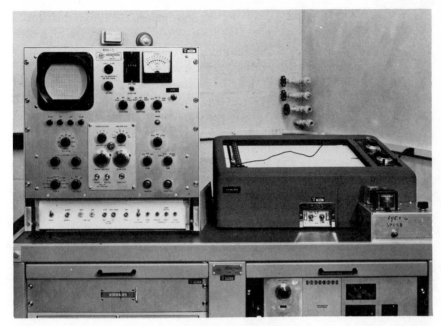

Fig. 98.4.   Typical multichannel pulse-height analyzer. Courtesy of RIDL Division, Nuclear-Chicago Corp.

of 400 data points. While the radioactive sample is being counted (usually for just 1 or a few minutes), one observes the spectrum accumulating, on the face of the oscilloscope. When the counting is stopped, the static 400-point spectrum (counts per channel versus channel number) is observed on the oscilloscope. The pulse-height data can then be plotted out on a precise X-Y or strip-chart recorder, printed out in digital form on paper tape or an electric typewriter, or transferred onto punched-paper tape, punched cards, or magnetic tape, as desired. When the data from one sample have been suitably recorded, the magnetic-core memory of the analyzer can be erased in preparation for the counting of the next sample.

Because of the finite time required by the analyzer to measure the magnitude of each pulse (more time, the larger the pulse), there can be an appreciable "dead-time loss" at high counting rates (typically, negligible at counting rates of less than about 50,000 cpm). The counting period can be preset for any desired "clock time" or "live time." A "dead-time" meter on the analyzer indicates the percentage of dead time while the sample is being counted. In general, counting rates much greater than about 100,000 cpm can lead to undesirable gain-shift problems, unless an electronic gain stabilizer is employed. Before a sample or a series of samples is to be counted, the gain of the analyzer system is carefully adjusted to the exact value desired, for example, 3.75, 7.50, or 15.0 keV/channel. This is done by adjusting the gain setting so that the photopeaks of one or more long-lived standard radionuclides fall exactly in the appropriate channels. Common calibration radionuclides used for this purpose are 270-day ^{57}Co (0.122 and 0.136 MeV), 2.62-year ^{22}Na (0.511 MeV from β^+, and 1.275 MeV), 30.0-year ^{137}Cs (0.662 MeV), and 5.263-year ^{60}Co (1.173 MeV and 1.332 MeV, in cascade). For example, if a gain setting of exactly 7.50 keV/channel is desired (400 channels thus covering the spectrum from 0 to 3.0 MeV), the gain should be adjusted so that the 0.662-MeV photopeak of ^{137}Cs centers in channel 88. A typical multichannel pulse-height analyzer, with one or two readout devices, costs about $10,000.

An illustrative pulse-height spectrum, that of 3.75-min ^{52}V, is shown in Fig. 98.5. Vanadium-52 emits a β^- particle and a 1.434-MeV gamma-ray photon in each disintegration. The main features of its NaI(Tl)

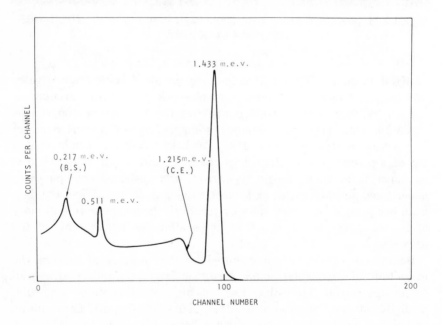

A SIMPLE PULSE-HEIGHT SPECTRUM

Fig. 98.5. Illustrative gamma-ray pulse-height spectrum.

(3 × 3-in.) pulse-height spectrum are thus the Compton-continuum region (0–1.215 MeV), the Compton edge (at 1.215 MeV), and the photopeak (at 1.434 MeV). Less prominent features are a small peak at 0.511 MeV (the results of β^+ annihilation photons generated in the lead shield by the 1.434-MeV gamma rays striking the crystal), and a small backscatter peak at 0.217 MeV. With higher-energy gamma rays, single- and double-escape peaks (at 0.511 MeV and 1.022 MeV less than the gamma-ray photon energy) are also observed. Also, with β^+ emitters and with high-energy gamma-ray emitters a small peak is sometimes observable at 0.681 MeV (0.511 MeV plus a 0.170-MeV back-scattered photon from the other 0.511-MeV annihilation photon); and with radionuclides which decay via a gamma-ray cascade, small "sum" peaks are observed (e.g., the ^{60}Co 1.173-MeV and 1.332-MeV cascade gamma-ray photons result in major photopeaks at these two energies, and a smaller peak at their sum, 2.505 MeV, the result of coincident gamma-ray photons striking the crystal simultaneously). With very low-energy gamma rays (particularly those with energies of less than 100 keV), a satellite "iodine escape peak" is observed at about 29 keV less than the photopeak; this is due to the escape, in some of the absorption events, of an iodine K X-ray photon (29 keV) from the NaI(Tl) crystal (which can occur fairly frequently with low-energy gamma rays because of their very slight penetration into the crystal).

In some instances, where improved geometrical counting efficiency is desired, a well-type NaI(Tl) crystal is used. In this case the sample vial is placed down in the cylindrical well of the crystal, thereby increasing the geometry from a typical ~30% value (sample on top of beta absorber, on top of solid crystal) to a value of about 90%. For a given set of external crystal dimensions (e.g., 3 in. × 3 in.), the effective NaI(Tl) path length of the well-type crystal is much less than that of the corresponding solid crystal; hence both the interaction efficiency and the photofraction are poorer for the well-type crystal, at least for gamma-ray energies above about 0.5 MeV. With well-type crystals, sum peaks are more pronounced. A special case is that of a β^+ emitter, for which the 1.022-MeV peak may be about the same size as the 0.511-MeV peak, whereas with a solid crystal the 0.511-MeV peak is essentially the only one observed.

If it is desirable to preferentially detect a radionuclide that emits gamma rays in cascade, in the presence of considerable gamma radiation of higher energy, one can sometimes utilize a well-type crystal and a sum peak, as mentioned above. Alternatively, a two-crystal coincidence technique, with the sample placed between the two crystals, can be employed. If one detector is set on a pulse-height channel which encompasses one of the cascade gamma-ray energies, the other is set to encompass the other cascade gamma-ray energy, and the circuitry is set to record only coincident pulses from the two detectors, detection of that activity is enhanced relative to detection of the interfering activities. For optimum efficiency, (1) the two detectors should be large in area relative to the size of the sample, (2) they should be thick enough to give good total-absorption efficiencies for the two gamma-ray energies involved, and (3) they should be as close together as possible. For selective detection of β^+ emitters, both channels are set to encompass the same photopeak (0.511 MeV). In some cases it is better to make one channel very wide in energy and to set the other just in the region of one of the photopeaks. Similarly, a $\beta^-\text{--}\gamma$ coincidence technique can be used effectively in some cases, using a β^- detector (organic crystal, plastic phosphor, or thin silicon semiconductor detector) and a gamma detector [NaI(Tl)] in coincidence. One can also utilize anticoincidence circuitry to preferentially reject pulses from an interfering activity that decays with the emission of cascade gamma rays.

b. With Semiconductor Detectors

Much interest is now evident in the use of lithium-drifted germanium, Ge(Li), semiconductor detectors, instead of, or in addition to, NaI(Tl) detectors, for gamma-ray spectrometry measurements. These are markedly superior in resolution to NaI(Tl) detectors, percentage resolutions [per cent full width at half-maximum (FWHM) of the photopeak] being typically about 15-fold better than for NaI(Tl). The relatively shallow effective-depth lithium–germanium detectors thus far available (maximum of about 2 cm), however, are not as efficient [compared with a 3-in.-thick NaI(Tl) detector] for gamma rays having energies greater than several tenths of 1 MeV. In addition, they must be maintained at liquid-nitrogen temperature at all times, they are considerably more expensive, and (for maximum resolution) they require a high-resolution amplifier and a more elaborate pulse-height analyzer—one with a few thousand, rather than just a few hundred, channels. Detectors of this type with a depletion layer of just a few millimeters (rather than 1 cm) are not as expensive and are very useful in the X-ray region ($\sim$10–100 keV). Much larger-area Ge(Li) detectors, having deeper depletion zones and costing less per unit of effective volume, are rapidly becoming available, making these detectors more generally useful. For maximum resolution, a field-effect transistor (FET) preamplifier, in some instances also operated at liquid-nitrogen temperature, should be used with a Ge(Li) detector. Whereas photopeaks in, for example, the region around 1 MeV that differ in energy by only about 50 keV or less are not resolved from one another by a NaI(Tl) detector, they are essentially completely resolved by a Ge(Li) detector even if their energies differ by as little as 5 keV.

For the lower-energy part of the X-ray region, a similar lithium-drifted silicon Si(Li), semiconductor detector is also very good (high resolution, good efficiency). This type of detector does not need to be maintained at liquid-nitrogen temperature except when in use (for maximum resolution).

4. Calculations

Typically, in instrumental activation analysis work, one uses a particular gamma-ray spectrometer to count a series of activated samples and standards (all activated under identical conditions) under exactly the same conditions, but at different decay times since t_0 (the end of the irradiation) and perhaps for different lengths of time. In order to make proper calculations, one usually computes the net photopeak-area (above the Compton continuum, using a linear base-line approximation) counting rates of the photopeak of interest, and then corrects all of these counting rates to t_0, or to any other desired reference time, using equation (7a). In making the decay corrections, one usually takes the calculated counting rate (net photopeak counts divided by live counting time) to be the same as the true net photopeak counting rate at the midpoint of the counting period (clock time). This is a fairly accurate assumption so long as the percentage of dead time during the counting period is small ($<5\%$), and the sample is not counted for a period of time longer than the half-life of the radionuclide of interest (otherwise corrections must be applied). If the dead time is negligible, one may prefer to use the rigorous equation, $A' = 0.693C/T(1 - e^{-0.693t/T})$, to relate the counts, C, obtained during the counting period, t, to the counting rate, A', at the start of the counting period. Knowing the exact amount of the element in question in the standard sample, one can readily compute the amount in each unknown sample, using equation (9a) or (9b). If desired, a computer program can be used to process the data, making all of the necessary calculations from the raw data and printing out an answer (in micrograms or parts per million), along with its standard deviation (based on its counting statistics), for each sample. Some multichannel analyzers have the capability of automatically adding together the counts in any group of channels selected, such as those included within a photopeak.

In many cases one can also analyze a series of samples for a number of elements of interest all at the same time, instead of one at a time. Also, by employing carefully standardized irradiation and counting conditions, one can develop a collection of sensitivity and gamma-ray spectral information for each radionuclide formed from the various elements. Then one can also analyze samples instrumentally for all elements that show up, even though some may not have been anticipated. Well-developed computer programs are available to (1) search the raw pulse-height digital data for all statistically significant photopeaks present; (2) from these photopeak energies, the counter resolution, and other criteria (such as irradiation time and flux, decay time, and counting conditions), determine which radionuclides might possibly be contributing significantly in each peak; (3) set up a matrix and solve by weighted least squares the combination of pure radionuclide reference spectra which best fits the data from the sample; and (4) calculate and print out the amount (in micrograms) or the concentration (in parts per million) of each element detected, along with its standard deviation (from counting statistics only). Another computer program may also be used, if desired, to calculate a firm (3σ) upper limit to the possible amount or concentration of each of many other elements which do not show up in the spectra. Thus either actual values or upper limits can be obtained for up to about 70 elements in a sample. A weighted least-squares computer program is also available for resolving a single photopeak (such as the 0.511-MeV β^+ annihilation peak) into its possible contributors of different half-lives.

The pulse-height spectrum of a mixture of radionuclides is, when measured properly, simply the sum of the spectra of the various components. The multichannel analyzers include the possibility of subtracting one spectrum from another, if desired. This feature can be of use for (1) subtracting out the counts due to background radiation, (2) subtracting out the spectra of one or more interfering activities, or (3) subtracting out the sample spectrum at a longer decay time from that obtained at a shorter decay time, in order to obtain a better representation of the short-lived species present. These operations, which are termed "spectrum stripping," may be carried out by live counting in subtract mode, or by subtraction of spectra stored in a second group of channels in the analyzer memory, on punched-paper, or magnetic tape. By operating on a time-sharing basis, one analyzer can simultaneously handle the input from two or four detectors, if desired—each going into a different group of channels. Most multichannel analyzers can also be operated in a "multiscaler" mode, that is, with all the incoming pulses being stored in the first channel for a preset period of time and then advancing regularly to the following channels. In this way, a gross decay curve is readily obtained fully automatically.

B. THE RADIOCHEMICAL-SEPARATION FORM

If (1) the induced activity of interest emits only β^- particles in all or almost all of its disintegrations, or (2) the interferences from other gamma-emitting induced activities prevent the purely instrumental detection of the activity of interest, one can often profitably resort to a post-irradiation radiochemical-separation procedure. Once the activity of interest is separated from all other induced activities, it can be counted on a beta-sensitive counter or, if it is a gamma emitter, on a gamma-ray detector.

1. Procedure

After activation, the sample is dissolved and chemically equilibrated with a relatively large amount (accurately known, and typically about 10 mg) of the element (or elements) of interest, as a carrier. Frequently, the equilibration is accomplished by alternate oxidation and reduction in acidic solution. If high levels of high-specific-activity interfering nuclides of other elements are present, one usually dilutes them with similar amounts (which, however, do not need to be known accurately) of holdback carriers of these elements. After the element of interest is separated out and purified by any suitable separation procedures (e.g., precipitation, volatilization, solvent extraction, electrodeposition, or ion exchange), it is then counted in an appropriate fashion. The amount of carrier element recovered is measured quantitatively by any suitable means (e.g., gravimetrically, volumetrically, by atomic absorption, or by reactivation), so that the results can be normalized to 100% recovery. It should be noted that, whenever a nucleus undergoes a nuclear reaction, there is imparted to the product nucleus, in the prompt disintegration of the compound nucleus, a recoil energy sufficient to eject it from the molecule in which it was originally present. The energetic recoil nuclei slow down in the medium and undergo various chemical reactions. As a result, most of the radionuclides formed by activation end up in a variety of chemical forms, mostly different from the original ones, and many of them virtually carrier-free. This must be kept in mind when carrying out the later chemical equilibration with added carrier. Recoils can also lead to losses of induced activity in the halogens and other sometimes volatile elements, and to losses by adsorption on the wall of the polyvials.

2. Freedom from Reagent Blank

Since any elemental contamination of the sample after activation can-not affect the results, no reagent-blank correction is necessary. For exam-ple, if one wished to analyze for arsenic a sample that contained only 0.001 μg of this element, the reagents used for the radiochemical separa-tion of the thermal-neutron induced ^{76}As activity, after activation, could contain even thousands of times more arsenic than the sample itself, without affecting the accuracy of the result. This is true because only the arsenic originally present in the sample has been made radioactive, and because the arsenic present as impurity in the reagents is negligible compared to the 10 mg of arsenic added as carrier.

3. Freedom from Loss Errors

Because of the equilibration of the induced activity of interest with a gross amount of carrier element, losses due to adsorption, coprecipita-tion, etc., become of negligible importance. Without such carrier dilution, of course, such losses could be very serious when one is determining microgram and even smaller amounts. Once equilibration with the carrier has been established, even sizable losses do not affect the accuracy of the result. This fact makes it possible to employ fairly rapid, nonquanti-tative separations, since the recovery of the carrier is measured and the results are normalized to complete recovery. In practice, of course, one still strives for fairly good recoveries—in the range of 50–100%, if possible. The higher the recovery, the better will be the counting statistics.

4. Beta Counting

If the separated activity must be beta-counted, or is best counted in this fashion, some precautions are necessary. One of these precautions differs appreciably from those involved in gamma-ray spectrometry, namely, the need to consider the effect of self-absorption. Because of the high penetrability of gamma rays (at least of energies above a few tenths of 1 MeV) in most materials (except those of very high density and high atomic number), activated samples with volumes of the order of 1 cm^3 can be counted by gamma-ray spectrometry with negligible difficulty from self-scattering or self-absorption of the gamma rays within the sample itself. This is not true for beta particles. The β^- particles, unlike the gamma rays, emitted by a particular radionuclide are not monoenergetic but instead exhibit an energy spectrum ranging from zero to a maximum value of (E_{max}) that is equal to the energy release (Q) in the β^- transition. The average and most probable β^- energies are typically about one-third of the E_{max} (in each β^- emission event, the total energy release, Q, is distributed between the readily detected β^- particle and the simultaneously emitted, virtually undetectable neutrino, ν). Tables such as the "Table of Isotopes" (9) and the General Electric *Chart of the Nuclides* (5), list the E_{max} values of the β^- particles emitted by the various radionuclides.

Beta particles are nowhere nearly as penetrating as gamma-ray photons of comparable energy, and, unlike gamma rays, they exhibit a maximum range in a given medium, the range depending on the energy. For beta particles, the range, R, in grams per square centimeter of aluminum, is given by the following equations (E in millions of electron volts):

$$\text{For } E < 0.8 \text{ MeV, } R = 0.407E^{1.38} \tag{12}$$

$$\text{For } E > 0.8 \text{ MeV, } R = 0.542E - 0.133 \tag{13}$$

Thus, the ranges in aluminum of β^- particles having energies of 0.5, 1, 2, 3, and 4 MeV are 0.16, 0.41, 0.95, 1.49, and 2.03 g/cm², respectively. Since the density of aluminum is 2.70 g/cm³, these ranges correspond to only 0.058, 0.15, 0.35, 0.55, and 0.75 cm, respectively. The range in most other materials is quite close to that in aluminum, in units of grams per square centimeter.

The relatively low penetration of β^- particles requires that there be very little material between the radioactive sample to be counted and the sensitive region of the detector. It also requires, where the β^- counting rates of two or more samples are to be accurately compared, not only that the overall counting geometry be exactly the same, but also that the chemical composition and thickness (expressed in grams per square centimeter) of the samples be exactly the same (or all be at least "infinitely thick"). For an "infinitely thin" sample, self-absorption of the β^- particles within the sample itself is zero (or at least negligible). For a sample of a thickness that is less than the range of the maximum-energy β^- particles emitted by the radionuclide, however, the self-absorption of the β^- particles will be appreciable. The fraction of the β^- particles generated within the sample that can escape decreases as the sample thickness is increased. An "infinitely thick" sample is one whose thickness (g/cm²) is greater than the range of the β^- particles. For a series of samples of a given specific activity (β^- dpm of a particular E_{max}, per gram), counted at a fixed geometry and detector efficiency, the β^-

counting rate is constant if all of the samples are infinitely thick or thicker—even if they are of different thicknesses.

For β^- counting, one usually employs a counter that has a high detection efficiency for charged particles, but a very low detection efficiency for gamma radiation. Suitable counters are Geiger-Müller and gas-proportional counters, organic crystal or plastic phosphor scintillation detectors, Cerenkov detectors, and thin silicon semiconductor detectors.

In β^- counting, after radiochemical separation of the activity of interest, it is usually desirable to verify that the activity being measured is, in fact, the one that it is supposed to be. This is done in two ways: (1) by counting the sample at two or more decay times, appropriately spaced, to see whether the observed half-life agrees with that of the radionuclide in question, and (2) by checking that the β^- E_{max} value is the same as that of the radionuclide in question. The energy check is readily made by counting the sample with and without an aluminum absorber of a suitable thickness placed between the sample and the detector. The same measurements are also made on the reference sample of radionuclide—a sample of the same chemical composition and thickness (g/cm^2). An aluminum absorber that decreases the counting rate by a factor of about 2 or 3 is quite suitable. The fraction, net counts per minute with absorber divided by net counts per minute without absorber, should be closely the same for sample and standard. It the half-life and β^- energy measurements do not indicate that the sample activity is quite pure and is that of the radionuclide in question, the sample will have to be purified further. It is also a good idea to check for the presence of any gamma-emitting impurity in the sample, by counting it also on the gamma-ray spectrometer.

Positrons emitted by radionuclides are also emitted with a spectrum of energies (rather than as monoenergetic particles), and exhibit an average or most probable energy of about one-third of the E_{max}. Their energy-range dependence is essentially the same as for β^- particles, that is, they are also related by equations (12) and (13). However, since positrons readily undergo an annihilation reaction with electrons once they have slowed down sufficiently, β^+ emitters are usually counted by gamma-ray spectrometry instead of by beta-particle counting. Also, if the separated activity of interest is an X-ray or gamma emitter, it can often best be counted by gamma-ray spectrometry.

In some cases, a sample matrix can become so highly radioactive that it is difficult or even hazardous to handle during the radiochemical-separation steps. If the activity of interest is relatively long-lived, and most of the interfering activities are comparatively short-lived, one can of course postpone handling the sample until the interferences have decayed away to a large extent. If this is not possible, the radiochemical separations must be carried out under "semi-hot-lab" conditions, that is, using some protective lead shielding, and long-handled tongs, remote-control pipets, etc. The subject of radiation hazards is mentioned only briefly in this chapter, because of limitations of space. In most activation analysis work the actual samples handled are not dangerously radioactive, but this is not always the case. In practice one estimates in advance, if possible, the probable radioactivity level of a sample to be activated and then monitors it before handling it. All operations are carried out to ensure that the chemist does not receive a whole-body exposure of more than 100 milliroentgens (mR)/week, the generally accepted tolerance level. Cumulative doses are checked regularly with film badges and pocket dosimeters.

In some instances, especially when the induced activity of interest is rather short-lived (minutes to hours), it may be preferable to yield the advantage of freedom from reagent blank and instead employ a preirradiation chemical separation. In this case, one must use very pure reagents, check every step of the procedure, and run a reagent blank, if very small amounts or very low concentrations are to be determined accurately. An example of this procedure is the preirradiation separation of vanadium in blood-serum samples, used in the author's laboratory. The half-life of the thermal-neutron activated vanadium, ^{52}V, is only 3.75 min. In this case, the reagent blank is typically only about 0.001 μg vanadium.

IV. SENSITIVITY OF ACTIVATION ANALYSIS FOR VARIOUS ELEMENTS

The discussion here will be limited to the interference-free limits of detection for various elements when exposed to (1) the high (10^{13} n/cm²-sec) thermal-neutron flux of a research reactor, and (2) the moderate (10^9 n/cm²-sec) 14-MeV flux of a small neutron generator. Some general comments were made previously concerning the limits of detection attainable in photonuclear and charged-particle activation analysis.

A. HIGH-FLUX THERMAL-NEUTRON ACTIVATION ANALYSIS

As can be seen from equations (5), (6), and (8), the amount of a particular radioactivity induced per unit mass of a particular element, when exposed to a flux of thermal neutrons, is (1) directly proportional to the thermal-neutron flux, ϕ, to the fractional isotopic abundance, a, and to the isotopic thermal-neutron capture cross section, σ; (2) inversely proportional to the chemical atomic weight of the element, AW; and (3) dependent in a more complex manner on the irradiation time, t_i, and the half-life of the radionuclide, T (both involved in the saturation term, S). The counting rate attainable from a given disintegration rate of a particular radionuclide depends on (1) the decay scheme of the radionuclide, (2) the counting geometry, and (3) the efficiency of the detector.

Thus, to compile a useful table of limits of detection, one may select or define values for each of these variables and then make arbitrary, but reasonable, assumptions regarding the minimum counting rate that is still measurable with some accuracy. This has been done in somewhat different ways by various authors, but the one developed in the author's former laboratory, by J. D. Buchanan (3), will be used here. Using his same assumptions but calculating limits of detection for a thermal-neutron flux of 10^{13}, instead of his 1.8×10^{12}, n/cm²-sec, and making a few other changes, one obtains the sensitivity values shown in Table 98.II (listed in the order of increasing atomic number). The values shown are calculated detection limits (in the absence of interferences) for a 1-hr (maximum) irradiation at 10^{13} flux, with radiochemical separation and β^- counting assumed for the β^- detection limits (allowing for decay and yield), and with gamma-ray spectrometry, using the photopeak counting rate of the gamma-ray photopeak listed, for the gamma detection limits. A 3×3-in. solid NaI(Tl) detector is assumed, with a mean sample-to-crystal distance of 2 cm.

For the β^- counting, the defined minimum detectable net counting rate is taken as 100 cpm if the half-life is less than 1 hr, and 10 cpm otherwise. For the gamma-ray counting, the defined minimum detectable photopeak counting rate is taken as 1000 cpm for $T \leq 1$ min, 100 cpm for $T > 1$ min but < 1 hr, and 10 cpm for $T \geq 1$ hr. The decay schemes, half-lives, and gamma-ray energies used or cited are taken from the "Table of Isotopes" (9), and the photopeak counting efficiencies from Heath's *Scintillation Spectrometry Gamma-Ray Spectrum Catalogue* (7). Although the sensitivities listed in Table 98.II are calculated ones, many of them have been checked experimentally in various laboratories, including the author's. In most cases, the experimental and calculated values agree within a factor of 2 or 3. Only rarely is there a discrepancy as large as a factor of 10. In most instances where no gamma limit is shown, the radionuclide is a pure β^- emitter. In most instances where

TABLE 98.II
Limits of Detection at a Thermal-Neutron Flux of 10^{13}
(1 hr maximum irradiation time)

Element	Radio-nuclide	Half-life	γ-Ray energy, MeV	Limit of detection, μg	
				β^- counting	γ-Ray spectrometry
F	^{20}F	11.56 sec	1.632	—	0.2
Ne	^{23}Ne	37.6 sec	0.439	—	0.4
Na	^{24}Na	14.96 hr	1.369	0.0004	0.0009
Mg	^{27}Mg	9.46 min	0.843	0.04	0.05
Al	^{28}Al	2.31 min	1.780	0.02	0.001
Si	^{31}Si	2.62 hr	1.264	0.009	20.0
P	^{32}P	14.28 days	—	0.02	—
S	^{35}S	87.9 days	—	2.0	—
	^{37}S	5.07 min	3.09	10.0	20.0
Cl	^{38}Cl	37.29 min	1.60	0.002	0.02
Ar	^{41}Ar	110.0 min	1.293	—	0.0001
K	^{42}K	12.36 hr	1.524	0.002	0.04
Ca	^{49}Ca	8.8 min	3.10	0.1	0.5
Sc	^{46}Sc	83.9 days	0.889	0.002	0.004
Ti	^{51}Ti	5.79 min	0.320	0.07	0.01
V	^{52}V	3.75 min	1.434	0.0004	0.0001
Cr	^{51}Cr	27.8 days	0.320	—	0.2
Mn	^{56}Mn	2.576 hr	0.847	0.000004	0.000005
Fe	^{59}Fe	45.6 days	1.095	10.0	40.0
Co	^{60m}Co	10.47 min	0.059	0.00002	0.0009
	^{60}Co	5.263 yr	1.173	0.04	0.07
Ni	^{65}Ni	2.564 hr	1.481	0.004	0.05
Cu	^{64}Cu	12.80 hr	$(0.511)\beta^+$	0.0002	0.0002
	^{66}Cu	5.10 min	1.039	0.002	0.009
Zn	^{65}Zn	245.0 days	1.115	0.9	5.0
	^{69m}Zn	13.8 hr	0.439	0.02	0.02
Ga	^{70}Ga	21.1 min	1.040	0.0004	0.09
	^{72}Ga	14.12 hr	0.835	0.0002	0.0004
Ge	^{75}Ge	82.0 min	0.265	0.0004	0.002
As	^{76}As	26.4 hr	0.559	0.0002	0.0005
Se	^{75}Se	120.4 days	0.265	—	0.2
	^{77m}Se	17.5 sec	0.161	—	0.05
	^{81}Se	18.6 min	—	0.001	
Br	^{80}Br	17.6 min	0.618	0.00005	0.0005
	^{82}Br	35.34 hr	0.554	0.0009	0.001
Kr	^{79}Kr	34.92 hr	0.261	—	0.5
Rb	^{86}Rb	18.66 days	1.078	0.04	0.7
	^{88}Rb	17.8 min	1.863	0.009	0.1
Sr	^{87m}Sr	2.83 hr	0.388	0.0009	0.0009

TABLE 98.II (*continued*)

Element	Radio-nuclide	Half-life	γ-Ray energy, MeV	Limit of detection, μg	
				β^- counting	γ-Ray spectrometry
Y	^{89m}Y	16.1 sec	0.915	—	1.0
	^{90}Y	64.0 hr	—	0.002	—
Zr	^{97}Zr	17.0 hr	0.747	0.2	0.2
Nb	^{94m}Nb	6.29 min	0.871	0.0009	0.1
Mo	^{99}Mo	66.7 hr	0.181	0.02	0.1
	^{101}Mo	14.6 min	0.191	0.02	0.03
Ru	^{103}Ru	39.5 days	0.497	0.09	0.1
	^{105}Ru	4.44 hr	0.726	0.002	0.004
Rh	^{104m}Rh	4.41 min	0.051	0.0007	0.00002
Pd	^{109m}Pd	4.69 min	0.188	0.1	0.009
	^{109}Pd	13.47 hr	0.307	0.0002	0.2
Ag	^{108}Ag	2.42 min	0.632	0.0009	0.002
	^{110}Ag	24.4 sec	0.658	—	0.0001
Cd	^{111m}Cd	48.6 min	0.247	0.1	0.02
	^{115}Cd	53.5 hr	0.53	0.009	0.04
In	^{116m1}In	54.0 min	1.293	0.000005	0.00001
Sn	^{125m}Sn	9.5 min	0.325	0.05	0.02
Sb	^{122m}Sb	4.2 min	0.0607	—	0.01
	^{122}Sb	2.80 days	0.564	0.0009	0.002
Te	^{131}Te	24.8 min	0.150	0.005	0.009
	^{131}I	8.05 days	0.364	0.4	0.2
I	^{128}I	24.99 min	0.441	0.0002	0.001
Xe	^{135}Xe	9.14 hr	0.250	—	0.01
Cs	^{134m}Cs	2.895 hr	0.128	0.0004	0.0004
	^{134}Cs	2.046 yr	0.605	0.05	0.04
Ba	^{139}Ba	82.9 min	0.166	0.005	0.01
La	^{140}La	40.22 hr	1.596	0.0004	0.001
Ce	^{141}Ce	32.5 days	0.145	0.2	0.2
	^{143}Ce	33.0 hr	0.293	0.02	0.04
Pr	^{142}Pr	19.2 hr	1.57	0.0001	0.009
Nd	^{147}Nd	11.06 days	0.091	0.02	0.02
Sm	^{153}Sm	46.8 hr	0.103	0.00009	0.0002
Eu	^{152m1}Eu	9.3 hr	0.963	0.0000009	0.00009
Gd	^{159}Gd	18.0 hr	0.363	0.001	0.009
Tb	^{160}Tb	72.1 days	0.966	0.005	0.02
Dy	^{165}Dy	139.2 min	0.095	0.0000004	0.0000002
Ho	^{166}Ho	26.9 hr	0.0806	0.00004	0.00004
Er	^{171}Er	7.52 hr	0.308	0.0005	0.0004
Tm	^{170}Tm	134.0 days	0.084	0.002	0.04
Yb	^{175}Yb	101.0 hr	0.396	0.0004	0.002
Lu	^{176m}Lu	3.69 hr	0.088	0.000009	0.000009

TABLE 98.II (*continued*)

Element	Radio-nuclide	Half-life	γ-Ray energy, MeV	Limit of detection, μg	
				β^- counting	γ-Ray spectrometry
Hf	^{179m}Hf	18.6 sec	0.217	—	0.007
	^{181}Hf	42.5 days	0.482	0.02	0.03
Ta	^{182m}Ta	16.5 min	0.172	0.02	0.04
	^{182}Ta	115.1 days	1.122	0.009	0.05
W	^{187}W	23.9 hr	0.479	0.0002	0.0007
Re	^{186}Re	88.9 hr	0.137	0.0002	0.004
	^{188m}Re	18.7 min	0.106	0.0007	0.004
	^{188}Re	16.7 hr	0.155	0.00004	0.0002
Os	^{190m}Os	9.9 min	0.361	—	0.4
	^{191}Os	15.0 days	0.129	0.02	0.04
	^{193}Os	31.5 hr	0.139	0.004	0.05
Ir	^{192m1}Ir	1.42 min	0.058	—	0.007
	^{192}Ir	74.2 days	0.317	0.0005	0.0004
	^{194}Ir	17.4 hr	0.328	0.00002	0.0001
Pt	^{199}Pt	31.0 min	0.197	0.004	0.02
	^{199}Au	3.15 days	0.158	0.02	0.01
Au	^{198}Au	2.697 days	0.412	0.00005	0.00005
Hg	^{197}Hg	65.0 hr	0.0776	—	0.002
	^{199m}Hg	43.0 min	0.158	—	0.1
Tl	^{204}Tl	3.81 yr	—	0.5	—
	^{206}Tl	4.19 min	—	0.04	—
Pb	^{209}Pb	3.30 hr	—	2.0	—
Bi	^{210}Bi	5.013 days	—	0.05	—
Th	^{233}Pa	27.0 days	0.313	0.007	0.01
U	^{239}Np	2.346 days	0.106	0.0007	0.0004

no β^- limit is shown, either the half-life is too short ($<$1 min), or the radionuclide decays by electron capture or isomeric transition. In a few instances, detection via a radioactive daughter is listed (e.g., tellurium via the ^{131}I daughter of ^{131}Te, and thorium via the ^{233}Pa daughter of ^{233}Th). For many elements, not all of the (n, γ) products are listed in the table—just the one, two, or three predominant ones, the ones of most analytical usefulness, are given. [Note: Twenty of the elements of the periodic system are monoisotopic in nature and hence form only a single (n, γ) product—except for a few elements, like cobalt, which form an isomeric state of the product radionuclide also. The remaining stable elements exist in nature in the form of from two to as many as ten stable nuclides, of various abundances. Each of these elements, in general, can form a number of different (n, γ) radionuclide products.] For some elements, the fast-neutron flux in the reactor can also generate quite significant activities, via (n, n'), (n, 2n), (n, p), or (n, α) reactions. These have been discussed in an earlier section and are not included in Table 98.II.

Taking the best β^- sensitivity for each element listed in the table, one finds that the values for the 68 elements range from as low as 4×10^{-7} μg (Dy) to as high as 10 μg (Fe), with a median of about 0.001 μg. Similarly, when the best gamma sensitivity is taken for each element, the values for the 71 elements range from as low as 2×10^{-7} μg (Dy) to as high as 40 μg (Fe), with a median of about 0.005 μg. If the best sensitivity (either β^- or gamma) is taken for each element, the values for the 75 elements range from as low as 2×10^{-7} μg (Dy) to as high as 10 μg (Fe), with a median of about 0.001 μg. Thus, a typical element can be detected down to about 1 ng, or 0.001 ppm in a 1-g sample.

It should be remarked that, where needed, all of the sensitivities shown in Table 98.II can be improved further by employing an even higher neutron flux, if available. Also, the sensitivities of the elements that form rather long-lived activities (half-lives in excess of a few hours) can all be improved by using an irradiation time longer than 1 hr, and also by counting for a longer-than-usual period of time.

B. MODERATE-FLUX 14-MeV-NEUTRON ACTIVATION ANALYSIS

With presently available 14-MeV-neutron generators, the maximum 14-MeV-neutron flux to which samples of appreciable size (1–5 cm³) can be exposed for any length of time is about 10^9 n/cm²-sec. The 14-MeV-neutron limits of detection for the 41 elements shown in Table 98.III are experimentally determined values, measured in the author's former laboratory by E. L. Steele and G. H. Andersen. They are for a maximum irradiation time of 5 min (in view of the target-decline problem), at a 14-MeV-neutron flux of 10^9 n/cm²-sec. The limits of detection in this case are defined as the amount of the element, under these conditions, that will give 100 photopeak counts in a counting period of 10 min or less, again assuming a 3 $\times$ 3-in. solid NaI(Tl) detector and a mean sample-to-crystal distance of 2 cm. The elements are listed in the order of increasing atomic number. In this experimental study, the rare-earth and inert-gas elements were not studied, nor were some of the more unusual elements, such as ruthenium, rhodium, and osmium.

TABLE 98.III
Limits of Detection at a 14-MeV-Neutron Flux of 10^9
(5 min maximum irradiation time)

Element	Reaction	Product	Half-life	γ-Ray energy, MeV	Limit of detection, μg
N	$^{14}N(n, 2n)$	^{13}N	9.96 min	$(0.511)\beta^+$	90
O	$^{16}O(n, p)$	^{16}N	7.14 sec	6.13	30
F	$^{19}F(n, p)$	^{19}O	29.1 sec	0.197	20
Na	$^{23}Na(n, p)$	^{23}Ne	37.6 sec	0.439	20
Mg	$^{24}Mg(n, p)$	^{24}Na	14.96 hr	1.369	80
Al	$^{27}Al(n, p)$	^{27}Mg	9.46 min	0.843	6
Si	$^{28}Si(n, p)$	^{28}Al	2.31 min	1.780	2
P	$^{31}P(n, \alpha)$	^{28}Al	2.31 min	1.780	8
K	$^{39}K(n, 2n)$	^{38}K	7.71 min	$(0.511)\beta^+$	90
Sc	$^{45}Sc(n, 2n)$	^{44}Sc	3.92 hr	$(0.511)\beta^+$	20
Ti	$^{46}Ti(n, 2n)$	^{45}Ti	3.09 hr	$(0.511)\beta^+$	90
V	$^{51}V(n, p)$	^{51}Ti	5.79 min	0.320	7
Cr	$^{52}Cr(n, p)$	^{52}V	3.75 min	1.434	10
Mn	$^{55}Mn(n, \alpha)$	^{52}V	3.75 min	1.434	40
Fe	$^{56}Fe(n, p)$	^{56}Mn	2.576 hr	0.847	30
Co	$^{59}Co(n, \alpha)$	^{56}Mn	2.576 hr	0.847	50
Ni	$^{60}Ni(n, p)$	^{60m}Co	10.47 min	0.059	300
Cu	$^{63}Cu(n, 2n)$	^{62}Cu	9.76 min	$(0.511)\beta^+$	9
Zn	$^{64}Zn(n, 2n)$	^{63}Zn	38.4 min	$(0.511)\beta^+$	30
Ga	$^{69}Ga(n, 2n)$	^{68}Ga	68.3 min	1.078	20
Ge	$^{76}Ge(n, 2n)$	^{75m}Ge	48.0 sec	0.139	5
As	$^{75}As(n, p)$	^{75m}Ge	48.0 sec	0.139	4
Se	$^{78}Se(n, 2n)$	^{77m}Se	17.5 sec	0.161	20
Br	$^{79}Br(n, n')$	^{79m}Br	4.8 sec	0.21	60
Rb	$^{85}Rb(n, 2n)$	^{84m}Rb	20.0 min	0.250	1
Sr	$^{88}Sr(n, 2n)$	^{87m}Sr	2.83 hr	0.388	1
Zr	$^{90}Zr(n, 2n)$	^{89m}Zr	4.18 min	0.588	4
Mo	$^{92}Mo(n, 2n)$	^{91}Mo	15.49 min	$(0.511)\beta^+$	30
Pd	$^{110}Pd(n, 2n)$	^{109m}Pd	4.69 min	0.188	4
In	$^{113}In(n, 2n)$	^{112m}In	20.7 min	0.156	30
Sb	$^{121}Sb(n, 2n)$	^{120}Sb	15.89 min	$(0.511)\beta^+$	7
Te	$^{130}Te(n, 2n)$	^{129}Te	68.7 min	0.455	60
I	$^{127}I(n, 2n)$	^{126}I	12.8 days	0.386	80
Ba	$^{138}Ba(n, 2n)$	^{137m}Ba	2.554 min	0.662	1
Ce	$^{140}Ce(n, 2n)$	^{139m}Ce	54.0 min	0.746	9
Hf	$^{180}Hf(n, 2n)$	^{179m}Hf	18.6 sec	0.217	80
Ta	$^{181}Ta(n, 2n)$	^{180m}Ta	8.15 hr	0.093	20
W	$^{186}W(n, 2n)$	^{185m}W	1.62 min	0.130	20
Pt	$^{198}Pt(n, 2n)$	^{197m}Pt	78.0 min	0.346	200
Hg	$^{200}Hg(n, 2n)$	^{199m}Hg	43.0 min	0.158	20
Pb	$^{208}Pb(n, 2n)$	^{207m}Pb	0.80 sec	0.570	100

The sensitivities are seen to range from as low as 1 μg (Rb, Sr, Ba) to as high as 300 μg (Ni), with a median of about 20 μg.

The same 14-MeV-neutron generators can, with a moderator, provide thermal-neutron fluxes of about 10^8 n/cm²-sec. Thus, all of the thermal-neutron limits of detection would be 100,000 times higher than the reactor 10^{13} flux limits shown in Table 98.II, if a 1-hr irradiation at 10^8 flux were employed. Because of the target-lifetime problem, however, it is more reasonable to consider a maximum irradiation time of about 5 min. Thus, for the elements of Table 98.II that form an (n, γ) product with a half-life of the order of 1 hr or longer, the 10^8-flux thermal-neutron limits are about 10^6 times those given in the table. For those with half-lives of a few minutes or less, the factor is 10^5.

V. SOURCES OF ERROR, ACCURACY, AND PRECISION

Since a number of the potential sources of error in the activation analysis method were mentioned in previous sections, this discussion will be brief and will be limited mostly to factors involved in thermal-neutron activation analysis. In general, if care is taken, and good counting statistics are obtained, absolute accuracies of the order of $\pm 1\%$ of the value are attainable in NAA work. It is very difficult, except by extreme care and measurement replication and averaging, to attain accuracies of the order of $\pm 0.1\%$. Moreover, if one becomes careless, the accuracies can be as poor as $\pm 10\%$.

A. GEOMETRICAL FACTORS

When a sample analyzed by NAA is compared with a standard sample of the element being determined, it is assumed that both were exposed to exactly the same neutron flux for the same period of time and were counted in exactly the same way. If any of these assumptions is not valid to a high degree of accuracy, a source of error has been introduced. If deviations from these assumptions are known to have occurred, accurate corrections can usually be made. For example, differences in neutron flux at the various positions where samples and standards are placed can be measured by means of flux-monitor samples placed at these locations (often, thin gold foil for thermal-NAA, thin copper foil for 14-MeV-NAA). Differences in counting efficiency between counters, and for different sample-to-detector distances, can be determined by calibration measurements. It is usually necessary to make the sample shape and volume exactly the same as that of the standard, in order to avoid difficultly correctable effects of flux gradients. For most work, carefully prepared aqueous standard solutions can be used (some exceptions are noted below). These are prepared at accurately known concentrations— high enough to give very good counting statistics under the irradiation and counting conditions employed. Some such reference solutions (e.g., gold and mercury solutions) are unstable in storage and therefore must be made up fresh.

B. THERMAL-NEUTRON SELF-SHIELDING ERRORS

Since some elements have very large cross sections (10 barns and higher) for the absorption of thermal neutrons, it is possible, unless due care is taken, to introduce into the thermal-NAA results errors due to "self shielding" within the sample. For the comparator method to be accurate, either (1) the sample and the standard must be exposed, throughout the volume of each, to the same average neutron flux, or (2) suitable corrections must be applied if this is not the case.

With aqueous standard samples even several cubic centimeters in volume, self-shielding is quite negligible, since oxygen has a very small (n, γ) cross section (<0.0002 b), and that of hydrogen is also small (0.33 b). With 1-cm^3 unknown samples, self-shielding may or may not be appreciable, depending on the overall (n, γ) cross section of the matrix. For a matrix that is essentially a pure element, the extent of self-shielding will depend on the shape and volume of the sample (increasing with increasing sample volume and symmetry) and will, for a given volume, increase with increasing density and atomic (n, γ) cross section, and decrease with increasing atomic weight. With low-Z, low-density elements of low (n, γ) σ (such as aluminum: σ of 0.23 b, and potassium: σ of 2.1 b), self-shielding is negligible even for a 1-cm-cube sample. However, for germanium, which has a modest σ (2.4 b) but a higher density (5.35 g/cm^3), self-shielding in a 1-cm-cube sample can be several per cent ($\sim 5\%$).

In order to avoid errors due to thermal-neutron self-shielding, when analyzing samples of appreciable (n, γ) cross section, one must do one of three things: (1) use a sample much smaller than 1 cm^3, (2) prepare the standard in a matrix of the same (n, γ) cross section (also, of course, matching the shape and volume of the sample), or (3) measure the extent of self-shielding in the sample and make a correction. The third alternative can be employed by measuring the attenuation of a thermal-neutron beam caused by the sample, or by placing a tiny bead or rod of a suitable detector element at the center of the sample. With careful calibrations, fairly accurate corrections for self-shielding can then be made. For a sample of a given volume, self-shielding can also be reduced by spreading the sample out in a thin layer.

With fast neutrons, the neutron absorption and scattering within a sample having a volume of the order of 1 cm^3 are quite small, since none of the cross sections is anywhere nearly as large as many of the thermal-neutron (n, γ) cross sections.

C. GAMMA-RAY SELF-ABSORPTION ERRORS

When one employs gamma-ray spectrometry to compare the net photopeak counting rate of a particular radionuclide with that of a standard sample of the element in question, it is assumed that the absorption and Compton scattering of these particular gamma-ray photons are either negligible in the sample and the standard or very closely the same. For a sample of low to medium density ($d < 3$ g/cm^3), having a volume of about 1 cm^3 or less, and gamma rays of energies above about 0.3 MeV, self-absorption of the gamma-ray photons within the sample is quite negligible. However, even in such cases, Compton scattering within the sample can be appreciable, and its extent will depend on the matrix. Since photons that are Compton-scattered within the sample have reduced energies, they cannot possibly contribute to the total-absorption peak observed with the detector; hence errors are possible unless suitable steps are taken. As an example, the percentages of the photons that undergo Compton scattering within a 1-cm^3 sample are, very roughly, 5%, 11%, and 28%, in H_2O, aluminum, and iron, respectively, for 0.5-MeV photons; 3%, 8%, and 21%, respectively, for 1-MeV photons; and 2%, 4%, and 12%, respectively, for 3-MeV photons. With high-density, high-Z matrices, both self-absorption and self-scattering can be quite large. To avoid errors in the results due to these effects, one either (1) reduces the sample size to such a point that the effects are negligible, (2) prepares the standard in a matrix that matches the sample in these respects, or (3) makes suitable corrections for these effects.

D. ERRORS DUE TO VARIOUS OTHER EFFECTS

The corrections needed if an (n, γ) product of interest is also formed appreciably in the sample by an (n, p) fast-neutron reaction on the element one unit above it in Z, and /or by an (n, α) fast-neutron reaction on the element two units above it in Z, have already been discussed. Fast neutrons impinging upon a sample containing an appreciable amount of hydrogen can produce many "knock-on" protons, which in turn can generate some induced activity, particularly via (p, n) reactions. For example, in analyzing samples which contain hydrogen and either carbon or oxygen (or both) for low levels of nitrogen, via the ^{14}N(n, 2n)^{13}N reaction, one must apply a correction for the small amount of ^{13}N formed by the ^{13}C(p, n)^{13}N and ^{16}O(p, α)^{13}N recoil-proton reactions. [Note: This effect can also be put to use, as has been done in the determination of ^{18}O in water or organic materials, using the ^{18}O(p, n)^{18}F recoil-proton reaction. A somewhat related technique is that of using recoil tritons to determine oxygen, via the ^{16}O(t, n)^{18}F reaction, generating the recoil tritons *in situ* by mixing the sample with finely powdered LiF and exposing this to a high thermal-neutron flux; the recoil tritons ($\sim$2.7 MeV in energy) are produced by the high-cross-section ^{6}Li(n, t)^{4}He reaction.]

Since ^{235}U is readily fissioned by thermal neutrons, and ^{238}U and ^{232}Th by fast neutrons, the various fission-product radionuclides thus formed, if the sample contains appreciable amounts of either or both of these elements, must not be confused with the (n, γ) products of the elements being determined. (Note: In some cases, the fission products of uranium and thorium are used for the activation analysis determination of these elements. Also, they can be determined via the very short-lived delayed-neutron fission products which they form—unusual radionuclides that have half-lives in the range of seconds to about 1 min and decay partially by neutron emission, which can be detected by neutron-sensitive counters, such as BF_3 or ^{3}He counters.) A more unusual possible type of error is illustrated by the thermal-neutron determination of very low levels of phosphorus in high-silicon matrices. Here, silicon forms 2.62-hr ^{31}Si via the ^{30}Si(n, γ)^{31}Si reaction. This decays by β^- emission to form stable ^{31}P.

E. THE EFFECT OF COUNTING STATISTICS ON PRECISION

Since radioactive decay is a statistically random process, the counts obtained when one employs a detector to measure and record a fraction of the emitted particles or photons also exhibit statistical fluctuations. With low-activity samples, the statistical uncertainty of the final analytical results is usually due mainly to the effect of the counting statistics. Because of limitations of space, the treatment of counting statistics here can be only a brief one, mentioning merely the principal factors and cases.

When a radioactive sample counted on some kind of detector produces N counts during the counting period, t, the standard deviation (σ) of N is simply $\pm N^{1/2}$. This means that, if the same sample were counted in exactly the same way time and time again (assuming that the decline in the disintegration rate of the sample, due to decay, is completely negligible during the whole course of the measurements), 68.3% of the observed values for N would fall within the limits of the average (or "true") value, $\bar{N}$, $\pm 1\sigma$; some 15.85% would be less than $\bar{N} - 1\sigma$; and some 15.85% would be greater than $\bar{N} + 1\sigma$. It is relevant to note that counts (N) of 100, 10,000, and 1,000,000 have σ's of ± 10, ± 100, and ± 1000 counts, respectively, or %σ's of $\pm 10\%$, $\pm 1\%$, and $\pm 0.1\%$, respectively.

Since measurements of samples, standards, and background are often carried out for different counting periods, the results of each are usually expressed as counts per unit time, for example, counts per minute (cpm). Since the counting period can usually be measured very accurately, its contribution to the statistical uncertainty of the answer is generally negligible, and the value for t is thus treated as an absolute value. Hence, if σ_N is $\pm N^{1/2}$, $\sigma_{N/t}$ is $\pm N^{1/2}/t$.

When even a pure radionuclide is counted, one automatically is acquiring counts because of the overall "background" radiation at the same time. Thus, the observed N is really N_{s+b}, that is, $N_s + N_b$, where the subscripts s and b refer to sample and background, respectively. In order to obtain the desired N_s, one must subtract a separately measured N_b from the observed N_{s+b}. If the background counting rate, B, is measured during a counting period, t_b, the standard deviation of the value obtained for the background counting rate, σ_B, is simply $\pm N_b^{1/2}/t_b$. If the sample is counted for a period of time, t, the standard deviation of the gross counting rate, σ_{S+B}, is $\pm N_{s+b}^{1/2}/t$, where S denotes the net sample counting rate.

When one adds two or more quantities together (in the same units), each with its own standard deviation, or subtracts one number from another, the standard deviation of the result is equal to the square root of the sum of the squares of the individual standard deviations; that is, in addition (with n being the number of terms added):

$$\sigma_{\text{sum}} = \pm \left(\sum_{1}^{n} \sigma_i{}^2 \right)^{1/2} \tag{14}$$

and in subtraction:

$$\sigma_{\text{differerce}} = \pm (\sigma_1{}^2 + \sigma_2{}^2)^{1/2} \tag{15}$$

Thus the standard deviation of the net sample counting rate, σ_S, is given by the equation

$$\sigma_S = \pm \, (\sigma_{S+B}^2 + \sigma_B{}^2)^{1/2} \tag{16}$$

or

$$\sigma_S = \pm \left(\frac{N_{s+b}}{t^2} + \frac{N_b}{t_b{}^2} \right)^{1/2} \tag{17}$$

or

$$\sigma_S = \pm \left(\frac{S + B}{t} + \frac{B}{t_b} \right)^{1/2} \tag{18}$$

In the special case where $t = t_b$,

$$\sigma_S = \pm \left(\frac{S + 2B}{t} \right)^{1/2} \tag{19}$$

In gamma-ray spectrometry measurements, one usually determines a particular net photopeak counting rate by subtracting the counts in the underlying Compton + background continuum from the gross counts in the analyzer channels that define the peak, and then dividing by the live counting time. A linear base line is assumed, going from the channel located at the trough of the Compton-edge dip to the left (lower-energy) side of the photopeak to an arbitrarily selected channel on the descending right (higher-energy) side of the photopeak. The area of this base is then $n(N_L + N_R)/2$, where in this case N_L and N_R are the numbers of counts in these two channels, and n is the number of analyzer channels included in the photopeak. The precision of the value obtained for the Compton + background region is determined by the σ's of these two channels, σ_{N_L} and σ_{N_R}. The standard deviation of the Compton + background region is thus $\pm n(N_L + N_R)^{1/2}/2$. The standard deviation of the gross photopeak counts, N_{gpp}, is simply $\pm N_{gpp}^{1/2}$, where N_{gpp} is the sum of the observed counts in the channels included in the photopeak. The desired net photopeak counts, N_{npp}, are obtained by subtracting the base counts from the gross counts, that is,

$$N_{npp} = N_{gpp} - \frac{n(N_L + N_R)}{2} \tag{20}$$

The standard deviation of N_{npp} is equal to the square root of the sum of the squares of the standard deviations of the two quantities:

$$\sigma_{Nnpp} = \pm \left[N_{gpp} + \frac{n^2(N_L + N_R)}{4} \right]^{1/2} \tag{21}$$

Then, N_{npp} and σ_{Nnpp} can both be converted to the corresponding count-ing-rate values by dividing each by the live counting time.

When the net counting rate of a sample is compared with the net counting rate of a standard (same radionuclide, corrected to the same decay time, same counting conditions), the standard deviation of the final answer will depend on the standard deviations of the two quantities. However, in the comparator method, one divides the sample net counts per minute by the standard net counts per minute and then multiplies this quotient by the number of micrograms of the element in the stan-dard, in order to obtain the micrograms of the element in the sample. In the multiplication of two or more quantities, each having its own standard deviation, or in the division of one quantity by another, the fractional or percentage standard deviation of the result is equal to the square root of the sum of the squares of the individual fractional or percentage standard deviations (with n being the number of terms multiplied together):

$$\%\sigma_{\text{product}} = \pm \left[\sum_1^n (\%\sigma_i)^2 \right]^{\frac{1}{2}} \tag{22}$$

and

$$\%\sigma_{\text{quotient}} = \pm[(\%\sigma_1)^2 + (\%\sigma_2)^2]^{\frac{1}{2}} \tag{23}$$

Often, of course, the $\%\sigma$ of the standard is negligible, compared with that of the sample (if the sample is of very low activity, and the stan-dard is of high activity). The assumption is that the amount of the element present in the standard is known very accurately, so that it can be taken as an absolute value (i.e., one of negligible uncertainty).

It should be noted that the σ obtained from a single counting of a sample merely means that the counting statistics indicate that the true value probably falls within $\pm 1\sigma$ of the value found—if all other sources of error are negligible, compared with the uncertainty introduced by the counting statistics. In other words, the true uncertainty, expressed as a standard deviation, is at least that large, and may be larger if other sources of variation are significant. One can multiply the σ obtained by various factors, to express the result to greater confidence levels (C. L.), that is, 90% C. L. = value $\pm 1.645\sigma$, 95% C. L. = value $\pm 1.960\sigma$, 99% C. L. = value $\pm 2.576\sigma$, 99.9% C. L. = value $\pm 3.291\sigma$.

If one repeats the analysis of a given sample a number of times, the uncertainty, $\sigma_{\bar{R}}$, of the mean value, $\bar{R}$—including all sources of variation (not just the counting statistics)—can be determined by taking the square root of the sum of the squares of the individual deviations from the mean value, and dividing this by the square root of $(n - 1)$, where n is in this case the number of individual determinations:

$$\sigma_{\bar{R}} = \pm \frac{\left[\sum_{1}^{n} (\bar{R} - R_i)^2\right]^{1/2}}{(n - 1)^{1/2}} \tag{24}$$

VI. APPLICATIONS OF ACTIVATION ANALYSIS

For a suitably high flux, activation analysis in one or more of its forms is capable of quantitatively determining almost every element in the periodic system at levels ranging all the way from gross levels down to submicrogram amounts. For about half of these elements, even subnanogram amounts can be determined. As a result of this ultrasensitivity, its wide dynamic range, and the wide range of sample sizes that can be accommodated, the method has found or is finding extensive use in almost every branch of science, industry, and medicine. When the method can be employed in its purely instrumental form, the speed of analysis and the nondestructive feature can also be of real value in its applications to various fields.

It is beyond the scope of this condensed treatment of the subject of activation analysis to review the large number of applications of the method that have been made to date in various fields. Especially the high-flux (reactor) thermal-NAA method has found many uses in such fields as chemistry (organic, inorganic, polymer), physics (very pure materials and semiconductors), metallurgy, industry (petroleum, chemical, mining, plastics, metals, rubber, solvents, paper, lumber, cement, etc.), biology, medicine, geochemistry, oceanography, archeology, and criminalistics (scientific crime investigation). Since activation analysis really determines specific stable nuclides of a polyisotopic element, forming different products from each one, the method can also be used in studies of isotopic fractionation and in stable-isotope tracer studies.

An extremely valuable reference source on activation analysis and its applications is *Activation Analysis: A Bibliography*, edited by G. J. Lutz et al. (12). This bibliography gives complete identifications of some 4000 publications in this field (papers, books, reports, etc.). Although all forms of activation analysis are included, of course the great majority of the references listed are concerned with high-flux thermal-NAA, since this is the most generally powerful and most extensively developed and applied form of the method. The bibliography is excellently cross-indexed according to elements, matrices, authors, techniques, and so on.

For more complete discussions of many of the topics mentioned briefly in this chapter, the reader is referred to the books by Bowen and Gibbons (2), Lyon (13), Lenihan and Thomson (10), Taylor (15), Koch (8), and Albert (1). Biomedical applications are discussed in detail in the book edited by Comar (4), and forensic (crime investigation) applications in the Proceedings edited by Guinn (6). The *Proceedings of the 1961 [and 1965] International Conference on Modern Trends in Activation Analysis*, edited by Wainerdi and Gibbons (16,17), contain a large number of excellent papers on the subject of activation analysis and its applications. An excellent compilation of all possible reactor thermal-neutron and fission-spectrum neutron reactions that form gamma-emitting products, and their yields, has been prepared by Lukens (11). The General Electric *Chart of the Nuclides* (5) is a very convenient and useful source of information. Valuable information on radiochemical-separation procedures is contained in the National Research Council *Radiochemistry Monographs* series (14). The very useful *Gamma-Ray Spectrum Catalogue* prepared by Heath (7) has already been cited, as has the invaluable "Table of Isotopes" prepared by Lederer et al. (9).

REFERENCES

1. Albert, P., *l'Analyse par Radioactivation,* Gauthier-Villars, Paris, 1964 (in French).
2. Bowen, H. J. M., and D. Gibbons, *Radioactivation Analysis,* Oxford University Press, London, 1963.
3. Buchanan, J. D., *Atompraxis,* **8,** 272 (1962) (also available as *Gen. At. Rept.* GA-2662, General Atomic, San Diego, Calif., 1961).
4. Comar, D., ed., *l'Analyse par Radioactivation et Ses Applications aux Sciences Biologiques,* Presses Universitaires de France, Paris, 1964 (mostly in English).
5. General Electric Company, *Chart of the Nuclides,* San Jose, Calif.
6. Guinn, V. P., ed., "Proceedings of the First International Conference on Forensic Activation Analysis," General Atomic Report GA-8171, General Atomic, San Diego, Calif., 1967.
7. Heath, R. L., *Scintillation Spectrometry Gamma-Ray Spectrum Catalogue,* 2nd Ed. (2 volumes, IDO-16880-1 and -2), Office of Technical Services, U.S. Department of Commerce, Washington, D.C., 1964.
8. Koch, R. C., *Activation Analysis Handbook,* Academic, New York, 1960.
9. Lederer, C. M., J. M. Hollander, and I. Perlman, *Table of Isotopes. Sixth Edition,* Wiley, New York, 1967.
10. Lenihan, J. M. A., and S. J. Thomson, eds., *Activation Analysis: Principles and Applications,* Academic, London, 1965.

11. Lukens, H. R., "Estimated Photopeak Specific Activities in Reactor Irradiations," *Gen. At. Rept.* GA-5073, General Atomic, San Diego, Calif., 1964.

12. Lutz, G. J., R. J. Boreni, R. S. Maddock, and W. W. Meinke, *Activation Analysis: A Bibliography,* National Bureau of Standards Technical Note 467, 1968.

13. Lyon, W. S., Jr., ed., *Guide to Activation Analysis,* Van Nostrand, Princeton, N.J., 1964.

14. National Research Council, *Radiochemistry Monographs* (series of monographs by different authors in the *NRC Nucl. Sci. Ser.*), Washington, D.C.

15. Taylor, D., *Neutron Irradiation and Activation Analysis,* Van Nostrand, Princeton, N.J., 1964.

16. Wainerdi, R. E., and D. Gibbons, eds., *Proceedings of the 1961 International Conference on Modern Trends in Activation Analysis,* Texas A & M University, College Station, Tex., 1961.

17. Wainerdi, R. E., and D. Gibbons, eds., *Proceedings of the 1965 International Conference on Modern Trends in Activation Analysis,* Texas A & M University, College Station, Tex., 1965.

18. Yule, H. P., and V. P. Guinn, in *Radiochemical Methods of Analysis,* International Atomic Energy Agency, Vienna, 1965, pp. 111–122.

SECTION E: Application of Measurement

Part I
Section E

Chapter 99

COMBINATION OF PHYSICAL AND CHEMICAL PROPERTIES FOR CHARACTERIZATION AND ANALYSIS

BY R. W. KING, *Sun Oil Company, Research and Development Division, Marcus Hook, Pennsylvania*

Contents

I. INTRODUCTION

Over the years the analytical applicability of physical property measurements, either alone or in combination with chemical methods, has been amply demonstrated. The use of physical properties in analytical chemistry has been practiced primarily in two connections, namely, for the rapid analysis of relatively simple systems such as binary or ternary mixtures, and for the analysis of extremely complex mixtures of molecules of similar type and size. The first application is one with which most chemists are fairly well acquainted. For example, we are all familiar with refractive index–composition diagrams and aware that the concentration of alcohol–water mixtures may be determined by using a hydrometer. In the analysis of these simple systems, physical property methods contribute rapidity, convenience, and, in many cases, a high order of accuracy. However, physical methods can be used to greatest advantage for the characterization and analysis of relatively unreactive, complex systems for which chemical techniques are for the most part inapplicable or incapable of distinguishing the subtle differences in structure that exist. For this reason it seems desirable to refresh the reader in regard to some of the elementary principles involved by a brief treatment of simple systems, and then plunge at once into the more formidable, but basically similar, applications of physical properties for the analysis of complex mixtures.

A. ANALYSIS OF SIMPLE SYSTEMS

The analysis of simple binary and ternary systems using physical properties is based on the observation that many of the conveniently measured properties of liquids are more or less additive; that is to say, a plot of the composition of a simple binary mixture against a physical property such as refractive index or density can often be represented by a smooth curve or straight line connecting the values of the chosen property for the pure components. For mixtures of two unassociated liquids the relation between composition and the measured property is usually very nearly linear. Mixtures of two associated liquids, however, are likely to exhibit substantial departures from linearity, as are mixtures involving an associated and an unassociated liquid. In some cases the relation may actually show a point of inflection, being concave upward over a portion of the composition range, and concave downward over the remainder. This type of behavior seems to be characteristic of the relations between composition and refractive index, and composition and density, for mixtures of glycols and water (53). Since the behavior of a given system is generally unpredictable, it is necessary, in such applications of physical properties, to establish a calibration curve by measuring the desired property for a series of blends carefully prepared by weight and covering the composition range of interest. Typical calibration curves for the analysis of several binary systems by means of refractive index are shown in Fig. 99.1 (30).

The accuracy of analyses based on property measurements is limited by the precision with which the property may be experimentally determined, and the difference between the values of the property for the

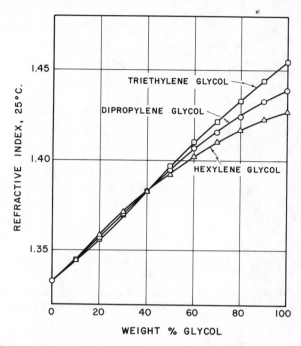

Fig. 99.1. Relation between refractive index and composition for aqueous solutions of glycols. Reproduced with the permission of *Analytical Chemistry*.

pure components, assuming that departures from linearity are not excessive. For example, a binary system in which the refractive indices of the components differ by 1 in the third decimal could hardly be analyzed with adequate accuracy if the refractometer available permitted reading to 1 or 2 units in the fourth decimal. On the other hand, a differential refractometer capable of measuring differences to within 1 or 2 units in the sixth decimal would make analysis of this system by refractive index measurement perfectly feasible.

The choice of the property to be used for analysis is affected by three considerations. First, it should be capable of being determined conveniently, rapidly, and precisely. Second, it must be chosen so that the desired accuracy is attainable. Finally, it must exhibit sufficient additivity that the accuracy is not impaired for certain areas of composition. For example, although refractive index is useful for analysis of aqueous mixtures of the three glycols shown in Fig. 99.1 over the entire range, the density–composition relationship is considerably less attractive for analytical purposes. The curve for dipropylene glycol shown in Fig.

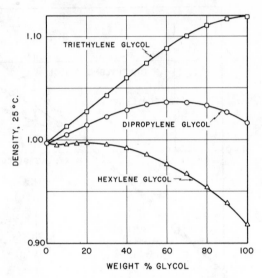

Fig. 99.2. Relation between density and composition for aqueous solutions of glycols. Reproduced with the permission of *Analytical Chemistry*.

99.2 (30) exhibits a pronounced maximum. In the vicinity of the maximum the change in density with composition is relatively slight; hence analysis by density is not satisfactory. However, below about 25% dipropylene glycol, density may be used to determine the glycol content to within about 0.15%. Triethylene glycol and hexylene glycol suffer similar limitations in the region above 70% and below 30%, respectively.

Although density and refractive index are the properties most commonly employed for the analysis of binary mixtures of liquids, there are obviously no limitations, save those already set forth, on the physical constant that can be chosen. Refractive dispersion, viscosity, surface tension, and boiling point are other possibilities. Often functions based on combinations of properties may be used to advantage since many of these more nearly exhibit true additivity than do the individual physical properties. Specific dispersion, refractivity intercept, specific refraction, and the specific parachor are all useful candidates for analytical application.

The analysis of binary mixtures of solids can often be accomplished by simple measurement of the melting point if the phase diagram of the system has been previously determined. The behavior of naphthalene–biphenyl mixtures shown in Fig. 99.3 is typical of many organic compounds. The two components form practically ideal solutions over

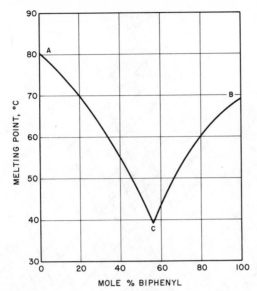

Fig. 99.3. Melting point–composition diagram for mixtures of naphthalene and biphenyl.

the entire range of composition; that is, the freezing point lowering is proportional to the mole fraction of the solute present. The curve *AC* shows the lowering of the melting point of naphthalene by admixture with biphenyl, and *BC* the lowering of the melting point of biphenyl by naphthalene. The intersection *C* is a eutectic point at which temperature both crystalline solids, *A* and *B*, can exist in equilibrium with a melt of fixed composition. Once the melting point–composition diagram has been established the system is amenable to analysis by melting point measurement. Ambiguities can be resolved by redetermining the melting point after adding a small amount of one of the pure components.

Unfortunately, the accuracy of this type of analysis is often rather poor since the capillary melting point is, in many instances, affected by the previous thermal history of the mixture. Sophisticated apparatus for the measurement of the freezing point, wherein equilibrium between solid and liquid phases is attained, can of course be used to improve the accuracy. However, considerably larger samples are ordinarily required, and much of the advantage afforded by a rapid determination of the capillary melting point is lost. If the proper equipment is available, the freezing point method can also serve for the analysis of mixtures of materials that are normally liquids at room temperature. For example,

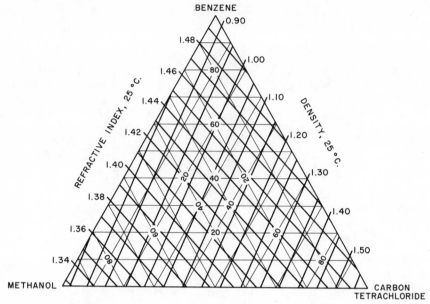

Fig. 99.4.   Ternary analytical diagram for the system carbon tetrachloride–methanol–benzene.

the concentration of *p*-xylene in mixtures of certain other aromatic hydrocarbons can be determined with reasonable accuracy by using a platinum resistance thermometer and the equipment described by Glasgow, Streiff, and Rossini (64).

The analysis of ternary mixtures by physical property methods is also quite feasible. In these cases it often happens, however, that several avenues need to be considered. If the three components of the system are completely miscible in all proportions, it is frequently possible to choose two properties for which the individual components exhibit sufficiently different values to permit an analysis. For example, in the ternary system carbon tetrachloride–methanol–benzene (197), the density is largely a measure of the fraction of carbon tetrachloride and the refractive index a measure of the fraction of methanol. The relation between composition and the density and refractive index may be established by measuring these properties for solutions of known composition.

These data may be used to develop simultaneous equations from which the composition of an unknown mixture may be unambiguously derived, or they may be used to construct a ternary composition diagram on triangular coordinate paper. An example of the latter is shown in Fig.

99.4. An alternative mode of presentation is to construct a composition grid on rectangular paper with the two measured properties as coordinates. An analytical chart of this type for the system acetone–methyl ethyl ketone–water is shown in Fig. 99.5 (175). In constructing this chart, plots were first made of refractive index against weight per cent methyl ethyl ketone with weight per cent water as the third parameter, and of density against weight per cent water, with weight per cent methyl ethyl ketone as the third parameter. By referring to these two plots, the grid of Fig. 99.5 was ruled in. If the spread in index and density is insufficient to produce the accuracy desired, or if there is evidence of paralleling of these properties as composition is changed, other combinations of physical constants should be investigated. For example, the system n-hexadecene–naphthalene–n-tetradecane cannot be analyzed by refractive index and density, but yields readily to a combination of density and boiling point (227).

In the analysis of ternary systems, advantage can sometimes be taken of the fact that one of the components may be conveniently removed by extraction and determined by difference, and the binary remaining analyzed by a single property measurement. For example, the system ethylbenzene–ethylcyclohexane–hexylene glycol can be analyzed by ex-

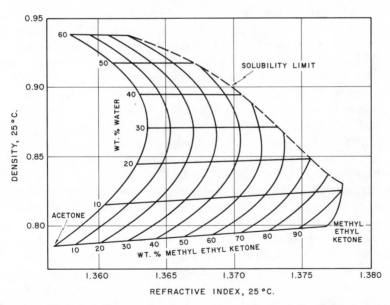

Fig. 99.5. Analytical diagram for the system acetone–methyl ethyl ketone–water. Reproduced with the permission of *Industrial and Engineering Chemistry.*

tracting the glycol with water, carefully drying the hydrocarbon phase, and determining its composition with respect to ethylbenzene and ethyl-cyclohexane by refractive index measurement (181).

Many times chemical methods may be used in combination with physical methods for the analysis of ternary mixtures. By way of illustration, many ternary systems made up of a paraffin, a cycloparaffin, and an aromatic hydrocarbon may be analyzed by quantitative sulfonation of the aromatic with a mixture of phosphoric and sulfuric acids. The sul-fonated aromatic is retained in the acid layer, and if the reaction is carried out using carefully measured quantities of acid and sample in a vessel similar to a Babcock bottle, the percentage of the aromatic may be determined by centrifuging and reading the volume of the two layers. An aliquot of the unreacted saturated hydrocarbon phase may then be withdrawn and analyzed by refractive index or some other appropriate physical property.

In combining chemical with physical methods it is not mandatory that the system be reduced to a binary by the chemical removal of one component in order to use a single physical property for the remainder of the analysis. In many cases a family of curves with the chemically measured component as a parameter may be prepared from measurements on mixtures of known composition. Any single curve in this family describes the change in a measured property with a change in the concentration of two of the three components in the system. Once the concentration of the chemically measured component has been determined, the curve corresponding to that for the value established chemically may be entered with the measured property, and the concentration of the second component read off. The concentration of the third component is obtained, of course, by the difference from 100%. For example, Fig. 99.6 illustrates such a family of curves for the system cyclohex-anol–phenol–cyclohexanone (35). In analyzing this system, the concentration of cyclohexanone is first established by chemical means and the refractive index curve corresponding to this value used to obtain the concentration of cyclohexanol. Interpolation between curves may, of course, be necessary. The concentration of phenol may be obtained by difference from 100%.

Ternary systems that are not miscible over the entire range of composition are often encountered in investigations of phase equilibria. In these cases it is generally possible to obtain sufficient analytical information for solution of the problem at hand by determining the value of the chosen property for ternary mixtures of known composition that are saturated with respect to one of the components. These data may

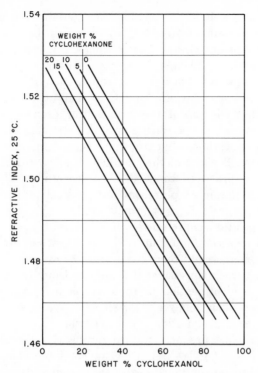

Fig. 99.6. Diagram for the analysis of the system cyclohexanol–phenol–cyclo-hexanone by a combination of chemical and physical methods.

then be used to determine the composition of the conjugate layers that separate when the three components are equilibrated at any composition falling within the area of incomplete miscibility. This device is often employed to establish tie-line compositions in studies of ternary phase equilibria, or for the evaluation of distribution coefficients.

Although physical property analyses of these relatively simple systems have been much practiced by physical chemists, a survey of the recent literature reveals that gas chromatography is becoming an increasingly popular tool for the rapid and precise analysis of such mixtures and may eventually displace physical methods entirely, except for the most conveniently handled binaries.

B. ANALYSIS OF COMPLEX MIXTURES

The same principles that serve for the analysis of simple systems can be used to analyze mixtures that are extremely complex. In this application, however, it is not possible to determine the concentrations of individual compounds. Instead the analysis must be reported in terms of the concentrations of classes of compounds, or certain structural features of the molecules. Although it might appear that chemical techniques would be the methods of choice for the examination of complex mixtures, in certain applications they are singularly inappropriate. The classical methods provide little useful information, for example, when applied to a complex material like petroleum or coal. Petroleum is fundamentally a mixture of hydrocarbons that are quite similar in their physical and chemical behavior. Because of their limited chemical reactivity, physical methods must be practiced to effect even the crudest separation. The exhaustive fractionation of petroleum by physical methods to produce very nearly ·pure hydrocarbons that may be identified from their physical properties or spectral characteristics has met with some success when applied to mixtures containing up to 15 or 20 carbon atoms. However, above this molecular weight there have, with the exception of n-paraffins, been very few hydrocarbons isolated from petroleum. Even after successful isolation, the identification of these compounds is beset with difficulties because of their relatively low chemical reactivity. Degradation techniques are also of questionable value when applied to petroleum or coal. The primary products formed during the degradation process are mixtures of more or less related compounds too complicated to be isolated, which are, on continued degradation, converted into very simple compounds that give little information regarding the original structure. Thus the oxidative degradation of coal produces carbon dioxide, acetic acid, and relatively simple aromatic acids and water.

It is apparent, then, that for a number of complicated mixtures that must be dealt with in modern technology the application of chemical analytical methods is insufficient to provide the necessary information for satisfactory scientific research. Reliable chemical methods for a detailed analysis of many complex mixtures are simply not available, and therefore the composition of such materials will remain unknown if we limit ourselves to the application of chemical techniques. It is for this reason that physical methods of analysis have been introduced into many fields of scientific endeavor. Advantage is taken of the fact that relationships exist between the physical constants (and functions thereof) and the chemical structure and elemental composition. In many cases, physical methods can be combined with chemical methods to provide substantially more information than can be derived from the application of either approach by itself.

In considering the mass of work that has been carried out to relate physical constants to chemical structure, it is immediately apparent that systems that are predominantly hydrocarbon in nature have been those most thoroughly studied. This is very probably a natural consequence of the fact that such systems are the most difficult to analyze by chemical methods. Only in recent years has the power of physical techniques been brought to bear on diverse materials such as acids, silicone oils, and glasses. It is therefore appropriate that any consideration of basic principles should reflect in scope and character the large amount of work related to hydrocarbon systems. The outstanding success with which physical methods have been applied to these materials makes them particularly useful as examples of the way in which analytically significant relationships between physical constants and composition may be developed. The reader should remember, however, that in the majority of cases the underlying principles are applicable to other organic and inorganic mixtures as well.

II. PHYSICAL PROPERTIES OF HYDROCARBONS

It is an almost impossible task to cover thoroughly all of the physical constants of hydrocarbons that have been measured. In this section only the properties that are potentially useful for characterization and analysis and are measurable in the liquid state are reviewed. The purpose of such a review is not particularly to present a compilation of accumulated data, but rather to show how such data may be related to chemical structure by means of theoretical or empirical relationships. The derivation of such relationships is extremely important to progress in the analysis of hydrocarbon mixtures. Synthesis and analysis go hand in hand, yet only a very small proportion of the theoretically possible hydrocarbons has ever been synthesized. Nevertheless, substantial progress has been made, especially in regard to the lower-boiling members of homologous series (190). A great many high-molecular-weight hydrocarbons have also been prepared, and their properties measured (203). However, the correlation of physical constants with structure has played, and will continue to play, a major role in (1) predicting the properties of non-synthesized members of a series, and (2) providing a reliable basis for analyzing hydrocarbon mixtures, especially those of high molecular weight, for which only a few data have been determined experimentally.

The approaches that have been taken to quantitatively express the variation of a physical property as a function of structural parameters can be conveniently separated into three categories (66): (1) total property group contribution methods, (2) homologous series methods, and (3) isomeric variation methods, in which the isomeric increment in the value of a property is correlated, rather than the total value of the property for each isomer.

Total Property Group Contribution Methods. An important type of correlation is the generalized one in which variations in properties due to isomer effects are neglected. These correlations usually express the value of a property as a function of the number and kinds of atoms and functional groups in the molecule. These methods are most useful for properties that vary linearly with molecular weight, such as molar refraction and molar volume.

Homologous Series Methods. A number of investigators have developed equations to express the physical properties of normal paraffin hydrocarbons and other normal series. These equations exhibit a wide variety of forms but are generally functions of a single variable—the number of carbon atoms in the molecule or a closely related quantity. Since only a single variable is involved, the variation in the property studied for members of an homologous series is amenable to simple graphical representation that can be employed for interpolation or extrapolation.

Isomeric Variation Methods. In this approach the incremental difference in the value of the property for each isomer is correlated, rather than the total value of the property. The incremental method has generally been used to evaluate the effect of geometry on the properties of a set of isomers.

The difficulty in developing relations between the properties of hydrocarbons and their structure lies in the fact that the simple relations are not sufficiently accurate and the accurate ones are too cumbersome to be applied routinely. Undoubtedly the increased use of high-speed computers will provide methods of calculation that are both accurate and convenient to apply.

A. DENSITY

It has long been assumed that the molecular volume of hydrocarbons was an additive function and could be built up of structural increments. A number of investigators have attempted to develop simple equations, based on the total property group contribution concept, that would permit the calculation of molecular volume, and therefore density, from structural parameters. The molecular volumes of the higher-molecular-weight n-paraffins (C_{10}–C_{40}) may be satisfactorily represented by the equation

$$\text{Mol. vol. } (20°C; 1 \text{ atm}) = 16.38N_1 + 30.8 \qquad \text{ml/g-mol}$$

where N_1 is the number of carbon atoms per molecule (104). American Petroleum Institute data for the n-paraffins from C_{10} to C_{40} were used to derive the chain carbon increment 16.38 shown in the equation. However, if one extrapolates to very high-molecular-weight paraffins, the sum of the CH_2 increments becomes so much greater than the constant term in this equation that the latter may be neglected. This leads to the concept of a limiting density, which can be expressed numerically as 14.026 divided by the limiting increment of volume per CH_2 group. Smittenberg has estimated the limiting density as 0.8513, which corresponds to a limiting CH_2 increment of volume of 16.47. Kurtz and Lipkin (109) have given the following formula for the calculation of the molecular volume of average paraffins:

$$MV \ (20°C) = 16.28N_1 + 31.2$$

These authors have suggested the following equation as a relationship of general utility for saturated hydrocarbons:

$$MV \ (20°C) = 16.28N_1 + 13.15N_2 + 9.7N_3 + 31.2$$

where N_1 = the number of chain carbons, N_2 = the number of cyclo-paraffin ring carbons, and N_3 = the number of ring junction carbons. For all hydrocarbons, Kurtz and Sankin (112) have proposed the formula

$$MV \ (20°C) = 16.28N_1 + 13.15N_2 + 9.7N_3 - 6.2N_4 + 31.2$$

where N_1 = the number of chain carbons, N_2 = the number of ring carbons, N_3 = the number of ring junction carbons, and N_4 = the number of double bonds.

It should be emphasized that, although this equation is quite useful, especially for the prediction of the molecular volume and density of the higher-molecular-weight hydrocarbons, it cannot take isomer effects into account. This simple linear formula often yields identical values for certain branched hydrocarbons that are in disagreement with experimental data. Fortunately, isomer effects become of less consequence at high molecular weights. A somewhat modified form of this equation has been reported to give more accurate results for certain classes of hydrocarbons (15).

Homologous series methods yield attractively simple and accurate equations for n-alkyl series of hydrocarbons. Smittenberg and Mulder

(210) have suggested that, in many cases, the physical constants of hydrocarbons belonging to an homologous series may be represented by an empirical relationship of the form

$$x = x_\infty + \frac{k'}{M + z'} \quad \text{or} \quad x = x_\infty + \frac{k}{c + z}$$

In these equations, x is the physical constant of a hydrocarbon of c carbon atoms or M molecular weight, x_∞ is the limit of this physical constant at an infinite number of carbon atoms, and k and z are empirical constants characteristic of a particular homologous series. These equations represent the line defined by plotting the physical constant considered against the reciprocal of the number of carbon atoms (or molecular weight), where x_∞ is the intercept at $1/c = 0$ (infinite number of carbon atoms), k is the slope, and z the curvature. This form of equation has also been found to represent adequately the properties of homologous series of alkanethiols and thiaalkanes (247). Values of these constants for predicting the densities of several homologous series of hydrocarbons are shown in Table 99.I.

Li and his associates (117) have proposed similar equations for the molar volume at 25°C for various homologous series of normal alkyl compounds. The general formula for such homologous series may be expressed as:

$$Y—(CH_2)_m—H$$

where m represents the number of carbon atoms in the normal alkyl chain, and Y represents a given end group containing one or more carbon atoms. For example, Y may be a methyl group, a cyclopentyl group,

TABLE 99.I

Values of the Constants in the Equation $d = d_\infty + [k/(c + z)]$
for the Density at 20°C of n-Alkyl Series of Hydrocarbons
as a Function of the Number of Carbon Atoms (210)

Type of hydrocarbon series	Values of constants		
	d_∞	k	z
n-Alkanes	0.8513	−1.3100	0.82
n-1-Alkenes	0.8513	−1.1465	0.44
n-Alkylcyclopentanes	0.8513	−0.5984	0.00
n-Alkylcyclohexanes	0.8513	−0.5248	0.00
n-Alkylbenzenes	0.8513	0.0535	−4.00

TABLE 99.II

Values of the Constants in the Equation $MV = V_0 + a_v m + b_v/(m - 1)$
$+ c_v/(m - 1)^2$ for Molar Volume at 25°C for n-Alkyl Series of
Hydrocarbons Represented by the Formula Y—$(CH_2)_m$—H (117)

Type of hydrocarbon series	Values of constants, ml/mole			
	V_0	a_v	b_v	c_v
n-Alkanes	45.82233	16.4841	14.56329	-4.56336
n-1-Alkenes	57.08054	16.4841	10.37057	-5.33246
n-Alkylcyclopentanes	95.80176	16.4841	-0.74372	1.64148
n-Alkylcyclohexanes	110.53675	16.4841	-0.81676	1.02295
n-Alkylbenzenes	91.99335	16.4841	-5.03136	4.71845

a phenyl group, a vinyl group, and so on, depending on the series. On
the basis of this description of such series, the equation for molecular
volume takes the form

$$MV \ (25°C) = V_0 + a_v m + \frac{b_v}{m - 1} + \frac{c_v}{(m - 1)^2}$$

where m is the number of carbon atoms in the normal alkyl chain,
and V_0, a_v, b_v, and c_v are constants. The constant a_v has the value
16.4841 ml/mole for all normal alkyl series. Values of the constants
for five normal alkyl series of hydrocarbons are given in Table 99.II.
This equation has also been applied to n-alkyl series of thiols, alcohols,
and carboxylic acids. Errors in the densities calculated using this equation are of the same order of magnitude as uncertainties in the experimental data.

For approximate work, a simple graphical representation of the data
for homologous series will often permit adequate interpolation or extrapolation. A convenient way to present such data is to plot the value
of the property against the reciprocal of the number of carbon atoms
plus 1. Data plotted in this way for a number of hydrocarbon series
are shown in Fig. 99.7. Generally, almost straight lines are obtained
that converge at a common point of intersection, the value for an
n-paraffin of infinite molecular weight. The number of reliable data on
homologous series is somewhat limited. However, if the densities of a
few of the lower-molecular-weight members of a series are available,
it is possible to use the concept of a limiting density to estimate in
an approximate way the densities of higher-molecular-weight members
of the series. This can be done by plotting the known values against

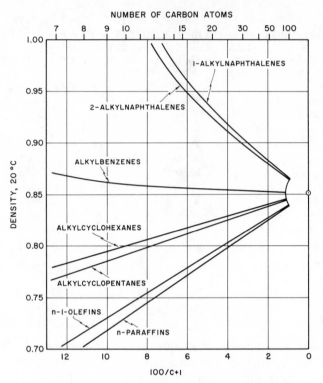

Fig. 99.7. Density of n-alkyl series of hydrocarbons.

$1/(c+1)$ and connecting them, as well as one can, with a straight line passing through the limiting density (0.8513 g/ml) at $1/(c+1) = 0$.

The relationship between the molecular volume (or density) of the normal paraffins and that of their branched isomers is not, as yet, well understood. It has been established that the introduction of a methyl group in a paraffin chain has a marked effect on the molecular volume, and that the magnitude and direction of this effect depend on the position of the substituent in the chain. Substituents near the end of a chain increase volume; substitutions near the center decrease volume (relative to the normal compound). These effects are reasonable if considered in the light of rotation of carbon–carbon bonds. If free rotation is interfered with, there will be a decrease relative to the normal compound. However, a branch on the second carbon atom can rotate with this carbon atom to sweep out more volume than if it were on the end of the chain. The isomer variation approach has been the most successful in predicting the properties of branched paraffins. Francis (57,58) has calculated the densities of all possible octanes, nonanes, decanes, and undecanes from the densities of the next lower paraffins. Wiener (237,238) developed a simple procedure for calculating the magnitude of internal interference by various structures. His procedure has been used by Platt (177) for the calculation of molecular volume. The Wiener-Platt concept emphasizes the importance of the number of pairs (p) of carbon atoms three bonds apart and of the Wiener number (w), which is obtained by summing the products of the number of carbon atoms on each side of each bond in the molecule. The Wiener number consequently is a measure of the number of bonds between *all* pairs of carbon atoms in the molecule.

The Wiener-Platt equation for calculating the difference in molecular volume between the n-hydrocarbon and its corresponding branched isomer is

$$\Delta V = -2.256 \, \Delta p - 10.95 \, \Delta w/c^2$$

where Δp and Δw are the Wiener increments, that is, the difference in the values of p and w between the normal hydrocarbon and its branched homolog, and c is the number of carbon atoms. Application of this equation to the branched hexanes and heptanes is shown in Table 99.III.

Recently the Wiener-Platt concept has been applied to the calculation of the molar volume (and consequently the density) of a number of higher-molecular-weight branched paraffins (66). Although the calculation can be tedious when applied to molecules containing large numbers of carbon atoms, it represents the best correlative approach available at the present time.

Few investigations into the density of multisubstituted cycloparaffins have been reported. The density of complex cycloparaffins is probably best determined by calculation of the molar volume, using the general equation of Kurtz and Lipkin for saturated hydrocarbons. The values so calculated will in general agree within 1 or 2 ml/mole with experimental data.

Relations for calculating the density of olefins and aromatics have also received little attention. The following simple formula has been proposed for alkylbenzenes (55):

$$d_4^{20} = 0.850 + (a + b + 1) \cdot \frac{0.12}{c}$$

where c = the total number of carbon atoms in the compound, a = the number of pairs of adjacent alkyl groups, and $b = 1$ for mono-*tert*-alkyl-

TABLE 99.III

Calculation of the Difference in Molecular Volume between Branched
and Normal Paraffins Using the Wiener-Platt Equation (104)

Compound	Molecular volume	w	Δw	p	Δp	Branched minus normal	
						ΔV(calc'd)	ΔV(expt'l)
n-Hexane	130.69	35	0	3	0	—	—
2-Methylpentane	131.93	32	−3	3	0	+0.91	+1.24
3-Methylpentane	129.72	31	−4	4	+1	−1.04	−0.97
2,2-Dimethylbutane	132.74	28	−7	3	0	+2.13	+2.06
2,3-Dimethylbutane	130.24	29	−6	4	+1	−0.43	−0.45
n-Heptane	146.54	56	0	4	0	—	—
2-Methylhexane	147.66	52	−4	4	0	+0.89	+1.12
3-Methylhexane	145.82	50	−6	5	+1	−0.92	−0.72
3-Ethylpentane	143.52	48	−8	6	+2	−2.72	−3.02
2,2-Dimethylpentane	148.70	46	−10	4	0	+2.23	+2.16
2,3-Dimethylpentane	144.15	46	−10	6	+2	−2.28	−2.39
2,4-Dimethylpentane	148.95	48	−8	4	0	+1.78	+2.41
3,3-Dimethylpentane	144.53	44	−12	6	+2	−1.83	−2.01
2,2,3-Trimethylbutane	145.19	42	−14	6	+2	−1.38	−1.35

benzenes, and 0 for other aromatics, including those with a tertiary alkyl group.

The accuracy of this equation is reasonably good for n-alkylbenzenes but leaves much to be desired when applied to other structures. Kurtz has reviewed the problems of expressing molecular volume as a simple additive function for these classes of hydrocarbons (104) and concludes that, considering the present state of our knowledge, application of the general equation may be the best approach, especially for structures of high molecular weight.

B. REFRACTIVE INDEX

Refractive index is an extremely attractive constant for use in the characterization of hydrocarbons. Its measurement is simple and accurate, and only small quantities of material are required. The refractive index is normally given for the wavelength of sodium light at 20°C.

As long ago as 1893, Eykman (51) reported the additivity of the molar refraction of CH_2 groups in homologous series, and showed that this property could be expressed as a simple function of carbon number. Following Eykman's work, a great many linear equations for the molar refraction of homologous series were proposed. A number of these utilized the Gladstone-Dale molar refraction, $M(n-1)/d$, probably because of its relative simplicity. For example, Huggins (83) suggested

$$\text{MR (Gladstone-Dale)} = 2.12 + 7.815n$$

where n = number of carbon atoms, for the sodium D line molar refraction of n-paraffin hydrocarbons at 20°C. Slightly different values for the constants in this equation have been suggested periodically (148,236).

Recent work based on accurate data for normal paraffins suggests that the Lorentz-Lorenz molar refraction

$$\text{MR} = \text{MV}\,\frac{n^2 - 1}{n^2 + 2}$$

is linear with the number of carbon atoms within the significance of the data. This relation may be expressed simply as

$$\text{MR} = R_0 + a_R m$$

where m is the number of carbon atoms in the normal alkyl chain, and a_R is the constant increment per CH_2 group and is the same for different homologous series. Li and his associates have used this equation (117) to represent the sodium D line molar refractions at 25°C of a number of n-alkyl series of hydrocarbons. From consideration of data for n-alkanes and 1-alkenes, a value of 4.64187 ml/mole was selected for a_R and the value of R_0 calculated for each of the homologous series by least-squares methods. The values of R_0 obtained in this fashion are shown in Table 99.IV. The series are considered to be represented by the general formula

$$Y\text{---}(CH_2)_m\text{---}H$$

where m is the number of carbon atoms in the normal alkyl chain and Y represents a given end group containing one or more carbon atoms. The equation is suitable not only for hydrocarbons but also for n-alkyl series of thiols, alcohols, and carboxylic acids.

The index of refraction is, of course, accessible from molar refraction if the molar volume is known or can be calculated. Li's equations for molar volume and molar refraction provide a mutually consistent set of relations that can be used to calculate these two properties. The refractive index can then be obtained by using the relation

$$\frac{n^2 - 1}{n^2 + 2} = \frac{\text{MR}}{\text{MV}}$$

TABLE 99.IV
Values of the Constants in the Equation $MR = R_0 + a_R m$
for Lorentz-Lorenz Molar Refraction at 25°C for n-Alkyl
Series of Hydrocarbons Represented by the Formula
$Y—(CH_2)_m—H$ (117)

Type of hydrocarbon series	Values of constants, ml/mole	
	R_0	a_R
n-Alkanes	6.72066	4.64187
n-1-Alkenes	10.93704	4.64187
n-Alkylcyclopentanes	23.14251	4.64187
n-Alkylcyclohexanes	27.76551	4.64187
n-Alkylbenzenes	26.55060	4.64187

For complex molecules of high molecular weight, the Lorentz-Lorenz molar refraction may be calculated to a reasonable approximation by using the atomic refraction constants tabulated by Eisenlohr (47). The general equation for hydrocarbons given by Kurtz can then be used to calculate the molar volume, and the refractive index derived from these two quantities.

Other equations for molar refraction based on total group property contribution methods have been developed. The following equation for Lorentz-Lorenz molar refraction gives values for hydrocarbons that are very nearly the same as those calculated using the Eisenlohr increments (104):

$$MR = 4.618(N_1 + N_2) - 2.2R + 3.518N_3 - 0.467N_4 - 0.902N_5 + 2.2$$

where N_1 = the number of chain carbons, N_2 = the number of ring carbons, R = the number of independent ring structures per molecule, N_3 = the number of ring junction carbons, N_4 = the number of double bonds, and N_5 = the number of triple bonds.

Equations for homologous series that give refractive index directly have also received considerable attention. Schoorl (204) derived the following relation for n-alkanes:

$$n_D^{20} = 1.4758 - 9.68/M + 87/M^2$$

where M is the molecular weight. Smittenberg and Mulder (210) have proposed an equation for refractive index similar in form to that proposed by them for density. The constants in this equation for several

TABLE 99.V

Values of the Constants in the Equation $n = n_\infty + [k/(c + z)]$
for the Refractive Index at 20°C of n-Alkyl Series of Hydrocarbons as a Function of the Number of Carbon Atoms (210)

Type of hydrocarbon series	Values of constants		
	n_∞	k	z
n-Alkanes	1.4752	−0.6838	0.82
n-1-Alkenes	1.4752	−0.5610	0.44
n-Alkylcyclopentanes	1.4752	−0.3920	0.00
n-Alkylcyclohexanes	1.4752	−0.3438	0.00
n-Alkylbenzenes	1.4752	0.1125	−2.30

homologous series of hydrocarbons are shown in Table 99.V. The deviations from experimental data of values calculated using the formula are of the same order of magnitude as uncertainties in the data. The Smittenberg equation has also been found to adequately represent the data for homologous series of alkanethiols and thiaalkanes.

If the equations of these authors for refractive index and density are compared, it will be seen that, except for n-alkylbenzenes, the denominators in the formulas are identical. A linear relationship must therefore exist between refractive index and density for these series of hydrocarbons. This linear relationship has proved to be quite useful for characterization and analysis and is discussed later in some detail.

The form of many of the equations that have been proposed for refractive index and molar refraction suggests that, if the values of refractive index for homologous series are plotted against the reciprocal of the molecular weight or number of carbon atoms, very nearly straight lines ought to be obtained that intersect at a limiting value for an n-alkyl chain of infinite number of carbon atoms. Figure 99.8, which was prepared by plotting data for a number of homologous series against the reciprocal of the number of carbon atoms plus 1, demonstrates that this is indeed the case. A practical consequence of this fact is that it is possible to estimate the refractive indices of higher-molecular-weight members of homologous series for which only limited data exist. The available data for lower-molecular-weight members are plotted against $1/(c + 1)$, and the best straight line is drawn through these points and the limiting refractive index (1.4752) at an infinite number of carbon atoms $[1/(c + 1) = 0]$.

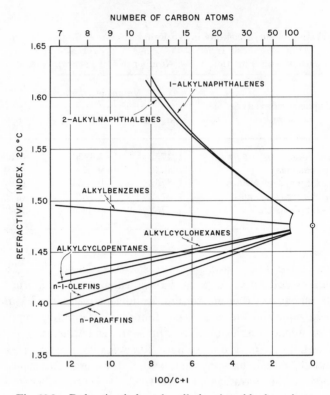

Fig. 99.8.   Refractive index of n-alkyl series of hydrocarbons.

Work related to the calculation of the refractive index of other than homologous series has been extremely limited. The Wiener-Platt concepts of isomeric variation seem to be the most promising approach to the calculation of the refractive indices of the individual paraffin isomers. Greenshields and Rossini have applied this approach to isomers in the 10–30 carbon atom range with marked success (66).

C. REFRACTIVE DISPERSION

The refractive dispersion is defined as the difference between refractive indices measured at two different wavelengths of incident radiation. The data most frequently used represent the F-C dispersion; that is, the difference between the refractive index for the hydrogen blue (F) line at 4861.3 Å and that for the hydrogen red (C) line at 6562.8 Å. For convenience, these values are multiplied by 10^4. It is sometimes necessary to measure the dispersion over other wavelength intervals because of absorption of radiation at the blue end of the spectrum by highly colored samples. If refractive indices are known at other wavelengths, the values for n_C and n_F can be approximated using the simplified Cauchy dispersion equation:

$$n = A + B/\lambda^2 = A + B\nu^2$$

where ν is the wavelength in reciprocal microns. If n_x, n_y, n_u, and n_v are the refractive indices when ν has the values ν_x, ν_y, ν_u, and ν_v, respectively, then

$$n_x - n_y = \frac{\nu_x{}^2 - \nu_y{}^2}{\nu_u{}^2 - \nu_v{}^2}\,(n_u - n_v)$$

This form of the Cauchy equation has no empirical constants that must be evaluated and is therefore quite useful (226). However, it does not adequately represent the properties of samples having relatively high dispersions (219). The Hartmann equation (71).

$$n_\lambda = n_0 + \frac{C}{(\lambda - \lambda_0)^a}$$

is entirely empirical. In this equation, n_λ is the refractive index of the substance for light of wavelength λ, and n_0, C, λ_0, and a are constants. The Hartmann equation is applicable to a large number of hydrocarbons if the exponent a is kept equal to 1.6 (54). Some of the parameters in the Hartmann equation have been found to be constant for certain groups of liquids. However, the calculation of one dispersion from another using this equation necessitates empirical evaluation of the constants for individual liquids, or some knowledge of the composition of the liquids being considered.

Sankin, Martin, and Lipkin (193) have described an empirical relationship between refractive index and wavelength that can be used for the calculation of one refractive dispersion from any other. No evaluation of constants is necessary, and the relation may be applied to both hydrocarbon and nonhydrocarbon liquids. This equation takes the form

$$(n_x - n_y) \times 10^4 = K(n_u - n_v) \times 10^4 + K'$$

where
$$K = \frac{\nu_x{}^3 - \nu_y{}^3}{\nu_u{}^3 - \nu_v{}^3}$$

and
$$K' = 38.0[\nu_x{}^2 - \nu_u{}^2 - K(\nu_u{}^2 - \nu_v{}^2)]$$

For conversion between the two most commonly used wavelength intervals, this reduces to

$$(n_F - n_C) \times 10^4 = 0.718(n_g - n_D) \times 10^4 + 7.5$$

By using this equation, it is generally possible to calculate the F-C dispersion from data obtained for the more easily and conveniently measured g-D interval.

Eby and Klett (46) applied this equation for the calculation of F-C dispersion from the dispersion measured for the mercury g and e lines. They reported that, when checked against Forziati's data for 60 hydrocarbons, the equation gave values that were systematically low by about 1.5 dispersion units for aromatic hydrocarbons and high by about the same amount for saturated hydrocarbons, and they proposed a slight empirical modification. This behavior when the equation was applied to Forziati's data had already been observed by the authors, who felt readjustment unwarranted since these data constituted but a small fraction of those for over 2000 liquids used in deriving the equation. Nevertheless, for very accurate work with low boiling hydrocarbons and their mixtures in the F-C dispersion range of 70–160, the modification of Sankin's equation proposed by Eby and Klett can be recommended. Its application is confined, however, to conversions between the F-C and g-e dispersion intervals. Its use for this purpose is facilitated by the very convenient table supplied by these authors.

The introduction of unsaturation into a molecule generally results in a marked increase in the refractive dispersion. The F-C dispersions of saturated hydrocarbons generally lie between 65 and 80, those of monoolefins between 90 and 95, and those of alkylbenzenes between 140 and 160, while alkylnaphthalenes may have dispersions as high as 300. Phenanthrenes and anthracenes have F-C refractive dispersions in the range of 400–600. This phenomenon makes dispersion an extremely useful tool for the characterization and analysis of hydrocarbons. The change in dispersion with molecular weight for n-alkyl series of hydrocarbons can be represented reasonably well as a straight line if dispersion is plotted against the reciprocal of molecular weight or number of carbon atoms. Figure 99.9 shows the result of plotting the F-C refractive dispersion against $1/(c + 1)$ for a number of homologous series. The common point of intersection of the lines represents the dispersion of a paraffin hydrocarbon of infinite molecular weight. The slopes of the curves are characteristic of each homologous series; the type of molecule can generally be estimated if molecular weight and dispersion are known. The refractive dispersion is frequently divided by the density in order to

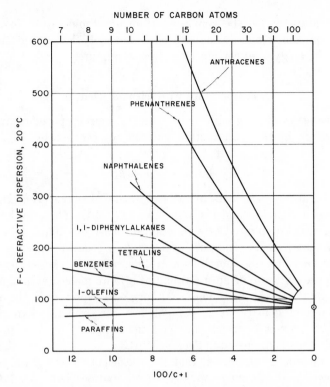

Fig. 99.9.   Refractive dispersion of n-alkyl series of hydrocarbons.

eliminate the effect of this variable. The quotient is termed the specific dispersion. This function is generally preferred in analytical applications and is discussed later.

D. TEMPERATURE COEFFICIENTS OF DENSITY AND REFRACTIVE INDEX

The change in the density and refractive index of hydrocarbons that occurs with a change in the temperature of measurement has been the subject of a number of investigations. Studies of the temperature coefficient of density have assumed a position of importance because petroleum products are sold largely by volume and some means must be provided to reduce the volume measurements to a standard reference temperature. The relationships that have been developed for hydrocarbons and petroleum fractions are of practical utility in the laboratory since they allow experimental observations to be corrected for variations in temperature and to be converted from one temperature to another. In addition, the temperature coefficient of density has proved to be of value in the analysis of certain hydrocarbon mixtures.

1. Temperature Coefficient of Density

The change of density with temperature for normal paraffins and other hydrocarbons in the liquid state can be represented by the general relation.

$$d_4^t = d_4^{20} + \alpha(t - 20) + \beta(t - 20)^2$$

The temperature coefficients of density, α and β, have been studied in some detail by Lipkin and his associates (120,124). These authors have shown that, if α is plotted as a function of density, a linear relationship exists for pure paraffins and pure cycloparaffins. Addition of an n-paraffin side chain of increasing length to a cycloparaffin ring causes the properties to approach those of an n-paraffin of infinite molecular weight along a reasonably straight line. These authors also represented the relation between α and density as a straight line for aromatic hydrocarbons (123). The manner in which these relationships can be used in hydrocarbon analysis is discussed in Section IV.B.3.c.

It has also been established that the coefficients α and β are a reciprocal function of molecular weight for all hydrocarbons (67,120). For molecular weights between 72 and 200, the average values of α and β shown in Table 99.VI have been proposed (120). For molecular weights greater than 200, values of α and β may be derived from the following relations (104,145):

$$\alpha \times 10^5 = -55.3 - \frac{3516}{M + 12} \qquad \text{(for } M > 100)$$

$$\beta \times 10^7 = +4.0 - \frac{941.2}{M} \qquad \text{(for } M > 200)$$

2. Temperature Coefficient of Refractive Index

The refractive index of a liquid depends on the number, charge, and mass of vibrating particles in the matter traveled by the light wave. Since for liquids the primary effect of a change in temperature is a change in density, and consequently a change in the number of vibrating particles per unit volume, a relationship must exist between the temperature coefficient of density and that of refractive index. Thus, if the varia-

TABLE 99.VI
Temperature Coefficients of Density for
Hydrocarbons (120)

Molecular weight	$\alpha \times 10^5$	$\beta \times 10^7$
72	−97.4	−8.1
85	−91.5	−6.4
100	−86.7	−5.0
120	−82.0	−3.7
140	−78.5	−2.6
160	−75.8	−1.8
180	−73.7	−1.2
200	−71.9	−0.7
225	−70.3	−0.2
250	−69.0	+0.2

tion of density is known, the variation of refractive index can be calculated and presents no new aspects. For hydrocarbons, Ward and Kurtz (226) proposed the linear relationship

$$\Delta n/\Delta t = 0.6 \, \Delta d/\Delta t$$

or its equivalent

$$\Delta d/\Delta t = 1.67 \, \Delta n/\Delta t$$

After a study of a number of pure hydrocarbons in the molecular weight range 200–400, van Nes and van Westen (161) proposed a modification of this relationship to

$$\Delta d/\Delta t = 1.71 \, \Delta n/\Delta t$$

The Eykman function (52)

$$(n^2 - 1)/(n + 0.4)d = k$$

represents the experimental data for the effect of temperature on density and refractive index quite accurately for a wide variety of hydrocarbons (105). If the constant k is carefully evaluated, the Eykman function can be used to extrapolate refractive index or density data over temperature intervals as large as 75°C (115). If the constant in Eykman's equation is temperature independent, it follows that the equation has inherent in it terms that define the relation between the temperature coefficients of refractive index and density. Differentiation yields

$$\Delta d/\Delta t = k(n^2 - 1 + 0.8n) \, \Delta n/\Delta t$$

which, upon the insertion of values for k and n for a number of hydrocarbons, gives the approximate relation

$$\Delta d/\Delta t = 1.67 \; \Delta n/\Delta t$$

as proposed by Ward and Kurtz.

It should be emphasized that, although these relations have been found quite useful for converting the refractive indices of mixtures of petroleum hydrocarbons from one temperature to another, they should not be indiscriminately applied to materials containing substantial quantities of nonhydrocarbons or used over extremely wide temperature intervals.

E. HYDROGEN CONTENT OF HYDROCARBONS

A great deal about the structure of an unknown hydrocarbon can often be deduced from its hydrogen content and molecular weight. It is possible from rather simple considerations to develop formulas for homologous series of hydrocarbons that express the relationship between hydrogen content, molecular weight, and a few structural factors characteristic of the type of series under consideration.

We can express the empirical formula for a hydrocarbon in its simplest terms as C_cH_h, where c is the number of carbon atoms and h the number of hydrogen atoms. If the molecule contains no rings or double bonds, that is, is a paraffin, we can write

$$h = 2c + 2 \tag{1}$$

Every ring or unsaturated bond that is introduced into an otherwise saturated structure results in the introduction of a deficiency of $2h$ atoms in the empirical formula. We can express this as

$$h = 2c + 2 - 2(R + Q) \tag{2}$$

where R and Q are used to denote rings and double bonds, respectively. The molecular weight of a hydrocarbon can be expressed generally as

$$M = 12.010c + 1.008h \tag{3}$$

Solving this for c and substituting in equation (2) gives

$$h = \frac{2M - 24.020[(R + Q) - 1]}{14.026} \tag{4}$$

The percentage of hydrogen may be written as

$$\%\mathrm{H} = 1.008h \times \frac{100}{M} \tag{5}$$

Substituting (4) in this equation and reducing the coefficients to their simplest terms gives

$$\%\mathrm{H} = 14.37 - 172.6\frac{[(R + Q) - 1]}{M} \tag{6}$$

which is a general relation for the hydrogen content of a hydrocarbon. Application of this relation is not limited to pure compounds. It is valid for the "hypothetical mean molecule" of a mixture of hydrocarbons, as often encountered in petroleum fractions. A number of interesting conclusions can be drawn from equation (6). These are discussed separately below for saturated and aromatic molecules.

1. Saturated Molecules

For a saturated hydrocarbon, the hydrogen content is an exact measure of the number of rings, because each ring closure involves the loss of two hydrogen atoms. This is true regardless of the number of carbons per ring and irrespective of whether the ring systems are condensed or noncondensed. For saturated hydrocarbons, the term $R + Q$ in equation (6) reduces to R_N, the number of cycloparaffin (naphthene) rings, which yields

$$\%\mathrm{H} = 14.37 - 172.6\frac{R_\mathrm{N} - 1}{M} \tag{7}$$

This implies that for each series of hydrocarbons with the same number of rings there is a linear relationship between the hydrogen content and the reciprocal of molecular weight. Figure 99.10 shows $\%\mathrm{H}$ plotted against $1000/M$ for homologous series of saturated hydrocarbons of different ring numbers. These homologous series all converge at the percentage hydrogen for a paraffin of infinite molecular weight. When $\%\mathrm{H}$ and molecular weight can be measured with sufficient accuracy, this figure provides a reliable way of estimating the mean number of rings of saturated hydrocarbon mixtures. It is of course possible to use equation (7) for this purpose by solving for R_N. For convenience this equation may be rewritten as

$$R_\mathrm{N} = 1 + M(0.08326 - 0.005793\%\mathrm{H}) \tag{8}$$

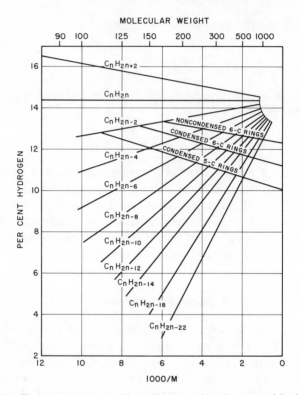

Fig. 99.10. Hydrogen content of homologous series of saturated hydrocarbons.

In addition to homologous series, repetitive series of increasing ring number may also be considered. Several series of this type are also shown in Fig. 99.10.

Reasoning directly analogous to that discussed above is valid also for olefinic unsaturation in noncyclic hydrocarbons. In this case $R + Q$ reduces to Q, the number of olefinic double bonds, and equation (8) becomes

$$Q = 1 + M(0.08326 - 0.005793\,\%\text{H})$$

2. Aromatic Molecules

In aromatic hydrocarbons both rings (R) and unsaturated bonds (Q) are present. However, if an aromatic hydrocarbon sample contains no olefinic unsaturation and no saturated rings, the hydrogen content is, to a degree, an indication of the type of aromatic. In this case, the general equation for hydrogen content [equation (6)] implies that a linear relationship exists between $R + Q$, the total number of rings plus double bonds, and the reciprocal of molecular weight. Figure 99.11 illustrates this relationship for several homologous and repetitive series of condensed aromatic hydrocarbons. Condensed aromatic systems may be built up in several ways, as shown below.

| linear | angular | peri-condensed |

kata-condensed

Figure 99.11 shows that the addition of aromatic nuclei in the linear or angular mode leads to a regular decrease in hydrogen content that approaches the value for an infinitely long chain of condensed benzene rings. On the other hand, creation of homologous series by the addition of n-alkyl groups causes the hydrogen content to approach that of the limiting paraffin along lines of constant $R + Q$. However, addition of nuclei to produce the most compact structures possible results in the gradual production of aromatic clusters whose hydrogen content approaches zero in an asymptotic fashion. The use of this figure obviously cannot lead to unambiguous identification except in very simple or special cases. For example, anthracenes and phenanthrenes are indistinguishable on the basis of $R + Q$, as are chrysenes and benzanthracenes. The figure is of utility in certain applications, however, and similar but more elegant use of hydrogen content has proved of some value in the analysis of coal (103).

F. BOILING POINT

The boiling points of hydrocarbons vary with molecular weight and structure. For a given homologous series the boiling points increase with increasing molecular weight. The influence of structure on boiling point for a series of isomers is, however, a great deal more subtle.

The relationships involving the normal boiling points of compounds of homologous series have been extensively studied (36,76,98). Many of the most successful formulas that have been developed involve a logarithmic function of the number of carbon atoms or molecular weight

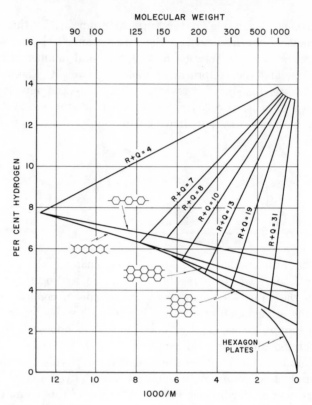

Fig. 99.11. Hydrogen content of series of aromatic hydrocarbons.

(36,76,245). Li (117) proposed a very accurate equation derived from API 44 data for n-alkyl series of hydrocarbons:

$$t_b = t_0 + c_t' \log (m - 1) - d_t \left[\int_{b_t}^{\infty} \exp (-u) d \ln u - \int_{u}^{\infty} \exp (-u) d \ln u \right]$$

where $u = b_t(m - 1)$ and b_t is 0.1505 for all normal alkyl series. The series is represented by the general formula $Y-(CH_2)_m-H$, where Y is an end group containing one or more carbon atoms. Values of the constants t_0, c_t', and d_t are given by the author for six normal alkyl series of hydrocarbons. To facilitate use of the equation, a table of values of the definite integrals for various values of m is also furnished. The equation has also been applied to n-alkyl series of thiols.

Equations of this complexity provide a degree of sophistication that can often be disregarded in ordinary work. It is possible, for example, to fit the data for normal alkyl series reasonably well by means of rather simple relations of the form used successfully by Smittenberg for refractive index and density. Least-squares evaluation of the constants for n-alkanes gives the following relation (162):

$$t_b = 813.1 - \frac{18012}{18.188 + c}$$

where c is the number of carbon atoms and t_b the boiling point in degrees centigrade.

The relation between structure and boiling point has not as yet been satisfactorily explained in terms of general concepts. Attempts have been made to relate boiling point to structure for a number of isomeric series. Francis (57) has calculated with reasonable success the boiling points of the C_8–C_{11} isoparaffins from the boiling points of the next lower paraffin by replacing a hydrogen atom with a methyl group. He has also derived approximate formulas for the boiling points of various series of alkylbenzenes (55).

Incremental difference methods appear to hold the most promise for the calculation of boiling point relationships for groups of isomers. The Wiener-Platt relationships have been used with moderate success for the calculation of the boiling points of paraffin isomers in the range of 5–30 carbon atoms (66). The accuracy to be expected from this type of calculation is indicated in Table 99.VII.

Possibly the most lucid generalized overall picture of the relationship between structure and boiling point for hydrocarbons has been presented by van Nes and van Westen (163). These authors plotted the boiling points for 1000 hydrocarbons against the reciprocal of molecular weight.

TABLE 99.VII

Deviation from the True Value of Boiling Points for Paraffin Hydrocarbons, Calculated Using Isomeric Variation Methods (66)

Range of carbon number	No. of compounds	Property	Avg. dev., °C	Std. dev., °C	Max. dev., °C
C_5–C_9	36	B.p. at 10 mm	±1.5	±1.9	4.9
C_{13}–C_{30}	31	B.p. at 10 mm	±2.8	±3.6	−7.4
C_5–C_{11}	103	B.p. at 760 mm	±1.1	±1.5	−5.7
C_{12}–C_{15}	17	B.p. at 760 mm	±2.5	±3.4	7.9

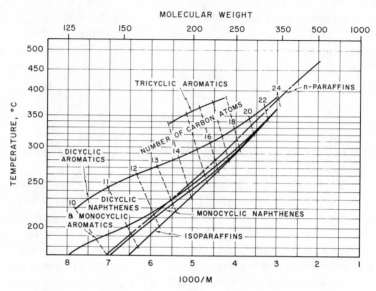

Fig. 99.12. Boiling points of hydrocarbons. Reproduced with the permission of Elsevier Publishing Company.

By judicious averaging and pooling of data, the relationships shown in Fig. 99.12 emerged. Although individual hydrocarbons may exhibit considerable deviation from these curves, this figure makes it possible to draw the following conclusions:

1. The boiling points for average isoparaffins and average mono-cycloparaffins are lower than those of the normal paraffins of equal molecular weight.

2. The curves for the boiling points of average dicycloparaffins and mononuclear aromatics very nearly coincide. Both have somewhat higher boiling points than the corresponding paraffin at low molecular weights. At high molecular weights this situation is reversed, the intersection with the *n*-paraffin curve occurring at about 180–190 molecular weight.

3. The position of the aromatics having three condensed rings is exceptional. Hydrogenation of such compounds to tricycloparaffins causes a decrease of nearly 100°C in boiling point.

4. The figure shows clearly that even very narrow-boiling fractions of hydrocarbons can contain molecules of highly divergent molecular weights, especially if aromatics containing two or more condensed rings are present.

G. VAPOR PRESSURE

The relation between vapor pressure and temperature for hydrocarbons has received the attention of numerous investigators over a period of years (21,36,39,218,240). The Antoine equation (218)

$$\log p = A - B/(t + C)$$

where A, B, and C are constants, p is vapor pressure in millimeters of mercury, and t is temperature in degrees centigrade, has been found to represent experimental data for hydrocarbons with fair accuracy. From the analysts' point of view, use of the relations between vapor pressure and temperature is limited almost entirely to the calculation of the boiling point at atmospheric pressure from the boiling point at reduced pressure. A great many methods for accomplishing this have been suggested (21,39,146,240). For routine work with hydrocarbon mixtures, the nomograph shown in Fig. 99.13 is satisfactory.

H. MELTING POINT

No satisfactory general relationship has been developed as yet for predicting the melting points of hydrocarbons. In general, melting points of members of homologous series tend to rise with increasing molecular weight (49). However, branching has an extremely large influence on melting point. As a general rule, the molecule with the most symmetrical branched structure will exhibit the highest melting point, presumably because of the fact that, when the symmetry of the molecule is high, there are more ways of fitting into a crystal lattice. The molecular weight effect is often totally overcome by the effect of symmetry. It is also true that the less symmetrical a molecule, the more difficult it is to crystallize. Some hydrocarbons exhibit the tendency to produce only glasslike solids upon cooling. Polarizability and dipole moment also affect the melting point because of their effect on the adhesive forces between molecules. The introduction of unsaturated bonds generally results in an increase in these factors, and, a priori, one could predict an increase in melting point. The introduction of unsaturation, however, often results in a decrease in symmetry that far outweighs the effect of adhesive forces. Only if the symmetry is retained does the introduction of unsaturation raise the melting point.

Although differences in melting point can be used to advantage for the separation of certain hydrocarbons, melting point has proved of limited utility for the identification or characterization of hydrocarbon mixtures, with the exception of the n-paraffins.

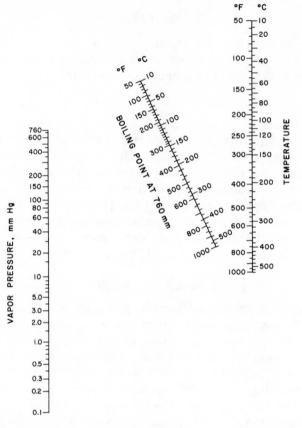

Fig. 99.13.   Boiling point–pressure nomograph for hydrocarbons.

1. Viscosity

Viscosity is a singularly important property of hydrocarbons, especially those of high molecular weight. Data on the viscosities of pure hydrocarbons reveal that this property in general is greatly influenced by the dimensions and shape of the molecule. In discussing viscosity, a distinction must be made between the dynamic viscosity η in centipoises (1 centipoise = 10^{-3} kg/m-sec) and the kinematic viscosity η/d in centistokes (1 centistoke = 10^{-6} m^2/sec). The dynamic viscosity is useful to express absolute forces between liquid layers. However, the use of kinematic viscosity is convenient when examining the motion of liquid substances. Unless otherwise specifically stated, it is the kinematic viscosity that will be referred to in the following discussion.

The viscosities of the normal and branched paraffins have been the subject of work by a number of investigators (82,158,200). Doolittle (44) has measured the viscosity of normal paraffins over a wide range of temperatures and has related viscosity to "free volume." The viscosities of normal and branched paraffins have been discussed by Schiessler (200) and by Cosby and Sutherland (34). For the normal paraffins a close relationship exists between the vapor pressure and the viscosity (221). At their atmospheric boiling points, all n-paraffins have approximately the same absolute viscosity.

A number of attempts have been made to relate viscosity with structure. Kramers (99) found that for solutions of macromolecules the viscosity contribution of a given molecule was proportional to the sum of the squares of the distances between the atoms constituting the molecule and the "center of gravity" of the molecule, if the square of the distance was taken as the average of all possible places. If this concept is applied to the viscosities of pure hydrocarbons, the viscosities of branched paraffins are in fact related to this sum. For molecules with internal branching, the viscosity calculated is in fairly good agreement with the actual value. For hydrocarbons with branching toward the end of the molecule, however, the calculated value is slightly too high. The direction of these results can be explained if we speculate that the shortening of the chain caused by branching decreases the viscosity. However, branching also results in a concentration of carbon atoms near tertiary and quaternary carbons that increases viscosity. As the tertiary or quaternary carbon approaches the center of the molecule, the branching effect is less pronounced and the influence of shortening predominates and is calculable.

Koelbel and his associates (97,130) found that plotting the viscosities at the same temperature of members of homologous series against the number of carbon atoms in the side chain yielded curves which could be represented by "parabolas." In Fig. 99.14, viscosities at 60°C of several homologous series for which data are available in the literature have been plotted in this fashion. For normal paraffins and normal olefins the number of carbon atoms in the "side chain" is predicated on taking butane and pentene, respectively, as a basis. From such curves it is possible to derive the relation

$$\eta = ac^2 + b$$

where a and b are constants, and c is the number of carbon atoms in the side chain. The values predicted by these relationships are in excellent agreement with the experimental data.

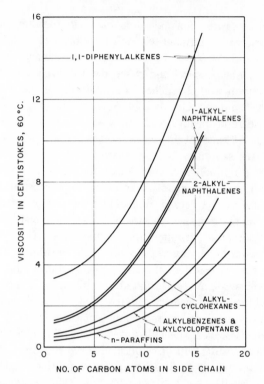

Fig. 99.14.   Viscosity of *n*-alkyl series of hydrocarbons.

Mikeska (149) and Schiessler (200,202,203) have presented considerable data on the viscosities of hydrocarbons of high molecular weight.
Consideration of their data together with Fig. 99.14 makes it possible
to draw the following broad conclusions:

1. The greater the complexity of the ring, the higher the viscosity
(for a given molecular weight).

2. Viscosity increases as the number of carbon atoms in linear side
chains is increased.

3. Viscosity also increases as side-chain branching increases, provided
the number of paraffinic carbon atoms attached to a given nucleus remains unchanged.

4. An olefinic linkage in the side chain results in a decrease in
viscosity.

Most aromatic hydrocarbons exhibit an increase in viscosity upon
hydrogenation (Fig. 99.14). However, when certain alkyl-substituted
fused-ring aromatics, notably phenanthrene and anthracene derivatives,

are hydrogenated, the viscosity decreases (189). Certain highly substituted benzenes have also been observed to behave in a similar manner (43). These molecules have in common a low degree of molecular flexibility. Table 99.VIII shows the magnitude of the changes in viscosity that occur for a number of aromatic hydrocarbons that exhibit this exceptional behavior. An n-alkylbenzene and an n-alkylnaphthalene have been included as representative of aromatics that show a viscosity increase upon saturation. Hydrogenation of petroleum fractions of high molecular weight almost always results in a marked decrease in viscosity. This fact, in conjunction with pure hydrocarbon data, has been cited (189) as evidence that the aromatic hydrocarbons in the lubricating oil fractions are predominantly of the fused-ring type. Although this view is undoubtedly substantially correct, the data in Table 99.VIII demonstrate that the results obtained on hydrogenating oil fractions cannot be used to exclude the possible presence of significant amounts of nonfused multisubstituted benzenes.

J. VISCOSITY–TEMPERATURE DEPENDENCE

The manner in which viscosity depends on temperature varies considerably for various types of hydrocarbons. A number of empirical relationships for the temperature dependency of viscosity have been proposed since this characteristic is of great practical importance for lubricating oils. The "viscosity index" is currently in wide industrial use. This term was originally defined by Dean and Davis (40) on the basis of their study of the change in Saybolt viscosities with temperature of distillates from the extremes of domestic crude types. The viscosity indices of the Pennsylvanian and Gulf Coast distillates were arbitrarily assigned the values 100 and 0, respectively. The viscosity index of an unknown oil may be obtained by comparing the ratio of its viscosities at 100 and 210°F with the ratios of the viscosities of the two reference oil fractions having the same 210°F viscosity as the sample. The formula commonly used is

$$VI = \frac{L - U}{L - H} \times 100$$

where L and H are the 100°F viscosities of the reference oils of 0 and 100 VI respectively and U is the 100°F viscosity of the unknown. The range is not limited to 0–100. If the value of U is greater than L, the viscosity index will be negative; if it is less than H, values of over 100 may be obtained. Tabular values of L and H which may be entered with the 210°F viscosity of the unknown are available for use

TABLE 99.VIII
Effect of Hydrogenation on the Viscosity of
Some Aromatic Hydrocarbons (43)

Structure	Per cent change in viscosity at $100°F^a$
	43
	3.5
	−1.3
	−15
	−22
	−24
	−40

^{a} $\dfrac{\eta_{cs}(\text{cycloparaffin}) - \eta_{cs}(\text{aromatic})}{\eta_{cs}(\text{aromatic})} \times 100.$

5684

with this equation. Modern viscosity index tables are given in terms of kinematic rather than Saybolt viscosity (5).

The use of the viscosity index has several important drawbacks that can be summarized as follows: (1) irrationalities exist at both the low and the high end of the VI scale; this becomes especially apparent at viscosity indices exceeding 140, and (2) the reference temperatures are fixed at 100 and 210°F. Proposals to correct these deficiencies have been made by a great many investigators (10,70,184,216,243). The "viscosity–temperature function" (VTF) suggested by Wright (243) is attractive because it encompasses many of the desirable features of previously suggested systems and at the same time possesses some unique advantages:

1. It is a single number which is characteristic of the oil composition. No auxiliary reference to actual viscosity level is necessary.

2. It is independent of any empirical relation that expresses the change of viscosity with temperature.

3. Specialized data or tables are not required, and the VTF can be obtained by a simple and rapid calculation.

The viscosity–temperature function is defined by the following equation:

$$\mu_2{}^m = \mu_1/2.25\mu_2$$

where the exponent m is the characterizing constant VTF, and μ_1 and μ_2 are the kinematic viscosities at 100 and 210°F, respectively. Figure 99.15 illustrates the application of actual data to this equation. The relationship between VTF and VI is simple enough to allow conversion from one system to another if irregularities in the viscosity index are ignored. Calculation of VTF and subsequent conversion to VI using Fig. 99.16 yields numbers in which ambiguities due to viscosity or viscosity index level are eliminated. Interpretation of the effect of molecular structure on viscosity–temperature behavior is considerably simplified when rational systems such as the viscosity–temperature function are used.

K. CRITICAL SOLUTION TEMPERATURE

The solution temperatures of hydrocarbons in certain solvents have been used for many years as an aid in the characterization of pure substances and in the analysis of hydrocarbon mixtures. Benzyl alcohol, furfural, and nitrobenzene have all been mentioned in the literature. Francis has reported the critical solution temperatures of a large number

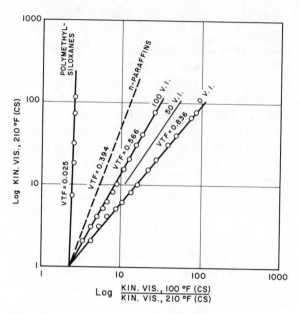

Fig. 99.15. Viscosity–temperature function of liquids. Reproduced with the permission of the American Society for Testing and Materials.

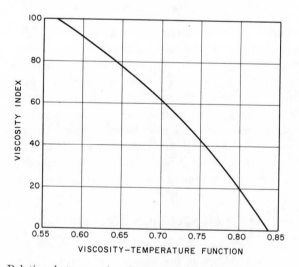

Fig. 99.16. Relation between viscosity–temperature function and viscosity index. Reproduced with the permission of the American Society for Testing and Materials.

of aromatic hydrocarbons in 60 organic solvents (59). However, the most generally convenient liquid proposed, and the one that has been investigated most, is aniline. Chavanne and Simon (28,29) determined the critical solution temperatures of a number of hydrocarbons in aniline and suggested its use for the estimation of the aromatic hydrocarbon content of mixtures. Tizard and Marshall (223) proposed that, instead of the critical solution temperature, the temperature of complete miscibility of equal volumes of aniline and hydrocarbon be used. The minimum temperature of complete miscibility is more conveniently measured than the critical solution temperature, and the term "aniline point" normally refers to this technique. Because of the flatness of the solubility curve, the aniline point is generally quite close to the critical solution temperature. Evans (50) has compared aniline point and critical solution temperature for a number of pure hydrocarbons, with the results shown in Table 99.IX. The aniline point is, in general, from 0.5 to 2° lower than the critical solution temperature.

TABLE 99.IX

Comparison of Critical Solution Temperature in Aniline with "Aniline Point" for Hydrocarbons (50)

Hydrocarbon	Critical solution temperature, °C	Aniline point, °C	Difference, °C
Paraffins			
n-Pentane	71.4	70.7	−0.7
n-Hexane	69.0	68.6	−0.4
n-Heptane	69.9	69.4	−0.5
n-Octane	71.8	71.7	−0.1
n-Nonane	74.4	74.4	0.0
n-Dodecane	83.7	83.7	0.0
Cycloparaffins			
Cyclopentane	18.0	16.8	−1.2
Methylcyclopentane	34.7	33.0	−1.7
Ethylcyclopentane	38.7	36.7	−2.0
n-Propylcyclopentane	50.5	48.7	−1.8
n-Butylcyclopentane	48.8	46.4	−2.4
Cyclohexane	31.0	30.2	−0.8
Methylcyclohexane	41.0	39.5	−1.5
Olefins			
Pentene-1	19.3	19.0	−0.3
Hexene-1	22.9	22.8	−0.1
Heptene-1	26.6	27.2	+0.6
Nonene-1	38.6	38.0	−0.6

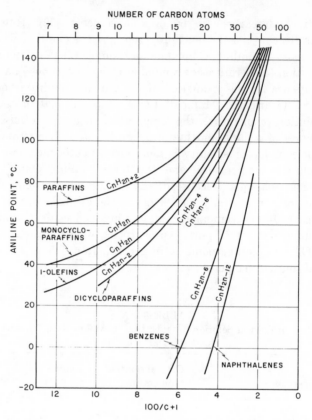

Fig. 99.17. Aniline point of *n*-alkyl series of hydrocarbons.

The various classes of hydrocarbons have different aniline points. Figure 99.17 has been constructed by combining available literature data for several *n*-alkyl series of hydrocarbons. This figure permits the following generalizations: in homologous series the aniline points increase with molecular weight; aromatic hydrocarbons exhibit the lowest, and paraffins the highest, values; cycloparaffins and olefins exhibit values that lie between those for paraffins and aromatics. Aniline point has played an important part in the development of procedures for hydrocarbon analysis, although its use in recent years has steadily been displaced by other methods. The associated phenomenon of demixing upon titration with water has been used as the basis of several methods for the analysis of binary systems composed of aromatic hydrocarbons and alcohols of relatively low molecular weight (13,20).

III. TYPES OF HYDROCARBONS IN PETROLEUM

In discussing the types of hydrocarbons that occur in petroleum fractions, it is necessary to distinguish between straight-run distillates on the one hand, and products produced by thermal or catalytic cracking or reforming on the other. The composition of cracked products is rendered more complicated than that of the straight-run distillates by the presence of substantial quantities of hydrocarbons containing olefinic linkages. It has been established with reasonable certainty that, with the possible exception of certain Pennsylvania oils (61), hydrocarbons containing olefinic double bonds do not occur in virgin petroleum.

It is not possible in this section to acknowledge all of the researches into the composition of petroleum and its products that have added to our knowledge of the subject. The purpose is rather to present a summary of the present state of our knowledge and to emphasize those observations regarding assemblages of structures that have to a certain extent simplified the analysis of these complicated mixtures.

A. HYDROCARBONS IN STRAIGHT-RUN DISTILLATES

The isolation of pure compounds from petroleum is an exceedingly difficult task. The early investigations of Young (244) and of Mabery (131), although they contributed to our knowledge, were largely qualitative in character. The most outstanding and comprehensive work in this field has been performed by Rossini and his co-workers (187) under the auspices of the American Petroleum Institute. Since the initiation of this work in 1927, a total of 230 hydrocarbons has been isolated from the gaseous, gasoline, kerosine, light gas oil, heavy gas oil, and lubricant portions of a Ponca Oklahoma crude. Table 99.X summarizes the distribution of these compounds according to type and carbon number. In the following discussion, the occurrences of the four classes of hydrocarbons represented in this table will be considered separately.

1. Normal Paraffins

The normal paraffins occur in most crude oils, but in varying proportions. Crudes of older geologic age are generally likely to contain larger quantities of these hydrocarbons. All of the normal paraffins up to *n*-tritriacontane have been isolated from the Ponca Oklahoma crude studied by Rossini. In the same crude, the proportions of the individual normal

TABLE 99.X

Distribution, by Class and Carbon Number, of the Hydrocarbons
Isolated from Ponca Oklahoma Crude

Carbon number	4	5	6	7	8	9	10	11	12	13	14	15	16	17	18	Total
Branched paraffins	1	1	4	6	15	7	5				1	1				41
Alkylcyclopentanes		1	1	5	13	2										22
Alkylcyclohexanes			1	1	8	3	1									14
Alkylcycloheptanes				1												1
Bicycloparaffins					3	3	5	1								12
Tricycloparaffins								1								1
Alkylbenzenes			1	1	4	8	22	4								40
Aromatic cyclo-paraffins						1	4	3		1				2	1	12
Fluorenes											1	2	3	1		7
Dinuclear aro-matics								1	2	12	15	5	1	1		37
Trinuclear aro-matics												1	4	1	1	7
Tetranuclear aro-matics													1	1	1	3
Total	1	2	7	14	43	24	39	10	12	16	8	8	6	5	2	197
Normal paraffins C_1–C_{33}																33
Grand total																230

paraffins are very nearly the same, that is to say, no one hydrocarbon
is conspicuous by its absence or predominance. The content of normal
paraffins relative to that of other hydrocarbon classes generally decreases
with increasing boiling point of the distillate fraction.

2. Branched Paraffins (Isoparaffins)

Rossini has isolated a large number of branched paraffins from the
gasoline fraction of several representative crude oils. Substantially all
of the branched paraffins boiling up to 132°C have been isolated from
Ponca crude. The singly branched isomers are more abundant than the
triply branched isomers. Until several years ago there was some question
as to whether branched paraffins persisted in any significant quantity
in the higher-boiling fractions. There is now, however, little doubt that
branched paraffins are present in some quantity even in the lubricating
oil portion of petroleum. The evidence for their presence in the lubricant

portion is indirect, consisting as it does of information accumulated on extensively fractionated samples, but can nevertheless be considered conclusive. Recently, Mair reported the isolation of 2,6,10-trimethylundecane and 2,6,10-trimethyldodecane from the light gas oil portion of the API representative petroleum (136). Bendoraitis and his co-workers have reported the isolation of 2,6,10,14-tetramethylpentadecane from both East Texas and Midcontinent crudes (8). The proportion of branched paraffins does, however, diminish with increase in the boiling point of the fraction.

3. Cycloparaffins

The proportion of cycloparaffins varies considerably with the type of crude. As a general rule, however, cycloparaffins constitute a substantial portion of petroleum. Most of the individual cycloparaffin hydrocarbons that have been isolated have boiling points in the gasoline range. The individual monocycloparaffins have been substantially fully isolated up to 132°C. In the higher-boiling fractions the rapid increase in the number of isomers, caused by the possibility of variation in the number of carbon atoms per ring, the occurrence of *cis-trans* isomerism, and the possibilities in regard to the manner in which individual rings may be linked to one another, makes the isolation of individual compounds a difficult task. However, one monocycloparaffin and six bicycloparaffins have been isolated from the kerosine fraction of Ponca crude. Mair and others (114,142) have also reported the isolation of the tricycloparaffin, adamantane, from petroleum.

Although only a relatively small number of representative cycloparaffins have been isolated so far, cyclopentane, cyclohexane, and decahydronaphthalene derivatives appear to be the predominant types. As a general rule in alkyl substitution it appears that the derivatives of the parent ring system with alkyl substituents of the smallest number of carbon atoms predominate over structures with longer chains. For example, dimethyl isomers are favored over ethyl isomers, and trimethyl isomers over methyl ethyl isomers. As the boiling point of fractions is raised, the contribution of dicycloparaffin and tricycloparaffin structures increases. All of the information thus far accumulated suggests that 5- and 6-membered ring structures are the predominant types in the cycloparaffin portion of petroleum. Little unequivocal information is as yet available concerning the ratio of 5- to 6-membered rings and the ratio of condensed to noncondensed cycloparaffins in the heavier petroleum fractions. Mass spectral data have been reported (139) that suggest that 5-membered rings predominate in the heavier fractions and that the ratio of 5- to 6-membered rings not in condensed structure may be on the order of 2:1. Interpretations of infrared spectra, however, are not in complete accord with this view. Calculation of the average ring size from physical properties, although admittedly subject to error, generally indicates a ratio of 5- to 6-membered rings closer to 1:1 for the lubricant portion of petroleum (192). There is also considerable evidence from mass spectra that a substantial proportion of the cyclo-paraffins in the heavier fractions are present as noncondensed structures (129,139,147). The ratio of noncondensed to condensed cycloparaffins is probably about 1:1. In general, the cycloparaffin portion of lubricant fractions can be thought of as a complex mixture of mono- and poly-cycloparaffins containing both condensed and noncondensed 5- and 6-membered rings. As pointed out before, the relative proportion of poly-cycloparaffins increases with increase in the boiling point.

4. Aromatics

All of the known aromatic hydrocarbons through nine carbon atoms have been isolated from the gasoline portion of Ponca crude. As a result of the examination of other representative petroleums, Rossini concluded that with a few exceptions these hydrocarbons are present in the same relative proportions in different crudes. As in the case of the cyclo-paraffins, where several possibilities for alkyl substitution exist, the pre-dominant isomers are generally those containing the substituents with the smallest number of carbon atoms. The individual alkylbenzenes have been fully isolated to 205°C, and eight additional mononuclear aromatics have been isolated from higher boiling material. Naphthalene and its two methyl derivatives have been isolated by Rossini from the kerosine and light gas oil fractions. The dinuclear aromatics appear first in the kerosine fraction and become increasingly important contributors in the light and heavy gas oil portions. The presence of various other simple alkylnaphthalenes has been established by Mair and Rossini (139) and also by other investigators (25,157). Reports in the literature suggest that eleven of the twelve 12-carbon naphthalenes are present in most petroleum gas oils (1,92,157). However, in the heavier fractions, the simple alkyl aromatics are probably present in only minor amounts. Most of the aromatics in the lubricant portion contain at least one cycloparaffin ring that is in all probability condensed to the aromatic nucleus. The predominance of these mixed aromatic–cycloparaffin hydro-carbons has been clearly demonstrated by Rossini and Mair in their examination of the lubricant portion of Ponca crude (139,140,143,186).

By a deliberate combination of fractionation processes, the bulk of this material was reduced to fractions containing molecules of very similar size and type, and representing 1/40,000 part of the original crude. From a study of the composition of these fractions by a comparison of their properties to those of known hydrocarbons of similar molecular weight, and by careful hydrogenatiòn of the aromatic nuclei, Rossini and Mair concluded that most of the aromatic hydrocarbons were of the following types.

1. Mononuclear aromatic mixed aromatic–cycloparaffin hydrocarbons consisting of one aromatic ring with one, two, or three cycloparaffin rings, probably most frequently condensed, together with paraffin side chains or connecting groups as appropriate to the molecular weight.

2. Dinuclear aromatic mixed aromatic–cycloparaffin hydrocarbons, consisting of two aromatic rings, probably condensed, with one or two cycloparaffin rings, the latter probably condensed, together with paraffin side chains or connecting groups as appropriate.

3. Trinuclear aromatic mixed aromatic–cycloparaffin hydrocarbons consisting of three aromatic rings, probably condensed, with one cycloparaffin ring together with paraffin side chains or connecting groups as appropriate.

The decrease in viscosity observed upon hydrogenation of the di- and trinuclear aromatics supports the supposition that, when two or more aromatic rings occur in the same molecule, they are invariably condensed together. Lipkin and his associates have studied the properties of the fractions separated by Rossini using their "double-bond index," and have confirmed that the aromatic rings are condensed (127). Moreover, these authors found no evidence of appreciable quantities of anthracenes, indicating that when three aromatic rings occur in the same molecule they are condensed in the angular rather than the linear fashion. It is therefore generally believed at the present time that the aromatic hydrocarbons in viscous oils consist largely of mixed aromatic–cycloparaffin hydrocarbons of condensed structure. The correctness of this general picture received support when Mair and Martinéz-Picó (137) reported the isolation of a number of aromatic and aromatic–cycloparaffin hydrocarbons from the 305–405°C portion of Ponca crude. These compounds are listed in Table 99.XI. The trinuclear aromatics are all phenanthrene derivatives, and eight of the eighteen hydrocarbons contain associated cycloparaffin rings. Carruthers has reported the isolation of several hydrocarbons of similar structure from a Kuwait petroleum (26).

TABLE 99.XI

Aromatic Compounds Isolated from the Heavy Gas Oil–Light Lubricating
Distillate (305–405°C) from Ponca Oklahoma Crude Oil (137)

Compound	Formula		Melting point, °C
	General	Molecular	
2,3,6,7-Tetramethylnaphthalene	C_nH_{2n-12}	$C_{14}H_{16}$	192–193
Monomethylfluorene	C_nH_{2n-16}	$C_{14}H_{12}$	84–84.5
Dimethylfluorene	C_nH_{2n-16}	$C_{15}H_{14}$	109–110
Dimethylfluorene	C_nH_{2n-16}	$C_{15}H_{14}$	157.5–158
Trimethylfluorene	C_nH_{2n-16}	$C_{16}H_{16}$	97–98
Trimethylfluorene	C_nH_{2n-16}	$C_{16}H_{16}$	139–140
Trimethylfluorene	C_nH_{2n-16}	$C_{16}H_{16}$	126–128
Phenanthrene	C_nH_{2n-18}	$C_{14}H_{10}$	94–95
1-Methylphenanthrene	C_nH_{2n-18}	$C_{15}H_{12}$	117–118
2-Methylphenanthrene	C_nH_{2n-18}	$C_{15}H_{12}$	—
3-Methylphenanthrene	C_nH_{2n-18}	$C_{15}H_{12}$	—
9-Methylphenanthrene	C_nH_{2n-18}	$C_{15}H_{12}$	—
1,8-Dimethylphenanthrene	C_nH_{2n-18}	$C_{16}H_{14}$	195.5–196
Cyclopentanophenanthrene	C_nH_{2n-18}	$C_{17}H_{14}$	142–143
1'- or 3'-Methyl-1,2-cyclo-pentanophenanthrene	C_nH_{2n-20}	$C_{18}H_{16}$	109–111
Pyrene	C_nH_{2n-22}	$C_{16}H_{10}$	123–128
4-Methylpyrene	C_nH_{2n-22}	$C_{17}H_{12}$	107–114
Dimethylpyrene	C_nH_{2n-22}	$C_{18}H_{14}$	127–140

The content of condensed polynuclear aromatics generally increases with
increase in the boiling point of the distillate fraction.

B. HYDROCARBONS IN PRODUCTS FROM CRACKING OPERATIONS

Rossini has studied the hydrocarbons that occur in gasolines produced
by various commercial cracking processes (187). All of the hydrocarbons
found in virgin gasolines occur also in gasolines produced by cracking
operations, although their relative proportions may be drastically different. In addition, cracking produces substantial quantities of olefins. Only
a relatively small number of the possible individual olefins that can
occur in such products have been identified. Knight (94) has studied
the 5- and 6-carbon olefins present in a catalytic gasoline and has identified the compounds shown in Table 99.XII. It appears likely from consideration of these data that the olefin portion of hydrocarbons produced

by cracking can be expected to contain monoolefins, conjugated diolefins, and cyclic monoolefins, and that, at least in the lower-boiling fractions, monoolefins predominate. The presence of nonconjugated diolefins and cyclic diolefins is probable, although direct evidence is scarce.

In addition to producing straight-chain, branched-chain, and cyclic olefins, cracking results in the production of significant quantities of mixed aromatic–olefin structures. Low-voltage mass spectrometry indicates the presence of such molecules as styrenes, indenes, dihydronaphthalenes, and acenaphthalenes (91).

The higher-boiling residues from severe cracking are generally rich in aromatic hydrocarbons and considerably depleted in paraffin and cycloparaffin structures. The polycycloparaffin systems that contribute heavily to the composition of virgin lubricant fractions have been largely eliminated by ring-opening reactions or converted to aromatics by

TABLE 99.XII

Relative Amounts of Individual Hydrocarbons in Pentene and Hexene Aggregates Isolated from a Catalytic Gasoline (94)

Pentenes		Hexenes	
Compound	Relative concentration, wt %	Compound	Relative concentration, wt %
1-Pentene	6	4-Methyl-1-pentene⎫	10
2-Methyl-1-butene	18	3-Methyl-1-pentene⎭	
trans-2-Pentene	10	2,3-Dimethyl-1-butene	5
cis-2-Pentene	20	4-Methyl-cis-2-pentene	8
2-Methyl-2-butene	42	4-Methyl-trans-2-pentene	2
2-Methyl-1,3-butadiene	0.5	2-Methyl-1-pentene	11
Cyclopentene	3	1-Hexene	4
1,3-Pentadiene	1	2-Ethyl-1-butene	3
		trans-3-Hexene	4
		cis-3-Hexene	6
		2-Methyl-2-pentene	13
		trans-2-Hexene	2
		cis-2-Hexene	4
		3-Methyl-trans-2-pentene	8
		3-Methyl-cis-2-pentene	9
		2,3-Dimethyl-2-butene	3
		3-Methylcyclopentene	3
		1-Methylcyclopentene	5
		Cyclohexene	Trace
		Dienes	Trace

thermal dehydrogenation. In addition, the mixed aromatic–cycloparaffin types that comprise the bulk of the aromatic portion of the heavier virgin distillates generally will have been converted, by opening and cracking of the associated cycloparaffin rings, to the simpler alkyl-substituted nuclei. Advantage has occasionally been taken of these ring-opening reactions to improve the viscosity–temperature characteristics of viscous fractions by relatively mild cracking in the presence of hydrogen (170).

IV. CHARACTERIZATION AND ANALYSIS

The characterization of hydrocarbons is hampered by their relatively low chemical reactivity. The isolation of substantially pure hydrocarbons from complex mixtures of natural products is also rendered exceedingly difficult, even at moderate molecular weights, by the presence of isomeric compounds of similar physical and chemical properties. When exhaustive fractionation of complex mixtures has been carried out using a deliberate combination of separation methods until further fractionation is ineffective, the material isolated is generally either one compound or a mixture of molecules of the same size and type—which may be called a "uniform" fraction. Identification or characterization of such extensively fractionated samples is possible by comparison of physical constants. In general, application of certain chemical reactions is best made to uniform fractions.

Since such extensive separation is so tedious and time consuming as to be economically unattractive in the industrial laboratory, a great deal of effort has been expended on the problem of characterizing and analyzing nonuniform samples. Investigation of complex mixtures of natural products has led to analytical methods that provide a substantial amount of information about composition and at the same time are reasonably rapid. Such investigations have also shown that, by combination of a few physical properties, simple functions can be defined that vary more or less regularly with composition or content of certain structural features. Such functions can supply valuable information for identification and characterization purposes. In this section, the application of physical and chemical methods to relatively uniform fractions is discussed. This is followed by a consideration of characterizing functions which embody various combinations of physical constants and are related to composition. The discussion is concluded with an example of the judicious combination of chemical and physical methods to produce a procedure for the analysis of mixtures of low-boiling hydrocarbons.

A. CHARACTERIZATION OF PURE SUBSTANCES AND UNIFORM FRACTIONS

1. Physical Methods

In ascertaining the character of a uniform fraction, the formulas and graphs of the various physical constants covered in Section II are valuable aids. Generally, one or more physical constants are determined for comparison with data for known compounds. A knowledge of the molecular weight is also useful, since it can then be ascertained whether or not the unknown hydrocarbons belong to one of the known series. Application of suitable chemical reactions can often supply additional information concerning structure.

However, the comparison of a few physical constants is generally effective only for low-boiling samples, since the physical properties of only the simpler hydrocarbons have been systematically investigated. For materials of relatively high molecular weight, the following procedure has been suggested (139).

Determine the boiling point, density, refractive index, dispersion, viscosity, critical solution temperature in one or more solvents, elemental composition, and molecular weight. From these data the molecular formula can be calculated, and a comparison of physical constants with those approximated by extrapolation from lower molecular weights, using the techniques of Section II, will often serve to determine the type of hydrocarbons (paraffin, cycloparaffin, benzene or naphthalene derivative, etc.) present.

2. Chemical Methods

Chemical reactions are used only in certain cases in the analysis of hydrocarbon mixtures. This is true partly because many hydrocarbons are not particularly reactive, and partly because physical techniques are often more sensitive to the minor differences in structure that must be distinguished. In general, chemical methods are most useful when applied to uniform fractions. The reactions of saturated hydrocarbons are few, and the reactions of aromatic hydrocarbons, such as sulfonation, halogenation, or nitration, are generally far from quantitative. They often result in products of ill-defined composition because of secondary reactions. There are, however, a few reactions that can provide important information. The most useful of these, the hydrogenation of aromatics, is discussed in some detail in Section VI.A.2.

The analysis of mixtures of normal paraffins and isoparaffins is generally quite difficult to accomplish on the basis of differences in their physical properties. Consequently, chemical methods that take advantage of the relative reactivity of tertiary hydrogens in saturated molecules have received some attention. Chlorosulfonic acid has been proposed as a reagent for isoparaffins, since it will not attack normal paraffins but reacts violently with branched hydrocarbons (206). Antimony penta-chloride will react with branched paraffins containing hydrogen attached to a tertiary carbon (198), and has been used to remove isoparaffin impurities from n-paraffins. The application of these reagents for quanti-tative work has been relatively little practiced, and their use has largely been displaced by physical separation techniques. Mixtures of normal paraffins and isoparaffins are generally analyzed by taking advantage of the fact that n-paraffins are quantitatively adsorbed by certain syn-thetic zeolites (159,173).

The determination of olefins, on the other hand, has for many years been predicated on the chemical reactivity of this class of hydrocarbons. Although hydrogenation, reactions with silver and mercury salts, and sulfonation have all received attention, the most generally satisfactory and most widely used method is bromination. There is, however, no bromination method available that gives results reliable in all respects. For very accurate work, the method of McIlhiney (212) is useful, since it provides a way of correcting for substitution reactions.

A satisfactory method for routine application has been described by Dubois and Skoog (45). It involves direct titration with stable potassium bromide–bromate reagent (56) and electrometric detection of the end point. A mercury catalyst is used to ensure complete addition, and substi-tution reactions are minimized by conducting the titration at a reduced temperature. It has been reported (242) that elimination of the catalyst does not adversely affect the accuracy of the method for a large number of olefins and in a number of instances gives values that are closer to the theoretical bromine number.

Bond (14) has reported a method for determining olefins by treatment with nitrogen tetroxide. The method is based on the fact that nitrogen tetroxide reacts quantitatively with olefins to produce ill-defined high-boiling addition products from which the other hydrocarbons may be separated by steam distillation or chromatography using solid adsorbents.

It is extremely difficult to distinguish the derivatives of cyclohexane and cyclopentane on the basis of differences in their physical properties. Under the proper conditions, however, cyclohexane derivatives may be catalytically dehydrogenated to their benzene analogs, while cyclopentane derivatives remain unchanged. A platinum or palladium catalyst is gen-erally used at temperatures around 350°C (179,183). This method has been used for analysis of the 6- and 7-carbon cycloparaffins in straight-run naphthas. However, the application of dehydrogenation to saturated hydrocarbon mixtures containing more than 7 carbon atoms is complicated by the fact that cyclohexanes that are geminally substituted—that is, contain two substituents on one ring carbon—are not readily dehydrogenated. The likelihood of geminal substitution increases rapidly with boiling point. It should be remembered also that the isomerization of cyclopentane derivatives to cyclohexanes is not impossible thermodynamically. Despite these deficiencies, dehydrogenation is a promising tool for examination of the higher-boiling cycloparaffins, especially in conjunction with some of the newer instrumental methods.

A number of chemical reactions are of potential use in the determination or identification of aromatic hydrocarbons. Aromatics as a class may be conveniently determined in olefin-free mixtures by sulfonation, using a reagent of sulfuric acid and phosphorus pentoxide (155). This procedure is generally reliable for materials boiling up to 600°F. Treatment with sulfuric acid mixtures of increasing concentration, including a final reaction with oleum, has been reported as suitable for the analysis of hydrocarbons in the lubricating-oil molecular weight range (191). Since, when present, olefins as well as aromatics react with sulfuric acid, the application of sulfonation methods to mixtures containing olefins results in the determination of the total content of these two classes of hydrocarbons.

Chromic acid oxidation has been suggested for the determination of methyl groups in mixtures of alkylbenzenes and mononuclear aromatic concentrates from petroleum (17). If oxidation is carried out at 130°C with a mixture of 2 parts of 5N chromic acid and 1 part of concentrated sulfuric acid, C-methyl groups can be determined as acetic acid without interference from benzoic acid.

A number of distinctive reactions have been proposed for the identification of condensed di- and trinuclear aromatics. Among the most useful are the molecular addition compounds that these materials form with picric and styphnic acid. The picrates and styphnates of condensed aromatics are generally quite crystalline and can often be used to separate such compounds from mixtures with other hydrocarbons. The aromatics may be regenerated, further fractionated, and often identified from the melting point of the picrate or styphnate.

A number of color reactions for the qualitative identification of aromatic hydrocarbons have also been described (48). In this connection the work of Sawicki (195,196) deserves especial mention.

Polarographic procedures for the determination of naphthalenes have been suggested (19). Paraffins, monoolefins, cycloparaffins, and alkylbenzenes do not interfere. Use of the method is not recommended, however, when polynuclear aromatics or aromatics containing an olefinic side chain are likely to be present.

B. CHARACTERIZATION OF MIXTURES BY PHYSICAL METHODS

Many of the methods discussed in the preceding section are largely applicable only to relatively low-molecular-weight, narrow-boiling, or well-fractionated samples. Such uniform fractions are encountered only infrequently by the analyst because of the time and expense that fractionation of this type entails. Usually, samples are very complicated in regard to size and type of molecule. In choosing physical properties which provide information on the chemical composition of such mixtures, it is necessary to discard the idea that we will ever be able to determine individual compounds by these methods, and to concentrate instead on selecting the content of certain structural elements as components. We then can find certain properties or combinations of properties that are sensitive to these structural elements to use as analytical functions. In the selection of properties it is also important to bear in mind whether or not a property is additive, that is, whether the property can be calculated for a mixture from the contributions of the components. This kind of additivity is obviously important, since when it exists, simple linear interpolation can be used to calculate composition.

For complicated mixtures the physical constants considered singly do not give unambiguous information about the composition. In such cases, functions built up from combinations of physical properties are often quite useful. Unfortunately, only rarely are such functions wholly additive. Nevertheless it is generally possible, if a mass of sufficiently diverse and representative data is available, to derive relationships between composition and combinations of physical properties that are analytically useful.

1. Boiling Point Relationships

Boiling point and density (or gravity) have been used for many years to characterize complex hydrocarbon mixtures obtained from the distillation of petroleum. The functions most often used are the characterization factor and the correlation index. Both of these functions have proved to be of considerable practical utility. However, the numerical values themselves have little analytical significance, for they by no means provide any unambiguous information concerning the composition of the mixture.

a. CHARACTERIZATION FACTOR

The characterization factor was developed from data on pure compounds (233,234) and is defined as

$$K = \frac{T_B^{\frac{1}{3}}}{g}$$

where T_B is the "cubic average boiling point" in degrees Rankine (°F + 460), and g is the specific gravity at 60°F. The characterization factor is calculated from the Engler distillation curve plotted so that the per cent distilled equals distillate recovered plus loss. The 10, 30, 50, 70, and 90% temperatures are first averaged to give the volumetric average boiling point in degrees Fahrenheit. A correction is then applied to give the cubic average boiling point. The correction is a function of the volumetric average boiling point and the slope of the distillation curve between 10 and 90% distilled. The cubic average boiling point is then converted to degrees Rankine and inserted in the above equation along with the specific gravity. Graphs have been developed from which the characterization factor may be read directly if the cubic average boiling point and gravity are known. Charts which permit the estimation of characterization factor from viscosity and gravity have also been published (235).

The characterization factor of the paraffin hydrocarbons is substantially constant, varying between 12.5 and 13.0. Cycloparaffins have K values between 11 and 12, and aromatics from about 9.8 to 12, depending on the percentage of the total carbon atoms that are present in paraffin chain structures. The entire range of hydrocarbon types is therefore covered by a relatively narrow numerical span. Characterization factors close to 12.5 generally indicate that a hydrocarbon mixture is largely paraffinic. Lower values of the K factor denote increased concentrations of cycloparaffin and/or aromatic hydrocarbons.

The characterization factor has been related to other useful properties and functions such as aniline point, viscosity, viscosity index, and viscosity–gravity constant (235). The relationship of K factor to the composition of oil fractions as given in terms of certain structural elements has been investigated by Leendertse and Smittenberg (165). These authors reported that the characterization factor was dependent on the percentages of the total number of carbons that were present in paraffinic

($\%C_P$) and in aromatic ($\%C_A$) structures, and could be calculated from the following formula:

$$K = 10.60 + (0.0266)\ \%C_P - (0.0191)\ \%C_A$$

b. Correlation Index

A somewhat different relation between specific gravity and boiling point has been suggested as a characterizing function by Smith (208). He found that, if the reciprocal values of the boiling point in degrees Kelvin for series of pure hydrocarbons were plotted against the specific gravity at 60°F, a reasonably straight line could be drawn through the points for the normal paraffin series. If a similar parallel line is drawn through the point for benzene, the extremes of 100% paraffin and 100% aromatic hydrocarbon composition have been delineated. Smith subdivided the interval between these two lines into ten equal parts and assigned the numbers 0 and 100 to the paraffin and benzene lines, respectively. The function defined in this fashion he termed the "correlation index." The index number of a hydrocarbon sample can be read from a graph such as that shown in Fig. 99.18 if the boiling point and gravity

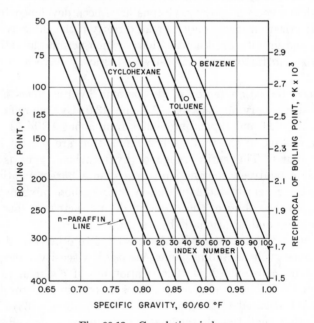

Fig. 99.18. Correlation index.

are known. The correlation index is an arbitrary way of expressing the characteristics of hydrocarbon mixtures. It is particularly useful for evaluating certain relatively low-boiling samples. Mixtures with index values from 0 to 15 are predominantly paraffinic. Values from 15 to 20 or 30 indicate cycloparaffins or mixtures of paraffins, cycloparaffins, and aromatics. Values above 60 generally suggest a predominance of aromatic structures.

2. Viscosity Functions

Viscosity functions are attractive, since the viscosity of the heavier hydrocarbons is not only an important but also a conveniently measured property. Viscosity has played a significant part in the development of a number of physical property methods for hydrocarbon analyses. It has also been related to a number of other physical constants of oil fractions. However, these relationships do not produce functions primarily related to composition, but generally are expressions useful for estimating a more difficultly measured quantity from data that are more convenient to obtain. For example, relations have been developed between viscosity and refractive index, surface tension, and ultrasonic velocity for saturated petroleum oils (228). It is possible that a study of the factors that permit such relations to exist will contribute to a better understanding of the effect of molecular configuration on viscosity.

a. VISCOSITY–GRAVITY CONSTANT

One of the most important viscosity functions related to composition is the viscosity–gravity constant proposed by Hill and Coats (75). Using data for a series of distillate fractions from crude oils of widely different types, these authors established a relation between specific gravity at 60°F and the Saybolt viscosity, V, at 100°F of the form

$$\text{Specific gravity} = a + b \log (V - 38)$$

where a and b are constants. Since they were able to show that b is a function of a, it followed that a, the viscosity–gravity constant, could be expressed as a function of gravity (g) and viscosity (V) in the following manner:

$$\text{Viscosity gravity constant (VGC)} = \frac{10g - 1.0752 \log (V - 38)}{10 - \log (V - 38)}$$

Other equations have been developed for materials whose 100°F viscosities fall outside the range of this expression. An alternative formula has been proposed for use with more viscous oils, in which the Saybolt viscosity at 210°F is used (75), and a modified form for application to materials of very low viscosity at 100°F has been reported (154). A similar function based on kinematic viscosity has been developed by E. Smith from data on pure compounds (207).

The viscosity–gravity constant has proved to be an extremely useful tool for the characterization of the more viscous fractions of petroleum. It is relatively insensitive to molecular weight, and a series of distillate oils from a single crude exhibit very nearly the same values of VGC. Values of VGC near 0.800 indicate samples of extremely paraffinic character, whereas values close to 1.00 indicate a preponderance of aromatic hydrocarbons. The following nomenclature has been proposed (108) to provide an approximate notion of the composition as it relates to certain values of VGC:

Paraffinic	0.790–0.819
Relatively paraffinic	0.820–0.849
Cycloparaffinic	0.850–0.899
Relatively aromatic	0.900–0.949
Aromatic	0.95 –0.99
Very aromatic	1.00 –1.05
Extremely aromatic	Above 1.05

That VGC is intimately bound up with the composition (as it can be expressed in terms of certain structural elements) has been amply demonstrated by van Nes and van Westen (166) and by Kurtz and his associates (106). Viscosity–gravity constant and characterization factor bear a relationship to one another, since both are functions of the same structural parameters. If both the kinematic viscosity at 210°F and the specific gravity of a sample are known, either the VGC or the K factor may be read from Fig. 99.19 with reasonable accuracy.

b. Viscosity–Molecular Weight Relations

It seems appropriate to consider in this section several extremely useful relationships that involve the use of viscosity but are not, in the strictest sense of the term, characterizing functions. These are the relationships between viscosity and molecular weight. Since the accurate experimental determination of molecular weight is rather difficult and time consuming, the use of functions based on more conveniently measured physical properties related to molecular weight is attractive in the industrial labora-

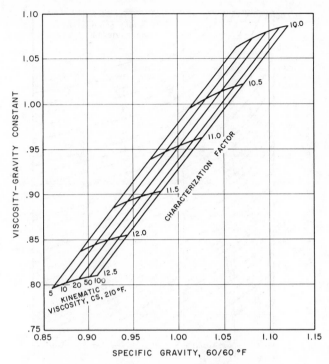

Fig. 99.19. Chart for the estimation of viscosity–gravity constant and characterization factor from specific gravity and viscosity.

tory. Hirschler (77) studied the relationship between viscosity and molecular weight for mixtures of petroleum hydrocarbons, and concluded that molecular weights between 240 and 700 could be estimated with an accuracy sufficient for most purposes from kinematic viscosities at 100 and 210°F. The final correlation was presented in the form of the chart shown in Fig. 99.20. For petroleum fractions, molecular weights read from this figure have been found to agree with experimental data to within 2–5%.

The derivation of the relationships pertinent to the development of the viscosity–temperature function indicates that plotting the log of the 100°F viscosity against the log of the 210°F viscosity should give straight lines of isomolecular weight. This has been found to be the case for a number of petroleum oils covering a wide range of compositions. However, certain pure hydrocarbons or oils containing a substantial proportion of condensed polynuclear aromatics can exhibit large deviations (243). Charts relating specific gravity and viscosity with

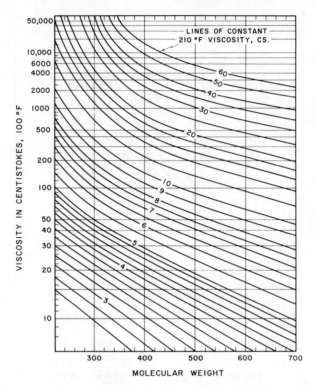

Fig. 99.20. Viscosity–molecular weight correlation for petroleum oils.

molecular weight have also been developed that are applicable to petroleum oils having molecular weights between 240 and 680 (77,150).

3. Refractive Index and Density Relations

Many relations involving refractive index and density have been proposed for the characterization of hydrocarbon mixtures. For the most part, such relationships have been extremely useful. That this has proved to be the case is due, first of all, to the high sensitivity of these physical constants to differences in molecular structure, and, second, to the fact that quite accurate measurements of these properties may be made conveniently with relatively simple and inexpensive equipment.

As has been previously mentioned, the formulas derived by Smittenberg for refractive index and density predict the existence of a linear relationship between these two properties for homologous series of satu-

rated hydrocarbons. This is illustrated in Fig. **99.21**. Graphs similar
to this figure are often useful in determining whether or not a given
hydrocarbon belongs to any one of the known series.

The similar slope and proximity of the lines for saturated hydrocarbons suggest that it should be possible to develop a general equation
relating refractive index, density, and molecular weight. Lipkin and Martin (121) developed the following formula relating refractive index to
density and the temperature coefficient of density (which is a function
of molecular weight):

$$n = \frac{69.878d - 0.4044Ad - 0.797A + 136.566}{5.543d - 0.746A + 126.683}$$

where n = the refractive index at 20°C for the D line of sodium, d = the
density at 20°C, and $A = -10^5 \times$ temperature coefficient of density,
α. As shown in Section II.D.1, α may be estimated from the molecular
weight.

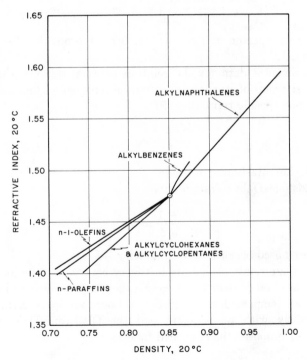

Fig. 99.21. Relation between refractive index and density for n-alkyl series of
hydrocarbons.

For over 500 saturated hydrocarbons for which data are available in the literature, the average deviation between the observed and calculated refractive index was 0.0019. For over 100 saturated petroleum fractions, the average deviation was about one-half this magnitude. This equation can also be arranged so as to allow the calculation of density, although the accuracy of this calculation is considerably less than that for refractive index.

a. SPECIFIC REFRACTION

A number of so-called specific refractions have from time to time been described in the literature. These refractions are generally alleged to give factors independent of temperature and pressure and to be additive in the true sense of the word, that is, they can be calculated by summing up the values for certain structural elements that comprise the molecule. The best known of these is the function proposed by Lorentz and Lorenz (128):

$$\frac{n^2 - 1}{(n^2 + 2)d} = k_1 = r$$

Lorentz-Lorenz specific refraction has been thoroughly discussed by Kurtz (104).

The experimental data for the relation between refractive index and density for a group of isomers can often be represented more accurately by the Newton specific refraction (9):

$$\frac{n^2 - 1}{d} = k_2$$

than by the Lorentz-Lorenz formula.

The Gladstone-Dale refraction (63)

$$\frac{n - 1}{d} = k_3$$

has often been used because of its simplicity.

Although these relationships are in theory contrived to be independent of temperature and pressure, in practice they have proved to be slightly dependent on temperature. Because the Lorentz-Lorenz refraction increases slightly with temperature and the Gladstone-Dale decreases, Eykman (51) proposed the empirical function:

$$\frac{n^2 - 1}{(n + 0.4)d} = k_4$$

The Eykman function represents accurately the effect of temperature and pressure on the relationship between refractive index and density for a wide variety of organic materials.

In spite of its several deficiencies the Lorentz-Lorenz specific refraction has been the most generally applied of these four functions.

If the specific refraction functions are multiplied by molecular weight, the following molecular refractions are obtained:

Lorentz-Lorenz:
$$\frac{M(n^2 - 1)}{(n^2 + 2)d} = K_1 = \text{MR}$$

Berthelot (Newton):
$$\frac{M(n^2 - 1)}{d} = K_2$$

Gladstone-Dale:
$$\frac{M(n - 1)}{d} = K_3$$

Eykman:
$$\frac{M(n^2 - 1)}{(n + 0.4)d} = K_4$$

For hydrocarbons, an additive property such as refraction can be expressed as the sum of certain atomic and structural increments. The atomic constants based on the Lorentz-Lorenz equation have been most widely used. The values of these constants derived by Eisenlohr (47) many years ago are still in general use, although slightly different values have been proposed periodically by other investigators. In general, molecular refractions calculated from the additive constants show acceptable agreement with experimental values and may be used to predict the molar refraction of unsynthesized compounds. However, more precise predictions of molar refraction can be made if specialized equations are developed for certain groups of compounds. This can often be done conveniently by considering data for homologous series. For a series of homologous hydrocarbons the summation of the atomic and structural increments to give molar refraction may be simplified to

$$\text{MR} = M \cdot r = C'r_{\text{CH}_2} + K$$

where C' is the number of CH_2 groups, r_{CH_2} is the refraction increment for one CH_2 group, and K is simply a constant that represents the contribution to the molecular refraction of all atoms and structural elements not in CH_2 groups.

For an homologous series the molecular weight may be expressed in a similar form:

$$M = 14.026C' + L$$

where L is a constant for any particular series and represents that portion of the molecular weight not in CH_2 groups. By combining these two equations, it follows that

$$r = r_\infty + q/M$$

where $r_\infty = r_{CH_2}/14.026$, and $q = K - Lr_\infty$. This equation predicts that r is a linear function of the reciprocal molecular weight and approaches the limiting value r_∞ as M becomes infinitely large. The constant r_∞ is the same for all normal alkyl series, whereas the constant q is characteristic of the type of series. Smittenberg (210) has applied this kind of reasoning to the Lorentz-Lorenz refraction. From data for homologous series of n-alkanes, n-1-alkenes, n-alkylcyclopentanes, n-alkylcyclohexanes, and n-alkylbenzenes, he was able to develop the group refractions shown in Table 99.XIII. The similarity between this treatment and that used by Li (117) in deriving equations for the molar refraction of a number of normal alkyl series of hydrocarbons and nonhydrocarbons (Section II.B) is obvious. Group refractions analogous to those presented by Smittenberg were derived from Li's data and are shown for comparison in the second column of the table. This general approach is almost universally applicable for the development of specialized equations or increments from data on homologous series, and permits the accurate prediction of the refraction of unsynthesized members.

The Lorentz-Lorenz refraction has received the particular attention of petroleum chemists because it can be shown that a linear relationship exists between it and the hydrogen content of saturated oil samples. Since molecular refraction is an additive property, this constant for the hydrocarbon C_cH_h can be calculated from the relation

$$MR = M \cdot r_D^{20} = cr_C + hr_H \tag{1}$$

TABLE 99.XIII
Numerical Values of Group Refractions

Group or function	Lorentz-Lorenz refraction	
	Smittenberg (210)	Calc'd from Li (117)
CH_2 group	+4.640	+4.642
Terminal H atom	+1.04	+1.04
Terminal olefinic bond	+1.59	+1.65
Terminal cyclopentane ring closure	−0.04	−0.07
Terminal cyclohexane ring closure	−0.03	−0.09
Terminal benzene ring	+25.60	+26.55

where r_C and r_H represent the atomic refraction of one carbon atom and one hydrogen atom, respectively, and M is the molecular weight. The numbers of hydrogen atoms, h, and of carbon atoms, c, in the empirical formula $C_c H_h$ are calculable from the percentage of hydrogen, H, as follows:

$$h = \frac{MH}{100.80} \quad \text{and} \quad c = \frac{M(100 - H)}{1201.0} \tag{2}$$

If these functions are substituted in equation (1), it follows that

$$\%H = \frac{100.8 \times 1201.0 r_D^{20} - 10080 r_C}{1201.0 r_H - 100.8 r_C} \tag{3}$$

If Eisenlohr's atomic increments are inserted for r_C and r_H, the following equation is obtained (224):

$$\%H = 112.37 r_D^{20} - 22.623 \tag{4}$$

This equation permits the calculation of hydrogen content from specific refraction. Since the determination of refractive index and density can readily be carried out with precision to the fourth decimal place, this equation also infers that the hydrogen content of a saturated oil can be derived with greater precision from the specific refraction than by elemental analysis. It must be realized, however, that the use of equation (4) for complex mixtures of saturated hydrocarbons involves three assumptions:

1. The specific refraction is additive on a weight basis.
2. The atomic increments for carbon and hydrogen are independent of the type of linkage in which they occur.
3. The atomic refractions of Eisenlohr are reliable.

When the linear relation between hydrogen content and specific refraction was first proposed by Vlugter, the available data suggested that these conditions were adequately fulfilled. An extensive reinvestigation of the relationship later undertaken by van Nes and van Westen (169) confirms that this is indeed the case. By the application of least-squares methods to experimental data on a great many saturated petroleum oils, they obtained the following linear relation:

$$\%H = 110.48 r_D^{20} - 22.078$$

which differs slightly from that developed using the Eisenlohr constants. This is hardly surprising since these constants were derived from a lim-

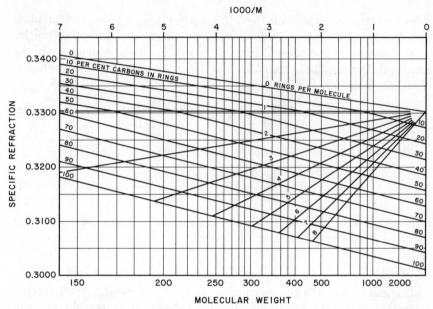

Fig. 99.22.   Specific refraction versus reciprocal molecular weight for saturated hydro-
crabons. Reproduced with the permission of Elsevier Publishing Company.

ited number of data of divergent character, whereas the data of van
Nes and van Westen were restricted to saturated mineral oil fractions.
The difference between these two equations is hardly great enough to
evoke concern, although it can be argued that the van Nes and van
Westen relationship is to be preferred since it was derived from data
on petroleum hydrocarbons.

The linear relation between hydrogen content and Lorentz-Lorenz specific refraction permits the construction of a graph similar to that given
in Fig. 99.10 with specific refraction replacing per cent hydrogen. A
graph constructed on this basis, using the van Nes and van Westen
relationship, is represented in Fig. 99.22. By using this figure, the average
number of rings, R_N, for saturated oils can be obtained without recourse
to elemental analysis. Alternatively, the relations for per cent hydrogen
can be substituted in equation (8), Section II.E, to give equations for
ring number based on specific refraction and molecular weight rather
than per cent hydrogen and molecular weight. Substitution of the relation
developed by using the Eisenlohr increments leads to the equation

$$R_N = 1 + 0.2145M - 0.6511Mr$$

of Vlugter, Waterman, and van Westen (224), whereas application of the van Nes-van Westen relation leads to the formula

$$R_N = 1 + 0.2112M - 0.6400Mr$$

Differences in the ring numbers calculated by means of these two equations are generally on the order of only 0.1–0.2.

b. REFRACTIVITY INTERCEPT

To express the relationship between refractive index and density, Kurtz and Ward (113) proposed the refractivity intercept:

$$RI = n - d/2$$

They demonstrated that, if n is plotted against d for repetitive hydrocarbon series, straight lines of very nearly the same slope are obtained whose intercepts are characteristic of the composition of the type of molecules comprising the series. The slope of 0.5 used in the derivation of the equation is based on data for normal paraffins and paraffin isomers. However, it is a sufficiently close approximation to the slopes exhibited by other hydrocarbon series that the equation is of wide applicability. Values of the refractivity intercept for members of a number of classes of hydrocarbons boiling below 250°C are given in Table 99.XIV.

TABLE 99.XIV
Refractivity Intercept Values for Hydrocarbons with
Boiling Points Less than 250°C (104)

Class of hydrocarbon[a]	Refractivity intercept, $n - d/2$
Paraffins	1.0458
Monocycloparaffins	1.0395
Dicycloparaffins	1.0326
Tricycloparaffins	1.0242
Monoolefins	1.0521
Diolefins	1.0613
Conjugated diolefins	1.0809
Cyclic monoolefins	1.0453
Cyclic conjugated diolefins	1.0580
Monocyclic aromatics	1.0626
Tetralins	1.0554
Naphthalenes	1.1082

[a] The values for cyclic compounds represent average values, including those for compounds with alkyl side chains typical of this boiling range.

In the original publication describing the refractivity intercept, the authors used the term "homologous series" in the broad sense to cover repetitive as well as homologous series. This has caused considerable misunderstanding because as Fig. 99.21 shows, the relation between n and d for n-alkyl series is represented, not by a series of parallel lines, but rather by a series of lines with a common point of intersection. However, Kurtz and Ward emphasized that, if a repetitive series is considered (105,113) in which the proportion of ring carbons to chain carbons remains constant, the intercept will remain practically constant. For example, natural rubber, which can be considered $(C_5H_8)_n$, has a refractivity intercept of about 1.0600. This corresponds fairly well with the average for nonconjugated diolefins of lower molecular weight. Hydrogenated rubber has an intercept of 1.0474, which may be compared to the values 1.0458 for average paraffins and 1.0470 for higher-molecular-weight paraffins. Hydrogenated cyclorubber has an intercept of 1.0332, compared to the average value of 1.0326 for dicycloparaffins. The usefulness of the refractivity intercept, despite the confusion that has tended to obscure the validity of its basic premise, cannot be seriously questioned. The refractivity intercept may be used not only for characterization but also for the quantitative analysis of mixtures of paraffins and cycloparaffins containing from 6 to 10 carbon atoms. For this purpose, graphs of intercept versus density similar to that shown in Fig. 99.23 are employed (68).

The relationships between refractive index and density that have proved useful for the analysis of mixtures of hydrocarbons appear to hold considerable promise when applied to other materials as well. For example, van Zijll Langhout has successfully used graphs based on specific refraction and elemental analysis for the determination of the composition of mixtures of sulfur compounds (247).

c. Density and Temperature Coefficient of Density

As has already been mentioned, a linear relation exists between the temperature coefficient of density and the density for various series of hydrocarbons. These linear relationships have been used by Lipkin, Martin, and Kurtz as the basis of methods for the analysis of special hydrocarbon mixtures (123,124). The first method is intended for the determination of the weight per cent of ring carbons in mixtures of paraffins and cycloparaffins. Per cent ring is defined as 100 times the ratio of the ring carbon atoms to the total carbon atoms. These authors plotted density versus the temperature coefficient of density for paraffins and noncondensed cycloparaffins containing no carbons in chains and found

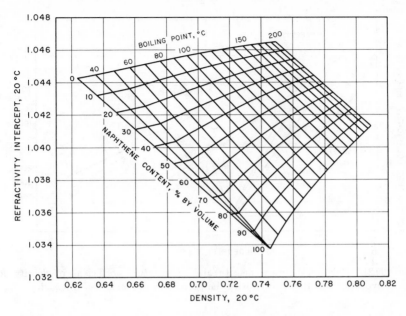

Fig. 99.23. Refractivity intercept graph. Reproduced with the permission of the American Society for Testing and Materials.

the linear relation shown in Fig. 99.24. A third straight line could be drawn through the data for condensed cycloparaffins. As also indicated in Fig. 99.24, the addition of a side chain to a cycloparaffin ring causes the properties to approach those of a paraffin of infinite molecular weight. The extent of approach is roughly proportional to the percentage of chain carbons. It was assumed that, for mixtures of paraffins and cycloparaffins in the region below the density of the limiting paraffin (~0.861), the total percentage of paraffinic chains can be obtained by linear interpolation between the paraffin line and the cycloparaffin ring line along a line of constant density. To avoid the use of graphs, the following equation was derived for $d \leqq 0.861$:

$$\text{Wt \% ring} = \frac{A + 190.0d - 217.9}{0.593d - 0.249}$$

where d = density at 20°C, and $A = -10^5 \times$ temperature coefficient of density, α.

A second equation was derived for the region above a density of 0.861 by assuming that the total percentage of paraffinic chains can be obtained by interpolation between the cycloparaffin ring line and the limiting

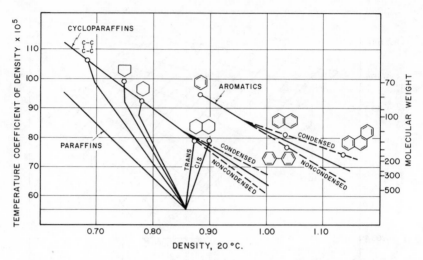

Fig. 99.24. Relation between density and temperature coefficient of density for hydrocarbons. Reproduced with the permission of *Analytical Chemistry*.

paraffin point along a straight line drawn through the point representing the sample. For $d \geqq 0.861$,

$$\text{Wt \% ring} = \frac{A + 102.8d - 142.8}{0.262}$$

In this region the cycloparaffin ring line was constructed midway between the condensed and noncondensed compounds and therefore assumes an equal distribution of the two types. In view of the limited extent of our knowledge, these equations seem to give quite reliable results when used for the analysis of saturated hydrocarbon mixtures from high-boiling petroleum oils.

Lipkin and Martin also considered that the relation between density and temperature coefficient of density for aromatic hydrocarbons could be represented by straight lines (Fig. 99.24). They developed a method analogous to that for saturated hydrocarbons which permitted the estimation of the proportion of aromatic rings to paraffin chains in certain more or less special mixtures of aromatics. Although this method proved useful for a short space of time, it is no longer of interest.

d. Specific Dispersion

For several reasons, dispersion is an attractive physical constant for the characterization of hydrocarbons. It is quite indicative of structure for aromatic hydrocarbons, is independent of errors in absolute units of measurement since it is a differential value, and is affected by temperature to only a minor degree over reasonable intervals. Although the experimental measurement of dispersion is not difficult for transparent samples, highly colored samples become a problem because of the limited transmission of incident radiation.

In the analysis of hydrocarbons the specific dispersion

$$\delta = 10^4(n_F - n_C)/d$$

has received considerably more attention than the dispersion itself. The specific dispersions of saturated hydrocarbons are very nearly the same (98–102) irrespective of molecular weight or structure. The specific dispersion of aromatic hydrocarbons varies from approximately 110 for alkylbenzenes of high molecular weight to over 500 for molecules with condensed ring systems. The dispersions of nonconjugated olefins are intermediate between those of the saturated and the aromatic hydrocarbons. The dispersions of various types of hydrocarbons have been extensively studied (6,225).

The specific dispersions of unsynthesized molecules can be predicted by using the quantitative relationships between dispersion and constitution developed by Thorpe and Larsen (220). The system developed by these investigators allows calculation of the dispersion of a pure hydrocarbon from its molecular structure with a high degree of accuracy. The molecule is considered in terms of the portion that is paraffinic, cycloparaffinic, or unsaturated. Specific dispersion is then calculated using an equation of the general form:

$$\delta = \frac{p}{C} X + \frac{n}{C} Y + \frac{u}{C} Z$$

where

p = the number of paraffinic carbon atoms;
n = the number of cycloparaffinic carbon atoms;
u = the apparent number of carbon atoms associated with ethylenic linkages;
C = the total number of carbon atoms per molecule;
X, Y, and Z = coefficients representing the dispersion contributions of a paraffinic, saturated cyclic system and an unsaturated linkage, respectively.

Corrections that must be made for conjugation and resonance have been detailed by the authors.

Lipkin and Martin (122) found that, if the equation developed for calculation of the refractive index of saturated hydrocarbons from density and temperature coefficient of density (Section IV.B.3) was applied to hydrocarbons containing double bonds, the calculated refractive index was lower than the observed refractive index, and the difference was a linear function of the specific dispersion. They proposed the following equation for calculation of the specific dispersion at 20°C:

$$\delta_{calc} = 98.3 + 2450(n - n_{calc})$$

where n = refractive index for the sodium D line at 20°C, and n_{calc} = refractive index calculated from density and temperature coefficient of density.

For pure hydrocarbons and mixtures of hydrocarbons of nearly the same specific dispersion, such as aromatic extracts, this equation is satisfactory. For mixtures such as ordinary petroleum fractions that contain hydrocarbons of appreciably different dispersions, however, the equation tends to give low results and a small correction (approximately 0.02 times the molecular weight) must be added to the calculated value. By means of these equations the specific dispersion of petroleum oils may be calculated from refractive index, density, and molecular weight with an accuracy almost equivalent to that realized by routine experimental measurement.

If the specific dispersion of homologous series of aromatic hydrocarbons is plotted as a function of the reciprocal number of carbon atoms plus 1, a series of straight lines is obtained that converge at the value of dispersion for the limiting paraffin. The slope of the lines is characteristic of the series. Data for several series of aromatic hydrocarbons are shown in Fig. 99.25. Diagrams of this sort can be quite useful in characterizing relatively homogeneous fractions of high-molecular-weight hydrocarbons.

Lipkin and his associates (127) have described a classification system for aromatics and olefins that they call the "double-bond index." The DBI is a function of the F-C specific dispersion, δ, the molecular weight, M, and the average number of double bonds per molecule, Q:

$$\text{DBI} = \frac{(\delta - 98)(M + 17)}{3190Q}$$

The number of double bonds refers to both aromatic and olefinic unsaturation and is determined by the quantity of hydrogen necessary to saturate the molecule. The DBI sorts most unsaturated hydrocarbons into

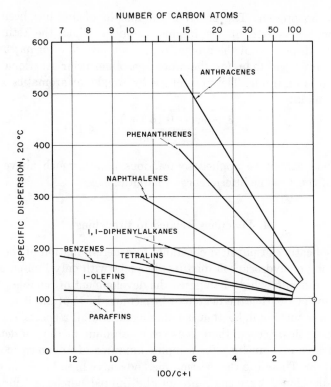

Fig. 99.25.   Specific dispersion of *n*-alkyl series of hydrocarbons.

definite classes. For a given unsaturated system, the DBI is independent of molecular weight. Cyclic and noncyclic monoolefins, isolated polyolefins, and isolated benzenes have a DBI of about 1. Conjugated diolefins and naphthalenes have a value of 2, while phenanthrenes and anthracenes exhibit values of 2.5 and 4, respectively. The authors have described the application of the DBI to a number of petroleum distillate fractions.

Grosse and Wackher (69) have described a method for the quantitative determination of aromatics in gasoline and naphtha fractions that is based on specific dispersion. Since all saturated hydrocarbons have a specific dispersion of about 98, the concentration of aromatic hydrocarbons present can be determined from a measurement of the specific dispersion of the sample if the average specific dispersion of the pure aromatics can be estimated with sufficient accuracy. A correction must also be employed to take account of the dispersion contribution of any olefins that may be present. The contribution due to olefins has been found to be about one-sixth of the bromine number. Since the estimate of the specific dispersion of the aromatics is more accurate for narrow-boiling-range fractions than for the whole gasoline, prior fractionation by distillation is desirable. The percentage by weight of aromatics is given by the formula:

$$\%A = \frac{\delta - 99 - 0.16 \text{ Br No.}}{\delta_{\text{arom}} - 99} \times 100$$

The method cannot be applied to fractions boiling much above 200°C because of the possible interference of naphthalenes.

4. Functions Related to Branching

The application of physical constants to a determination of the degree of branching of hydrocarbon mixtures has met with only limited success. Two main difficulties have hampered the development of reliable physical methods for evaluating the degree of branching. First of all, it is difficult to define a branching index that is unambiguous in all respects, especially for cycloparaffins. Second, there is as yet no absolute method of determining the degree of branching for complex mixtures of unknown molecular configuration. Therefore the existing methods have been based on general relations developed by using data for the limited number of individual hydrocarbons that have been synthesized. These methods are, as a consequence, somewhat limited in scope when applied to complex mixtures, and are at best only semiquantitative. The two most widely explored physical constants that possess the necessary sensitivity to the presence of branchings are the magneto-optical rotation and the parachor.

When placed in a magnetic field, substances that may not normally be optically active are able to rotate the plane of polarization of light. The angle of rotation depends on the nature of the substance and may be expressed by the following formula:

$$\alpha = VHL \cos \beta$$

Here V is called the Verdet constant. It is only slightly dependent on temperature but varies with the wavelength of the light. H is the strength of the magnetic field in gauss, L is the length of the path traveled by light through the substance, and β is the angle between the beam of light and the magnetic field lines. The Verdet constant of saturated hydrocarbons has been related to the degree of branching as defined by Westerdijk (232). However, the experimental determination of the Verdet constant is extremely difficult. Even when relative methods are used, rather complicated apparatus and skillful operation are required to obtain reasonable accuracy. For this reason, the use of the Verdet constant will not be discussed further.

The molecular parachor, P, was introduced as an additive function by Sugden (214) and takes the form:

$$P = \frac{\sigma^{\frac{1}{4}} M}{d - D_v}$$

where σ = the surface tension in dynes per centimeter, d = the density of the liquid, M = the molecular weight, and D_v the density of the vapor in equilibrium with the liquid under the conditions of measurement. When P is determined at 20°C, the value of D_v may be neglected for values of molecular weight greater than 150. This function has been used successfully by Waterman, Leendertse, and their associates (229) for estimating chain branching in terms of

φ = the number of n-alkyl groups, minus 1

Using API Project 44 data for normal alkanes, n-alkylcyclopentanes, and n-alkylcyclohexanes, these investigators constructed diagrams of specific refraction and specific parachor as a function of the reciprocal molecular weight. They also included lines representing bicyclic naphthenes, tricyclic naphthenes, etc., by making use of a general equation proposed by Geleen (62) for relating a specific additive function with the number of rings and the molecular weight of saturated hydrocarbon mixtures. These diagrams are shown in Figs. 99.26 and 99.27. For nonbranched saturated hydrocarbon mixtures, the average number of rings per molecule can be read directly from either of these diagrams. Chain branching, however, lowers both the specific refraction and the specific parachor, and results in an apparent ring number approximately 0.05 ring too high per branching unit for the specific refraction diagram and 0.14 ring too high in the specific parachor diagram. Consequently, it can be shown that the degree of branching is given approximately by the relation

$$\varphi = 10(R_T{}^p - R_T{}^r)$$

and R_T, the true average number of rings per molecule, is given by

$$R_T = R_T{}^r - \tfrac{1}{2}(R_T{}^p - R_T{}^r)$$

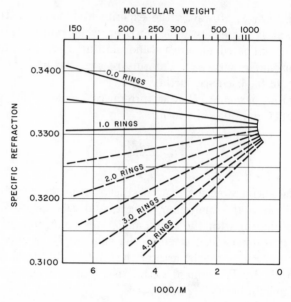

Fig. 99.26. Specific refraction versus reciprocal molecular weight for saturated hydrocarbons. Reproduced with the permission of Elsevier Publishing Company.

in which $R_T{}^r$ and $R_T{}^p$ are the apparent ring numbers estimated from the specific refraction and the specific parachor diagrams, respectively.

Since this method of branching analysis is based predominantly on reliable data for pure individual hydrocarbons, it holds for widely differing types of branched as well as unbranched saturated hydrocarbon mixtures. It is particularly useful for the analysis of saturated polymers. In spite of the fact that physical methods for branching number have been rather extensively investigated, it is likely that instrumental methods such as infrared and nuclear magnetic resonance spectroscopy will prove, in the long run, to be the most fruitful approach to the problem.

C. HYDROCARBON ANALYSIS OF THE LOWER-BOILING FRACTIONS OF PETROLEUM

As was emphasized earlier, the compound-by-compound analysis of even the fractions of petroleum boiling below 200°C, as practiced by API Project 6, is impractical in the industrial laboratory. Substantial progress has been made, however, toward the goal of inexpensive individual hydrocarbon analysis by the application of mass spectrometry and gas chromatography. Indeed, it is now possible to determine the indi-

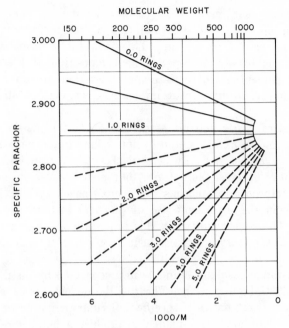

Fig. 99.27. Specific parachor versus reciprocal molecular weight for saturated hydro-carbons. Reproduced with the permission of Elsevier Publishing Company.

vidual hydrocarbons in mixtures of as many as 40 or 50 components in a matter of a few hours. Although the contributions that instrumental methods have made and will undoubtedly continue to make cannot be ignored, the obvious fact is that this kind of instrumentation is beyond the reach of many laboratories. In addition, such detailed analyses are, in many instances, not required.

For these reasons, combinations of physical and chemical properties will continue to play an important role in the analysis of the lower-molecular-weight hydrocarbons. We have seen that there are a number of physical constants that can serve as indicators of composition. It follows that there are a great many ways in which such physical constants, or functions thereof, may be combined with chemical reactivity for analytical purposes. As one might anticipate, a number of methods have been put forward for the analysis of low-boiling hydrocarbons. The purpose of this section is not to debate the relative merits of these proposals, but rather to illustrate the practical application of physical property methods by a discussion of one or two firmly established procedures.

Most methods that have been proposed as routine analytical tools have as their objective the determination of composition in terms of the various classes of compounds. This is a form of molecular type analysis; that is to say, it reports the composition of the sample in terms of the percentages of certain molecules that are alike in regard to structure and reactivity—for example, paraffins, cycloparaffins, olefins, and aromatics.

The method of Kurtz and his associates (111) for the determination of olefins, aromatics, paraffins, and cycloparaffins (naphthenes) has been selected as an example, not only because the general principles on which it is based have received acceptance, but also because it incorporates a system of internal cross checks. Such cross-check systems are indispensable aids in the intelligent use of physical methods in analytical applications. The steps involved in this procedure are diagramed in Fig. 99.28. A brief summary of the salient features of the analysis is as follows:

1. The full-range gasoline sample is first separated by distillation into fractions of 25°C boiling range. The distillation cut points are chosen in such a way that aromatics of specific molecular weight occur in certain fractions. The distillation is intended to produce a number of fractions that are more homogeneous in composition than the original sample and therefore lend themselves more readily to analysis by physical methods.

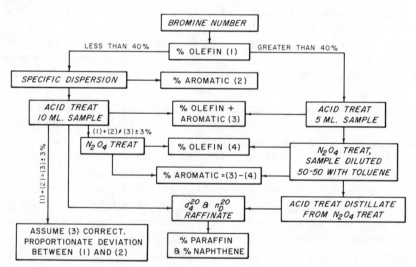

Fig. 99.28.   Scheme of analysis for olefinic gasolines. Reproduced with the permission of *Analytical Chemistry.*

2. The percentage of unsaturated (olefinic and aromatic) hydrocarbons is determined by reaction with a mixture of phosphoric and sulfuric acids under controlled conditions at ice temperature (Section IV.A.2).

3. The olefins are determined by bromination, using a bromide–bromate reagent as titrant (Section IV.A.2). The following relation is used to calculate the per cent olefin from the bromine number:

$$\% \text{ Olefin} = \frac{\text{Br No.} \times \text{MW}}{160}$$

The authors give a table from which the molecular weight of the olefins in each fraction may be estimated.

4. The aromatic content of each fraction is calculated from specific dispersion and bromine number, using the procedure described by Grosse and Wackher (Section IV.B.3.d):

$$\text{Vol} \% \text{ aromatics} = \frac{\delta - 99 - 0.16 \text{ Br No.}}{\delta_{\text{arom}} - 99} \times \frac{d}{0.868} \times 100$$

In this equation, δ is the specific dispersion, and d the density of the fraction. The value 0.868 is taken as the average density of monocyclic aromatics. The authors give values for the specific dispersion of the aromatics in each fraction that can be substituted in the denominator of the formula.

5. A cross check of the data obtained in steps 2, 3, and 4 is carried out by adding the olefins as determined in step 3 to the aromatics determined in step 4 and comparing this value to the sum of the olefins and aromatics determined in step 2. If the analysis is correct, the two values should agree within 2–3%.

6. If the values do not agree, the olefin content is redetermined, using the nitrogen tetroxide method of Bond (Section IV.A.2). The new value is then subtracted from the total of olefins and aromatics determined in step 2, and the difference taken as the percentage of aromatic hydrocarbons.

7. The unreacted material from step 2 (saturated hydrocarbons) is analyzed for cycloparaffins by means of physical properties. A graph of refractivity intercept versus density (Fig. 99.23) is used. A check of the complete removal of unsaturated components can be obtained by comparing the average boiling point estimated from this figure with the mid-boiling point of the fraction. If the estimated boiling point is more than 10° higher than that observed, it is indicative of incomplete removal of aromatics and olefins.

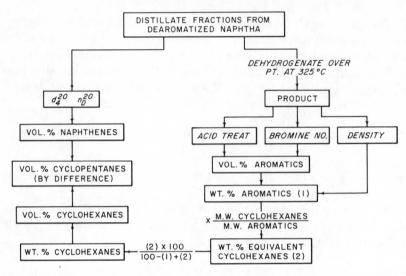

Fig. 99.29. Scheme of analysis for the determination of cyclopentanes and cyclohexanes in light naphthas.

This procedure is an excellent example of the way in which physical and chemical methods may be combined in a complementary fashion to define composition in terms of the concentration of hydrocarbons of similar structure and reactivity. In discussing their procedure, the authors suggest checking the experimental determination of dispersion by comparing it to the value calculated from density, refractive index, and molecular weight, using the method of Lipkin and Martin (Section IV.B.3.d). They suggest also that a comparison of the refractive index calculated from the density and molecular weight of the saturated portion (Section IV.B.3) with that observed is a fairly sensitive test for incomplete removal of aromatics and olefins. The cross checks cited by Kurtz and his co-workers are extremely valuable aids and serve to emphasize the care with which physical methods must be applied in order to obtain meaningful results when one is dealing with complex mixtures.

A procedure that was used to advantage for a time in the author's laboratory to distinguish between cyclohexane and cyclopentane rings in light naphtha (boiling range 145–220°F) fractions is shown schematically in Fig. 99.29. In this scheme a combination of catalytic dehydrogenation, distillation, acid absorption, and refractivity intercept allows the determination of methylcyclopentane, cyclohexane, methylcyclohexane, and 7-carbon cyclopentanes.

These are but two examples of the many possible ways in which physical and chemical methods may be used to characterize mixtures of the lower-boiling hydrocarbons. The exploitation of useful combinations is limited only by the imagination of the investigator and the nature of the problem at hand.

V. METHODS OF SEPARATION

In the analysis of a complex mixture it is often desirable to attempt some simplification of the mixture by the application of suitable separation techniques. Some prior segregation of molecules in regard to size, type, or reactivity serves two useful purposes. First, it often makes the application of physical methods a simpler task, and, second, almost without exception it provides access to more information about constitution than could be obtained from examination of the unfractionated material. Because of the important place that physical separation techniques have assumed in the analysis of hydrocarbon mixtures, and because of their close association with physical property methods of analysis, no discussion of this subject would be complete without some brief description of the separation techniques that have been most frequently and most successfully applied.

It is the purpose of this section, not to deal with the topic of separation principles in detail, but to demonstrate to the reader, by discussion and illustration, the way in which these methods, either singly or in combination, can be brought to bear on the problems peculiar to hydrocarbon analysis. For a comprehensive treatment of these topics the reader should consult Section C: "Separation: Principles and Technics" of Part I, Volumes 2 and 3, of this Treatise. Chapters 27, 29, 31, 32, and 34 will be particularly useful.

A. DISTILLATION

A thorough discussion of the principles of distillation and its application for analytical purposes is given in Volume 2, Chapter 29.

1. Straight Rectification

Separation by distillation takes place according to volatility. Although for a given homologous series the boiling point rises with increasing molecular weight, the distillation of complex mixtures of molecules of different types does not necessarily produce fractions containing compounds of similar molecular weights. As a matter of fact, in the case of hydrocarbons it is possible for a narrow fraction with a boiling point of approximately 350°C to contain a paraffin with a molecular weight of about 310, as well as a trinuclear aromatic of molecular weight 180.

Distillation is effective as a separation tool only if adequate rectification of the distillate is provided for. The rectifying section of distillation equipment is generally constructed as a vertical cylinder (distillation column) between the boiling vessel and the condenser. Part of the condensed distillate (reflux) is returned to this column in a manner countercurrent to the ascending vapor from the boiling vessel. The ratio between the amount of distillate returned to the column and the amount withdrawn as product is generally termed the "reflux ratio." The descending liquid should come into intimate contact with the ascending vapor. Discrete plates or trays may be constructed as an integral part of the column, or it may simply be filled with a suitable packing material, such as glass or steel helices or gauze packing, to promote close contact between the vapor and returning liquid. The amount of material that resides in the rectifying section is termed the "holdup."

To express the separating power of the rectification section, it is generally compared to an idealized plate column (vapor and liquid in equilibrium on each plate) that produces the same degree of separation for the same mixture. The separating power of a column is therefore expressed as a number of "equivalent theoretical plates," that is to say, as the number of stages in an idealized column to which it would be equivalent. The number of theoretical plates is generally determined at total reflux, using a binary mixture that obeys Raoult's law and whose components exhibit a constant ratio of vapor pressure in the temperature range between their boiling points. A convenient mixture for the calibration of laboratory columns is n-heptane–methylcyclohexane (18).

For most efficient operation it is desirable to minimize heat exchange between the rectifying section and its surroundings. This may be done by using silvered vacuum jackets or electrically heated jackets with sections that may be independently controlled so as to adapt the ambient temperature to the gradient that exists in the column.

The primary factors that determine how well a given separation can be accomplished by distillation are the separating power and holdup of the apparatus, the volume to be distilled, and the reflux ratio applied. The highest separating efficiency is realized at infinite reflux ratio, that is, by returning all condensed vapor to the column without withdrawing any product. Estimation of the reflux ratio necessary to produce a given separation is rather difficult. In any case, the reflux ratio should not be chosen lower than the theoretical plate rating of the column; for example, for a 50-plate column the reflux ratio should be at least 50.

In batch distillation, as the amount of holdup relative to the volume in the boiling flask increases, the sharpness of the separation decreases. When relatively small amounts of liquid must be distilled, the apparatus must be constructed so as to have as low a holdup as possible.

As high a boilup rate as is practicable is generally to be preferred in laboratory distillations, since it reduces the time required to complete a distillation. The limit of boilup rate in packed columns occurs when "flooding" takes place, that is, when columns of liquid are formed by the returning condensate through which the ascending vapor actually bubbles.

The factors discussed so far are obviously interdependent. An examination of their relationship to one another has led Podbielniak to suggest a simple expression that is useful in comparing one distillation column with another (178).

It is often impossible to separate compounds of similar boiling point by straight rectification. In such cases recourse may be had to azeotropic distillation, extractive distillation, or distillation at reduced pressures. It may occasionally be necessary to resort to various combinations of these techniques in order to effect the desired fractionation.

2. Azeotropic Distillation

Azeotropic distillation of hydrocarbons depends on the addition of an agent which has the power of forming azeotropes with some of the components of the mixture. In azeotropic distillation the added compound is of the same general volatility as the compounds to be separated and can be introduced into the boiling flask with the original charge. In general, polar organic compounds such as alcohols or nitriles are used. With few exceptions, these form minimum-boiling azeotropic mixtures with paraffinic, cycloparaffinic, aromatic, and olefinic hydrocarbons.

For the binary minimum-boiling azeotropic mixtures formed between a polar organic compound and a paraffin, cycloparaffin, and aromatic hydrocarbon of the same boiling point, the departure of the partial and total vapor pressures from ideality will be greatest for the paraffin azeotrope, intermediate for the cycloparaffin, and least for the aromatic azeotrope; that is to say, the boiling point lowering is greatest for the paraffin azeotrope and least for the aromatic azeotrope. By way of illustration, small quantities of saturated hydrocarbons may be separated from aromatics of about the same boiling point by azeotropic distillation with a polar organic compound.

Certain fluorocarbons have been found to behave in a way exactly opposite from the polar azeotrope agents (133). They produce the greatest boiling point lowering for aromatic hydrocarbons, and the least for paraffins. Thus, alternate azeotropic distillation with polar organic compounds and with fluorocarbons is often a very effective way to achieve separations that cannot be accomplished in any other way. Azeotropic distillation is, for the most part, more effective if applied to fractions that have first been segregated into molecules of roughly similar size by simple rectification. Rossini has discussed the azeotropic distillation of hydrocarbons in considerable detail and has summarized in a general way the separations that may be accomplished (188).

3. Extractive Distillation

Extractive distillation, which is often useful in separating different types of hydrocarbons of similar boiling point, is in principle a vapor–liquid partition. In practice, a polar liquid of considerably higher boiling point than the mixture to be distilled is allowed to flow down through the rectification section along with the returning reflux. The presence of the solvent results in an increase in the volatility differences between certain components and permits their separation by rectification. Extractive distillation has been little applied as a laboratory tool because proper control of the operation is rather difficult and necessitates the use of somewhat complicated equipment. The relative merits of extractive and azeotropic distillation have been discussed by Carlson (24).

4. Distillation at Reduced Pressures

When a mixture of hydrocarbons contains components that have substantially the same boiling point at the normal distilling pressure but significantly different boiling points at a different pressure, a separation can often be achieved by distilling the mixture alternately at the two pressures. Rossini has found that this procedure is particularly useful in separating cycloparaffins from paraffins and mononuclear from dinuclear aromatics. As pressure is decreased, the average cycloparaffin becomes relatively more volatile than an average paraffin of the same volatility at the initial pressure. The behavior of dinuclear relative to mononuclear aromatics is similar.

B. SOLVENT EXTRACTION

The simplest example of solvent extraction is seen when a sample is brought into contact with a solvent in which it is not completely miscible, with the result that, on standing, two phases are formed. The action of the solvent is to distribute the components of the sample between the two phases. The part that is dissolved in the solvent is generally referred to as the extract, and the undissolved part as the raffinate. The effectiveness of extraction is dependent largely on solubility. The ratio of the concentration of a certain component in the two phases (distribution coefficient) is in many cases nearly constant and depends only slightly on the concentration of other components.

The basis of the use of solvent extraction as a separation method for hydrocarbons is that various components of a mixture exhibit different distribution coefficients, so that the ratio of their concentrations in the solvent phase differs from that in the original mixture. Since the type and not the size of a hydrocarbon molecule is mainly responsible for its solubility, solvent extraction separates primarily according to type. However, some fractionation according to molecular size may also occur, and it is therefore advisable to apply the extraction to rather narrow-boiling mixtures. The stringency of this requirement depends on the change in solubility with molecular weight. Some idea of the extent to which this may vary may be gained by inspection of Fig. 99.17, which is a plot of the aniline point as a function of the number of carbon atoms for several different types of hydrocarbons. Aniline point is indicative of the solubility in aniline, with the hydrocarbon of lowest aniline point being the most soluble and that of highest aniline point the least soluble. The figure shows that there is a decrease in the solubility of a given type of hydrocarbon as the number of carbon atoms in paraffinic chains is increased. It can be seen that a mixture of paraffins and cycloparaffins that is intended for separation by extraction with aniline should be homogeneous in regard to molecular size. Differences in the number of carbon atoms per molecule between the two types should not exceed 1 at the 25-carbon-atom level. If, however, the separation of aromatics from paraffins and cycloparaffins is intended, variation in molecular size of 5–10 carbon atoms may be tolerated.

The separation achieved after a single treatment with solvent is never complete. The simple batch extraction of naphthalenes from a concentrate of catalytic cycle oil aromatics is illustrated in Fig. 99.30. In this figure a triangular phase diagram is represented, the apices of which indicate pure solvent, A, naphthalenes, B, and cycle stock aromatics,

C. If we start with material of composition *S* (33% naphthalenes) and add solvent, the composition moves along the line *SA*. When a sufficient quantity of *A* is added, the gross composition *O* comes within the demixing area. After being shaken, this system separates into an extract layer of composition *T*, and a raffinate layer of composition *L*. If, after this first stage the layers are segregated and the solvent is removed from each, the composition follows the lines *ATE* and *ALR*. The result is a partial separation of *S* into an extract, *E*, and a raffinate, *R*, which are richer and poorer, respectively, in naphthalenes.

A number of procedures have been devised to increase the efficiency of batch extractions. For example, instead of removing the solvent completely, it can be removed only partially from *T* to give composition *P*, which separates into two layers of compositions *H* and *J*. Segregation of these layers and removal of solvent gives an extract and a raffinate of composition *F* and *G*, respectively. Alternately, more charge stock, *S*, may be added to the extract layer *T* and more solvent to the raffinate layer *L*, giving two new pairs of layers. The difficulty with such repetitive (cross-current) batch extractions is that the desired portion of the sample is generally obtained in very small yield. Countercurrent extraction, on the other hand, accomplishes such repetition almost automatically and generally leads to much better yields. In this process the two liquid phases are passed in opposite directions through the extraction system.

For separating a normally liquid mixture it is often useful to use only one solvent and to let the mixture, together with a small amount of solvent, constitute the other liquid phase. One of the disadvantages of using a single solvent, however, is that only one component of a binary mixture may be obtained in the pure state. By using two mutually immiscible solvents in a countercurrent fashion, it is theoretically possible to obtain both components of a binary mixture in reasonable purity. Rossini (187) has described several convenient continuous laboratory extractors for use with a single solvent or with two solvents. The results obtainable with continuous batch extraction apparatus can generally be approached by the application of a series of mixing–settling operations performed manually with a number of separatory funnels. Various schemes for both cross-current and countercurrent extraction have been described (37).

It is difficult to put forward any hard and fast rules for the selection of the most suitable solvent or pair of solvents for a given separation. Consideration of a list of common solvents arranged in increasing order of polarity is often helpful in selecting solvent pairs, since the compounds at either extreme of such a list generally meet the requirement of im-

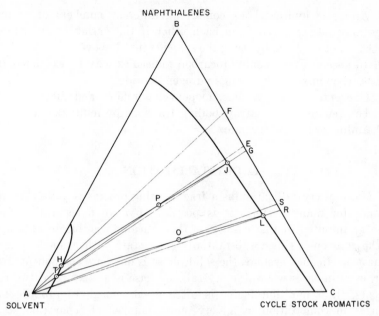

Fig. 99.30.   Diagram of simple batch extraction.

miscibility. Often it is necessary to determine some ternary diagrams (Fig. 99.30) or to evaluate the distribution coefficients of model compounds before the proper choice can be made. The difference between the critical solution temperatures of the solvent with the two pure components or groups of constituents may often be used as a rough measure of the selectivity of which a solvent is capable, provided the extent of solubility is satisfactory. Extensive compilations of the critical solution temperatures of hydrocarbons in various solvents have been published (60). For a solvent to manifest its maximum selectivity, the operating temperature should be chosen in the vicinity of the lower of the two critical solution temperatures of the two components with the solvent. A more detailed treatment of the principles and practice of solvent extraction is contained in Chapter 31, Volume 3, of this Treatise.

With regard to hydrocarbons, the following separations can be performed by extraction:

1. Aromatic hydrocarbons can be separated from paraffins and cycloparaffins with polar solvents such as furfural, liquid sulfur dioxide, glycols, nitrobenzene, and methylcyanide. About six ideal stages are required for essentially quantitative separation.

2. Aromatic hydrocarbons containing different numbers of aromatic rings may be separated from each other if the variation in molecular weight is not too large by using the solvents listed above.

3. In special cases, nonhydrocarbon molecules may be extracted from hydrocarbon mixtures by using hydrogen fluoride.

4. Concentrates of paraffins, monocycloparaffins, and dicycloparaffins may be prepared from narrow-boiling fractions by multistage extraction with aniline or methyl formate.

C. CRYSTALLIZATION

The use of crystallization as a fractionating process is quite attractive because for many systems it is possible to have a separation factor close to infinity. In this case, it is theoretically possible to obtain one of the components in a substantially pure state with a single stage of separation. In such systems the solid phase is one pure component. However, even in cases where the crystalline phase is a mixture of two components, the separation factor is often great enough to produce large changes in composition in a very small number of separation stages.

Before the process of crystallization can be applied, it is desirable to have the component to be crystallized in as high a concentration as possible. Such a concentration is usually achieved by a suitable combination of other fractionating processes. In contrast to distillation, which separates largely by molecular size or boiling point, and to azeotropic distillation, extraction, and adsorption, which separate largely according to molecular type, crystallization separates largely by freezing point. For unassociated liquids the freezing point is a property that depends to a great extent on molecular size and molecular symmetry. For a given molecular weight, the freezing point is generally higher for molecules of greater symmetry and lower for those of lower molecular symmetry. The use of multistage crystallization on a laboratory scale has been discussed thoroughly by Tipson (222).

In crystallization it is often desirable to add a solvent which will remain liquid and nonviscous at the temperatures likely to be encountered, and which can be removed readily by evaporation or simple extraction with water. For purposes of characterization it is often desirable to remove the maximum amount of a given material of high freezing point from its mixture with other components. In this case it is frequently advantageous to conduct the initial crystallization without solvent, and then to carry out purification of this concentrate by subsequent crystallizations from the appropriate solvent.

In certain cases the solvent(s) is chosen in such a way that advantage may be taken of its ability to preferentially dissolve certain components of the mixture during the crystallization process. In this case, both solubility and melting point are responsible for the separation obtained and the process is therefore one of "extractive crystallization."

Fractional crystallization is often an extremely useful way of concentrating small quantities of impurities from substantially pure materials for further characterization and analysis. A very convenient way of performing multistage crystallizations for this purpose (and of course for the purpose of purification as well) is to employ the technique of zone melting. In this procedure, a narrow molten zone is caused to pass repeatedly along a solid column of the material at a very slow rate. If the mixture obeys Raoult's law, the liquid zone is gradually enriched in the impurities, which are therefore eventually concentrated at one end of the column. The theory and practice of zone refining have been discussed in detail by Pfann (176).

The use of crystallization for analytical purposes is discussed in further detail in Chapter 32, Volume 3, of this Treatise.

D. FORMATION OF SOLID MOLECULAR COMPOUNDS

When two or more different kinds of molecules are firmly held together without the influence of strong chemical bonding forces such as ionic or homopolar bonds, we may call the result a "molecular compound." The formation of solid molecular compounds, as well as their decomposition into the constituent molecules, is strongly influenced by the environment and temperature. High concentrations of the components and low temperatures favor the formation of solid molecular compounds, whereas low concentrations and somewhat elevated temperatures favor regeneration of the constituents. The formation of solid molecular compounds and subsequent regeneration of the components can often be used for the effective separation of certain classes of compounds.

Powell (180) has described three types of molecular compounds, which he classifies as follows:

Type I. Powell attributes the formation of a compound of this type to some residual form of attraction between the components that may possibly involve specific parts or features of the molecules. The presence of certain groups in both of the constituent molecules is required, but the size and shape of the molecules may vary considerably. An example of this type consists of the compounds formed by the interaction of aromatic hydrocarbons with certain polynitro compounds. Alkylbenzenes, naphthalenes, anthracenes and phenanthrenes, and other condensed poly-nuclear aromatics form solid molecular compounds with polynitro com-pounds such as picric acid, trinitrobenzene, and trinitroresorcinol.

The stability of the compounds formed between alkylbenzenes and picric acid increases with increasing alkyl substitution. The compounds formed between picric acid and pentamethyl- or hexamethylbenzene are relatively stable. Compounds formed by picric acid with naphthalene and other condensed ring aromatics are also quite stable. The formation of these molecular compounds appears to depend primarily on the pres-ence of the aromatic ring system in the one component and the three nitro groups in the other.

Type II. The molecular compounds of this type consist of a framework of one component within which the molecules of the second component are laterally enclosed in channels. These compounds are therefore held together to a greater extent by the geometry of the arrangement of molecules than by attractive forces between the two kinds of molecules. The channels of the arrangement formed by the molecules of one com-ponent are such that the molecules of the other component may have different lengths but must have approximately the same cross section. The number of molecules that may be contained within the channels depends on the length of the channels and of the molecules. Usually there is not a simple ratio between the number of molecules in channels and the number of molecules of the channel-forming component. Exam-ples of this type of compound are the products formed by urea with straight-chain paraffin hydrocarbons, alcohols, esters, aldehydes, ketones, and carboxylic acids. Branched or cyclic compounds generally have cross sections too large to permit them to enter into such compound formation with urea. However, they are able to form analogous compounds with thiourea. Another example of the type II compounds are the products that desoxycholic acid forms with normal paraffin hydrocarbons and normal alkyl carboxylic acids.

Type III. The compounds of this type are similar to those of type II in that the geometry of the molecules is a far more important consid-eration than the existence of attractive forces. A molecular compound of this type consists of a framework of one component within which the molecules of the second component are trapped. Powell has suggested the name clathrate or cage compound for this type of structure. The second component must consist of molecules that are of an appropriate size to fit into the cavities formed by the first component. In this case, all of the molecular dimensions are important in determining whether a molecular compound will be formed. There is a limiting composition, which is determined by the number of cavities formed relative to the number of molecules forming the framework. However, since all of the cavities may not be occupied, a simple ratio will not always exist between the number of trapped molecules and the number of molecules forming the cage. Examples of compounds of this type are those which the nickel cyanide ammonia complex forms with benzene, thiophene, pyrrole, aniline, and phenol. A considerable number of these so-called Werner complexes have been disclosed (182,199). The complexes formed between 1 mole of nickel thiocyanate and 4 moles of a substituted primary aromatic amine are capable of forming clathrates with a great many aromatic hydrocarbons. Most of these complexes are extremely selective for molecules of specific dimensions and can produce separations of certain isomers that are rather difficult to achieve by other means.

The application of molecular compound formation as a separation tool is more or less self-evident. The mixture to be separated is brought into contact with an excess of the compound-forming reagent (urea, thiourea, etc.) under conditions favorable for the reaction. When the reaction is complete, the solid molecular compound is removed from the mixture by filtration and washed, and the separated material recovered by decomposition of the molecular compound. This can generally be accomplished by raising the temperature or introducing a solvent in which the molecules forming the framework are soluble but the trapped molecules are not. The conditions necessary for the effective application of urea have been discussed by Zimmerschied and Redlich and their co-workers (185,248). Schiessler and Flitter (201) have discussed the formation of compounds with thiourea in some detail. Details of the preparation and use of Werner complexes have been described by de Radzitsky and Hanotier (182).

E. LIQUID CHROMATOGRAPHY

1. Adsorption Chromatography

A number of porous solids, such as clays, charcoal, silica gel, or alumina, possess the property of being able to take up large amounts of organic substances both from solution and from gases. Solids that exhibit this property are generally termed "adsorbents"; they owe this behavior to the presence of millions of submicroscopic pores which produce an extremely large available surface area. When an adsorbent is brought into contact with a mixture, the surface of the adsorbent often exerts a selective action that results in an enrichment of the adsorbed molecules in certain components of the mixture. This effect may be extremely sensitive to small differences in molecular structure. The separating effect of adsorption can be greatly enhanced by percolation of the liquid mixture through a vertical column filled with adsorbent. The adsorption column is basically a method for countercurrent application of adsorption. Although the adsorbent is held motionless, it is obvious that the liquid and the adsorbent move in opposite directions relative to each other. Adsorption is a distribution process and hence is analogous to other distribution processes, such as fractional distillation and countercurrent liquid–liquid extraction.

The commonly practiced procedure for the resolution of a mixture is to introduce a small quantity of the sample into the adsorption column and to follow it with a suitable solvent to effect transport of the material out the bottom of the column. During such "development," the less strongly adsorbed compounds move downward at a faster rate than those that are more strongly adsorbed, and in many cases complete separation may be attained. Development is continued until the components appear in the percolate, from which they are recovered by removal of the solvent. This procedure is generally known as "separation by elution."

It is also possible to carry out the development with a solvent that is more strongly adsorbed than any of the components of the mixture. When this is done, the whole sample is forced downward toward the end of the adsorbent bed. Under these conditions the least adsorbed molecules migrate most rapidly and emerge first, while the most strongly adsorbed molecules emerge just before the solvent front. This technique, which is termed "displacement," is generally somewhat less versatile than the elution procedure. Usually the separation that can be obtained by displacement is less than that possible by development with a weakly adsorbed solvent. The displacement method does have the advantage that it is more rapid than elution and if carefully carried out does not require the removal of solvent from the fractions obtained.

Adsorption has been used with considerable success for the separation, purification, and analysis of hydrocarbons (79,132,134,241). Although the use of a large number of adsorbents has been reported, silica gel and alumina have proved the most effective for the separation of mixtures of liquid hydrocarbons. As a general rule, hydrocarbons containing larger numbers of double bonds are more strongly adsorbed than those containing fewer double bonds. Hirschler (81) states that adsorption affinity on silica gel increases in the following order: saturates, monoolefins, diolefins, monocyclic aromatics, polycyclic aromatics. As a rule, the more cyclic the molecule the more strongly it is adsorbed. However, the difference in adsorbability between paraffins and cycloparaffins is far less than the difference between cycloparaffins and aromatics. Within a given homologous series, the adsorbability is a function of molecular weight. Silica gel preferentially adsorbs the lower-molecular-weight members, whereas the reverse is true for activated carbon and alumina.

Molecules that exhibit very subtle differences in structure are often susceptible to separation by adsorption. For example, a polyalkylbenzene is more strongly adsorbed on silica gel than a normal alkylbenzene with the same number of carbon atoms. In the case of mononuclear aromatics with two alkyl groups attached to the benzene ring, the *ortho* isomer is more strongly adsorbed than the *para* isomer (135). These differences in adsorbability make adsorption a powerful tool for the separation of complex mixtures. For a comprehensive treatment of its analytical applications, the reader should consult Chapter 34, Volume 3, of this Treatise.

The utility of adsorption for the separation and determination of aromatic hydrocarbons in oil fractions was demonstrated by Mair and Forziati (134). Lipkin (119) has described a convenient elution procedure for the determination of aromatics in hydrocarbon mixtures boiling above 200°C. The accuracy is about 1%, and both fractions are recovered for further investigation. Criddle and LeTourneau (38) developed a rapid displacement procedure for gasoline hydrocarbons in which fluorescent dyes are used to locate zones in the column corresponding to the saturated, olefinic, and aromatic hydrocarbons. This procedure has been modified so as to be applicable to the analysis of higher-molecular-weight oils (96) and mixtures containing oxygenated solvents (95). Snyder (211) has described the use of linear elution adsorption chromatography for the efficient separation of compound classes over alumina and silica gel. Jezl (85) has used calcined alumina to concentrate sulfur compounds from gas oils and lubricating oils. Nitrogen compounds have been separated from shale oil by Smith and his co-workers using Florisil, a synthetic magnesium silicate (209). O'Donnell has investigated the composition of asphalts, using adsorption on Attapulgus clay (174).

In the laboratory application of adsorption, the simple displacement procedure is generally quite useful for the separation of relatively uncomplicated mixtures of low or moderate molecular weight. For complex mixtures of high molecular weight, such as are encountered in oil constitution research, however, the elution technique has generally proved superior. In some cases it is possible to accomplish the desired separation by using high ratios of adsorbent to sample and employing a single eluent. As an example, saturated hydrocarbon mixtures of narrow boiling

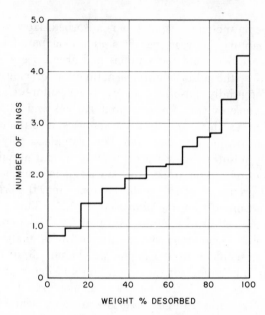

Fig. 99.31. Separation of a saturated hydrocarbon oil by pentane elution.

range can be separated according to increasing number of cycloparaffin rings by careful elution from silica gel with pentane (79). Figure 99.31 shows the separation, by means of this technique, of the saturated portion of a 700–800°F boiling-range fraction from a Gulf Coast crude. The ratio of adsorbent to oil was approximately 40:1. The least cyclic hydrocarbons were eluted first, and the more polycyclic material last. The ring numbers were calculated from specific refraction and molecular weight, using the equation of Vlugter, Waterman, and van Westen (Section IV.B.3.a). The saturated and aromatic portions of this fraction were first separated according to Lipkin's procedure.

It is also possible to separate mixtures of aromatic hydrocarbons according to increasing number of aromatic rings by elution with a low-molecular-weight saturated hydrocarbon. However, the volumes of eluent required are impractically large. A much more convenient way to accomplish separations of aromatic hydrocarbons is to use a series of developers of increasing adsorbability. If the volume of each developer is chosen correctly, each desorbent will elute all the materials that are less strongly adsorbed than itself. Hirschler and James (80) have termed this mode of operation "consecutive multiple displacement development." Results of the application of this technique to an aromatic concentrate from a Gulf Coast distillate, using alumina as the adsorbent, are shown in Fig. 99.32. The separation according to number of aromatic rings is reasonably sharp, and the action of successive developers is clearly indicated by the peaks and valleys in the plot of effluent oil concentration versus per cent of charge desorbed. Similar separations employing other developers have also been described (27,139).

It is obvious that selection of the proper developer or series of developers is critical to the successful accomplishment of the separation at hand. Consideration of the "adsorption index" is often helpful in making this selection. The adsorption index of a compound is the apparent volume in milliliters adsorbed per kilogram of adsorbent at an equilibrium concentration of 0.2% by volume of the substance in a specified solvent, usually n-heptane. The higher the adsorption index, the more strongly the compound is adsorbed. Hirschler (81) has determined the adsorption

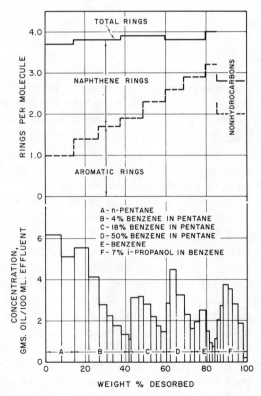

Fig. 99.32. Separation of a concentrate of petroleum aromatics by displacement development.

TABLE 99.XV
Adsorption Indices of Organic Liquids on Silica Gel

Compound	Adsorption index	Compound	Adsorption index
Saturated hydrocarbons	0	α-Methylnaphthalene	45
Octene-1	2	Dimethylnaphthalenes	53
Cyclohexene	3	Nitropropane	101
Trichloroethylene	6	Dioxane	145
Chloroform	12	Nitrobenzene	162
tert-Butylbenzene	20	Ethyl ether	178
Methylene chloride	21	Acetone	185
Thiophene	22	Ethyl acetate	195
Benzene	24	Pyridine	204
Ethylbenzene	27	Pyrrole	217
sym-Tetrachloroethane	28	Methanol	251
Toluene	31	Quinoline	273
Tetralin	36	Aniline	275
Ethylene dichloride	38	Morpholine	286

indices of a number of organic solvents on silica gel, as shown in Table 99.XV. The adsorption index for saturated hydrocarbons on silica gel is zero; for monoolefins the range is from 2 to 4. Monocyclic aromatics range from 22 to 31, and substituted naphthalenes from 45 to 53.

Instead of adding to a column, stepwise, a series of solvents of progressively increasing desorbing power, it is possible to apply a continuous concentration gradient of the developing mixture. This technique has been termed "gradient elution"; if properly applied, it is capable of producing results somewhat superior to multiple displacement development. Gradient elution has been used extensively for the separation of proteins, polysaccharides, and other natural products (2).

For some separations it is possible to take advantage of the special properties of certain adsorbents. Some natural and synthetic zeolites possess the property of being able to adsorb n-paraffins and n-olefins selectively from mixtures with other hydrocarbons. This is due to the fact that the channels in the crystal structure of these materials have a diameter of approximately 5 Å. Unbranched paraffin and olefin molecules can penetrate into these channels, whereas branched and cyclic molecules cannot. As a consequence, n-paraffins and n-olefins are strongly retained by these "molecular sieves" and may be quantitatively removed from mixtures as long as the capacity of the zeolite is not exceeded.

The quantitative determination of n-paraffins in hydrocarbon mixtures by the use of molecular sieves has been described by Norris and O'Connor (172,173).

Two other synthetic zeolites, in which the dimensions of the channels are approximately 10 and 13 Å, are commercially available. Mair and Shamaiengar have used these materials to produce some rather striking separations of high-molecular-weight hydrocarbons according to size and shape (141).

2. Partition Chromatography

Partition chromatography is fundamentally different from the adsorption chromatography discussed above. In this method of separation, an inert support (which may be an adsorbent) is coated with a thin layer of a solvent and arranged in a column so that a second solvent, immiscible with the first, may move over it. The rate of movement of a certain compound in a system of this sort is determined by its partition coefficient between the two liquid phases. Partition chromatography may therefore be considered a form of countercurrent extraction that permits the separation of small samples with exceptional efficiency in a fairly simple manner. Although the mechanism of separation is quite different, the laboratory practice of partition and of adsorption chromatography is essentially similar. Generally, one is restricted to the use of much smaller amounts of sample in partition than in adsorption chromatography.

In most of the useful separations of hydrocarbons by partition chromatography that have been reported (138,194), the stationary phase has been a polar liquid. The use of a lipophilic support to hold the nonpolar phase stationary would seem quite promising, for the partition coefficients of hydrocarbons would then be well in favor of the stationary liquid. Some preliminary experiments using this approach have been reported by Desty and Goldup (42) and indicate that this treatment has considerable merit for hydrocarbon separations.

F. THERMAL DIFFUSION

If a temperature gradient is applied to a homogeneous mixture, it will give rise to a concentration gradient of the constituents. This phenomenon has been termed "thermal diffusion." The degree to which the concentration gradient occurs is dependent on the ratio of the coefficient of thermal diffusion to that of normal (concentration) diffusion, because the difference in concentration caused by thermal diffusion tends to be counteracted by concentration diffusion. This ratio is highly susceptible to certain properties of the components of the mixture to be separated. In the thermal diffusion of liquids, the concentration gradient is small and must be multiplied a number of times to be of practical use. Such magnification can be achieved by confining the liquid mixture in the annular space between two vertical coaxial cylinders that are maintained at different temperatures (31). Under these conditions, convective flow of the mixture is established and a component that has a tendency to move to the cold wall will concentrate at the bottom of the column, because the cold liquid is usually more dense than the bulk mixture. Conversely, a component which tends to move toward the warm wall will be concentrated at the top of the column. After a period of time, the convective flow, which tends to neutralize the vertical concentration gradient, results in the establishment of a state of equilibrium. In the thermal diffusion of a mixture, the components will, when this state has been established, be arranged along the height of the column according to their tendency to migrate to the cold wall. As no column has infinite separating power, the transition between two successive components is continuous. Kramers and Broeder (100) and Jones and his coworkers (87,88) have studied the thermal diffusion of liquid hydrocarbons, using columns of this type. The separation of sulfur compounds by thermal diffusion has been described by Thompson and his associates (217). A comprehensive treatment of the analytical applicability of thermal diffusion is given in Chapter 27, Volume 2, of this Treatise.

The fundamental basis for the separation of gases by thermal diffusion is known to depend on mass differences. In the case of liquids, this principle does not apply rigorously and factors other than mass control the nature and magnitude of separation. From considerations based on the cage model of a liquid, Kramers and Broeder have concluded that for hydrocarbon systems molecular mass is less important in determining the direction of separation than the energy required for a molecule to escape from its "hole" in the liquid. They suggested that in a convective column the order of separation of hydrocarbons from top to bottom will be as follows: light n-paraffins, heavy n-paraffins, branched paraffins, cycloparaffins and mononuclear aromatics, and dinuclear aromatics. This order has been largely substantiated.

An analytical thermal diffusion column similar to that described in Chapter 27, Volume 2, is used in the author's laboratory. Figure 99.33 shows the results of the thermal diffusion of a mixture of n-hexadecane

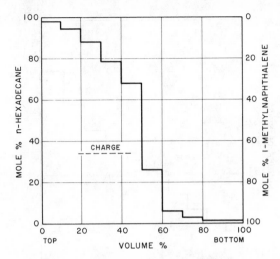

Fig. 99.33. Separation of a mixture of n-hexadecane and 1-methylnaphthalene by thermal diffusion.

and 1-methylnaphthalene by means of this apparatus. Ten per cent of the charge mixture was recovered as essentially 100% n-hexadecane and 20% as practically pure 1-methylnaphthalene.

The application of thermal diffusion to a saturated hydrocarbon mixture of high molecular weight is shown in Fig. 99.34. The ring numbers were calculated by the method of Vlugter, Waterman, and van Westen (Section IV.B.3.a). Generally speaking, it can be seen that the molecules are arranged in order of increasing ring number in going from the top to the bottom of the column. The thermal diffusion process does not distinguish between aromatic and saturated rings. Mixtures of n-heptyl-benzene–n-heptylcyclohexane, toluene–methylcyclohexane, and benzene–cyclohexane show no separation at all (164). The thermal diffusion of complex oil fractions containing both saturated and aromatic hydrocarbons produces a separation according to total number of rings only. Little segregation of saturated molecules from aromatic molecules takes place. For this reason, the technique is generally more useful when applied after some prior separation as to type has been achieved, perhaps by adsorption. By combining thermal diffusion and adsorption it should be possible to separate mixtures of high-boiling hydrocarbons into fractions containing a constant number of aromatic rings and varying numbers of cycloparaffin rings, and into fractions containing a constant number of cycloparaffin rings and varying numbers of aromatic rings.

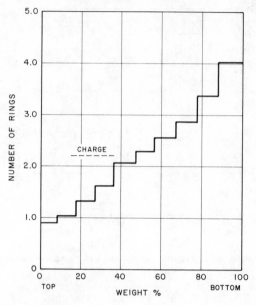

Fig. 99.34. Separation of a saturated hydrocarbon oil by thermal diffusion.

VI. PHYSICAL METHODS FOR THE ANALYSIS OF MATERIALS OF HIGH MOLECULAR WEIGHT

A. VISCOUS FRACTIONS OF PETROLEUM

1. The Choice of Fundamental Units of Composition

The manner in which the constitution of complex mixtures of high molecular weight is expressed depends largely on the features which one wishes to choose as component parts. For example, the composition can be given according to the amounts of the various types of molecules in the sample. This may be termed molecular type analysis. Alternatively, the different kinds of atoms present may be regarded as the components and the analysis expressed in terms of their relative abundance. This kind of information can be obtained from elemental analysis and a determination of the average molecular weight. Finally, certain structural elements can be considered as components of the mixture. For instance, petroleum oils can be considered to be built up of aromatic rings, cycloparaffin (naphthene) rings, and paraffin chains (either "free" or "combined"). Van Nes and van Westen have termed the analysis based on these concepts "structural group analysis" because it provides an estimate of the average proportions in which these structural groups occur, irrespective of the way in which they are combined into molecules.

In the application of molecular type analysis to hydrocarbon mixtures it is customary to consider four types of molecules: aromatics, olefins, cycloparaffins, and paraffins. To avoid confusion it can be assumed that a molecule is

Aromatic	if it contains at least one aromatic ring.
Olefinic	if it contains at least one olefinic double bond.
Cycloparaffinic	if it contains at least one cycloparaffin ring.
Paraffinic	if it contains neither an aromatic nor a cycloparaffin ring nor an olefinic double bond.

An obvious disadvantage of such simplified definitions is that if a certain molecule contains more than one classifying feature it will fit into more than one group. For example, styrenes and indenes can be classified as either aromatic or olefinic. As a consequence of the rapidly increasing frequency with which a multiplicity of classifying features may be found in a single hydrocarbon molecule as the molecular weight becomes greater, the viscous fractions of petroleum are in practice ill-suited to molecular type analysis. As we have seen, molecular type analysis is at present finding its main application in the lower-boiling ranges.

In expressing the composition of hydrocarbon mixtures in terms of the average proportion of structural groups, aromatic rings, naphthene (cycloparaffin) rings, and paraffin chains have for many years been chosen as the structural elements. It has become the practice to express the analysis in terms of the proportions of these structural elements that would comprise a hypothetical mean molecule. In applying the "average molecule" concept to lubricating oils, prior separation of the sample into aromatic and saturated fractions by adsorption has been recommended (110). Each of the fractions is then characterized in terms of the composition of the average molecule. This scheme is, in effect, a way of combining molecular type analysis and structural group analysis and yields considerably more information than when either method alone is applied.

There is considerable advantage in having a method of structural group analysis in which the structural elements are chosen in such a way that their sum is equal to 100%. This requirement can be met by expressing the composition in terms of the percentage of the total number of carbon atoms present in aromatic ring structures, naphthene ring structures, and paraffin chains. Thus "carbon type analysis" or "carbon distribution" indicates the relative amounts of carbon in aromatic, naphthenic, or paraffinic structure. If the molecular weight of the sample is known and if some assumptions can be made regarding the type and degree of condensation of the rings present, it is possible to recalculate the average molecule analysis as carbon distribution.

As an example that may help to clarify the concepts that have been discussed, consider the saturated and aromatic hydrocarbon molecules shown in Fig. 99.35. These molecules are illustrative of the complex hydrocarbons that may be present in the viscous fractions of petroleum. The saturated molecule contains 3 cycloparaffin rings and paraffin chains appropriate to bring the total number of carbon atoms to 27. Of these 27 carbon atoms, 13 are in cycloparaffin (naphthene) rings. The percentage of carbons in naphthenic structures, ($\%C_N$) is therefore 48. The remaining 14 carbon atoms are in paraffin chains. The percentage of carbons in paraffin chain structures ($\%C_P$) is therefore 52.

The aromatic molecule illustrated contains 2 aromatic rings, 2 cycloparaffin rings, and paraffin chains appropriate to bring the total number of carbon atoms to 27. Of the 27 carbon atoms, 10 are in aromatic

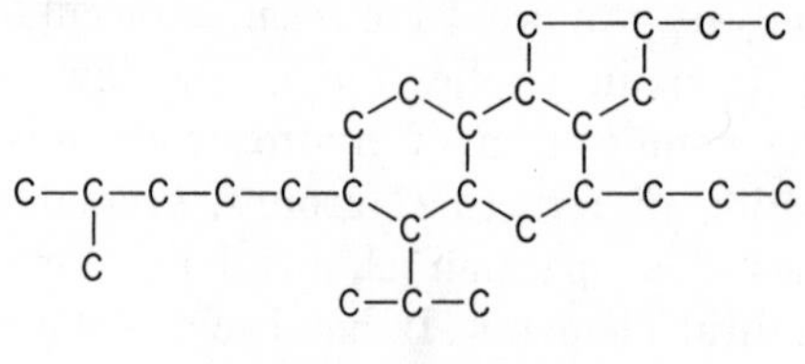

POSSIBLE STRUCTURE OF
NAPHTHENE HYDROCARBON

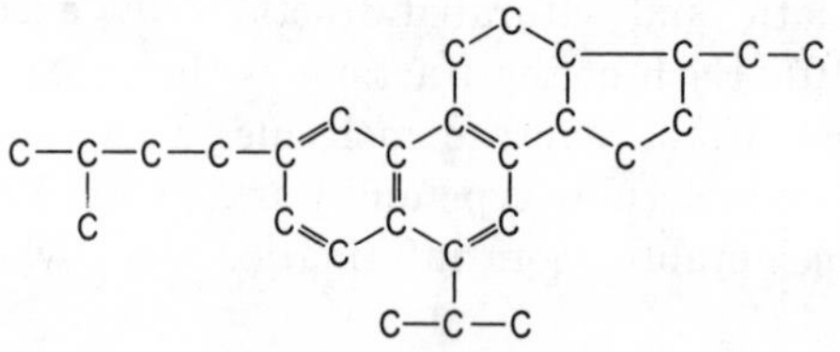

POSSIBLE STRUCTURE OF
AROMATIC HYDROCARBON

Fig. 99.35. Examples of hydrocarbon molecules representative of those that occur in the viscous fractions of petroleum.

rings. The percentage of carbons in aromatic structures ($\%C_A$) is therefore 37. There are 7 carbons in cycloparaffin (naphthene) rings; $\%C_N$ is therefore 26. The remaining 10 carbons are in paraffin chain structures; consequently $\%C_P$ is 37. If the two types of molecules were mixed in equimolar proportions, the carbon type composition of the mean molecule in the resulting blend would be reported as follows: $\%C_A$, 18.5; $\%C_N$, 37; and $\%C_P$, 44.5. The average molecule in such a blend would be visualized as having 1 aromatic ring (R_A), 2.5 naphthene rings (R_N), and 12 carbons in paraffin chains (C_P). On the basis of molecular type analysis, such a blend would of course contain 50% saturated and 50% aromatic hydrocarbons.

The determination of composition in terms of the hypothetical mean molecule and carbon distribution is the most important means presently available for the economical analysis of heavier petroleum fractions. Structural group analysis occupies a position midway between ultimate analysis, where atoms are the components, and molecular analysis, where molecules are the components.

The almost universal choice of these structural elements in oil constitution research stems from the fact that R_T, the total number of rings per molecule, and $\%C_A$, the percentage of carbons in aromatic structures, may be determined exactly by the so-called direct method involving an elemental analysis and molecular weight determination before and after complete hydrogenation of the sample. The related structural elements R_A, R_N, and C_P and $\%C_N$ and $\%C_P$ may be derived from R_T and $\%C_A$ if some assumptions can be made regarding the type and degree of condensation of rings. The direct method has never been well suited for routine work. The elemental analyses must be carried out with extreme accuracy, and the quantitative hydrogenation of oil fractions generally requires extended reaction periods in a high-pressure autoclave. For these reasons a great deal of attention has been directed toward the discovery of relations between simply measured physical constants and the constitution in terms of structural group analysis (116,213). In general, no simple relation exists between these physical constants and the composition. In developing methods based on physical constants, two different approaches are possible. First, observations on individual hydrocarbons may be taken as basic data and the relation between physical properties and structure estimated. Rules for the behavior of these properties in mixtures also must be worked out. Second, analytical data may be collected on a large number of oil fractions by the application of exact methods. On the basis of these data, empirical correlations between physical properties and composition may be sought.

It is questionable which avenue of attack can provide the more rapid advance in the development of methods for structural group analysis. Fortunately, both have been, and probably will continue to be, actively pursued. Each also has certain limitations. For example, if a structural group analysis has to be developed for a structural group element that cannot be determined exactly in oil fractions, recourse must be made to the study of synthesized hydrocarbons. A case in point is the determination of the number of chain branchings in oil fractions where only data on individual hydrocarbons of known structure can be used, since the value sought cannot be measured exactly for oil fractions. If data are desired that can be measured with exactness on the oil fraction itself, the second line of attack seems preferable. Petroleum fractions are invariably complicated mixtures, and the study of such fractions would appear to require less work than a study of all of the possible pure components.

In this section the majority of the physical property methods that are useful in oil constitution research are reviewed. The data and assumptions involved in their development are discussed, and their limitations pointed out. Since the basic data used, either in developing these relations or in proving their validity for oil fractions, are for the most part derived from application of the direct method, this technique is described first in some detail.

2. The Direct Method

The determination of the total number of rings, R_T, and the percentage of carbon in aromatic structure, $\%C_A$, by means of elemental analysis and molecular weight determination before and after complete hydrogenation, can be considered an almost classical method in oil analysis. When supplemented by an assumption of an average ring type, the procedure is called the "direct method," and has served to a great extent as a basis for other methods of structural group analysis. Upon hydrogenation of an oil fraction in such a way that only aromatic rings are converted to cycloparaffin (naphthene) rings, the increase in hydrogen content is an exact measure of the total number of aromatic carbon atoms present, for each aromatic carbon atom takes up one hydrogen atom. If the empirical formula of the hypothetical "mean molecule" of the original hydrocarbon mixture is written as $C_c H_h$ and after hydrogenation as $C_{c'} H_{h'}$, it follows that $h' - h$ hydrogen atoms are taken up per molecule and that $h' - h$ is equal to the mean number of aromatic carbon atoms per molecule in the original mixture. Therefore, the percentage of carbon atoms in aromatic structure ($\%C_A$) can be written as

$$\%C_A = \frac{h' - h}{c} \times 100 \qquad (1)$$

The empirical formulas are, of course, derived from the elemental analysis and molecular weight determination before and after hydrogenation. If the percentage of carbon is designated by C and the percentage of hydrogen by H and the assumption is made that only carbon and hydrogen are present, then $C = 100 - H$.

Since $c = MC/1201$, $h = MH/100.8$ and $h' = M'H'/100.8$, where M is the mean molecular weight, these values can be substituted in equation (1) to give

$$\%C_A = \frac{1201}{100.8} \times 100 \times \frac{-H + M'H'/M}{100 - H}$$

or

$$\%C_A = 1191 \times \frac{M'H' - MH}{M(100 - H)} \qquad (2)$$

The formula is applicable when no foreign elements are present and when no cracking or ring opening takes place during hydrogenation. Since, in an actual experiment, foreign elements are often present and a small amount of cracking may occur during hydrogenation, some corrections are required. These factors, as well as imperfections in the experimental measurements of H, H', M, and M', frequently cause an apparent decrease in the mean number of carbon atoms on hydrogenation. Consequently, the increase in hydrogen content is not reflected in a corresponding increase in molecular weight as required. Van Nes and van Westen (167) have suggested that this may be satisfactorily taken into account by assuming that the molecular weight decrease during hydrogenation is caused by the splitting off of CH_2 groups. In such a case, equation (1) becomes

$$\%C_A = \frac{(h' - h) + 2(c - c')}{c} \times 100$$

Substituting

$$c = \frac{MC}{1201}, \quad h = \frac{MH}{100.8}, \quad h' = \frac{M'H'}{100.8}, \quad \text{and } c' = \frac{M'C'}{1201}$$

in this equation and collecting terms gives

$$\%C_A = \frac{1201(M'H' - MH) - 201.6(M'C' - MC)}{1.008MC} \tag{3}$$

which may be applied in the practical case.

Since the hydrogenated sample consists only of saturated hydrocarbons, the mean number of naphthene rings per molecule can be calculated from the per cent hydrogen and the molecular weight by means of the formula given in Section II.E:

$$R_N = 1 + M'(0.08326 - 0.005793H') \tag{4}$$

Since this ring number is the sum of the mean number of aromatic rings and naphthene rings per molecule that were present in the original sample before hydrogenation, it should be designated as R_T, the mean total number of rings.

The conversion of percentage of carbon in aromatic structures, $\%C_A$, into the mean number of aromatic rings per molecule, R_A, requires an assumption as to the type and degree of condensation of aromatic rings present. Conversion of the mean total number of rings, R_T, to the percentage of carbon in ring structure, $\%C_R$, requires similar assumptions. Over the years a considerable weight of evidence has accumulated that suggests that no substantial amount of other than 5-membered or 6-membered rings occurs in petroleum. The evidence available also suggests that the cycloparaffins in high-molecular-weight saturated oils contain about equal quantities of 5- and 6-membered rings, and that the proportion of condensed to noncondensed structures is about 1:1 (129,147,192). Also, Rossini and his associates (140,143,186) concluded a number of years ago that whenever two or more aromatic rings occur in the same molecule they are condensed together in such a way that two carbon atoms are shared—as in naphthalene, phenanthrene, etc.

The simplest and most obvious assumption that is in reasonable accord with these findings is that of kata-condensed 6-membered rings. For aromatic molecules this choice agrees with the findings of Rossini; for cyclic saturated molecules the choice of kata-condensed 6-membered rings appears sound, since this molecule represents an average of the probable types of ring structures that are present (see Fig. 99.10). Thus, this assumption can be viewed as supposing that two groups of molecules will counterbalance one another—that is, the molecules with more than the number of ring carbon atoms in corresponding molecules having condensed 6-membered rings will counterbalance those having fewer numbers of ring carbon atoms.

In converting the ring number, R_T, to the percentage of carbon in ring structure, $\%C_R$, two cases must be distinguished—one where R_T is equal to or less than 1 and one where R_T is greater than 1. For the first case it is assumed that only monocyclic and noncyclic hydrocarbons are present. The number of carbon atoms per ring is therefore 6, and

$R_T \leqq 1$

$$\%C_R = \frac{6R_T}{M'C'/1201} \times 100 = \frac{720600R_T}{M'C'} \tag{5}$$

When R_T is greater than 1, it can be assumed that monocyclic and polycyclic structures are present. The condition of condensed 6-membered rings that is imposed requires that the first ring have 6 carbon atoms and subsequent rings contribute 4 additional carbon atoms. It follows that for

$R_T \geqq 1$

$$\%C_R = \frac{2 + 4R_T}{M'C'/1201} \times 100 = 240200 \frac{(1 + 2R_T)}{M'C'} \tag{6}$$

These two conditions are equivalent to saying that molecules with more than one ring are present only when the value of R_T exceeds 1.

Since $\%C_R = \%C_A + \%C_N$, the carbon in naphthenic structure in the original sample can now be calculated from $\%C_R$ and $\%C_A$. Finally, $\%C_P$, the percentage of carbon in paraffinic structure, is found from

$$100 = \%C_A + \%C_N + \%C_P \tag{7}$$

The percentage of carbon in aromatic structure calculated according to equation (3) may be used to derive R_A, the average number of aromatic rings per molecule. The same assumptions and reasoning used in deriving $\%C_R$ from R_T may be applied to give the following equations:

$R_A \leqq 1$

$$R_A = \%C_A \times \frac{MC}{720600} \tag{8}$$

$R_A \geqq 1$

$$R_A = \%C_A \times \frac{MC}{480400} - 0.500 \tag{9}$$

Substituting for $\%C_A$ its equivalent in terms of equation (3) gives:

$R_A \leqq 1$

$$R_A = 0.001653(M'H' - MH) - 0.0002775(M'C' - MC) \qquad (10)$$

$R_A \geqq 1$

$$R_A = 0.002479(M'H' - MH) - 0.0004163(M'C' - MC) - 0.500 \qquad (11)$$

Since R_A and R_T are now known, R_N, the mean number of naphthene rings in the original sample, may be calculated from the relation

$$R_N = R_T - R_A \qquad (12)$$

Finally, C_P, the number of carbons in paraffin chains, can be calculated from $\%C_P$

$$C_P = \%C_P \times \frac{MC}{120100} \qquad (13)$$

In applying the direct method it is often advantageous to separate the sample into a saturated and an aromatic hydrocarbon fraction by adsorption. Structural group analysis of the saturated portion may then be made directly from the elemental analysis and molecular weight of the fraction, using the relations given in equations (4)–(7) and (13), and remembering that in this case R_A and $\%C_A$ are zero and therefore $R_T = R_N$ and $\%C_R = \%C_N$. The composition of the aromatic fraction is determined from similar data obtained before and after complete hydrogenation, using equations (3)–(7) and (11)–(13). In this case, the value of R_A will always be equal to or greater than 1.

Martin (145) analyzed a number of aromatic fractions by the direct method and suggested the following equation for the calculation of R_A:

$$R_A = 0.00248M \left(\frac{H'C}{C'} - H \right) - 0.500 \qquad (14)$$

This equation is equivalent to that obtained by substituting the value for $\%C_A$ given by equation (2) (ideal case) in equation (9), save that the experimental molecular weight determination after complete hydrogenation has been eliminated by making use of the relation

$$M' = M(C/C')$$

The use of this relation has some merit, since the molecular weight is generally the least accurate of the experimental data. Martin also points out that when every molecule contains at least one aromatic ring (as in the case when aromatic fractions prepared by adsorption are being analyzed) and kata-condensed 6-membered rings are assumed:

$$3R_A + R_N = \frac{2c - h}{2}$$

Substituting

$$c = \frac{MC}{1201} \quad \text{and} \quad h = \frac{MH}{100.8}$$

gives

$$3R_A + R_N = \frac{M}{2}\left(\frac{2C}{1201} - \frac{H}{100.8}\right)$$

or

$$3R_A + R_N = M(0.000833C - 0.00496H)$$

It follows, therefore, that for aromatic fractions the calculation of R_N may be made directly once R_A is known:

$$R_N = M(0.000833C - 0.00496H) - 3R_A \qquad (15)$$

and need not be obtained by difference. The extent to which $R_A + R_N$ calculated by equations (14) and (15) agrees with R_T calculated from data on the completely hydrogenated aromatics by equation (4) is indicative of how well the assumption of 6-membered condensed rings represents the true ring type, since, as it may be recalled, R_T is exact for mixtures of saturated hydrocarbons. The fact that a check of this kind is possible is one advantage gained by a prior separation according to type before application of the direct method. In addition, considerably more information about the sample becomes available, since the separate structural group analysis of both fractions may be obtained with relatively small additional effort.

The laboratory hydrogenation of oils as practiced in connection with the direct method has been treated extensively by van Nes and van Westen (168), Lipkin and his co-workers (125), and Miron (152).

The precision of the direct method is determined to a large extent by the precision of the ultimate analysis and molecular weight measurements. It must be recognized, however, that systematic errors are possible because of the presence of foreign elements, the occurrence of cracking and ring opening during hydrogenation, and the assumption of kata-condensed 6-membered rings. The high degree of accuracy required in the determination of the hydrogen content and the lengthy time required for complete analytical hydrogenation make the direct method ill suited to routine work. Its use has generally been confined to research applications and as a method of verification in special cases. Its great advantage lies in the fact that it provides an exact way of determining $\%C_A$ and R_T. As a consequence, it has often been used to provide basic data for the development of other methods of structural group analysis that make use of certain properties of hydrocarbons that may be obtained conveniently, accurately, and quickly.

3. Rapid Physical Property Methods

Physical property methods for structural group analysis in existence today can be divided into two categories: those that are intended to be applied to the whole oil sample, and those that require a prior separation according to molecular type, that is, aromatic hydrocarbons from saturated hydrocarbons.

a. APPLICATION WITHOUT PRIOR TYPE SEPARATION

(1) The n-d-M Method

The n-d-M method described by van Nes and van Westen (160) provides a procedure whereby the carbon distribution and ring content of olefin-free petroleum oils can be calculated from measurements of refractive index, density, and molecular weight. The method is based on carefully obtained direct method data for 133 gas oil and lubricating oil distillate fractions from five crude oils. The crudes were selected so as to cover the range of types normally encountered in the petroleum industry.

The procedure for calculating the structural group analysis of an oil fraction from measurements of molecular weight and the refractive index and density at 20°C is as follows. From the observed density, d, and the observed refractive index, n, at 20°C, the factors v and w are calculated by means of the following equations:

$$v = 2.51(n_D^{20} - 1.4750) - (d_4^{20} - 0.8510)$$

$$w = (d_4^{20} - 0.8510) - 1.11(n_D^{20} - 1.4750)$$

The percentage of aromatic carbon is calculated from v and the molecular weight by means of one of the following equations:

If v is positive: $\%C_A = 430v + 3660/M$

If v is negative: $\%C_A = 670v + 3660/M$

The percentage of carbon in ring (aromatic + naphthenic) structures is calculated from w and the molecular weight:

If w is positive: $\%C_R = 820w - 3S + 10000/M$

If w is negative: $\%C_R = 1440w - 3S + 10600/M$

where S is the weight percent of sulfur. The percentage of naphthenic carbon, $\%C_N$, and the percentage of paraffinic carbon, $\%C_P$, are calculated as follows:

$$\%C_N = \%C_R - \%C_A$$

$$\%C_P = 100 - \%C_R$$

The average number of aromatic rings per molecule, R_A, is calculated from v and the molecular weight:

If v is positive: $R_A = 0.44 + 0.055Mv$

If v is negative: $R_A = 0.44 + 0.080Mv$

The average total number of rings per molecule, R_T, is calculated from w and the molecular weight:

If w is positive: $R_T = 1.33 + 0.146M(w - 0.005S)$

If w is negative: $R_T = 1.33 + 0.180M(w - 0.005S)$

The average number of naphthene rings per molecule, R_N, can be calculated by difference:

$$R_N = R_T - R_A$$

Since the assumption of kata-condensed 6-membered rings was made in the calculation of certain of the direct method data, this assumption is implicit in the equations of the n-d-M method. To eliminate the labor involved in these calculations the authors have developed a series of nomographs from which the analysis may be obtained simply and quickly.

The method is based upon the discovery that linear relations exist between the composition as found by the direct method and the refractive index, density, and reciprocal molecular weight. Equations expressing these relationships were found to have the general form

$$\%C = a/M + b\,\Delta d + c\,\Delta n$$

where $\%C$ is the percentage of carbon atoms in some structure (aromatic, naphthenic, or total rings); a, b, and c are constants; M is the molecular weight; and Δd and Δn are the differences between the measured values of d and n and those for the limiting paraffin. These formulas are also transposable into a form which expresses the number of rings per molecule, R (either aromatic, naphthenic, or total), as

$$R = a' + b'M \, \Delta d + c'M \, \Delta n$$

where a', b', and c' are constants different from those in the first equation. In evaluating the constants in these formulas by means of the basic data obtained by the direct method, it was found necessary to divide the range of compositions studied into two parts, each with somewhat different constants in the formulas. That one set of constants will not satisfy the whole range the authors attribute to the fact that the oils with a mean number of rings per molecule greater than 1 contain kinds of hydrocarbons somewhat different from those with an average of less than one ring. Since the constants were evaluated for refractive index and density measurements at both 20 and 70°C, two sets of final formulas have been given. The final equations also incorporate a small correction for sulfur content.

The composition range covered by the samples used in deriving the formulas may be described (1) in terms of carbon distribution—up to about 75% carbon atoms in ring structure, the percentage in aromatic rings, however, not being larger than 1.5 times the percentage present in naphthenic structure, and (2) in terms of ring number—up to four rings per molecule, not more than half of them being aromatic. Strictly interpreted, the limitations in composition impose similar restrictions regarding the compositions to which the method may be applied without extrapolation beyond the basic data.

As a result of their experience in applying this procedure to a wide variety of oil samples, the authors observed that (1) the n-d-M method gives reliable results on synthetic oils prepared by polymerization, (2) application to paraffin waxes yields accurate results, (3) application to residual oils has never given any indication of incorrectness, and (4) application to other products that fall outside the range of the basic data cannot be recommended because large differences between the actual and calculated values may occur. However, the method is useful for such extreme samples if only comparative figures are required. The n-d-M method cannot be applied to individual hydrocarbons.

(2) The Refractivity Intercept–Kinematic Viscosity–Gravity Constant (RI–KVGC) Method

The time-consuming and rather demanding experimental determination of molecular weight required by the n-d-M procedure can be eliminated if characterizing functions that are primarily dependent on composition but relatively insensitive to molecular weight are chosen as the basis for a structural group analysis method. Smith (207) has described a procedure for the structural group analysis of lubricating oils utilizing the refractivity intercept (RI) and a modified viscosity–gravity constant (KVGC) as characterizing constants. He found both of these factors to be essentially independent of molecular weight for a large number of hydrocarbons of high molecular weight. In application, the only physical constants that need be measured experimentally are the refractive index at 25°C, the specific gravity (25/25)°C and the kinematic viscosity at 210°F. The refractivity intercept is calculated according to the equation:

$$RI = n - g/2$$

where $n =$ the refractive index for the sodium D line at 25°C, and $g =$ the specific gravity at 25/25°C. The kinematic viscosity–gravity constant is calculated according to the equation:

$$KVGC = g - Y$$

where Y is a function of the kinematic viscosity, V, at 210°F, as follows:
For values of V from 1.5 to 15 centistokes

$$Y = 0.0915(\log V) - 0.02885(\log V)^2$$

For values of V from 15 to 50 centistokes

$$Y = 0.0221 + 0.0461(\log V) - 0.00625(\log V)^2$$

Values for the percentage of carbon atoms in ring structures, $\%C_R$, and the percentage of the ring carbon atoms occurring in aromatic ring structures, $\%A_R$, are deduced by entering the diagram shown in Fig. 99.36 with the calculated values of RI and KVGC. The carbon distribution is then calculated from the equations:

$$\%C_A = (\%C_R)(\%A_R/100)$$

$$\%C_N = \%C_R - \%C_A$$

$$\%C_P = 100 - \%C_R$$

Figure 99.36 was developed primarily from data on the pure hydrocarbons of high molecular weight reported by API Project 42. The values for the refractivity intercept and kinematic viscosity–gravity constant

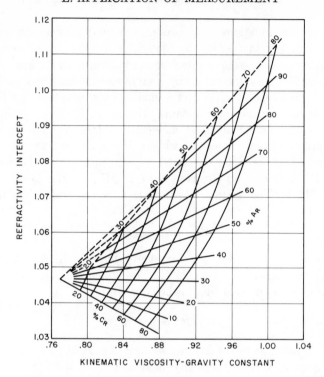

Fig. 99.36. Diagram for RI–KVGC method. Reproduced with the permission of the Ohio State University.

were calculated for each of the pure hydrocarbons for which the necessary physical constants were available and then plotted using these functions as coordinates. The points for the pure paraffin–naphthene hydrocarbons could be represented by a group of almost parallel lines, one corresponding to cyclopentyl ring structures, another to cyclohexyl ring structures, and others to condensed ring structures. The author found that a similar plot of RI and KVGC data for a series of dewaxed aromatic-free oil fractions could be represented by a single straight line. By superimposing this line on the graph derived from data on pure paraffin–naphthene hydrocarbons, it was possible to locate points along it corresponding to 20%, 30%, etc., carbon atoms in ring structure ($\%C_R$).

This single straight line is the paraffin–naphthene base line for the composition diagram in Fig. 99.36. In constructing the remainder of this figure, RI and KVGC data on pure paraffin–aromatic hydrocarbons were used exclusively, since the isolation from petroleum fractions of a mixture of alkyl aromatic hydrocarbons containing no associated cyclo-paraffin ring is extremely difficult. In order for the final diagram to be representative of the composition of natural lubricating oils, it was necessary to impose certain restrictions on the molecular weight range, and to make certain assumptions as to the predominant type of aromatic present for a given carbon distribution as follows:

1. Paraffin–aromatic hydrocarbons which contain less than 20% ring carbon atoms are of the benzene homolog type.

2. Paraffin–aromatic hydrocarbons which contain 40% ring carbon atoms are predominantly of the naphthalene homolog type.

3. Petroleum oils with an average molecule in which half of the ring carbons are in aromatic structures occupy a position corresponding to the tetrahydronaphthalene homologs having the same carbon distribution.

4. (*a*) A petroleum oil containing half of the ring carbons in aromatic structures contains 33% noncondensed (benzene or tetralin derivatives) aromatic ring types to 67% condensed ring types (naphthalene, phenanthrene, etc., derivatives).

(*b*) The initial slope of the %C_R lines leaving the paraffin–naphthene base line is the same as that of the lines joining the paraffin–naphthene base line and the benzene homolog line.

Assumptions *1* and *2* were used to fix the base line for natural paraffin–aromatic hydrocarbons as shown by the dashed line in Fig. 99.36. The values for %C_R along the paraffin–aromatic base line were located by reference to the data for pure hydrocarbons in a manner similar to that employed for the paraffin–naphthene base line. Assumption *3* was the determining factor in establishing the lines of constant "per cent ring carbons in aromatic rings" (%A_R). In accord with this assumption, the 50 %A_R line corresponds to the location of the pure hydrocarbon tetrahydronaphthalene homologs.

The slope of the %C_R lines is determined by conditions *4a* and *4b*. To satisfy condition *4a*, this line was drawn so as to cross the 50%A_R line at a point one-third of the distance from the intersection of a %C_R line for naphthalene homologs to the point of intersection of a similar line for benzene homologs.

The rest of the %A_R lines were established by graduating between the paraffin–naphthene and paraffin–aromatic base lines in such a way that the position of the 50 %A_R line still corresponds to that of the line representing tetrahydronaphthalene homologs.

It is obvious from this brief review of its derivation that the RI-KVGC method is based to a great extent on data for pure hydrocarbons of high molecular weight. In assessing the validity of the method, Smith compared the carbon distributions determined from RI and KVGC with those determined by the n-d-M method for 52 oil samples of divergent composition and found that the methods, in spite of their very different derivations, were in excellent agreement. He also concluded, from a comparison of both procedures to direct method data reported by Mair, that the RI-KVGC method is more suitable than the n-d-M for determining the carbon distribution of highly aromatic oils. In general, the method may be applied to any olefin-free distillate oil having a viscosity below 20 centistokes at 210°F. This viscosity range corresponds to about a 300–600 range of molecular weights. If $\%C_A$ is less than 5%, the viscosity may be as high as 30 or 40 centistokes at 210°F, before appreciable error is introduced.

The principal advantages of the method are the simplicity and speed of the analytical procedure and the excellent precision attainable. All of the physical properties required may be measured rapidly and precisely, and the carbon distribution determined in approximately 15 min. Disadvantages are the somewhat restricted molecular weight range over which the method is applicable, and the fact that only carbon distribution can be determined since no provision is made for the estimation of R_A, R_N, or R_T (ring analysis).

(3) The Refractivity Intercept–Viscosity–Gravity Constant Method

Kurtz and his associates (106) found the refractivity intercept and the viscosity–gravity constant described by Hill and Coats to be sufficiently molecular-weight-independent functions of composition to permit their use as the basis of a method for determining carbon distribution. The only experimental data required are the viscosity, gravity, and refractive index. The refractivity intercept is calculated from the following relation:

$$RI = n - d/2$$

where n = the refractive index at 20°C for the sodium D line, and d = the density at 20°C. The specific gravity at 60/60°F of petroleum fractions may be converted to density with an accuracy adequate for this calculation by using the relation given by Ward and Kurtz (226).

The viscosity–gravity constant is calculated by use of the following equation:

$$VGC = \frac{10g - 1.0752 \log (V - 38)}{10 - \log (V - 38)}$$

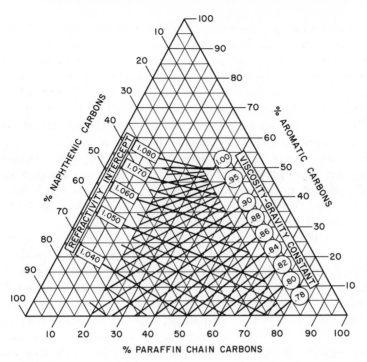

Fig. 99.37. RI-VGC diagram. Reproduced with the permission of *Analytical Chemistry*.

where g = the specific gravity at 60/60°F, and V = the Saybolt viscosity at 100°F. The point corresponding to the values of each of these functions is then located on the triangular diagram shown in Fig. 99.37 and the carbon distribution read off.

A correction must be applied to the values of $\%C_N$ and $\%C_P$ read from the diagram if the sample contains more than 0.5 wt % of sulfur. The corrections are given by the following formulas:

$$\text{For } \%C_N: \quad -\frac{\text{wt } \% \text{ S}}{0.288} \qquad \text{For } \%C_P: \quad +\frac{\text{wt } \% \text{ S}}{0.216}$$

The corrected values for $\%C_A$ can be obtained from the relation:

$$\%C_A = 100 - (\%C_N + \%C_P)$$

In developing this diagram, the authors used data for approximately 258 petroleum oils or fractions thereof for which the carbon distribution, viscosity–gravity constant, and refractivity intercept could be derived. Lines of constant VGC were located on the triangular coordinate system by first plotting VGC as a function of a single composition variable on rectangular coordinates. It was found that the relation between VGC and %C_P for aromatic-free oils, and between VGC and %C_P and VGC and %C_A for whole oils, could be represented by straight lines. The relation for saturated oils was used to establish the terminal locus of the VGC lines along the paraffin–naphthene carbon side of the triangular coordinate system. The relations for whole oils, when transferred to triangular coordinates, established additional points through which lines of constant viscosity–gravity constant were constructed covering the VGC range from 0.790 to 0.910. Above VGC values of 0.910 the lines were placed by inspection to fit the data available.

In locating the refractivity intercept lines, 1.0300 was chosen as the limiting value for 100% C_N and 1.0500 as the value corresponding to 100% C_P. Linear graduation between these two values along the paraffin–naphthene carbon side of the diagram was sufficient to establish the location of the terminus of the intercept lines on this axis. It was not possible to establish a single limiting value of the refractivity intercept for 100% C_A, since the intercept of aromatic carbons is influenced quite strongly by the type of aromatic structure in which they reside. The authors discovered, however, that if the equivalent refractivity intercept of the aromatic carbons was calculated according to the relation

$$RI_A = \frac{RI_{sample}(100) - 1.0300(\%C_N) - 1.0500(\%C_P)}{\%C_A}$$

for a large number of petroleum oils and plotted as a function of the observed refractivity intercept of the sample, the locus of the points could be described reasonably well by a smooth curve. By interpolating between 1.0300 and values of RI_A read from this curve for a given value of intercept, it was possible to fix the terminal points of the refractivity-intercept lines along the naphthene–aromatic carbon side of the diagram. Similarly, interpolation between 1.0500 and RI_A fixed corresponding points along the paraffin–aromatic carbon side.

In evaluating this method the authors compared carbon distribution data obtained by its use with that obtained by means of the n-d-M method and other more detailed hydrocarbon analysis procedures (110). They established that, within the range of composition recommended for the n-d-M method, agreement between the two procedures was quite satisfactory. However, the refractivity intercept–viscosity–gravity constant method gave more reliable results when applied to samples of high aromatic content.

For oils having a viscosity–gravity constant of 0.900 or less, the method gives the value of $\%C_A$ with a standard deviation of about 1%. The standard deviation for $\%C_N$ is about 1.5%. The carbon distribution of oils of high aromatic content having viscosity–gravity constants between 0.900 and 1.00 can be determined with a standard deviation in $\%C_A$ of about 2% and in $\%C_N$ and $\%C_P$ of 4.5%.

This method has the advantage of simplicity and a fairly broad range of applicability. The physical constants required may be measured quickly and precisely with relatively inexpensive equipment. The carbon distribution is obtained directly from the diagram, and no additional calculations are required. The chart itself has been demonstrated to be quite useful in following graphically the composition changes produced by certain refinery operations, and a nomograph has been developed that may be used in place of the chart if desired (144). The method suffers the disadvantage that no provision is made for the calculation of R_A, R_N, or R_T.

(4) The Refractivity Intercept–Density Method

Kurtz and his co-workers (107) have also proposed a method for the determination of carbon distribution in which the only physical constants that need be measured are the refractive index and the density. However, the average number of carbon atoms per molecule must also be estimated from an elemental analysis and the molecular weight. The refractivity intercept is calculated from refractive index and density measurements at 20°C. Since density is molecular weight dependent, the measured value and the average number of carbon atoms are used in conjunction with Fig. 99.38 to estimate an equivalent density for a molecule of 30 carbon atoms. The point corresponding to the values for the refractivity intercept and the equivalent density is then located on the triangular chart shown in Fig. 99.39, and the values of carbon distribution are read off.

A sulfur correction must be applied to the values of $\%C_N$ and $\%C_P$ read from the graph if the sulfur content is greater than 0.5 wt %. The corrections are as follows:

$$\text{For } \%C_N: \quad -\frac{\text{wt }\%S}{0.416} \qquad \text{For } \%C_P: \quad +\frac{\text{wt }\%S}{0.478}$$

The corrected values for $\%C_A$ can be obtained from the equation:

$$\%C_A = 100 - (\%C_N + \%C_P)$$

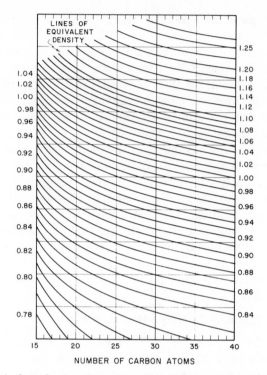

Fig. 99.38. Equivalent density chart. Reproduced with the permission of *Analytical Chemistry*.

In developing the intercept–density method the authors demonstrated that the general equation for molecular volume (Section II.A) could be used to calculate the density of high-molecular-weight hydrocarbons with sufficient accuracy to permit the use of such calculated densities for the analysis of petroleum fractions. The densities of all of the "basic structures" in the range of 15–40 carbon atoms were calculated using this equation. The term basic structures refers to the structural formulas that can be written for a given number of carbon atoms if isomer effects are ignored. For example, if molecules containing 20 carbon atoms are considered, the densities can be calculated for molecules containing no rings, for molecules containing one 5-membered saturated ring, one 6-membered saturated ring, and for molecules containing two saturated rings, both of them 5-membered, both of them 6-membered, and combinations thereof, etc. Further calculations also have to be made for molecules containing one, two, three, etc., aromatic rings along and in combination with 5- and 6-membered saturated rings. In developing the "basic structures," all multiring structures were assumed to be kata-condensed. Densities for molecules containing 15, 20, 26, 30, and 40 carbon atoms were calculated in this manner. By using the carbon distribution of the "basic structures" and the calculated densities, lines of constant density were located on triangular composition diagrams representing molecules of 15, 20, 26, 30, and 40 carbon atoms. The density lines on all of these diagrams possessed the same slope, and consequently lines which passed through any one point, such as the point of intersection with the paraffin–aromatic carbon side of the diagram, were the same at all other points of composition. The authors were able to take advantage of this property in constructing the equivalent density chart shown in Fig. 99.38. It was then possible to use a single triangular composition diagram for 30 carbon atoms for oils with as few as 15 and as many as 40 carbons, by applying this chart to convert the measured density to an equivalent density at 30 carbon atoms.

Refractivity intercept was used to provide a second physical property function which was relatively independent of molecular weight and would intersect the density lines in such a way as to uniquely define the carbon

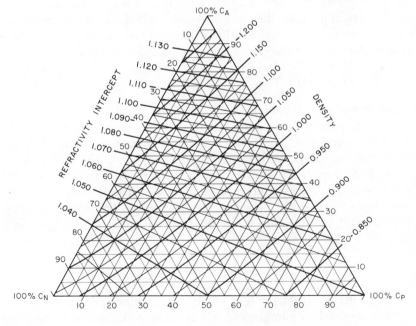

Fig. 99.39.   Diagram for refractivity intercept-density method. Reproduced with the permission of *Analytical Chemistry*.

distribution. The lines of constant refractivity intercept were located on Fig. 99.39 in the same way as for the RI-VGC method. The authors found it necessary to make some adjustment in the placement of the density lines on the final diagram, however, so as to take account of the presence of noncondensed cycloparaffins in actual oil samples.

Kurtz and his co-workers applied this method to 369 oil samples for which carbon distribution data were available. The standard deviation of the differences between the accepted analysis and that obtained using the intercept–density method was 1.45 for %C_A, 2.61 for %C_N, and 2.45 for %C_P. A comparison with direct method data for 83 oils produced substantially the same standard deviations. Since one carbon atom equals 5% in an oil with 20 carbon atoms, 4% in an oil with 25 carbon atoms, and 3.3% in an oil with 30 carbon atoms, the standard deviations correspond to one-half a carbon atom or less. The method is reliable over a very broad range of carbon distribution and exhibits excellent agreement with detailed analytical procedures on samples containing as much as 80% of the total carbons in aromatic structures.

The refractivity intercept–density method has the advantage of an extremely broad range of applicability. Since the determination of molecular weight and per cent carbon is required, however, the data necessary for its application are not so readily obtained as are those for the foregoing viscosity methods. Like these methods, it suffers the disadvantage that no calculation of the ring analysis is provided for.

(5) V_k-n-d Methods

Boelhouwer and Waterman (11) have proposed a method for the analysis of petroleum oils that requires the determination of only the kinematic viscosity at 20°C, the refractive index, and the density. For oils containing sulfur, a correction must be applied. The graph shown in Fig. 99.40 is entered with the log of the measured viscosity and the refractive index of the oil under examination, and a density, d', is read. From the difference, Δd, between this density and the measured value, the density d^* of a theoretical sample that is completely hydrogenated is estimated by the formula:

$$d^* = d - 0.62\,\Delta d - 0.008\%S$$

where %S is the weight per cent of sulfur. The kinematic viscosity, V_k^*, of the completely hydrogenated oil is estimated from the following relation:

$$\log V_k^* = (1 - 3\,\Delta d)\log V_k + \Delta d$$

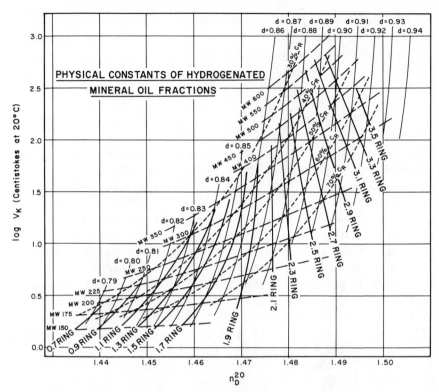

Fig. 99.40. Diagram for the V_k-n-d method. Reproduced with the permission of the *Journal of the Institute of Petroleum*.

Figure 99.40 is re-entered with d^* and $\log V_k^*$, and the molecular weight, M, and $\%C_R$ and R_T are read. The percentage of carbons in aromatic structures, $\%C_A$, is estimated directly from Δd:

$$\%C_A = 600\,\Delta d - (40\,\Delta d)^2$$

The average number of aromatic rings per molecule, R_A, is deduced from the following equations:

When $R_A \leqq 1$

$$3R_A = \frac{M \times \%C_A}{2800}$$

When $R_A \geqq 1$

$$2R_A + 1 = \frac{M \times \%C_A}{2800}$$

Finally, the percentages of carbon atoms in naphthenic and paraffinic structures, $\%C_N$ and $\%C_P$, and the average number of naphthenic rings per molecule, R_N, are calculated by the equations:

$$\%C_N = \%C_R - \%C_A$$

$$R_N = R_T - R_A$$

$$\%C_P = 100 - \%C_R$$

The method is based on the observations of Boelhouwer, Van Steenis, and Waterman that, when the viscosity or the logarithm of the viscosity of aromatic-free mineral oil fractions is plotted against molecular weight, a series of lines of constant ring number, R_T, results. By using such diagrams, it is possible to find the number of rings from molecular weight and viscosity. The authors subsequently discovered that a plot of the logarithm of the viscosity against the refractive index also produced a series of lines of constant ring number for aromatic-free oils. They also found it possible to locate lines of constant density, molecular weight, and $\%C_R$ on such a diagram to produce Fig. 99.40. Placement of these lines and development of the "prediction factors" for calculation of the physical constants of the "theoretical hydrogenated oil" are based on the ring analyses of a number of aromatic-free oils and on n-d-M analyses of approximately 100 oil fractions ranging from 12 to approximately 30% C_A. Consequently, the range of applicability of the method is somewhat limited. Agreement with the n-d-M method is good, since n-d-M analyses were used in the derivation.

The V_k-n-d method has the advantage that both ring analyses and carbon distribution analyses may be calculated from kinematic viscosity, refractive index, and density, all of which are constants that can be measured quickly and precisely. It suffers the disadvantage that the calculation is somewhat cumbersome and the range of applicability of the method restricted.

As a result of further investigation into the application of viscosity for the characterization of mineral oil fractions, Cornelissen and Waterman (32,33) developed a series of nomograms that permit the estimation of the carbon distribution, molecular weight, and ring analysis. Since the method was developed from data on whole oils, the "prediction factors" of the V_k-n-d method are eliminated. However, the method is not suitable for application to saturated fractions or completely hydrogenated samples. The kinematic viscosity at 20°C, the refractive index, and the density of the original oil are measured. Nomograms are provided for certain integral values of the logarithm of the kinematic viscosity over the range $\log V_k = 0.300$ to $\log V_k = 2.500$. Two nomograms that bracket the logarithm of the measured viscosity are entered with the measured refractive index and density, and values of $\%C_A$, $\%C_N$, R_A, and R_N and molecular weight read off. Linear interpolation between these values is then used to estimate the composition of the oil corresponding to the measured viscosity.

A correction must be applied to the values of R_N and $\%C_N$ when sulfur is present. The corrections to be deducted are given by the relations:

$$\text{for } R_N: \quad 0.0008 \times M \times \%S \qquad \text{for } \%C_N: \quad 3 \times \%S$$

where $\%S$ is the weight per cent of sulfur, and M is the molecular weight.

The remaining elements of composition are then calculated according to the formulas:

$$\%C_R = \%C_A + \%C_N$$

$$\%C_P = 100 - \%C_R$$

$$R_T = R_A + R_N$$

The authors consider this method an improvement over the procedure put forward by Boelhouwer and Waterman. The method is based on the concept of using a three-dimensional space model to represent the effect of composition on the three physical constants. The space model is considered to consist of the mutually perpendicular coordinates $\log V_k^{20}$, n_D^{20}, and d_4^{20}. The composition of a point located within the space model may be defined by $\%C_A$ and $\%C_N$. Other points with different values of the physical constants may have the same composition (if the molecular weight is different) or different compositions. Points of the same composition will form space lines which are the intersecting lines of surfaces of equal $\%C_A$ and $\%C_N$. Employing this concept, the authors constructed cross sections of the surfaces of equal $\%C_A$ and $\%C_N$ values parallel to the index–density coordinate plane, for certain different values of the logarithm of the viscosity. The fact that each cross section for a given viscosity produced a series of practically straight parallel lines for equal values of $\%C_A$ and $\%C_N$ when index and density were the rectangular coordinates suggested that the construction of nomograms to simplify the calculation of composition would be rather simple and straightforward. The authors consider the replacement of the cross sections with nomograms to be advantageous because of the fact that, in the nomograms, values of several structural elements and physical constants can be represented simultaneously.

The composition data used in constructing the cross sections and consequently the nomograms were based on n-d-M analyses of approximately 100 oil fractions. Since this collection of data is the same one as was used in deriving the V_k-n-d relationship, the method suffers similar limitations. Its use is further restricted by the fact that it cannot be applied to aromatic-free samples. Within the range of 12–30%C_A it agrees well with the n-d-M method.

This procedure is advantageous in that it circumvents the necessity of an experimental determination of molecular weight. It requires the measurement of only the kinematic viscosity, refractive index, and density. The elimination of "prediction factors" and the availability of simple nomograms provide a distinct improvement over the V_k-n-d method, although interpolation between nomograms is somewhat troublesome. It has the advantage of providing for the estimation of both ring analysis and carbon distribution, as well as for the estimation of molecular weight. It suffers the disadvantage that the range of composition over which it may be applied is quite limited.

(6) Applicability of Methods

Stout and his co-workers have made a rather extensive survey of the applicability of the six methods which have been described (213). They compared the carbon distribution determined by each method with that established by detailed analytical procedures for 44 oils covering a wide range of composition. The results of this work are summarized in Table 99.XVI, which tabulates the average and standard deviations from the referee analyses of data produced by each method. Examination of this table shows that each method gives almost equivalent results

TABLE 99.XVI

Deviation of Six Methods for Carbon Distribution
from Detailed Analyses for 44 Oils (213)

Method	Deviation of average			Average deviation			Standard deviation			Percentage of oils to which method is applicable
	%C_A	%C_N	%C_P	%C_A	%C_N	%C_P	%C_A	%C_N	%C_P	
n-d-M	+2.4	−3.1	+0.7	2.4	3.4	1.3	3.3	3.7	1.9	64
RI-KVGC	+1.9	−0.7	−1.2	2.1	3.0	2.6	2.7	3.7	3.2	86
V_k-n-d	+0.1	−0.5	+0.4	0.7	2.4	2.5	1.2	3.0	3.8	39
Cornelissen-Waterman	+1.1	−1.2	+0.1	1.5	2.2	2.9	1.9	2.6	4.0	27
RI-VGC	+0.3	−0.6	+0.3	1.5	2.5	2.6	2.0	3.3	3.3	86
RI-density	0	−1.1	+1.1	1.4	2.5	2.0	1.9	3.1	2.5	100

TABLE 99.XVII
Maximum VGC or %C_A of Oils to Which Methods
Are Applicable (213)

Method	Maximum VGC	Maximum %C_A
V_k-n-d	0.880	20%
Cornelissen-Waterman	0.880	30%
n-d-M	0.950	40%
RI-VGC	1.000	50%
RI-KVGC	1.050	55%
RI-density	1.080	85%

in the range of composition where it may be applied. The last column shows quite clearly that the methods developed by Waterman and his associates are rather severely restricted in range of applicability. As a further result of this study, it was suggested that the practical limits over which each of these methods can be used could be defined by the values of %C_A or viscosity–gravity constant given in Table 99.XVII.

(7) Other Methods

A number of other more or less complex methods for the analysis of oil samples have been proposed. In particular, the procedures described by Montgomery and Boyd (153) and by Deansley and Carleton (41) should be mentioned. Both possess features that make their use in certain specialized applications advantageous. Montgomery and Boyd proposed a system of structural group analysis wherein the carbon distribution was defined as follows:

C_1 = the number of carbons in paraffin chains
C_2 = the number of carbons in saturated ring structures
C_3 = the number of carbons in condensed saturated ring structures
C_4 = the number of carbons in noncondensed aromatic structures
C_5 = the number of carbons is condensed aromatic rings

Obtaining quantitative information concerning these five structural groups requires the experimental determination of molecular weight, elemental analysis, the determination of refractive index and density, and a spectroscopic estimate of the total number of aromatic carbon atoms. Three chemical and two physical property equations are then solved simultaneously by high-speed computer for C_1, C_2, C_3, C_4, and C_5. The method is based on data for pure hydrocarbons and is likely to be of particular utility for the characterization of well-fractionated samples.

The dispersion–refraction method of Deansley and Carleton is one of the few procedures proposed for higher-molecular-weight materials in which the presence of olefinic unsaturation may be taken into account. In practical application it has a number of drawbacks, not the least of which is the fact that it requires the experimental determination of molecular weight, refractive index, density, refractive dispersion, per cent carbon, per cent hydrogen, and bromine number. In addition, the calculations are rather complicated, and the estimation of the percentage of carbon atoms in rings is based on rather arbitrary considerations. The chief merit of the dispersion–refraction method is that it can be applied to cracking residues and similar materials. It also is quite suitable for the analysis of highly aromatic samples.

b. METHODS REQUIRING PRIOR SEPARATION ACCORDING TO MOLECULAR TYPE

The advantages of separating olefin-free petroleum oils according to molecular type and then applying structural group analysis methods to the saturated and aromatic fractions were pointed out in connection with the description of the direct method. It is possible to retain these advantages and at the same time avoid the time-consuming and exacting labor involved in the direct method if relatively rapid and reliable physical property procedures for the analysis of the separated saturated and aromatic portions can be found. Fortunately, analysis of the saturated portion may be carried out without recourse to elemental analysis by substituting the specific refraction for per cent hydrogen (Section IV.B.3.a), and several methods are now available that permit the analysis of the aromatic portion without resorting to hydrogenation.

(1) The Dispersion–Density Method

Martin and Sankin (145) have proposed a method for the analysis of petroleum aromatic fractions that eliminates the necessity for hydrogenation. The physical properties required for application of this procedure are density, F-C specific dispersion, and molecular weight. The density and F-C refractive dispersion are measured at 20°C. The specific dispersion is calculated using the formula:

$$\delta_{F-C} = \frac{(n_F^{20} - n_C^{20})(10^4)}{d}$$

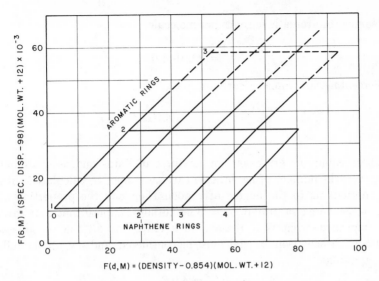

Fig. 99.41. Aromatic analysis graph. Reproduced with the permission of *Analytical Chemistry.*

Functions of dispersion and molecular weight, $F(\delta, M)$, and density and molecular weight, $F(d, M)$ are then calculated according to the equations

$$F(\delta, M) = (\delta - 98)(M + 12) \times 10^{-3}$$

and

$$F(d, M) = (d - 0.854)(M + 12)$$

The average number of aromatic rings, R_A, and the average number of naphthene rings, R_N are read from the graph shown in Fig. 99.41. Alternatively, the ring analysis may be calculated from the equations:

$$R_A = 0.042F(\delta, M) + 0.55$$

When $R_A \leqq 2$

$$R_N = 0.073F(d, M) + 0.64 - 0.073F(\delta, M)$$

When $R_A \geqq 2$

$$R_N = 0.073F(d, M) + 0.95 - 0.082F(\delta, M)$$

The following equations have been proposed by the authors to permit calculation of the carbon distribution from the ring analysis:

Total number of carbon atoms per molecule:

$$C_T = 0.071M + 0.4R_A + 0.1R_N$$

Number of aromatic carbons per molecule:

$$C_A = 2.0 + 4.0 R_A$$

Percentage of aromatic carbons, $\% C_A$:

$$\% C_A = \frac{100 C_A}{C_T}$$

These equations are exact for mixtures of kata-condensed aromatics. However, the number of naphthenic carbons cannot be calculated without assumptions as to the degree of condensation and the proportion of 5- and 6-membered saturated rings. The authors suppose that four carbons per naphthene ring is a reasonable average for all the possible aromatic–naphthene isomers that must be considered, and suggest the following equation:

$$C_N = 4.0 R_N$$

Then

$$\% C_N = \frac{100 C_N}{C_T}$$

and

$$\% C_P = 100 - (\% C_A + \% C_N)$$

The average number of chain carbons, C_P, can also be calculated:

$$C_P = C_T - (C_A + C_N)$$

This method was developed on the basis of data for individual hydrocarbons and petroleum fractions. The function of dispersion and molecular weight was derived from consideration of the "double-bond index" of Lipkin (127). Double-bond index, as given by the relation,

$$\text{DBI} = \frac{(\delta - 98)(M + 17)}{3190 Q}$$

has the drawback that the number of double bonds, Q, is difficult to determine. However, the product of DBI and Q can be calculated from dispersion and molecular weight and used to determine the number of aromatic rings if assumptions can be made concerning the type and degree of condensation. Multiplication of (DBI $\times$ Q) by 3.19 and replacement of $(M + 17)$ with $(M + 12)$ produced the equation for $F(\delta,$

M). From a study of data for pure hydrocarbons, the authors found that the value of this function depends primarily on aromatic type but is influenced to a minor extent by the number of substituents on the aromatic nucleus. Two compounds of the same aromatic type and number of substituents have the same value of $F(\delta, M)$ even if they are considerably different in molecular weight. A substituent has the same effect upon $F(\delta, M)$ whether it is part of an alkyl chain or part of a cycloparaffin ring. The relationship between $F(\delta, M)$ and the number of aromatic rings R_A (and consequently the position of the horizontal lines in Fig. 99.41) was established with the aid of direct method data for 29 aromatic fractions from petroleum. The number of aromatic rings calculated by the direct method was based on the assumption of kata-condensed 6-membered rings. Evaluation of the substituent effect upon $F(\delta, M)$ in conjunction with these data led the authors to conclude that an average of three substituents per molecule is likely for the benzenes and naphthalenes in petroleum.

The function of density and molecular weight was derived from a consideration of the method suggested by Lipkin for calculating the number of aromatic rings in mixtures of alkyl aromatics (123) from density and the temperature coefficient of density. The graphs presented by Lipkin incorporate straight lines representing constant numbers of aromatic rings. If the equation of any one of these lines is defined by

$$d = a\alpha + b$$

where d = density, α = temperature coefficient of density, and a and b are constants, it is possible to substitute for α an equivalent function of molecular weight:

$$-\alpha \times 10^5 = 55.3 + \frac{3516}{M + 12}$$

to give

$$d = 55.3a + b + \frac{3516a}{M + 12}$$

As M becomes large, the value of d approaches $(55.3a + b)$ so d_∞, the density of the limiting paraffin, may be inserted for this quantity. The value $3516a$ is constant for any single constant ring line, and, since it is a function of d and M, $F(d, M)$ may be inserted to replace it, giving

$$d = d_\infty + \frac{F(d, M)}{M + 12}$$

or

$$F(d, M) = (d - d_\infty)(M + 12)$$

If the value 0.8536 is used for the density of the limiting paraffin, the final equation results:

$$F(d, M) = (d - 0.854)(M + 12)$$

Since this function is related to both number of aromatic rings and number of naphthene rings, it can be used in conjunction with $F(\delta, M)$ to estimate the number of naphthene rings. Values of $F(d, M)$ were calculated from data for 84 pure hydrocarbons and 29 petroleum fractions. Values of R_N for the petroleum fractions were established by the direct method. These data were used to calculate the average values of $F(d, M)$ for alkyl aromatics and the average increment in $F(d, M)$ per naphthene ring necessary to establish the positions of the lines for constant numbers of naphthene rings in Fig. 99.41.

The dispersion–density method is useful over a wide range of molecular weights and compositions. It has been tested for aromatics in the gasoline, gas oil, and lubricating oil ranges of petroleum and for fractions whose average compositions exhibit as many as three aromatic rings per molecule. The average deviations of the analyses from direct method data are about 0.1 aromatic ring and 0.2 naphthene ring. Above three aromatic rings per molecule the accuracy is uncertain, and the calculated analyses should be used only for comparative purposes. Saturated hydrocarbons, olefins, some noncondensed polycylic aromatics, anthracenes, and nonhydrocarbons interfere, but as much as 5–20% of various of these may be present without introducing serious error. These interfering compounds, moreover, generally are absent or are present in only minor amounts in aromatic concentrates from petroleum. The "aromatic analysis graph" is convenient for the study of changes in composition of a series of fractions, as well as for the determination of average composition.

The dispersion–density method has been modified by Hazelwood (73) so as to eliminate the necessity of an experimental measurement of specific dispersion. Hazelwood succeeded in developing equations for R_A and R_T based on refractive index, density, and molecular weight. The differences between the measured refractive index and density and the values of these properties for the limiting paraffin are first calculated using the relations

$$\Delta d = d_4^{20} - 0.8510 \qquad \Delta n = n_D^{20} - 1.4750$$

These values are then substituted along with the molecular weight in the following formulas for R_A and R_T:

$$R_T = 0.080M(\Delta d - 0.50\,\Delta n) + 1.10$$

$$R_A = 0.040M(2.75\,\Delta n - \Delta d) + 0.58$$

The other structural elements may then be calculated by relations similar to those used in the dispersion–density method. Agreement with direct method data is equivalent to that for the dispersion–density method. The limitations in application are also similar.

(2) Comprehensive Examination of Oil Fractions

Kurtz and Martin (110,213) have suggested a complete procedure for the detailed structural group analysis of the heavier petroleum fractions. This sequence of analytical operations is outlined in Fig. 99.42. Pertinent details of the method are as follows:

1. The quantitative separation of the saturated hydrocarbons from aromatics is accomplished by percolating 100 ml of sample through 1000 g of 28–200 mesh silica gel. The saturated hydrocarbons are eluted with pentane, and the aromatics with a mixture of benzene and methanol. A complete description of the separation procedure has been given by Lipkin et al. (119).

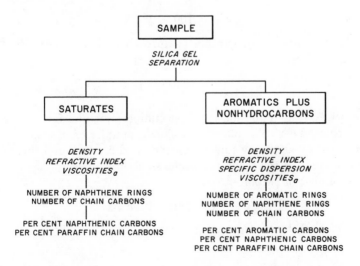

Fig. 99.42.   Scheme of analysis for viscous petroleum fractions.

2. The aromatic fraction is analyzed by means of the dispersion–density method just described.

3. The density, refractive index, and molecular weight of the saturated portion are determined. The number of rings per molecule, R_N, is calculated using the relation

$$R_N = 1 + 0.2145M - 0.6511Mr$$

suggested by Vlugter, Waterman, and van Westen.

In this equation M = molecular weight, and r = Lorentz-Lorenz specific refraction, $(n^2 - 1)/(n^2 + 2)d$.

The percentage of carbon atoms in saturated ring structures, $\%C_N$, is then calculated, using the "wt % ring" equation of Lipkin (Section IV.B.3.c):

When $d \leq 0.861$

$$\text{wt \% ring} = \frac{A + 190.0d - 217.9}{0.593d - 0.249}$$

When $d \geq 0.861$

$$\text{wt \% ring} = \frac{A + 102.8d - 142.8}{0.262}$$

The percentage of carbons in paraffin chain structures, $\%C_P$, is calculated by difference from 100%:

$$\%C_P = 100 - \%C_N$$

Finally, the number of paraffin chain carbons is estimated from the following relation:

$$C_P = \frac{\%C_P}{100} \times (0.071M + 0.1R_N)$$

The "wt % ring" equation is rather cumbersome and for convenience may be replaced with the following decidedly simpler relations, which can be developed by assuming that kata-condensed 6-membered rings represent the average of the saturated cyclic structures present.

Total number of carbon atoms, C_T:

$$C_T = 0.071M + 0.1R_N$$

Number of naphthene ring carbons:

When $R_N \leq 1$

$$C_N = 6R_N$$

When $R_N \geq 1$

$$C_N = 2 + 4R_N$$

Number of paraffin chain carbons:

$$C_P = C_T - C_N$$

The carbon distribution is calculated as follows:

$$\%C_N = \frac{100C_N}{C_T}$$

$$\%C_P = \frac{100C_P}{C_T}$$

The data in Table 99.XVIII for 12 samples covering a wide range of composition show that the actual differences between $\%C_N$ calculated from the "wt % ring" equation and that calculated using these equations are well within the limitations imposed by the assumptions involved in each series of equations.

This method has a number of advantages over procedures that are applied without prior separation. Since both ring analyses and carbon distributions are calculated independently on the saturated and aromatic portions of the sample, considerably more information is available to the analyst. In addition, the various physical property functions are applied to the analysis of the types of molecules for which they are best suited.

TABLE 99.XVIII

Comparison of $\%C_N$ for Saturated Oils Calculated According to
Lipkin (124), and Assuming Kata-Condensed Six-Membered Rings

Saturated oil	R_N	Wt % Ring A and d_4^{20}	$\%C_N$, assuming kata-condensed 6-membered rings	Difference
1	0.42	9.7	9.8	+0.1
2	0.69	16.5	15.9	−0.6
3	0.85	14.9	14.5	−0.5
4	1.33	22.4	20.8	−1.6
5	1.55	31.8	30.6	−1.2
6	1.77	35.3	33.7	−1.6
7	2.07	31.4	29.2	−2.2
8	2.30	42.2	41.0	−1.2
9	2.35	32.2	31.2	−1.0
10	3.21	38.3	40.0	+1.7
11	3.30	42.2	41.0	−1.2
12	3.63	62.1	63.2	+1.1

B. ANALYSIS OF OTHER NATURAL PRODUCTS

The application of physical property methods is not limited to hydrocarbon systems. Substantial insight into the composition of complex mixtures containing appreciable numbers of nonhydrocarbon molecules has been acquired in recent years by the application of physical constitution analysis.

1. Coal

The work of van Krevelen in the field of coal constitution research is an excellent demonstration of the potential of these techniques (102). The analysis of coal is considerably more difficult than that of petroleum. This is so primarily because, although petroleum is a complex mixture, the bulk of the molecules are of relatively low molecular weight, whereas coal exhibits the properties and behavior typical of macromolecular structures. In order to extend the principles of structural group analysis to the analysis of coal, van Krevelen found it useful to introduce the concept of the "mean structural unit," which is the assembly of atoms averaged over the number of aromatic clusters (the systems of aromatic rings with completely conjugated carbon–carbon double bonds). Thus the mean structural unit is neither a strictly defined molecule nor a basic molecular unit, but can be considered as a kind of mathematical model by means of which the properties of the system can be interpreted.

In view of the high carbon content and low reactivity of coal, van Krevelen concluded that it consists largely of systems of condensed aromatic rings. The aromaticity is therefore an important structural parameter and can be expressed in terms of the fraction of aromatic carbon, $f_a = C_A/C$. The hydrogen atoms may be classified as "aromatic" or "saturated," depending on whether they are bonded to aromatic or saturated carbon atoms. It follows that the fraction of "aromatic" hydrogen, $h_a = H_A/H$, is also an important structural parameter which permits conclusions to be drawn about the H/C ratio in the side chains and the ring character of the nonaromatic part of the mean structural unit. The latter is important because coal may contain saturated rings in addition to aromatic rings. H_A/H may be determined in approximate fashion by infrared spectroscopy.

We have already shown (Section II.E) that for a hydrocarbon containing no rings or double bonds we may write

$$H = 2C + 2$$

where H and C represent the number of atoms per structural unit. We have also seen that two hydrogen atoms are eliminated for every ring closure and for the formation of every double bond. However, if only aromatic double bonds are involved, two aromatic carbon atoms, C_A, are created by the introduction of every double bond. Therefore, the creation of every aromatic carbon atom involves the elimination of one hydrogen atom. Consequently, we may write

$$H = 2C + 2 - 2R - C_A$$

Dividing both sides of this equation by C, rearranging, and substituting f_a, the fraction of aromatic carbon, for C_A/C gives:

$$2\frac{R-1}{C} = 2 - f_a - \frac{H}{C} \tag{1}$$

The term $2(R-1)/C$ van Krevelen has called the "ring condensation index." This equation may be used for compounds containing elements other than carbon and hydrogen alone, provided the type of bond between the atoms of these elements is known. The H/C ratio must then be corrected in such a way that the compound is mathemtically transformed into a hydrocarbon with the same carbon skeleton as the original compound. Since coal is a macromolecular structure of largely unknown molecular weight, determination of the constitution must be made in terms of the mean structural unit. On the one hand, the structural unit may present a picture of the mean composition of the molecules in coal; on the other hand, it is equally possible that the mean structural unit is comparable to the "monomer" of a polymeric compound. In the latter case the number of hydrogen atoms in equation (1) must be further reduced by the average number of bridges per structural unit. If the average number of connecting bridges between monomer units is designated as B, the equation for the structural unit then becomes

$$2\frac{R-1}{C} = 2 - f_a - \frac{H}{C} - \frac{B}{C} \tag{2}$$

If B/C cannot be determined or be shown to be negligibly small compared with the ring condensation index, the calculation of the latter by means of equation (1) gives a value which must be considered only as an average for the whole molecule. Besides these structural parameters, the ring aromaticity, that is, the fraction of aromatic rings, R_A/R, is sometimes made use of, along with the fraction of the total carbon atoms that occur in cyclic structures. The latter is sometimes referred to as the "cyclicity." Cyclicity and ring condensation index permit conclusions to be drawn about the compactness of the total ring system. The greater the total number of rings formed from the cyclic carbon atoms available, the more compact the ring system will be. Ring condensation index and ring aromaticity together give an impression of the compactness of the *aromatic* clusters.

A common feature of these structural parameters is that all have the limiting values of zero and unity. For instance, the ring condensation index of the structural units becomes zero for noncondensed aromatic rings and unity for graphite. A negative value implies that no rings are present. An impression of the size and composition of the structural units can be obtained only if, in addition to the quantities in equation (1) or (2), the value of C_A may be determined. If this condition is satisfied, it is possible to calculate not only the size of the aromatic clusters but also the absolute values of the other quantities (H, O, N, R, etc.) from the structural parameters and data supplied by elemental analysis. However, since the molecular weight is unknown, it is obvious that even equation (1) or (2) cannot be solved without recourse to other relationships that provide access to f_a or C. Van Krevelen has made extensive use of additive functions in estimating f_a and C_A. Certain functions of physical constants (such as molecular volume) may be calculated with considerable accuracy for some structural assemblages of atoms by summing the atomic contributions. For other types of structures such a calculation will lead to differences between the calculated and the experimental values. In the latter case, these deviations can serve as a valuable source of information since they often allow conclusions to be reached regarding the particular structural feature responsible for the deviation.

These functions are commonly expressed per mole, and may be written in terms of the contributions of the atomic and structural increments as follows:

$$MF = C\varphi_c + H\varphi_h + O\varphi_o + \cdots + \Sigma X\varphi_x \tag{3}$$

where MF = additive molecular function; M = molecular weight; C, H, O, etc. = number of atoms of carbon, hydrogen, oxygen, etc.; X = number of structural factors; and φ_c, φ_h, φ_x = values of the atomic and structural increments.

Such functions may be used to provide data on structure even if the molecular weight is not known, because when equation (3) is divided by C, M/C in the corresponding reduced equation

$$\frac{M}{C} F = \varphi_c + \frac{H}{C} \varphi_h + \frac{O}{C} \varphi_o + \cdots + \frac{X}{C} \varphi_x \tag{4}$$

follows directly from the per cent carbon:

$$\%C = \frac{12C}{M} \times 100 \quad \text{or} \quad \frac{M}{C} = \frac{1200}{\%C} = M_c$$

Consequently, if C_A were designated as the structural feature X, equation (4) could be used in determining the aromaticity in terms of f_a ($= C_A/C$). In deciding what additive function would be useful for this purpose, it is necessary to consider the way in which aromatic and saturated carbon atoms differ. The bond distances are shorter and the bond energies greater in the former than in the latter. The shorter bond distance results in a smaller volume per carbon atom for aromatic structures, and suggests that molecular volume, M/d, would be a suitable choice as the additive function.

Armed with this information, van Krevelin and Chermin developed a graphical densimetric method for the determination of f_a (101). By plotting the molar volume per gram atom of carbon, M_c/d, against the atomic hydrogen to carbon ratio, H/C, for a large number of aromatic hydrocarbons, they found that the values of M_c/d for each series of compounds of constant aromaticity (f_a) were a linear function of the aromatic H/C ratio. However, since none of the pure compounds contained more than three aromatic rings, the extrapolation of the lines of constant aromaticity to lower values of H/C was somewhat doubtful. To overcome this deficiency van Krevelin and Chermin introduced data that had been obtained in an exhaustive investigation of the constitution of pitch. Basically, this graph, shown in Fig. 99.43, holds only for hydrocarbons. In general, however, it may be applied to organic substances containing foreign atoms provided that appropriate corrections are applied to the experimental values for M_c/d and H/C. Van Krevelen has provided rather complete tables of these corrections. However, if it is assumed that in hard coal (1) oxygen occurs mainly in —OH and —O— groups, (2) nitrogen is present chiefly in heterocyclic form, (3) sulfur is present mainly in —S— groups, and (4) all double bonds are aromatic, then the corrected molar volume of coal per gram atom of carbon is given by

$$\left(\frac{M_c}{d}\right)_{\text{corr}} = \left(\frac{M_c}{d}\right)_{\text{exp}} - \left(8.1\frac{O}{C} + 6.4\frac{N}{C} + 12.5\frac{S}{C}\right) \qquad (5)$$

and no corrections to the H/C ratio are required. The value of f_a may therefore be graphically derived from Fig. 99.43. Only the density and elemental analysis are required. The ring condensation index may be determined by using equation (1) or (2).

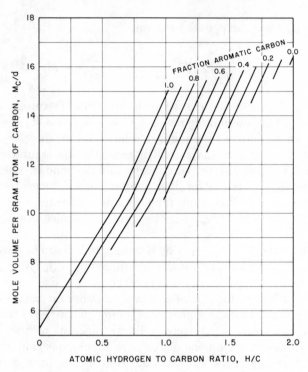

Fig. 99.43. van Krevelen's aromaticity diagram. Reproduced with the permission of Elsevier Publishing Company.

There remains the problem of the determination of C_A or C in order that the complete composition of the mean structural unit may be estimated. The number of physical constants that can be conveniently determined for a substance like coal is rather restricted. However, the refractive index can be measured fairly accurately by a simple, though indirect, method. The refractive index and the density can be used to calculate the Lorentz-Lorenz molar refraction, which is an additive function that obeys the general relationship of equation (4). The molar refraction is distinguished from the molar volume in that the contribution of the structural factors, φ_x, is not always constant. Thus in this case, if C_A is designated as the structural feature X, φ will prove to be a function of C_A. This relationship therefore provides a possible approach for determining the absolute value of C_A.

In pursuing this line of reasoning, van Krevelen was able to show that, for aromatic hydrocarbons, the difference between the experimental molar refraction per gram atom of aromatic carbon and the value calculated from the atomic increments was in fact a function of the number of aromatic carbons per structural unit. He also found the refractometric increment to be an unambiguous function of the aromatic surface area. He attached great importance to this discovery and consequently preferred to utilize the aromatic surface area as a means for ultimately arriving at an estimate of C_A. He found that, if the reciprocal of the aromatic surface area derived from molar refraction for a number of vitrinites was plotted as a function of $100 - \%C$, the points described a straight line that passed through the point for graphite at $100\%C$. He concluded from the equation of this line that, for the vitrinites, the aromatic surface area in square angstroms is given by

$$S \approx \frac{830}{100 - \%C}$$

Since the surface area of an aromatic carbon atom is 2.46 square angstroms, the number of aromatic carbons C_A is given by

$$C_A \approx \frac{340}{100 - \%C}$$

Thus for the vitrinites it is unnecessary to experimentally determine the refractometric increment. Instead, the foregoing empirical relations, which involve elemental analyses only, may be used with a fair degree of confidence.

Since $f_a = C_A/C$ and f_a and C_A can now be determined, the total number of carbon atoms in the mean structural unit can be calculated. The numbers of the other atoms can then be calculated from their atomic ratios and the number of carbons. The number of rings, R, can also be calculated from the ring condensation index. The compositions of a number of vitrinites calculated in this manner have been given by van Krevelen and are shown in Table 99.XIX. This table demonstrates very strikingly the progressive growth of the aromatic ring system during coalification.

Another approach to the analysis of coal and similarly condensed polyaromatic structures is, of course, the application of hydrogenation to convert the sample to a mixture of saturated hydrocarbons. The analysis of such a mixture may then be carried out by conventional methods with some confidence. However, in the case of coal and other substances that behave as macromolecules, hydrogenation is difficult and is always accompanied by cracking and depolymerization, so that large changes in molecular weight are the rule. Schuhmacher and his co-workers (205) have reported on the hydrogenolysis of a Dutch bright coal of 86.5%

TABLE 99.XIX

Composition of the Mean Structural Units for a Number of Vitrinites (102)

Coal No.	Per cent carbon	Arom. C	Total C	H	O	N	No. of rings	Molecular weight
		Number of atoms in the mean structural unit						
1	70.5	14	21	18	5	0.3	6	350
2	75.5	14	18	14	3	0.3	5	290
3	81.5	18	22	17	2	0.4	6	330
4	85.0	23	27	20	2	0.4	7	380
5	89.0	31	35	24	1	0.6	8	470
6	91.2	39	42	25	1	1	11	550
7	92.5	45	48	24	1	1	14	620
8	93.4	52	53	23	1	1	17	680
9	94.2	59	59	22	1	1	19	750
10	95.0	68	68	21	1	1	24	860
11	96.0	85	85	19	1	1	34	1060

carbon content. These investigators conducted the hydrogenation at 325°C and at about 400 atm. Stannous chloride was used as a catalyst, and tetralin as a hydrogen donor. The hydrogenolysis products were separated into a number of fractions by solvent extraction, and the ring content was determined from elemental analysis and molecular weight. The results are shown in Fig. 99.44. It can be concluded from this figure that a mean molecular weight of 400 corresponds to an average of about six rings per structural unit. This agrees fairly well with the result of the physical constitution analysis given for coal number 4 in Table 99.XIX. Hydrogenolysis has provided some valuable information on the skeletal structure of coal. In the first place, confirmation has been obtained concerning the abundance of cyclic structures. Furthermore, the results suggest that hydrogenolysis is accompanied by depolymerization reactions.

These approaches to the physical constitution analysis of coal are noteworthy because they are representative of two devices that are commonly employed to permit the application of physical property methods to nonhydrocarbon systems. The first of these is the mathematical transformation of the nonhydrocarbon molecule into its hydrocarbon analog. From this point, the problem is treated as one of hydrocarbon characterization. The second device is, of course, the actual chemical conversion of the nonhydrocarbon mixture into a mixture of hydrocarbons, which is then analyzed by the methods discussed for oil constitution analysis.

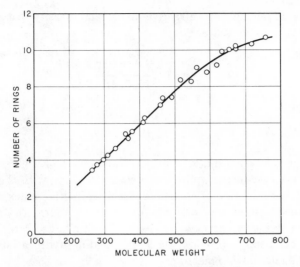

Fig. 99.44.   Ring numbers of fractions from the hydrogenolysis of coal. Reproduced with the permission of Elsevier Publishing Company.

2. Oxygenated Compounds

a. PHENOLS AND AROMATIC ETHERS

Karr and Comberiati (89) have proposed a method for the ring analysis of samples of high oxygen content where the oxygen is present principally in aromatic ether groups and/or phenolic hydroxyl groups. Such mixtures are often encountered in research on petroleum and shale oils and in the examination of low-temperature coal tars. The procedure depends on the mathematical transformation of the properties of oxygen-containing molecules to those of their hydrocarbon analogs, after which the ring analysis methods developed for hydrocarbons may be used.

In discussing the physical properties of hydrocarbons it was demonstrated that for n-alkyl series the refractive index and the density change in a manner that is approximately linear with respect to the reciprocal of the number of carbon atoms or molecular weight. It was also pointed out that these properties approach the value of refractive index (1.4751) or density (0.8513) for a paraffin of infinite molecular weight (Sections II.A and II.B). The corresponding properties of n-alkyl series of oxygenated compounds will behave in a similar manner, and Karr and Comberiati have taken advantage of this fact.

The rate of change of refractive index and density with molecular weight may be expressed as

$$\frac{dn}{dM} = \frac{n_D^{20} - 1.475}{1000/M} \tag{1}$$

$$\frac{dd}{dM} = \frac{d_4^{20} - 0.851}{1000/M} \tag{2}$$

The average refractive index or density for isomers of a given molecular weight, if inserted in the appropriate equation along with the molecular weight, will give the rate of change. By using refractive index and density data at several molecular weights for a given series, the consistency of the linear relationship may be observed and average rates for a class of oxygen compounds (O) and the corresponding hydrocarbons (H) derived. These may be designated as

$$\left(\frac{dn}{dM}\right)^{O} ; \left(\frac{dd}{dM}\right)^{O} \quad \text{and} \quad \left(\frac{dn}{dM}\right)^{H} ; \left(\frac{dd}{dM}\right)^{H}$$

It follows that the difference between the refractive index or density of a mixture of oxygen compounds and the analogous hydrocarbon mixture at a given average molecular weight is represented by the relations

$$\Delta n_D^{20} = \frac{1000}{M}\left[\left(\frac{dn}{dM}\right)^{O} - \left(\frac{dn}{dM}\right)^{H}\right] \tag{3}$$

$$\Delta d_4^{20} = \frac{1000}{M}\left[\left(\frac{dd}{dM}\right)^{O} - \left(\frac{dd}{dM}\right)^{H}\right] \tag{4}$$

A simpler notation is to express the differences in the rates of change as a constant k:

$$\left[\left(\frac{dn}{dM}\right)^{O} - \left(\frac{dn}{dM}\right)^{H}\right] = k_n$$

and

$$\left[\left(\frac{dd}{dM}\right)^{O} - \left(\frac{dd}{dM}\right)^{H}\right] = k_d$$

The differences in physical properties predicted by equations (3) and (4) will be excessively large when the mixtures considered consist only in part of oxygen compounds. In such cases the equations must be multiplied by the factor $OM/1600$, the average number of oxygen atoms per molecule. In this relation, O is the weight per cent of oxygen. Equations (3) and (4) can therefore be rewritten as

$$\Delta n_D^{20} = k_n \left(\frac{O}{1.6}\right) \tag{5}$$

and

$$\Delta d_4^{20} = k_d \left(\frac{O}{1.6}\right) \tag{6}$$

As a further simplification, the authors designate $k/1.6$ as k^*. The values of k_n^* and k_d^* therefore indicate the increase in refractive index and density for each 1% increase in oxygen content of the sample as compared to an analogous sample of hydrocarbons. Equations (5) and (6) then become simply

$$\Delta n_D^{20} = k_n^* \times \text{wt } \%O \tag{7}$$

$$\Delta d_4^{20} = k_d^* \times \text{wt } \%O \tag{8}$$

By using physical property data for over 1200 oxygen compounds and their hydrocarbon analogs, Karr and Comberiati derived values of k_n^* and k_d^* for a number of homologous series of aromatic ethers and phenols. They recommend the use of the following average values of the constants for samples containing aromatic ethers, phenolic compounds, and mixtures of these oxygenated compound types:

	k_n^*	k_d^*
Aromatic ethers	0.0007	0.007
Phenolic compounds	0.0023	0.011
Mixtures	0.0015	0.009

In application of the method, the refractive index, density, molecular weight, and oxygen content are determined. The values of Δn_D^{20} and Δd_4^{20} are calculated and subtracted from the observed values to obtain the approximate physical properties of the structurally analogous aromatic hydrocarbons. The ring analysis method of Hazelwood [Section VI.A.3.b.(1)] is then applied. This procedure gives results that agree well with the known values for individual oxygenated compounds and synthetic mixtures. It has been found useful in the examination of fractions from low-temperature coal tars, such as tar acids containing phenolic compounds, neutral oils containing aromatic ethers, and pitch resins containing both phenolic hydroxyl and aromatic ether groups.

b. Fatty Acid Derivatives

Waterman attacked the problem of the structural analysis of cyclic fatty acid derivatives (polymerized drying oils; copolymerization products of fatty oils with various hydrocarbons) by actual chemical deoxygenation to produce saturated hydrocarbons of corresponding structure (228). In this case, chemical transformation was necessary because of the limited amount of reliable physical data on cyclic or branched acids. The deoxygenation of the acid esters can be effected either by high-pressure hydrogenation:

$$R\text{—}CH_2\text{—}COOR' + H_2 \xrightarrow[\substack{330\text{-}350°C \\ 200\ atm}]{NiCu} RCH_3 + CO_2 + R'H$$

or by application of the Grignard reaction to produce a tertiary alcohol, dehydration of the alcohol, and saturation of the olefinic linkage to produce the saturated hydrocarbon. The direct hydrogenation results in hydrocarbons with one carbon atom less than the number present in the hydrocarbon chain of the original ester. In the Grignard method the hydrocarbon chain of the molecule is lengthened. The ring analysis of the saturated hydrocarbons produced may be carried out using specific refraction or elemental analysis and molecular weight. Waterman used these methods to confirm the formation of rings during the thermal polymerization of linseed oil and tung oil. He found that the thermal polymerization leads chiefly to the formation of dimers containing approximately two rings per molecule. The deoxygenation technique has also been used by Wijnands (12) in his investigation of the structure of novolaks, prepared by polycondensation of formaldehyde with phenol, p-cresol, and m-cresol. The novolaks were transformed into saturated hydrocarbon mixtures by direct hydrogenation. Ultimate analysis of the hydrocarbons confirmed the linear structure of the condensation products.

c. Cyclic (Naphthenic) Acids

Investigations into the constitution of the higher-molecular-weight naphthenic acids found in petroleum have frequently involved conversion to the analogous hydrocarbons. The general formula for naphthenic acids may be written as $R(CH_2)_nCOOH$, where R is an alkylated cyclic nucleus of one or more rings. The ring structures are generally cycloparaffinic, although there is evidence that aromatic rings may be present in the higher-molecular-weight naphthenic acids. It has been established that n may be 1, 2, 3, 4, or 5, but that compounds where $n = 1$ predominate. The procedure used by Zelinsky (246) for conversion of naphthenic acids to their corresponding hydrocarbons involves preparation of the ester, alcohol, and iodide, and finally, conversion of the iodide to the hydrocarbon. The resulting hydrocarbon mixture may then be characterized by structural group analysis. It is also possible to convert the acids directly to a mixture of saturated hydrocarbons by multiple hydrogenations using Raney nickel (78). The hydrocarbons prepared in this manner are very likely to contain one less carbon atom than the parent acid as the result of complete removal of the carboxyl group.

To avoid the necessity of converting acids to hydrocarbons, Jezl and Sankin (84) developed a method for the ring analysis of naphthenic acids based on elemental analysis and molar refraction. In principle this method involves (1) calculation of the numbers of carbon, hydrogen, and sulfur atoms (usually small) from elemental analysis and molecular weight; (2) subtraction of the sum of the atomic refractions for these atoms, plus that for the two oxygen atoms in a carboxyl group, from the experimental molar refraction; and (3) division of this difference by the molar refraction equivalent of an aromatic ring to give the number of aromatic rings per molecule. The number of rings plus double bonds $(R + Q)$ is calculated from the empirical formula. The number of aromatic rings, together with an estimate of the average number of double bonds per aromatic ring, is used to derive the number of cycloparaffin rings from $R + Q$. This series of calculations is expressed by the following two equations:

$$R_A = M(0.1724r - 0.0003471C - 0.001881H - 0.0004264S) - 0.64$$

$$(1)$$

where M = molecular weight; r = Lorentz-Lorenz specific refraction, $(n^2 - 1)/(n^2 + 2)d$; and C, H, and S = %C, %H, and %S.

$$R_N = M(0.000833C - 0.00496H) - 3.8R_A \qquad (2)$$

where M = molecular weight; C and H = %C and %H; and R_A = number of aromatic rings from equation (1).

Equation (2) assumes 2.8 double bonds per aromatic ring, which is an average of the correct values for benzenes and naphthalenes. The error caused by using this average is less than 0.2 naphthene ring for any sample containing less than 50% aromatic molecules. The accuracy of the overall ring analysis is about 0.1 aromatic ring and 0.3 naphthene ring. The analyses generally agree well with those obtained by actual preparation of the hydrocarbons by the Zelinsky route or by direct hydrogenation.

In order to avoid the necessity for time-consuming elemental analysis, Jezl and Sankin also examined the possibility of using a modified form of the n-d-M method to determine the number of aromatic and naphthenic rings in naphthenic acids. The n-d-M method utilizes the following function in the determination of aromatic ring content [Section VI.A.3.a.(1)]:

$$Mv = M[2.51(n - 1.475) - (d - 0.851)] \qquad (3)$$

where n is the refractive index for the sodium D line at 20°C, d is the density at the same temperature, and M is the molecular weight. The value of Mv is then used in the following equations to calculate the number of aromatic rings per molecule:

If Mv is positive: $R_A = 0.44 + 0.055Mv$ $\qquad (4)$

If Mv is negative: $R_A = 0.44 + 0.080Mv$ $\qquad (5)$

The authors tabulated the differences between the values of Mv for a large number of monocarboxylic acids and their analogous hydrocarbons. They concluded that, since naphthenic acid fractions probably contain no noncyclic compounds, the average difference in Mv between carboxylic cyclanes and cyclane hydrocarbons can be used to derive a revised form of the n-d-M relations for calculating the number of aromatic rings. The following equations are the result of introducing the average difference in Mv of 22.1 into equations (4) and (5):

$Mv \geqq -22.1$: $R_A = 1.66 + 0.055Mv$ $\qquad (6)$

$Mv \leqq -22.1$: $R_A = 2.21 + 0.080Mv$ $\qquad (7)$

When applied to naphthenic acid fractions, equations (6) and (7) give results for R_A that agree well with those calculated from molar refraction and elemental analysis.

Jezl and Sankin also explored the possibility of using a similar approach to revise the n-d-M equations for calculating the total number of rings per molecule, R_T, since the number of naphthenic rings would then be accessible by difference. They concluded that, although suitable equations could probably be developed, the available data on actual naphthenic acid samples did not cover a wide enough range of R_T to allow an adequate assessment of their accuracy.

3. Sulfur Compounds

The structural group analysis of high-molecular-weight sulfur compounds (generally prepared as concentrates from natural sources such as petroleum or coal tar) can be carried out by converting the sulfur compounds to their hydrocarbon analogs. Mozingo and his associates (155) have described the use of Raney nickel to effect almost quantitative conversion of sulfur compounds to hydrocarbons at relatively low temperatures.

By using techniques similar to those applied by Karr and Comberiati for oxygenated compounds (Section VI.B.2.a), Karr, Wendland, and Hansen (90) have developed proportionality constants that allow the refractive index and the density of sulfur-rich aggregates from mineral oils to be converted to the corresponding properties of analogous mixtures of hydrocarbons. Because the questionable reliability of some of the available physical property data on sulfur compounds made it necessary to exercise some selective judgment, the authors found it more convenient to prepare reciprocal molecular weight plots than to calculate the rate of change of the properties using linear equations. Plots were prepared for 12 homologous series of sulfur compounds. On each plot a similar curve was constructed for the analogous hydrocarbon series. When values for more than one isomer were available, the arithmetic average was taken as the representative value for the particular molecular weight. The term "analogous hydrocarbon" is defined as the hydrocarbon that would result if each sulfur atom in aliphatic combination were replaced by a methylenic ($-CH_2-$) or ethylenic ($-CH_2-CH_2-$) group and each sulfur atom in aromatic combination by an acetylenic ($-CH\!=\!CH-$) group. This is a rather arbitrary convention, but it permits one to relate the various classes of sulfur compounds with the major classes of petroleum hydrocarbons. Thus, thiacyclopentanes and thiacyclohexanes become cyclopentanes and cyclohexanes; thiophenes become benzene derivatives; benzothiophenes become naphthalenes; and so on.

The relationship between the physical properties and reciprocal molecular weight proved to be linear in all cases; consequently the proportionality constants k_n and k_d could be derived directly from the plots. Using the notations of Karr and Comberiati, we may write

$$\Delta n_D^{20} = \frac{1000}{M}\left[\left(\frac{dn}{dM}\right)^{\mathrm{S}} - \left(\frac{dn}{dM}\right)^{\mathrm{H}}\right] \tag{1}$$

and

$$\Delta d_4^{20} = \frac{1000}{M}\left[\left(\frac{dd}{dM}\right)^{\mathrm{S}} - \left(\frac{dd}{dM}\right)^{\mathrm{H}}\right] \tag{2}$$

If we designate the differences between the rates of change as k_n and k_d, equations (1) and (2) become

$$\Delta n_D^{20} = \frac{1000}{M} \times k_n \tag{3}$$

and

$$\Delta d_4^{20} = \frac{1000}{M} \times k_d \tag{4}$$

If these are transposed to solve for k_n or k_d, they demonstrate that to deduce k_n or k_d from the plots it is necessary only to read the value of the property difference at some convenient value of $1000/M$ and divide by this value of $1000/M$.

When the mixtures examined consist only in part of sulfur compounds, equations (3) and (4) must be multiplied by the factor $SM/3200$, equal to the average number of sulfur atoms per molecule, where S is the sulfur content in weight per cent, giving

$$\Delta n_D^{20} = k_n \left(\frac{S}{3.2}\right) \tag{5}$$

$$\Delta d_4^{20} = k_d \left(\frac{S}{3.2}\right) \tag{6}$$

By way of additional simplification, $k/3.2$ may be designated as k^*. Equations (5) and (6) then become simply

$$\Delta n_D^{20} = k_n^* \times \text{wt } \%\text{S} \tag{7}$$

$$\Delta d_4^{20} = k_d^* \times \text{wt } \%\text{S} \tag{8}$$

Thus if each sulfur atom in a sulfur-rich petroleum fraction is replaced by a methylenic, ethylenic, or acetylenic group, depending on whether the sulfur is in aliphatic or aromatic combination, the resulting decrease in density and refractive index is given simply by the product of the sulfur content in weight per cent and the appropriate constant. The authors list k^* values for 12 different structural types of sulfur compounds. For high-sulfur aggregates from petroleum, they recommend use of the average constants: $k_n^* = 0.0022$; $k_d^* = 0.0064$. In application of the method, the refractive index, density, molecular weight, and sulfur content are determined. The values of Δn_D^{20} and Δd_4^{20} are calculated and subtracted from the observed values to obtain the approximate physical properties of a mixture of structurally analogous hydrocarbons. The adjusted values of the physical properties can then be used to carry out the structural group analysis by any suitable method.

As a further convenience for the user, the authors have incorporated their sulfur correction term into the equations given by van Nes and van Westen [Section VI.A.3.a.(1)] and by Hazelwood [Section VI.A.3.-b.(1)]. The n-d-M relations for carbon distribution become:

If v is positive: $\%C_A = 430v + 3660/M + 0.38S$
If v is negative: $\%C_A = 670v + 3660/M + 0.59S$

If w is positive: $\%C_R = 820w + 10000/M - 3.2S$
If w is negative: $\%C_R = 1440w + 10600/M - 5.7S$

The modified ring analysis equations are as follows:

If v is positive: $R_A = 0.44 + 0.055M(v + 0.9 \times 10^{-3}S)$
If v is negative: $R_A = 0.44 + 0.080M(v + 0.9 \times 10^{-3}S)$

If w is positive: $R_T = 1.33 + 0.146M(w - 4.0 \times 10^{-3}S)$
If w is negative: $R_T = 1.33 + 0.180M(w - 4.0 \times 10^{-3}S)$

Hazelwood's equations take the following form when the sulfur correction term is incorporated:

$$R_T = 0.080M(\Delta d - 0.50\,\Delta n - 5.3 \times 10^{-3}S) + 1.10$$

$$R_A = 0.040M(2.75\,\Delta n - \Delta d + 0.3 \times 10^{-3}S) + 0.58$$

The equations for carbon distribution are unchanged, save that the value for the total number of carbon atoms per molecule must be estimated from elemental analysis and molecular weight, remembering that the determined value for carbon must be increased by the substitution of two carbon atoms for each sulfur atom present.

Karr, Wendland, and Hansen found it rather difficult to evaluate quantitatively the improvement afforded by use of the modified n-d-M and Hazelwood equations. From the results of their application to a series of test mixtures and high-sulfur concentrates from petroleum, the authors concluded that, in general, the equations incorporating the sulfur correction gave more rational and reliable results. They recommend that the following criteria be used in selecting the proper equations for application to the analysis of a particular sample:

% Sulfur	n_D^{20}	d_4^{20}	Recommended modified equation
<3%	<1.57	<0.99	n-d-M
<3%	>1.57	>0.99	Hazelwood
>3%	<1.57	<0.99	MW > 400, Hazelwood; MW < 400, n-d-M
>3%	>1.57	>0.99	Hazelwood

Mathematical transformation methods, of which this is an excellent example, will undoubtedly improve in accuracy as more reliable data on the physical properties of sulfur, oxygen, and nitrogen compounds become available.

In another approach, van Zijll Langhout (247) has developed graphs based on ultimate analysis that can be applied for the identification of pure compounds containing carbon, hydrogen, and sulfur. They can also be used to analyze simple mixtures of such compounds and to represent distinct chemical transformations in which these materials may be involved. Plots of specific refraction versus the weight per cent of sulfur also have some merit for characterization purposes, since homologous or repetitive series of sulfur compounds of different structural types produce straight lines of different slope for each series that approach the limit 0.3308 at 0% sulfur. Similar graphs based on other physical constants can be constructed. The accuracy and usefulness of such graphs will probably increase as more reliable physical constants for sulfur compounds become available. This kind of treatment might possibly be useful for the characterization of sulfur compounds in crude oil fractions, provided they could be adequately separated from the hydrocarbons with which they occur. This is extremely difficult, however, since the sulfur compounds in petroleum oils resemble the corresponding hydrocarbons very closely, both physically and chemically.

4. Inorganic Systems

Physical properties provide a potentially useful means of defining the composition of inorganic systems. The identification of minerals, alloys, and so forth on the basis of certain bulk properties has long been a common practice. The application of these methods for the quantitative analysis of such materials, however, has been employed to a considerably less degree. Although the scope of this chapter does not allow more than a perfunctory treatment of this subject, the work of Waterman deserves mention as an example of the utility of physical methods as applied to inorganic analysis.

a. Composition of Glasses

In investigating the behavior of certain glasses upon thermal treatment, Waterman (228) found that, if the specific refraction, $(n^2 - 1)/(n^2 + 2)d$, was plotted as a function of the composition for the sodium and potassium silicate glasses, two straight lines were obtained which intersected at the specific refraction value for molten quartz. These straight lines therefore make it possible to determine the composition of potassium and sodium glasses by determining their density and refractive index and calculating their specific refraction. Waterman has also reported exploratory work indicating that the viscosity behavior of molten glasses may prove to be a useful property for characterization.

VII. EXPERIMENTAL MEASUREMENT OF PHYSICAL PROPERTIES

The purpose of this brief section is not to provide the reader with a comprehensive survey of the techniques for measuring the necessary physical properties, for this has been the subject of a number of individual chapters in this volume and preceding ones. Neither is it intended to provide a set of instructions for laboratory practice. The purpose is, rather, to guide the reader to the sources of specific procedures that have been found practical for the routine examination of oil samples in the author's laboratory.

A. DENSITY

The density of a substance is the weight in vacuum of a unit volume of the material at a given temperature. In the methods commonly in use today the unit of mass is the gram and the unit of volume the milliliter. The test temperature is generally 20 or 25°C. For a detailed discussion of density measurement, Chapter 81 of this Treatise should be consulted.

The application of most physical property methods requires a density accurate to within 1–2 units in the fourth decimal. To achieve satisfactory accuracy in routine work, it is necessary to control the temperature within ±0.02°C, to correct for the effect of air buoyancy, and to compare the balance weights used so that their relative values are known to the nearest 0.05 mg.

The most common method of density determination involves finding the weight of liquid occupying a known volume defined by the shape of a given vessel. Since in most cases it is impracticable to determine the volume of the vessel from its geometry, the vessel is calibrated in terms of the weight of pure water which it will hold. A great many devices have been used for defining a volume so that the filling and weighing of the vessel are reproducible and convenient. In the author's laboratory the twin-arm bicapillary pycnometers described by Lipkin are used (118,126). They fill conveniently by a simple siphoning action, are easy to weigh and clean, and are available commercially. Detailed procedures for the use of these pycnometers are described in ASTM Methods of Test D941-55 and D1481-62 (3).

B. REFRACTIVE INDEX AND DISPERSION

For the application of many physical property methods, the refractive index must be measured with an accuracy of 1 or 2 units in the fourth decimal place. An ordinary Abbe refractometer is generally adequate for this purpose if the temperature of measurement is controlled to within 0.1°C and the instrument is carefully calibrated with liquids of certified refractive index. For calibration, standard samples of isooctane, methylcyclohexane, and toluene may be obtained from the American Petroleum Institute.

In this instrument the refractive index is measured by the critical angle method, using white light. In its usual form, the refractometer requires a sample of only about 0.05 ml of liquid and is designed for a maximum of speed, convenience, and simplicity. All transparent and light-colored liquids with refractive indices in the range of 1.30–1.70 can be examined. The refractive index of dark samples can often be determined by removing the back cover plate of the measuring prism and resorting to direct illumination. Detailed procedures for the use of Abbe refractometers have been published in ASTM Method of Test D1807-64 (4).

For work of greater accuracy and for the measurement of dispersion, the use of a Pulfrich or Precision Abbe refractometer is recommended. The Precision Abbe instrument has been used in the author's laboratory for a number of years for the measurement of the dispersion of hydrocarbon liquids and has proved quite satisfactory. It is a critical angle instrument identical in principle to the ordinary Abbe refractometer. However, improved accuracy has been obtained by dispensing with the compensating prisms and using unusually large and precise Abbe prisms mounted on a long vertical, taper bearing. The instrument is used with monochromatic light sources and can measure the refractive index for a variety of wavelengths. The sector is marked in equal, arbitrary divisions, and a table is used to convert scale readings to values of refractive index.

The Precision Abbe is theoretically capable of measuring the absolute value of n with an accuracy of 2 or 3 units in the fifth decimal place.

However, this accuracy is seldom realized when the instrument is used for liquids because of the difficulty of introducing the illumination at truly grazing incidence. Refractive indices of liquids are generally reliable to 6 units in the fifth decimal, and refractive dispersions to 12 units in the fifth decimal. To attain this kind of accuracy, however, the temperature of the prisms must be controlled to within 0.02°C of the test temperature, and the instrument fitted with a special thermometer holder and enclosed in a constant-temperature box. In addition, the refractometer must be calibrated periodically with certified reference liquids. A further complication that occurs in measurement of the widely used F-C dispersion is the inherent weakness of illumination produced by conventional hydrogen discharge lamps. With the Precision Abbe a lamp capable of producing light having an intensity of at least 3 foot-candles as measured on the entrance face of the illuminating prism is required. Several suitable designs have been described in the literature (22,93). A detailed description of the use of this refractometer is given in ASTM Method of Test D1218-61 (3).

A comprehensive treatment of refractometry is contained in this Treatise, Part I, Volume 6, Chapter 70.

C. VISCOSITY

In considering the determination of viscosity it is necessary first to distinguish between the dynamic or absolute viscosity and the kinematic viscosity. The cgs unit of dynamic viscosity (η) is the poise, which has the dimensions grams per centimeter per second. The kinematic viscosity is the quotient of the dynamic viscosity divided by the density, η/d, both at the same temperature. The unit of kinematic viscosity, V_k, is the stoke, which has the dimensions square centimeters per second. It has become common practice to use the hundredth part of a stoke, centistokes, in expressing the vicosity. It is with the measurement of kinematic viscosity, expressed in centistokes, that this section is primarily concerned.

In determining the kinematic viscosity, the time required for a fixed volume of sample, contained in a glass viscometer, to flow through a calibrated capillary under an accurately reproducible head of liquid, and at a closely controlled temperature, is measured. The viscosity is calculated from this efflux time. The viscometer is calibrated by using standard oils having viscosities established by reference to water in a master viscometer, or by direct comparison with another, carefully calibrated viscometer.

A number of suitable capillary viscometer designs have been described in the literature (86). In the author's laboratory the Cannon-Fenske (23) instrument is used for all determinations save those where the required 10 ml of sample is not available or the material is opaque. In these cases a Zeitfuchs cross-arm viscometer (86) is used. The latter design is suitable for either transparent or opaque liquids and requires only about 2 ml of sample. It has the disadvantage that check determinations cannot be made without completely emptying and cleaning the viscometer. It is for this reason that the Cannon-Fenske instrument is preferred for routine use, since replicate determinations may be made by simply repositioning the head level of the sample in the capillary arm of the instrument.

For accurate viscometry it is essential to control the temperature to within $\pm 0.02°F$ of the test temperature and to avoid gradients in the bath. It is also necessary to keep the viscometers scrupulously clean and to make sure that the sample is free from all solid particles, water, or lint. Detailed instructions for the determination of kinematic viscosity are given in ASTM Method of Test D445-65 (3). When required, the viscosity index may be calculated directly from the kinematic viscosity using the ASTM Viscosity Index Tables (5).

In several of the hydrocarbon analysis methods mentioned in Section VI, the Saybolt universal viscosity is required. The kinematic viscosity may be converted to Saybolt viscosity by means of the tables given in ASTM Method of Test D2161-66 (3).

For a detailed discussion of viscosity the reader should consult Chapter 83 of this volume.

D. SURFACE TENSION

The use of the molecular parachor to determine the degree of branching of hydrocarbon liquids (Section IV.B.4) requires the measurement of surface tension. Although the determination of this property is fundamentally simple, a great deal of care is required to produce data of acceptable quality. For work of highest accuracy the capillary height method can be recommended. Suitable apparatus for the application of this technique has been described by Harkins and Jordan (72). The capillary method has the disadvantage, however, of being quite slow. The ring method, on the other hand, is very fast and fundamentally is capable of high accuracy, especially if apparatus employing a chainomatic balance is used (72). For routine work intended for characterization, the use of a duNouy torsion balance is generally satisfactory. Modern duNouy tensiometers are capable of a reproducibility of ± 0.05

dyne and, if carefully calibrated and adjusted, an accuracy of about 1%. Details of the use of this instrument for the measurement of surface and interfacial tension are given in ASTM Method of Test D971-50 (3). The measurement of surface tension is discussed at considerable length in Chapter 82.

E. MOLECULAR WEIGHT

In many cases a satisfactory value for the molecular weight of oil fractions may be obtained by the use of physical property correlations (Section IV.B.2.b). However, for certain well-fractionated samples or relatively pure hydrocarbons, it may be necessary to measure the molecular weight experimentally.

The determination of molecular weight is generally rather time consuming and requires a careful technique if reliable results are to be obtained. The ebullioscopic method has been used with considerable success for hydrocarbons and has received rather general acceptance because of its convenience and precision. It is based on the measurement of the changes in the boiling point of a solvent that occur upon successive additions of sample. The molecular weight is calculated from the temperature and sample concentration measurements. The sample must be soluble and stable at the boiling point of the solvent, and it must have a relatively low vapor pressure at that temperature. The vapor pressure requirement is critical; in general, the sample should boil at least 100°C higher than the solvent. If the boiling point is too low, or the sample unstable, a cryoscopic method must be used. Rossini has discussed the application of both ebullioscopic and cryoscopic methods to hydrocarbons and has defined the areas where each is most useful (187).

A number of thermistor methods for molecular weight determination have been described (16,156) that make use of the "electro-osmotic" technique proposed by Hill (74) and Baldes and Johnson (7). In a typical apparatus a twin thermistor bridge is used to detect the small temperature difference that arises from the difference in solvent activity between a drop of pure solvent and one of solvent plus nonvolatile solute hanging on the thermistors. The dial reading required to balance the bridge can be related to the molality of the solution drop by calibration. Once the calibration has been performed for a given solvent, the molecular weight of an unknown solute can be calculated from the dial reading and its concentration in the solvent. The limitations are similar to those of the ebullioscopic technique. An instrument utilizing this principle is commercially available and has been used with excellent results in the author's laboratory. The method is rapid and quite accurate. As many

as eight to ten determinations may be made per day, and an accuracy
of 1–2% in the molecular weight range of 180–500 can be attained with
little difficulty. The practical application of the thermistor micro method
has been discussed by Wilson (239). Details of the experimental technique are contained in ASTM Method of Test D2503-67 (3).

F. ANILINE POINT

The aniline point is the minimum equilibrium solution temperature
for an equivolume mixture of aniline and sample. Equal volumes of
aniline and sample are placed in a tube and mixed mechanically. The
mixture is heated at a controlled rate until the two phases become miscible. The mixture is then cooled at a controlled rate, and the temperature
at which the two phases separate is recorded as the aniline point. Relatively simple equipment may be used. A test tube 25 mm in diameter
and 150 mm in length fitted with a thermometer and wire ring stirrer
may serve as the sample tube. This is inserted in a glass jacket about
40 mm in diameter and 175 mm in length. The whole apparatus is then
immersed just below the jacket lip in a transparent oil bath provided
with a means for controlled heating and cooling. Ten milliliters of sample
and 10 ml of freshly distilled aniline are generally used. The sample
must be carefully dried by shaking with a suitable desiccant, which
is then removed by filtration. The precision of the determination is about
0.2°C. A complete procedure for the measurement of aniline point is
given in ASTM Method of Test D611-64 (3).

VIII. APPLICATIONS

Up to this point we have dealt mainly with a description of the physical properties that have proved useful in analyzing complex mixtures,
their combination to produce simple functions related to composition,
and, finally, their integration into relatively rapid routine analytical
procedures for both hydrocarbon and nonhydrocarbon systems. From
time to time, certain brief illustrations of the utility of these methods
have been used to augment the discussion.

The purpose of this section is to present material drawn mainly from
the experience of the author and his associates, which demonstrates more
completely the usefulness of physical property methods in a number
of applications. Most of the examples are concerned with hydrocarbon
systems of high molecular weight, although several applications to other
materials are given. In this connection, it should be remembered that

many complex natural products can be rendered amenable to characterization by the methods described, provided they are converted to hydrocarbons. As has been pointed out, such conversion is eminently practical in many cases.

A. HYDROCARBONS

The problem of analyzing mixtures of hydrocarbons boiling up to about 120°C is yielding rapidly to instrumental techniques such as gas chromatography and mass spectrometry. As a consequence, molecular type analysis of the lower-boiling hydrocarbons by the application of physical properties is becoming less commonplace. The use of these techniques for the analysis of gasoline and naphtha fractions has been adequately illustrated (Section IV.C) and therefore is not considered in this section. It is in the analysis of the higher-boiling petroleum oils that methods for structural group analysis are most often applied. Indeed, it is unlikely that they will ever be entirely displaced for this purpose.

The use of physical properties (and structural group analysis) for the analysis of mixtures of high-molecular-weight hydrocarbons has found its greatest application in four connections, namely, (1) characterizing "uniform" fractions separated from petroleum; (2) analyzing oils and oil products, including the comparison of oils from different crudes; (3) following changes in composition occasioned by the application of physical separation processes such as solvent extraction, adsorption, and thermal diffusion, and elucidating the reactions involved in chemical conversion processes such as hydrogenation, dehydrogenation, and polymerization; and (4) providing additional information on molecular composition through the structural group analysis of a series of fractions.

1. Characterization of Uniform Fractions

Fundamental research on oil constitution frequently requires the characterization of fractions that are relatively homogeneous as to hydrocarbon type. Such fractions are usually produced only after the application of a series of rigorously applied physical separation techniques. The characterization of such samples is best accomplished by the careful measurement of a number of physical properties, the values of which are then compared to those predicted for various types of hydrocarbons of similar molecular weight. The relations between the various physical properties and molecular weight for homologous series discussed in Section II are often useful for this purpose.

As many different properties as is practicable should be determined, so as to leave no reasonable doubt concerning the conclusions drawn from the comparison of the properties of the fraction with those of pure compounds. As a minimum, determination of the following properties is recommended: (1) the molecular weight; (2) the content of carbon and hydrogen; (3) the content of sulfur, nitrogen, and oxygen; (4) the boiling point at 1 mm Hg; (5) the density; (6) the refractive index for the D, F, and C lines; and (g) the aniline point. From these properties the molecular formula, $C_nH_{2n+x}S_vN_yO_f$, the specific refraction, and the specific dispersion are calculated. The procedure for characterization is best illustrated by the following example, given by Mair (139), of the examination of a sample from the exhaustive fractionation of the lubricant portion of Ponca crude.

The values of the suggested properties of a sample separated from a petroleum oil by exhaustive fractionation were found to be as follows: elemental analysis, carbon 89.75 and hydrogen 10.246 per cent by weight; molecular weight, 334.5 g/mole; density, 1.004 g/ml; refractive index, 1.573; refractive dispersion, $n_F - n_C \times 10^4$, 202; boiling point at 1 mm, 217°C; aniline point, 6°C.

From the elemental analysis and molecular weight, the molecular formula is calculated to be $C_{25}H_{34}$, which gives the value of $x = -16$ in the formula C_nH_{2n+x}. From the refractive index and density and the refractive dispersion and density, the specific refraction and specific dispersion are calculated and found to be 0.3282 and 201, respectively.

The specific dispersion, 201, is higher than that which would be expected for aromatic-free material, that is, 98–100 (Fig. 99.25), and indicates the presence of aromatic rings. The specific refraction, 0.3282, would correspond to an aromatic-free fraction of $x = -0.3$, but not $x = -16$. This too indicates the presence of aromatic rings. The aniline point of 6°C is also indicative of a high proportion of aromatic structures (Fig. 99.17).

The fraction is carefully hydrogenated under conditions chosen to minimize the likelihood of rupture of C–C bonds. The hydrogenated material is found to have the following values of the selected physical properties: elemental analysis, carbon 87.13 and hydrogen 12.872 per cent by weight; molecular weight, 344.6 g/mole; density, 0.933 g/ml; refractive index, 1.502; refractive dispersion, $n_F - n_C \times 10^4$, 93; boiling point at 1 mm, 193°C; aniline point, 87°C.

From these data, the following properties are calculated: molecular formula, $C_{25}H_{44}$; specific refraction, 0.3163; specific dispersion, 100.

For the hydrogenated sample, the value of the specific dispersion,

100, shows that the material is aromatic free (Fig. 99.25). The molecular formula gives a value of $x = -6$ in the formula C_nH_{2n+x} and indicates a deficiency of 8 hydrogen atoms. In the absence of olefins, this is good evidence for a four-ring cycloparaffin. The values of the specific refraction and aniline point are both in fairly good agreement with those predicted (Figs. 99.22 and 99.17) for a cycloparaffin of four rings.

The increase in the number of hydrogen atoms per molecule upon hydrogenation is ten, and indicates that five aromatic double bonds have been saturated. It can be concluded, therefore, that in the original molecule there are two aromatic rings condensed together as in the naphthalene nucleus. This conclusion is further substantiated by the fact that the specific dispersion, 201, is in accord with values of 185–200 for molecules containing a naphthalene nucleus, but not with values of 150–160 for molecules containing two separate benzene rings (Fig. 99.22).

The selected fraction therefore consists of molecules composed of two aromatic rings, condensed as in the naphthalene nucleus, plus, on the average, two cycloparaffin rings, plus paraffin side chains as appropriate. The aromatic rings and the cycloparaffin rings are present in the same molecule and not in different molecules. Certain mixtures composed of aromatic molecules and cycloparaffin molecules having properties similar to those of the selected fraction, and exhibiting the same change in empirical formula on hydrogenation, are possible. However, the processes of fractionation always employed before characterization of this kind is attempted would readily have separated such mixtures.

A somewhat simpler example can be found in the characterization of an aromatic-free fraction separated from a dewaxed, solvent-refined, light lubricating distillate from Kuwait crude. In this case, the oil was separated into a paraffin–cycloparaffin fraction and an aromatic fraction by adsorption. The paraffin–cycloparaffin portion was then separated into cycloparaffin-lean and cycloparaffin-rich fractions by continuous thermal diffusion. A paraffin–isoparaffin fraction was isolated from the cycloparaffin-lean portion by extractive crystallization at $-40°C$. This fraction was then further separated by batch thermal diffusion.

The uppermost fraction from the latter separation was examined and found to have the following properties: elemental analysis, carbon 85.11 and hydrogen 14.89 per cent by weight; molecular weight, 345 g/mole; density, 0.8036 g/ml; refractive index, 1.4503; refractive dispersion, 80; aniline point, 116°C; solidification point, $-4°C$.

From the elemental analysis and molecular weight, the molecular formula is calculated to be $C_{24.5}H_{51.0}$, which gives the value of $x = 2$ in the formula C_nH_{2n+x}. From the refractive index and density, and the refractive dispersion and density, the specific refraction and specific dispersion are calculated and found to be 0.3346 and 99, respectively.

The specific dispersion, 99, confirms the absence of aromatic rings (Fig. 99.25). The specific refraction, 0.3346, is very close to the predicted value of 0.3347 for a normal paraffin or an isoparaffin of 345 molecular weight (Fig. 99.22). The aniline point also agrees with the predicted value of 115°C (Fig. 99.17). The molecular formula, specific refraction, and aniline point all indicate that the sample is composed entirely of paraffins and branched paraffins. The solidification point of —4°C suggests the absence of appreciable quantities of straight-chain paraffins, since the melting point of a C_{25} n-paraffin is 53.7°C. It thus can be concluded that this fraction is predominantly a mixture of high-molecular-weight branched paraffins.

Examination of the sample by infrared and nuclear magnetic resonance spectroscopy substantiated this conclusion. The spectroscopic data suggested that the average molecule contains approximately 2 tertiary carbon atoms, 4 methyl groups, and 19 methylene groups. The infrared spectrum also indicated the presence of isolated pendant methyl groups. Consequently, the average molecule can be visualized as containing approximately 23 carbon atoms in a linear chain with 2 pendant methyl branches.

2. Structural Group Analysis of Oil Products

From their inception, methods for structural group analysis have been extensively applied for the most obvious purpose—that is, for the analysis of oils and oil fractions from various crudes. In this way, a mass of comparative data has accumulated on the character of crude oils, variations in the composition of fractions from a single crude, and so forth. These data are often helpful in identifying products of unknown origin or in giving valuable indications as to their method of manufacture. The composition of petroleum oils, as deduced from structural group analysis, can often be related to a desirable property, suggesting end uses for new products or new applications for established ones. For example, Kurtz and his associates have been able to relate many of the properties of oil-extended rubber to the composition of the extender oil (108,215).

The variables in composition that may be encountered in domestic crude types are adequately demonstrated by the differences in carbon distribution shown in Table 99.XX for the light lubricant portions of Michigan, East Texas, Webster, and Mirando crudes. It should be observed that the differences in carbon distribution are not nearly as large

TABLE 99.XX
Carbon Type Analyses of the Light Lubricant
Portion of Four Crudes

Carbon distribution	Crude			
	Michigan	East Texas	Webster	Mirando
$\%C_A$	15	16.5	13.5	22.5
$\%C_N$	20.5	20.5	36.5	42.5
$\%C_P$	64.5	63	50	35

as might be expected. The two crudes commonly described as paraffinic, Michigan and East Texas, exhibit almost identical compositions. They differ significantly, however, from the Gulf Coast crudes, Webster and Mirando. The latter are often described as naphthenic, and do indeed contain a greater portion of the total carbon atoms in naphthenic structures. Carbon distribution data are often useful in identifying the type of crude used to produce a given lubricant, or conversly in selecting the crude best suited for the manufacture of oils for particular end uses.

While carbon distribution is convenient for comparing crudes of different types, the general observation can be made that the carbon distribution of lubricant fractions distilled from the same crude shows little change with boiling point. As a consequence, it is often necessary to apply other structural group methods that provide more information than can be obtained from carbon distribution alone.

For example, in studying the composition of the aromatics recovered from the manufacture of lubricating oils by the so-called split-extraction process, carbon distribution proved to be rather ineffective. Table 99.XXI shows the carbon distribution of a number of fractions prepared by vacuum distillation of the aromatic concentrates recovered from the furfural extraction of the lighter lube distillates, and from the Duo-Sol extraction of the heavier distillates. It is quite obvious that these data reveal little that is indicative of a change in the composition of the average molecule with increasing molecular weight. If, however, the fractions are separated into aromatic and paraffin–cycloparaffin portions by adsorption, and each of the resulting fractions is characterized by ring analysis, it is possible to reach some definitive conclusions concerning the types of structures that predominate. Table 99.XXII gives the values of properties for fractions separated in this fashion. Figure 99.45 shows

TABLE 99.XXI
Carbon Distribution of Fractions Prepared by Vacuum Distillation of Extracts from "Split-Extraction" of a Lube Stock

	Furfural Extract of Light Lubes				
Fraction	1	2	3	4	5
Boiling range, °F	Int.–704	704–735	735–760	760–788	788+
Vol % of extract	22.8	20.4	19.4	19.8	17.6
Molecular weight	284	321	332	347	382
Carbon distribution					
%C_A	27	30	29	33	39
%C_N	34	34	32	31	30
%C_P	39	36	39	36	31

	Duo-Sol Extract of Heavy Lubes					
Fraction	1	2	3	4	5	6
Boiling range, °F	Int.–750	750–830	830–890	890–960	960–980	980+
Vol % of extract	9.8	10.3	10.0	10.6	10.0	49.3
Molecular weight	337	375	399	442	465	—
Carbon distribution						
%C_A	29	29	30	31	—	—
%C_N	37	35	34	30	—	—
%C_P	34	36	36	39	—	—

the results of plotting the functions of density and molecular weight and dispersion and molecular weight for the aromatic portions on the graph described by Martin and Sankin [Section VI.A.3.b.(1)]. Points corresponding to the paraffin–cycloparaffin portions of the distillate fractions are shown on the graph of specific refraction versus reciprocal molecular weight (Section IV.B.3.a) in Fig. 99.46.

Such a treatment of the data reveals that the aromatics recovered by furfural extraction of the light lubricant portion are predominantly naphthalenes containing one and two cycloparaffin rings. Small quantities of benzene- and phenanthrene-type aromatics are also present. The aromatics recovered by Duo-Sol extraction of the heavy residue contain much larger quantities of phenanthrene types and, in addition, have a larger proportion of associated cycloparaffin rings. They consist predominantly of naphthalenes and phenanthrenes containing two associated cycloparaffin rings. The saturated hydrocarbons removed by furfural from the lower-boiling portion average a little over two cycloparaffin rings per molecule, whereas those removed by Duo-Sol extraction are much more polycyclic in character and average almost four rings per

TABLE 99.XXII

Physical Properties of Aromatic and Saturated Portions of Extract Fractions
from the "Split-Extraction" of a Lube Stock

Furfural Extract of Light Lubes

Fraction	1	2	3	4	5
Boiling range, °F	Int.–704	704–735	735–760	760–788	788+

Aromatic portion					
Wt % of whole fraction	57.7	63.6	69.4	75.6	84.4
d_4^{20}	0.9888	0.9916	0.9973	1.0051	1.0224
Specific dispersion	201	199	200	208	222
Molecular weight	266	292	317	325	332
$F(d, M)$	37.5	41.8	47.2	50.9	57.9
$F(\delta, M)$	28.6	30.7	33.6	37.1	42.7

Saturated portion					
Wt % of whole fraction	42.3	36.4	30.6	24.4	15.6
n_D^{20}	1.4697	1.4733	1.4758	1.4788	1.4899
d_4^{20}	0.8582	0.8630	0.8684	0.8729	0.8910
Molecular weight	303	359	380	410	428
Lorentz-Lorenz sp. refr.	0.3249	0.3252	0.3247	0.3247	0.3244
$1000/M$	3.30	2.78	2.63	2.44	2.34

Duo-Sol Extract of Heavy Lubes

Fraction	1	2	3	4	5	6
Boiling range, °F	Int.–750	750–830	830–890	890–960	960–980	980+

Aromatic portion						
Wt % of whole fraction	70.5	73.2	75.6	79.0	81.3	—
d_4^{20}	1.0041	1.0063	1.0072	1.0069	1.0180	—
Specific dispersion	199	201	200	205	208	—
Molecular weight	320	359	382	423	445	—
$F(d, M)$	49.8	56.5	60.4	66.5	75.0	—
$F(\delta, M)$	33.5	38.2	40.2	46.6	50.3	—

Saturated portion						
Wt % of whole fraction	29.5	26.8	24.4	21.0	18.7	—
n_D^{20}	1.4869	1.4942	1.4950	1.4948	—	—
d_4^{20}	0.8943	0.9089	0.9100	0.9087	—	—
Molecular weight	387	445	474	535	—	—
Lorentz-Lorenz sp. refr.	0.3215	0.3204	0.3205	0.3208	—	—
$1000/M$	2.58	2.25	2.11	1.87	—	—

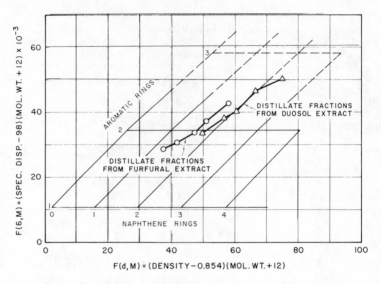

Fig. 99.45. Composition of the aromatic portion of fractions from the "split extraction" of a lube stock.

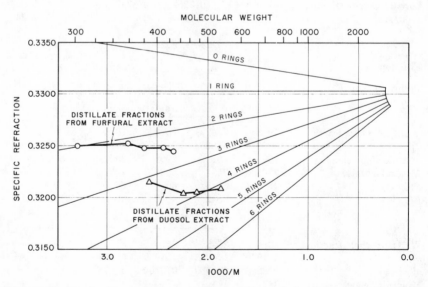

Fig. 99.46. Composition of the saturated portion of fractions from the "split extraction" of a lube stock.

molecule. The relative insensitivity of carbon distribution to these differences is due to the fact that the increase in the proportion of aromatic
and cycloparaffin rings as the molecular weight increases is just enough
to hold the carbon type composition essentially constant.

In examining the commercially significant behavior of complex hydrocarbon systems like the viscous fractions of petroleum, it is often possible
to discover relationships between the bulk properties of the oil and its
composition as defined by structural group analysis. For example, in
investigations into the viscosity–temperature behavior of oils, it can be
established that a fairly well-defined relationship exists between the
ring content, R_T, and the viscosity index of aromatic-free oils. This
relationship, shown in Fig. 99.47, was derived from the examination
of a large number of fractions prepared by the thermal diffusion of
aromatic-free oils and oil fractions from a number of crudes. The relation
is essentially independent of molecular weight, viscosity level, or type
of crude oil.

A study of Fig. 99.47 leads to some interesting conclusions. First
of all, it appears that, for hydrocarbon systems that exhibit New-
tonian behavior, there is a limit to the viscosity index that can be attained. This limit appears to be about 170 or 180 viscosity index and
corresponds to that of a mixture of high-molecular-weight n- or isopar-

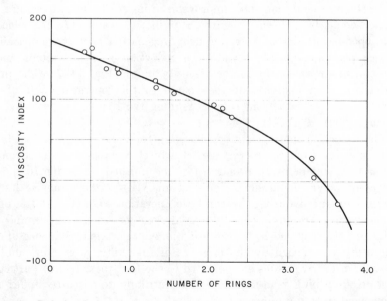

Fig. 99.47.   Relation between viscosity index and ring number for aromatic-free oils.

affins. Second, the fact that the relationship between ring number and viscosity index is a curve is of importance, since it leads to the conclusion that viscosity index is not an additive quantity. It follows that a blend of two oils of widely differing ring contents will exhibit a viscosity index higher than the arithmetic mean. In general, it may be concluded that improved viscosity indices can be obtained by blending oils of divergent character. Conversely, if an oil is separated into portions of divergent character, as, for example, by solvent extraction, the improvement is less than might be calculated from the very low viscosity index of the extract and the viscosity index of the original oil. These latter effects are due, at least in part, to the peculiarities of the Dean and Davis viscosity index system.

3. Following Chemical Reactions and Physical Separations

Physical property methods can also serve a useful purpose in following the course of chemical reactions such as cracking, hydrogenation, and dehydrogenation or physical separation processes such as distillation, solvent extraction, and thermal diffusion.

For example, it was demonstrated in laboratory experiments that the viscosity index of an essentially aromatic-free oil could be improved by mild hydrocracking over a platinum catalyst, followed by distillation and solvent extraction. The improvement in viscosity–temperature behavior could have been brought about in two ways: (1) by selective ring opening, or (2) by selective dehydrogenation of polycycloparaffins. Either possibility would result in a reduction in ring content and a consequent improvement in the viscosity index (Fig. 99.47). The properties and ring analyses of the charge to this operation and of the raffinate and extract from solvent refining the hydrolube produced are given in Table 99.XXIII.

The reduction in ring number, in going from the charge to the distillate-raffinates from hydrocracking, clearly shows that the improvement in viscosity–temperature properties of the finished oil is due to the removal of polycycloparaffins. The ring analyses of the aromatic hydrocarbons produced during the catalytic operation suggest that the polycycloparaffins are preferentially dehydrogenated to aromatics and are subsequently removed in the solvent refining operation. Evidence of this selectivity is afforded by the presence of substantially greater numbers of rings in the aromatics as compared to the feedstock. In this particular case, ring-opening reactions seem to contribute to viscosity index improvement to only a minor extent.

TABLE 99.XXIII

Hydrocracking of a 200 Viscosity Aromatic-Free Lubricating Distillate

	Charge	Hydrocracked at:		
		375°C	400°C	430°C
Yield, raw lube distillate, wt % of charge	—	65.6	62.0	46.1
Extract, wt % of raw lube distillate	—	5.4	19.8	21.0
Raffinate	—	94.6	80.2	79.0
		Raffinates		
Viscosity, SUS,100°F	184	213	152	126
Viscosity index	101	104	113	116
R_N	1.91	1.92	1.63	1.50
		Extracts		
R_A	—	1.39	1.25	1.54
R_N	—	1.21	1.10	0.80
R_T	—	2.60	2.35	2.34

The course of refinery and laboratory operations can often be followed satisfactorily without having to resort to ring analysis. When carbon distribution will suffice, the triangular charts developed by Kurtz and his co-workers (Section VI.A.3.a) are particularly convenient, as they allow a graphical representation of the course of the reaction or separation. The results of plotting the viscosity–gravity constant and refractivity intercept of fractions from a variety of chemical and physical processes are shown in Fig. 99.48. The solid points on the left indicate the changes in composition resulting from the analytical hydrogenation of a Gulf Coast distillate (125). The gradual decrease in aromatic carbon content and increase in naphthenic carbon content as aromatic rings are converted to cycloparaffin rings are clearly demonstrated. There is also evidence that the final hydrogenation treatments, which were continued in an effort to eliminate the last traces of ultraviolet absorption, may have resulted in some ring opening, because there is a decrease of between 2 and 4% in the naphthene ring carbons and a corresponding increase in the paraffin chain carbons during these final hydrogenations.

The progress of multistage extractions can be conveniently followed in a similar manner. The squares on the extreme right show the changes in composition produced by the repeated extraction of a Venezuelan lubricating distillate with furfural (171). As aromatic rings are preferentially removed, the percentage of carbon in aromatic structures decreases and that in naphthenic structures increases from 24% in the feedstock to 36% in the fifth raffinate. However, the ratio of carbon in naphthenic structure to that in paraffinic structure remains relatively constant, and the principal difference between the raffinate and the extract is therefore in the ratio between the quantities of aromatic and of naphthenic rings, and not in the amount of carbon in paraffinic structure. The tendency of successive raffinates to move along a line representing a constant ratio of %C_N to %C_P is a phenomenon generally observed during the extraction of viscous petroleum oils.

The cluster of triangular points shown in Fig. 99.48 demonstrates the type of data generally obtained when the VGC and the intercept of a series of fractions distilled from the lubricant portion of a single crude are plotted. Only very small changes in carbon distribution are in evidence. Other structural group methods that allow determination of "ring analysis" are often more useful for following changes in composition effected by distillation.

Data obtained by subjecting one of the distillate fractions discussed

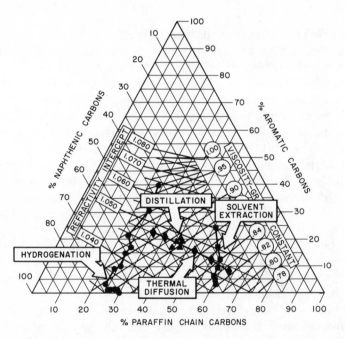

Fig. 99.48. Change in carbon distribution effected by chemical and physical processes.

above to thermal diffusion are indicated in the figure by the solid diamonds. The separation of the feedstock into fractions that tend to be richer and leaner in cyclic structures than the original is clearly demonstrated. The thermal diffusion of oils, when followed on a triangular diagram of this sort, generally produces lines that point in the direction of the apex representing 100% paraffin chain carbons.

4. Molecular Type Analysis by the Structural Group Analysis of a Series of Fractions

None of the applications of structural group analysis that have been illustrated supplies any information on the way in which the various structural groups that comprise the individual molecules are joined. Although in these applications good use has been made of information developed through the concept of the structure of an "average molecule," in no case has it been possible to estimate the proportions of the various types of individual molecules that contribute to the average structure. In many instances, structural group analysis can be of service in this endeavor, if it is combined with fractionation by special separation processes.

The use of structural group analysis to determine the proportions of various molecular types is based on the fact that it is possible to segregate oil fractions of narrow molecular weight range into a paraffin–cycloparaffin (saturated) portion and an aromatic portion. The paraffin–cycloparaffin portion may then be further separated according to number of cycloparaffin rings, and the aromatic portion according to number of aromatic rings. To permit the determination of the proportion of the types of molecules present in the original oil sample, it is generally assumed that the final fractions are made up of only two types of molecules. For example, a fraction obtained from the paraffin–cycloparaffin portion that exhibits a value of $R_N = 2.2$ is assumed to be composed of 80% dicycloparaffins and 20% tricycloparaffins, and to contain no tetracycloparaffins, monocycloparaffins, or paraffins at all. Similarly, an aromatic fraction containing an average of 1.2 aromatic rings per molecule is considered to be composed of 80% benzene-type and 20% naphthalene-type aromatics. If the separation of aromatics is carefully conducted, and a large number of fractions collected and examined by ultraviolet spectrophotometry before making up the final composites, it is almost always possible to satisfy the condition that a given portion of the separated material should contain no more than two nuclear types.

On the other hand, in the fractionation of a saturated oil there is considerably less assurance that this condition will always be met, since there is no convenient way of following the separation. Data accumulated in the author's laboratory suggest that, if the boiling range of the sample is no wider than 200°F, either thermal diffusion or adsorptive percolation will serve to provide fractions from which the proportions of paraffins, monocycloparaffins, etc., in the original sample may be estimated with an accuracy of 15–20% of the amount of each molecular type present. The following examples will serve to illustrate the application of these techniques.

a. Aromatic Hydrocarbons

The complete procedure for estimating the proportions of mono-, di-, trinuclear, etc., aromatic hydrocarbons in an oil sample may be summarized as follows:

1. The oil sample (which should not be wider boiling than 200°F) is first separated into aromatic and saturated portions by means of elution chromatography (Section V.E.1).

2. The aromatic portion is then further fractionated by multiple displacement development over activated alumina (Section V.E.1).

3. The large number of fractions collected are examined by ultraviolet spectrophotometry and combined in such a way that each final fraction will contain no more than two nuclear types.

4. The average number of aromatic rings in each final fraction is determined from the specific dispersion and molecular weight [Section VI.A.3.b.(1)].

5. The ring numbers are evaluated for each fraction in terms of the contribution of each molecular type. The analysis of the whole aromatic portion is calculated by summing the products of the mole percentage of each fraction and its content of each type of molecule. It is necessary to use mole per cent since the structural group analysis is in terms of rings per molecule.

As an example, the data contained in Fig. 99.32 can be used to estimate the composition of the original unfractionated aromatics. The method of calculation is outlined in Table 99.XXIV. These data also illustrate a complication that often occurs when aromatic fractions prepared by adsorptive percolation are subjected to this type of calculation. This complication arises from the fact that a substantial, if not the greater, part of the last fractions to be desorbed are nonhydrocarbon molecules containing sulfur, nitrogen, and oxygen. In this case, the last 17% of the material consists largely of nonhydrocarbons. Since the strong ad-

TABLE 99.XXIV

Molecular Type Analysis of Aromatics from a Gulf Coast Distillate
by Ring Analysis of a Series of Fractions

Fraction	Weight per cent of charge	Mole per cent of charge	No. of aromatic rings/ molecule	Aromatic type,[a] mole %			
				Mono-nuclear	Di-	Tri-	Tetra-
1–2	14.3	11.9	1.05	11.3	0.6	—	—
3–4	12.6	11.2	1.42	6.5	4.7	—	—
5–7	10.9	10.1	1.74	2.6	7.5	—	—
8–11	10.9	10.6	1.95	0.5	10.1	—	—
12–15	11.2	11.3	2.25	—	8.5	2.8	—
16–18	9.0	9.8	2.59	—	4.0	5.8	—
19–22	10.6	12.0	2.87	—	1.6	10.4	—
23–26	5.6	6.2	3.23	—	—	4.8	1.4
27–33	14.9	16.9	(1.96)[b]	(4.3)	(7.5)	(4.8)	(0.3)
				25	44	29	2

[a] $\Sigma(0.25 \times 1) + (0.44 \times 2) + (0.29 \times 3) + (0.02 \times 4) = 2.07$ rings/molecule. Experimental value on whole sample = 2.02 rings/molecule.

[b] Contains nonhydrocarbons.

sorbability of these compounds is due primarily to the presence of the heteroatom, it was assumed that the composition of the hydrocarbon portion of the molecules in this aggregate was the same as that for the rest of the sample. The internal consistency of the entire procedure may be checked by comparing the average number of aromatic rings calculated by summing the individual contributions of the types of molecules with the average number of rings calculated for the original material. The agreement should be within 0.1 to 0.2 ring. For the example shown in Table 99.XXIV, the agreement is satisfactory.

b. Saturated Hydrocarbons

The procedure for estimating the proportions of paraffins, monocycloparaffins, dicycloparaffins, etc., in the saturated portion of petroleum oils is similar in many respects to that described for the aromatic portion and may be summarized as follows:

1. A representative sample of saturated hydrocarbons is isolated from the oil fraction by elution chromatography (Section V.E.1).

2. The saturated portion is further fractionated according to ring number by careful adsorptive percolation over silica gel (Section V.E.1) or by thermal diffusion (Section V.F)

3. The average number of cycloparaffin rings per molecule is calculated for each fraction by a suitable procedure. The specific refraction–molecular weight relationships (Section IV.B.3.a) are generally the most convenient to apply.

4. The ring numbers are evaluated for each fraction in terms of the contribution of each molecular type. The analysis of the whole saturated portion is calculated by summing the products of the weight percentage of each fraction and its content of each type of molecule. Use of weight percentage is permissible in this case since fractionation of saturated oils by either pentane elution or thermal diffusion does not result in appreciable molecular weight differences between the final fractions.

The data presented in Fig. 99.34 can be used to illustrate application of the procedure to a saturated oil of about 250 moleculer weight. Details of the calculation are shown in Table 99.XXV. As in the case of the aromatic portion, a good check of internal consistency can be made by comparing the ring number calculated by summing the contributions of the individual types of molecules with that calculated from the properties of the whole saturate portion.

These examples demonstrate the type of information that can be obtained by the combination of ring analysis methods with certain fractionation processes. Analyses of this kind are somewhat time consuming and consequently are not generally applied routinely. However. the ad-

TABLE 99.XXV

Molecular Type Analysis of Saturates from a Light Naphthenic
Distillate by Ring Analysis of a Series of Fractions

Fraction	Weight per cent of charge	No. of cycloparaffin rings/ molecule	Saturated hydrocarbon type,[a] wt %					
			Paraffins	Monocyclo-paraffins	Di-	Tri-	Tetra-	Penta-
1	8.4	0.89	0.9	7.5	—	—	—	—
2	9.4	1.04	—	9.0	0.4	—	—	—
3	9.5	1.32	—	6.5	3.0	—	—	—
4	9.3	1.62	—	3.5	5.8	—	—	—
5	10.6	2.06	—	—	10.0	0.6	—	—
6	9.7	2.29	—	—	6.9	2.8	—	—
7	10.4	2.56	—	—	4.6	5.8	—	—
8	10.6	2.86	—	—	1.5	9.1	—	—
9	10.1	3.37	—	—	—	6.4	3.7	—
10	12.0	4.03	—	—	—	—	11.6	0.4
			1	26	32	25	15	0.5

[a] $\sum (0.01 \times 0) + (0.26 \times 1) + (0.32 \times 2) + (0.25 \times 3) + (0.15 \times 4) + (0.005 \times 5) = 2.27$ rings/molecule. Experimental value on whole sample = 2.21 rings/molecule.

TABLE 99.XXVI
Molecular Type Analysis of the Light Lubricant Portion
(650–850°F) of Four Crudes

Hydrocarbon type, vol %	Crude			
	Michigan	East Texas	Webster	Mirando
Aromatics	29	25.5	29.5	49.5
Paraffins	20	19.5	0.5	0
Monocyclic naphthenes	43.5	40	22.5	4
Dicyclic naphthenes	5.5	11	33.5	11.5
Tricyclic naphthenes	2	3.5	10	22
Tetracyclic naphthenes	—	0.5	4	11
Pentacyclic naphthenes	—	—	—	2

ditional information obtainable is often of substantial value in research investigations. For instance, although the carbon distribution data for four light lubricant fractions shown in Table 99.XX convey the impression that there are differences in composition among these oils, they do not provide any clear conception as to what the magnitude of these differences might be when the composition is expressed in terms of the concentrations of individual types of molecules. Results of the application of fractionation and ring analyses to these samples are shown in Table 99.XXVI. A comparison of the percentages of paraffin, cycloparaffin, and aromatic hydrocarbons in the light lubricant portions of these four crudes is now possible. The East Texas sample contains about 25% aromatic hydrocarbons, the Michigan and Webster oils about 30%, and the Mirando about 50%. Both the East Texas and Michigan fractions contain about 20% paraffin and 40% monocycloparaffin hydrocarbons and only small amounts of polycycloparaffins. On the other hand, the Webster and Mirando samples contain substantially no paraffins and exhibit a wide distribution of cycloparaffins, ranging from monocyclic to pentacyclic. The dicycloparaffins predominate in Webster, whereas tricycloparaffins are the predominant saturated hydrocarbon type in Mirando.

c. APPLICATION OF SUCCESSIVE FRACTIONATION PROCESSES

In addition to the foregoing, it is possible to apply successively two fractionation processes which effect separation of the oil in different ways, so that even more information may ultimately be obtained. For example, the oil may be first separated into an aromatic and a saturated

TABLE 99.XXVII

Molecular Type Analysis of a Solvent-Refined Light Lubricating
Distillate from Kuwait Crude

Type	Non-aromatic, wt %	Mononuclear aromatic, wt %	Dinuclear aromatic, wt %	Total wt %
Noncyclic	29	—	—	29
Monocyclic	44	5	—	49
Dicyclic	5	10	1	16
Tricyclic	—	3	2	5
Tetracyclic	—	Trace	1	1
Total	78	18	4	100

portion. The saturated portion, as we have seen, may then be fractionated by adsorption or thermal diffusion. The aromatic portion can be fractionated by multiple displacement over activated alumina to give fractions that are more or less uniform with regard to the number of aromatic rings. These may then be further separated by thermal diffusion according to the total number of rings (aromatic and naphthenic) since thermal diffusion is incapable of distinguishing between saturated and unsaturated ring systems. Ring analyses of all of the final fractions then permits the ultimate type analysis in terms of paraffins; cycloparaffins with 1,2,3, etc., rings; mononuclear aromatics with 0,1,2, etc., cycloparaffin rings; and so forth.

Results of the application of this series of operations to a dewaxed light lubricating distillate prepared by solvent extraction from Kuwait crude are shown in Table 99.XXVII. It is evident from the data that the oil is not highly cyclic in character. The saturated portion, which constitutes about 80% of the sample, consists largely of isoparaffins and monocycloparaffins. It is also clear that, although the majority of the aromatic molecules contain associated cycloparaffin rings, a small proportion of mono- and dinuclear aromatics appear to simply be substituted with alkyl chains. High-molecular-weight tetralins or indans are the predominant aromatic type.

B. NONHYDROCARBONS

The methods that have served so well for hydrocarbon analysis are often of use in investigating mixtures of nonhydrocarbons as well. Structural group analyses have frequently been employed in connection with investigations into the composition of fatty and cyclic acids and esters, drying oils, and so forth.

1. Fatty Acid Derivatives

An interesting example quoted by Waterman is the characterization of an addition compound formed by the reaction of methyl linoleate with styrene vapor at 280°C (230). The properties of this material are shown in the first column of Table 99.XXVIII. These data indicate the presence of olefinic unsaturation and ester groups. The compound in question was converted into the corresponding saturated hydrocarbon by using the Grignard reaction with methylmagnesium iodide, followed by dehydration and hydrogenation (Section VI.B.2.b). The properties and ring analysis of this hydrocarbon are compared in the second and third columns to those calculated for the $C_{28}H_{54}$ bicycloparaffin one would expect to obtain if the original compound resulted from a Diels-Alder condensation of styrene and the primarily conjugated linoleic ester. The agreement is excellent and provides convincing proof of the bicyclic nature of the addition compound.

In another phase of this investigation into the preparation of homogenous copolymerization products of drying oils, a rapid method for the determination of the percentage of styrene in the copolymerization products was required. A graphical method was developed (231)

TABLE 99.XXVIII

Analysis of Copolymerization Product of Styrene and Methyl Linoleate (230)

	Properties of addition compound	Properties of saturated hydrocarbon obtained from addition compound	
		Obs'd	Calc'd for $C_{28}H_{54}$
n_D^{20}	1.5067	1.4768	—
d_4^{20}	0.9631	0.8668	—
Lorentz-Lorenz sp. refr.	0.3088	0.3257	0.3253
Molecular weight	403	395	390.7
Saponification value	140.0	—	—
Iodine value	64.8	—	—
% Carbon	81.27	86.11	86.07
% Hydrogen	10.49	13.94	13.93
Average no. of rings ⎱ from %H and mol. wt.		1.99	2.00
per molecule ⎰ from n, d, and mol. wt.		2.07	2.00

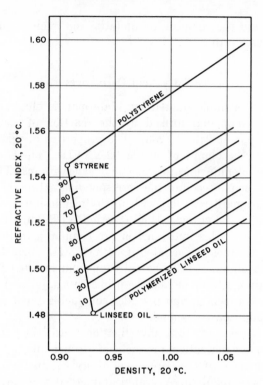

Fig. 99.49. Refractive index–density diagram for the analysis of styrenated oils. Reproduced with the permission of Elsevier Publishing Company.

based on the observation that the polymerization reactions of pure styrene and pure linseed oil could be represented by two parallel lines on a diagram of refractive index versus density. Analysis of a large number of styrenated linseed oils by another procedure showed that parallel lines may be located between the 100% styrene and 100% linseed oil extremities so as to permit rapid estimation of the percentage of styrene in the products from measurements of refractive index and density. The completed diagram is shown in Fig. 99.49.

2. Cyclic (Naphthenic) Acids

Physical property methods have also proved useful in investigations into the constitution of the cyclic (naphthenic) acids that are present in many crude petroleums. Evidence for the occurrence of polycyclic carboxylic acids in the higher-molecular-weight acid fractions has been

TABLE 99.XXIX
Properties of Naphthenic Acids and Hydrocarbons
Produced Therefrom by Deoxygenation (65)

Property	Naphthenic acids	Crude naphthene hydrocarbons	Naphthenes, hydrogenated 50 hr
Acid number	177	0	0
n_D^{20}	1.4965	1.4849	1.4781
d_4^{20}	0.9860	0.8920	0.8815
Specific dispersion	—	107	98
Molecular weight	317	288	286
Lorentz-Lorenz sp. refr.	0.2966	0.3212	0.3212
Aniline point, °C	—	77.8	84.2
% Carbon	78.20	86.51	86.46
% Hydrogen	11.17	13.08	13.41
% Oxygen	10.52	0.31	0.12
% Sulfur	0.11	0.10	0.01
Rings/molecule, R_T	—	—	2.6

obtained by converting these materials to the corresponding hydrocarbons and then subjecting them to ring analysis (65). For example, a sample of naphthenic acids of molecular weight 317 from a Gulf Coast petroleum was deoxygenated by way of the ethyl esters, alcohols, and iodides. The properties of the resulting hydrocarbons are shown in the second column of Table 99.XXIX. The aniline point and specific dispersion of this material indicate that a small amount of aromatic hydrocarbons was present. In order to eliminate aromatic unsaturation, the crude hydrocarbons were hydrogenated with Raney nickel for 50 hr at 230–250°C and 2700–3000 psi. The properties of the hydrogenated material are shown in the third column. The aniline point and specific dispersion indicate that the material was completely saturated. The change in aniline point from 77.8 to 84.2°C suggests that about 5% of aromatic hydrocarbons were present in the unhydrogenated material. The number of rings per molecule was calculated from the elemental analyses and the molecular weight. The value of 2.6 rings per molecule demonstrates rather conclusively that polycyclic carboxylic acids containing as many as three rings are present in substantial quantities.

Deoxygenation of naphthenic acids can also be accomplished by direct catalytic hydrogenation with Raney nickel at 280–290°C and 5000–6000 psi (78). This is less time consuming than the four-step chemical reduction procedure mentioned above, but has the disadvantage that the presence of aromatic rings will go undetected since the product of this reaction is the fully saturated hydrocarbon having, in all likelihood, one less carbon atom than the parent acid.

Additional information on the constitution of these acids can of course be obtained by fractionating the hydrocarbons produced from them according to ring number by adsorption or thermal diffusion. When this procedure was applied to the final hydrocarbons in Table 99.XXIX, fractions containing from 1.8 to 3.3 rings per molecule were obtained. The original naphthenic acids therefore contained molecules having from one to four rings, with two- and three-ring molecules predominating.

These examples are indicative of but a few of the many ways in which physical properties can be used for characterization and analysis. Although new separation techniques and instrumental methods will undoubtedly further our knowledge of the composition of the complex mixtures that constitute certain natural products, physical property methods will very probably continue to play an important role in the routine examination of such materials.

REFERENCES

1. Adams, N. G., and D. M. Richardson, *Anal. Chem.*, **23**, 129 (1951).
2. Alm, R. S., R. J. P. Williams, and A. Tiselius, *Acta. Chem. Scand.*, **6**, 826 (1952).
3. *ASTM Standards, 1970*, American Society for Testing and Materials, Philadelphia, Pa., 1970, Parts 17 and 18.
4. *ASTM Standards 1970*, American Society for Testing and Materials, Philadelphia, Pa., 1970, Part 29.
5. "ASTM Viscosity Index Calculated from Kinematic Viscosity," *ASTM Data Series*, DS 39a, American Society for Testing and Materials, Philadelphia, Pa., September 1965.
6. Auwers, K., von, *Ann.*, **415**, 98 (1918).
7. Baldes, E. J., and F. Johnson, *Biodynamica*, **46**, 1 (1939).
8. Bendoraitis, J. G., B. L. Brown, and L. S. Hepner, *Anal. Chem.*, **34**, 49 (1962).
9. Berthelot, M., *Ann. Chim. Phys.* (3) **48**, 22 (1856).
10. Blott, J. F. T., and C. G. Verver, *J. Inst. Petrol.*, **38**, 193 (1952).
11. Boelhouwer, C., and H. I. Waterman, *J. Inst. Petrol.*, **40**, 116 (1954).
12. Boelhouwer, C., H. I. Waterman, and P. H. W. Wijnands, *J. Polymer Sci.*, **17**, 411 (1955).
13. Bogin, C. D., *Ind. Eng. Chem.*, **16**, 380 (1924).
14. Bond, G. R., *Ind. Eng. Chem., Anal. Ed.*, **18**, 692 (1946).
15. Boyd, M. L., and D. S. Montgomery, *J. Inst. Petrol.*, **45**, 345 (1959).
16. Brady, A. P., H. Hoff, and J. W. McBain, *J. Phys. Colloid Chem.*, **55**, 304 (1951).
17. Brandenberger, S. G., L. W. Maas, and I. Dvoretzky, *Anal. Chem.*, **33**, 453 (1961).
18. Brooks, F. R., F. M. Nelson, and V. Zahn, *Petrol. Refiner*, **27**, 620 (1948).
19. Burdett, R. A., and B. E. Gordon, *Anal. Chem.*, **19**, 843 (1947).

20. Caley, E., and A. Habboush, *Anal. Chem.*, **33**, 1613 (1961).
21. Calingaert, G., and D. S. Davis, *Ind. Eng. Chem.*, **17**, 1287 (1925).
22. Campanile, V. A., and V. Lantz, *Anal. Chem.*, **26**, 1394 (1954).
23. Cannon, M. R., and M. R. Fenske, *Ind. Eng. Chem., Anal. Ed.*, **10**, 297 (1938).
24. Carlson, C. S., "Extractive and Azeotropic Distillation," in A. Weissberger, ed., *Technique of Organic Chemistry*, Vol. IV, Interscience, New York, 1951, Chap. III.
25. Carruthers, W., and A. G. Douglas, *J. Chem. Soc. II, 1955*, 1847.
26. Carruthers, W., and A. G. Douglas, *J. Chem. Soc., 1957*, 278.
27. Charlet, E. M., K. P. Lanneau, and F. B. Johnson, *Anal. Chem.*, **26**, 861 (1954).
28. Chavanne, G., and L. J. Simon, *Compt. Rend.*, **168**, 1111, 1324 (1919).
29. Chavanne, G., and L. J. Simon, *Compt. Rend.*, **169**, 185, 285 (1919).
30. Chiao, T.-T., and A. R. Thompson, *Anal. Chem.*, **29**, 1678 (1957).
31. Clusius, K., and G. Dickel, *Naturwissenschaften*, **26**, 546 (1938).
32. Cornelissen, J., and H. I. Waterman, *Anal. Chim. Acta*, **15**, 401 (1956).
33. Cornelissen, J., and H. I. Waterman, *J. Inst. Petrol.*, **43**, 48 (1957).
34. Cosby, J. N., and L. H. Sutherland, *Proc. Am. Petrol. Inst.*, Sect. III: *Refining*, **22**, 13 (1941).
35. Cova, D. R., *J. Chem. Eng. Data*, **5**, 282 (1960).
36. Cox, E. R., *Ind. Eng. Chem.*, **27**, 1423 (1935).
37. Craig, L. C., and D. Craig, "Extraction and Distribution," in A. Weissberger, ed., *Technique of Organic Chemistry*, Vol. III, Interscience, New York, 1950, Chap. IV.
38. Criddle, D. W., and R. L. LeTourneau, *Anal. Chem.*, **23**, 1620 (1951).
39. Davis, D. S., *Ind. Eng. Chem.*, **17**, 735 (1925).
40. Dean, E. W., and G. H. B. Davis, *Chem. Met. Eng.*, **36**, 618 (1929).
41. Deanesly, R. M., and L. T. Carleton, *Ind. Eng. Chem., Anal. Ed.*, **14**, 220 (1942).
42. Desty, D. H., and A. Goldup, "Chromatography of Hydrocarbons," in E. Heftmann, ed., *Chromatography*, Reinhold, New York, 1961, p. 663.
43. Dixon, J. A., and S. G. Clark, II, *J. Chem. Eng. Data*, **4**, 94 (1959).
44. Doolittle, A. K., *J. Appl. Phys.*, **22**, 1471 (1951).
45. Dubois, H. D., and D. A. Skoog, *Anal. Chem.*, **20**, 624 (1948).
46. Eby, H. M., and R. A. Klett, *Anal. Chem.*, **30**, 100 (1958).
47. Eisenlohr, F., "Spectrochemie Organischer Verbindungen," in J. Schmidt, ed., *Chemie in Einzeldarstellung*, Vol. III, Enke, Stuttgart, 1912.
48. Ekkert, L., *Pharm. Zentral.*, **72**, 51 (1931).
49. Etessam, A. H., and M. F. Sawijer, *J. Inst. Petrol.*, **25**, 253 (1939).
50. Evans, E. B., *J. Inst. Petrol. Technol.*, **23**, 220 (1937).
51. Eykman, J. F., *Rec. Trav. Chim.*, **12**, 157 (1893).
52. Eykman, J. F., *Rec. Trav. Chim.*, **14**, 185 (1895).
53. Fogg, E. T., A. N. Hixson, and A. R. Thompson, *Anal. Chem.*, **27**, 1609 (1955).
54. Forziati, A. F., *J. Res. Natl. Bur. Std.*, **44**, 373 (1950).
55. Francis, A. W., *Chem. Rev.*, **42**, 107 (1948).
56. Francis, A. W., *Ind. Eng. Chem.*, **18**, 821 (1926).
57. Francis, A. W., *Ind. Eng. Chem.*, **35**, 442 (1943).
58. Francis, A. W., *Ind. Eng. Chem.*, **36**, 256 (1944).

59. Francis, A. W., *Ind. Eng. Chem.*, **36**, 1096 (1944).
60. Francis, A. W., "Solvent Extraction of Hydrocarbons," in A. Farkas, ed., *Physical Chemistry of the Hydrocarbons*, Vol. I, Academic, New York, 1950, Chap. 7.
61. Fred, M., and R. Putscher, *Anal. Chem.*, **21**, 901 (1949).
62. Geleen, H., H. I. Waterman, and J. B. Westerdijk, *Riv. Combust.*, **10**, 1 (1956).
63. Gladstone, J. H., and T. P. Dale, *Phil. Trans.*, **153**, 317 (1863).
64. Glasgow, A. R., Jr., A. J. Streiff, and F. D. Rossini, *J. Res. Natl. Bur. Std.*, **35**, 355 (1945).
65. Goheen, G. E., *Ind. Eng. Chem.*, **32**, 503 (1940).
66. Greenshields, J. B., and F. D. Rossini, *J. Phys. Chem.*, **62**, 271 (1958).
67. Griswold, J., and J.-N. Chew, *Ind. Eng. Chem.*, **38**, 364 (1946).
68. Groennings, S., *ASTM Bull.*, January 1958, p. 64.
69. Grosse, A. V., and R. C. Wackher, *Ind. Eng. Chem., Anal. Ed.*, **11**, 614 (1939).
70. Hardiman, E. W., and A. H. Nissan, *J. Inst. Petrol.*, **31**, 255 (1945).
71. Hardy, A. C., and F. H. Perrin, *Principles of Optics*, McGraw-Hill, New York, 1932.
72. Harkins, W. D., and H. F. Jordan, *J. Am. Chem. Soc.*, **52**, 1751 (1930).
73. Hazelwood, R. N., *Anal. Chem.*, **26**, 1073 (1954).
74. Hill, A. V., *Proc. Roy. Soc. (London)*, **A127**, 9 (1930).
75. Hill, J. B., and H. B. Coats, *Ind. Eng. Chem.*, **20**, 641 (1928).
76. Hinrichs, G., *Z. Physik. Chem.*, **8**, 229 (1891).
77. Hirschler, A. E., *J. Inst. Petrol.*, **32**, 133 (1946).
78. Hirschler, A. E., Sun Oil Co., unpublished information.
79. Hirschler, A. E., and S. Amon, *Ind. Eng. Chem.*, **39**, 1585 (1947).
80. Hirschler, A. E., and R. L. James, Preprints, General Papers, Division of Petroleum Chemistry, ACS Meeting, Dallas, Tex., April 1956, p. 187.
81. Hirschler, A. E., and T. S. Mertes, "The Separation of Aromatics by Selective Adsorption," in B. T. Brooks, C. E. Boord, S. S. Kurtz, Jr., and L. Schmerling, eds., *The Chemistry of Petroleum Hydrocarbons*, Vol. I, Reinhold, New York, 1954, Chap. 8.
82. Hugel, G., *Chim. Ind.*, **26**, 1282 (1931).
83. Huggins, M. L., *J. Am. Chem. Soc.*, **63**, 116 (1941).
84. Jezl, J. L., and A. Sankin, Sun Oil Co., unpublished information.
85. Jezl, J. L., and A. P. Stuart, *Ind. Eng. Chem.*, **50**, 943 (1958).
86. Johnson, J. F., R. L. LeTourneau, and R. Matteson, *Anal. Chem.*, **24**, 1505 (1952).
87. Jones, A. L., and R. W. Foreman, *Ind. Eng. Chem.*, **44**, 2249 (1952).
88. Jones, A. L., and E. C. Milberger, *Ind. Eng. Chem.*, **45**, 2689 (1953).
89. Karr, C., Jr., and J. R. Comberiati, *Anal. Chem.*, **33**, 1420 (1961).
90. Karr, C., Jr., R. T. Wendland, and W. E. Hanson, *Anal. Chem.*, **34**, 249 (1962).
91. Kearns, G. L., N. C. Maranowski, and G. F. Crable, *Anal. Chem.*, **31**, 1646 (1959).
92. King, R. W., F. A. Fabrizio, and A. R. Donnell, "Application of Gas Chromatography to Some High-Boiling Compounds Present in Petroleum and Tar Oils," in N. Brenner, J. E. Callen, and M. D. Weiss, eds., *Gas Chromatography*, Academic, New York, 1962, Chap. 10.
93. King, R. W., and A. E. Hirschler, *Anal. Chem.*, **26**, 1397 (1954).
94. Knight, H. S., *Anal. Chem.*, **30**, 9 (1958).

95. Knight, H. S., and S. Groennings, *Anal. Chem.*, **26**, 1549 (1954).
96. Knight, H. S., and S. Groennings, *Anal. Chem.*, **28**, 1949 (1956).
97. Koelbel, H., W. Siemes, and H. Luther, *Brennstoff-Chem.*, **30**, 362 (1949).
98. Kopp, H., *Ann.*, **41**, 169 (1842).
99. Kramers, H. A., *Physica*, **4**, 1 (1944).
100. Kramers, H., and J. J. Broeder, *Anal. Chim. Acta*, **2**, 687 (1948).
101. Krevelen, D. W. van, and H. A. G. Chermin, *Fuel*, **33**, 338 (1954).
102. Krevelen, D. W. van, and J. Schuyer, *Coal Science*, Elsevier, New York, 1957.
103. *Ibid.*, p. 160.
104. Kurtz, S. S., Jr., "Physical Properties and Hydrocarbon Structure," in B. T. Brooks, C. E. Boord, S. S. Kurtz, Jr., and L. Schmerling, eds., *The Chemistry of Petroleum Hydrocarbons*, Vol. 1, Reinhold, New York, 1954, Chap. 11.
105. Kurtz, S. S., Jr., S. Amon, and A. Sankin, *Ind. Eng. Chem.*, **42**, 174 (1950).
106. Kurtz, S. S., Jr., R. W. King, W. J. Stout, D. G. Partikian, and E. A. Skrabek, *Anal. Chem.*, **28**, 1928 (1956).
107. Kurtz, S. S., Jr., R. W. King, W. J. Stout, and M. E. Peterkin, *Anal. Chem.*, **30**, 1224 (1958).
108. Kurtz, S. S., Jr., R. W. King, and J. S. Sweely, *Ind. Eng. Chem.*, **48**, 2232 (1956).
109. Kurtz, S. S., Jr., and M. R. Lipkin, *Ind. Eng. Chem.*, **33**, 779 (1941).
110. Kurtz, S. S., Jr., and C. C. Martin, *India Rubber World*, **126**, 495 (1952).
111. Kurtz, S. S., Jr., I. W. Mills, C. C. Martin, W. T. Harvey, and M. R. Lipkin, *Anal. Chem.*, **19**, 175 (1947).
112. Kurtz, S. S., Jr., and A. Sankin, *Ind. Eng. Chem.*, **46**, 2186 (1954).
113. Kurtz, S. S., Jr., and A. L. Ward, *J. Franklin Inst.*, **222**, 563 (1936).
114. Landa, S., and S. Hála, *Erdoel Kohle*, **11**, 698 (1958).
115. Lauer, J. L., and R. W. King, *Anal. Chem.*, **28**, 1697 (1956).
116. LeMaire, G. W., "A Study of Structures of Molecular Complexes Found in Heavy Petroleum Fractions," *Quart. Colo. School Mines*, **53**, No. 4, 1958.
117. Li, K., R. L. Arnett, M. B. Epstein, R. B. Ries, L. P. Bitler, J. M. Lynch, and F. D. Rossini, *J. Phys. Chem.*, **60**, 1400 (1956).
118. Lipkin, M. R., J. A. Davison, W. T. Harvey, and S. S. Kurtz, Jr., *Ind. Eng. Chem., Anal. Ed.*, **16**, 55 (1944).
119. Lipkin, M. R., W. A. Hoffecker, C. C. Martin, and R. E. Ledley, *Anal. Chem.*, **20**, 130 (1948).
120. Lipkin, M. R., and S. S. Kurtz, Jr., *Ind. Eng. Chem., Anal. Ed.*, **13**, 291 (1941).
121. Lipkin, M. R., and C. C. Martin, *Ind. Eng. Chem., Anal. Ed.*, **18**, 380 (1946).
122. Lipkin, M. R., and C. C. Martin, *Ind. Eng. Chem., Anal. Ed.*, **18**, 433 (1946).
123. Lipkin, M. R., and C. C. Martin, *Anal. Chem.*, **19**, 183 (1947).
124. Lipkin, M. R., C. C. Martin, and S. S. Kurtz, Jr., *Ind. Eng. Chem., Anal. Ed.*, **18**, 376 (1946).
125. Lipkin, M. R., C. C. Martin, and R. C. Worthing, *Proceedings, Third World Petroleum Congress*, Sect. VI, The Hague, 1951.
126. Lipkin, M. R., I. W. Mills, C. C. Martin, and W. T. Harvey, *Anal. Chem.*, **21**, 504 (1949).
127. Lipkin, M. R., A. Sankin, and C. C. Martin, *Anal. Chem.*, **20**, 598 (1948).
128. Lorentz, H. A., *Wied. Ann.*, **9**, 641 (1880).
129. Lumpkin, H. E., *Anal. Chem.*, **28**, 1946 (1956).

130. Luther, H., and H. Koelbel, *Brennstoff-Chem.*, **30**, 300 (1949).
131. Mabery, C. F., *Ind. Eng. Chem.*, **15**, 1233 (1923).
132. Mair, B. J., *J. Res. Natl. Bur. Std.*, **34**, 435 (1945).
133. Mair, B. J., *Anal. Chem.*, **28**, 52 (1956).
134. Mair, B. J., and A. F. Forziati, *J. Res. Natl. Bur. Std.*, **32**, 151, 165 (1944).
135. Mair, B. J., A. L. Gaboriault, and F. D. Rossini, *Ind. Eng. Chem.*, **39**, 1072 (1947).
136. Mair, B. J., N. C. Krouskop, and T. J. Mayer, *J. Chem. Eng. Data*, **7**, 420 (1962).
137. Mair, B. J., and J. L. Martinéz-Picó, *Proc. Am. Petrol. Inst.*, Sect. III: *Refining*, **42**, 173 (1962).
138. Mair, B. J., M. J. Montjar, and F. D. Rossini, *Anal. Chem.*, **28**, 56 (1956).
139. Mair, B. J., and F. D. Rossini, "Summary of Work of the American Petroleum Institute Research Project 6 on Hydrocarbons in the C_{13} to C_{38} Fraction of Petroleum," in "Symposium on Composition of Petroleum Oils, Determination and Evaluation," *ASTM Spec. Tech. Publ.* No. 224, American Society for Testing and Materials, Philadelphia, Pa., 1958, p. 9.
140. Mair, B. J., and F. D. Rossini, *Ind. Eng. Chem.*, **47**, 1062 (1955).
141. Mair, B. J., and M. Shamaiengar, *Anal. Chem.*, **30**, 276 (1958).
142. Mair, B. J., M. Shamaiengar, N. C. Krouskop, and F. D. Rossini, *Anal. Chem.*, **31**, 2082 (1959).
143. Mair, B. J., C. B. Willingham, and A. J. Streiff, *J. Res. Natl. Bur. Std.*, **21**, 581 (1938).
144. Mapstone, G. E., *Petrol. Refiner*, **40**, 156 (1961).
145. Martin, C. C., and A. Sankin, *Anal. Chem.*, **25**, 206 (1953).
146. Maxwell, J. B., and L. S. Bonnell, *Vapor Pressure Charts for Petroleum Hydrocarbons*, Esso Research and Engineering Co., Linden, N.J., 1955.
147. Melpolder, F. W., R. A. Brown, T. A. Washall, W. Doherty, and C. E. Headington, *Anal. Chem.*, **28**, 1936 (1956).
148. Mibashan, A., *Trans. Faraday Soc.*, **41**, 374 (1945).
149. Mikeska, L. A., *Ind. Eng. Chem.*, **28**, 970 (1936).
150. Mills, I. W., A. E. Hirschler, and S. S. Kurtz, Jr., *Ind. Eng. Chem.*, **38**, 442 (1946).
151. Mills, I. W., S. S. Kurtz, Jr., A. H. A. Heyn, and M. R. Lipkin, *Anal. Chem.*, **20**, 333 (1948).
152. Miron, S., *Anal. Chem.*, **27**, 1947 (1955).
153. Montgomery, D. S., and M. L. Boyd, *Anal. Chem.*, **31**, 1290 (1959).
154. Moore, C. C., and G. R. Kaye, *Oil Gas J.*, **33**, 108 (1934).
155. Mozingo, R., D. E. Wolf, S. A. Harris, and K. Folkers, *J. Am. Chem. Soc.*, **65**, 1013 (1943).
156. Muller, R. H., and H. J. Stolten, *Anal. Chem.*, **25**, 1103 (1953).
157. Mukhamedova, L. A., M. Baĭburova, and A. Robinzon, *Khim. i Tekhnol Topliva i Masel*, **3**, 18 (1958).
158. Nederbragt, G. W., and J. W. M. Boelhouwer, *Physica*, **13**, 305 (1947).
159. Nelson, K. H., M. D. Grimes, and B. J. Heinrich, *Anal. Chem.*, **29**, 1026 (1957).
160. Nes, K. van, and H. A. van Westen. *Aspects of the Constitution of Mineral Oils*, Elsevier, New York, 1951.
161. *Ibid.*, p. 89.

162. *Ibid.*, p. 96.
163. *Ibid.*, p. 99.
164. *Ibid.*, p. 161.
165. *Ibid.*, pp. 179–180.
166. *Ibid.*, pp. 200–202.
167. *Ibid.*, p. 260.
168. *Ibid.*, pp. 265–298.
169. *Ibid.*, p. 306.
170. *Ibid.*, p. 413.
171. *Ibid.*, p. 429.
172. Norris, M. S., and J. G. O'Connor, *Anal. Chem.*, **31**, 275 (1959).
173. O'Connor, J. G., F. H. Burow, and M. S. Norris, *Anal. Chem.*, **34**, 82 (1962).
174. O'Donnell, G., *Anal. Chem.*, **23**, 894 (1951).
175. Othmer, D. F., M. M. Chudgar, and S. L. Levy, *Ind. Eng. Chem.*, **44**, 1872 (1952).
176. Pfann, W. G., *Zone Melting*, Wiley, New York, 1958.
177. Platt, J. R., *J. Phys. Chem.*, **56**, 328 (1952).
178. Podbielniak, W. J., *Ind. Eng. Chem., Anal. Ed.*, **13**, 639 (1941).
179. Polishuk, A. T., C. K. Donnell, and J. C. S. Wood, *Anal. Chem.*, **26**, 1087 (1954).
180. Powell, H. M., *Endeavour*, **9**, 154 (1950).
181. Qozati, A., and M. van Winkle, *J. Chem. Eng. Data*, **5**, 269 (1960).
182. de Radzitsky, P., and J. Hanotier, *Ind. Eng. Chem., Proc. Design Develop.*, **1**, 10 (1962).
183. Rampton, H. C., *Anal. Chem.*, **21**, 1377 (1949).
184. Ramser, J. H., *Ind. Eng. Chem.*, **41**, 2053 (1949).
185. Redlich, O., C. W. Gable, A. K. Dunlop, and E. W. Millar, *J. Am. Chem. Soc.*, **72**, 4153 (1950).
186. Rossini, F. D., *Proc. Am. Petrol. Inst.*, Sect. III: *Refining*, **19**, 99 (1938).
187. Rossini, F. D., B. J. Mair, and A. J. Streiff, *Hydrocarbons from Petroleum*, Reinhold, New York, 1953.
188. *Ibid.*, Chap. 4.
189. *Ibid.*, p. 397.
190. Rossini, F. D., K. S. Pitzer, R. L. Arnett, R. M. Braun, and G. C. Pimental, *Selected Values of Physical and Thermodynamic Properties of Hydrocarbons and Related Compounds*, API Res. Proj. 44, Carnegie Press, Pittsburgh, Pa., 1953.
191. Rostler, F. S., and H. W. Sternberg, *Ind. Eng. Chem.*, **41**, 598 (1949).
192. Sachanan, A. N., *Chemical Constituents of Petroleum*, Reinhold, New York, 1945, p. 237.
193. Sankin, A., C. C. Martin, and M. R. Lipkin, *Anal. Chem.*, **22**, 643 (1950).
194. Sauer, R. W., T. A. Washall, and F. W. Melpolder, *Anal. Chem.*, **29**, 1327 (1957).
195. Sawicki, E., and R. R. Miller, *Anal. Chem.*, **30**, 109 (1958).
196. Sawicki, E., R. R. Miller, T. W. Stanley, and T. R. Hauser, *Anal. Chem.*, **30**, 1130 (1958).
197. Scatchard, G., and L. B. Ticknor, *J. Am. Chem. Soc.*, **74**, 3724 (1952).
198. Schaarschmidt, A., *Z. Agnew. Chem.*, **44**, 474 (1931).
199. Schaeffer, W. D., W. S. Dorsey, D. A. Skinner, and C. G. Christian, *J. Am. Chem. Soc.*, **79**, 5870 (1957).

200. Schiessler, R. W., J. N. Cosby, D. G. Clarke, C. S. Rowland, W. S. Sloatman, and C. H. Herr, *Proc. Am. Petrol. Inst.,* Sect. III: *Refining,* **23,** 15 (1942).
201. Schiessler, R. W., and D. Flitter, *J. Am. Chem. Soc.,* **74,** 1720 (1952).
202. Schiessler, R. W., C. H. Herr, A. W. Rytina, C. A. Weisel, F. Fischl, R. L. McLaughlin, and H. H. Keuhner, *Proc. Am. Petrol. Inst.,* Sect. III: *Refining,* **26,** 254 (1946).
203. Schiessler, R. W., and F. C. Whitmore, *Ind. Eng. Chem.,* **47,** 1660 (1955), Am. Doc. Inst., Doc. 4597.
204. Schoorl, N., *Organische Analyse,* Vol. II, 3rd Ed., D. B. Centen, Amsterdam, 1937.
205. Schuhmacher, J. P., H. A. van Vucht, M. P. Groenewege, L. Blom, and D. W. van Krevelen, *Fuel,* **35,** 281 (1956).
206. Shepard, A. F., and A. L. Henne, *Ind. Eng. Chem.,* **22,** 356 (1930).
207. Smith, E. E., *Ohio State Univ. Eng. Expt. Sta. Bull.* No. 152, May 1953.
208. Smith, H. M., *U.S. Bur. Mines Tech. Paper* No. 610, 1940.
209. Smith, J. R., C. R. Smith, Jr., and G. U. Dinneen, *Anal. Chem.,* **22,** 867 (1950).
210. Smittenberg, J., and D. Mulder, *Rec. Trav. Chim.,* **67,** 813, 826 (1948).
211. Snyder, L. R., *Anal. Chem.,* **33,** 1527 (1961).
212. *Standard Methods for Testing Petroleum and Its Products,* Institute of Petroleum, London, IP Method 9/42.
213. Stout, W. J., R. W. King, M. E. Peterkin, and S. S. Kurtz, Jr., "Adsorption and Physical Property Methods," in "Symposium on Composition of Petroleum Oils, Determination and Evaluation," *ASTM Spec. Tech. Publ.* No. 224, American Society for Testing and Materials, Philadelphia, Pa., 1958.
214. Sugden, S., *J. Chem. Soc.,* **125,** 1177 (1924).
215. Sweely, J. S., S. W. Ferris, M. E. Peterkin, and S. S. Kurtz, Jr., *Rev. Gen. Catchouc,* **34,** 170 (1957).
216. Tadayon, J., E. W. Hardiman, and A. H. Nissan, *J. Inst. Petrol.,* **35,** 28 (1949).
217. Thompson, C. J., H. J. Coleman, H. T. Rall, and H. M. Smith, *Anal. Chem.,* **27,** 175 (1955).
218. Thomson, G. W., *Chem. Rev.,* **38,** 1 (1946).
219. Thorne, H. M., W. Murphy, and J. S. Ball, *Ind. Eng. Chem., Anal. Ed.,* **17,** 481 (1945).
220. Thorpe, R. E., and R. G. Larsen, *Ind. Eng. Chem.,* **34,** 853 (1942).
221. Thorpe, T. E., and J. W. Rodger, *Phil. Trans.,* **A185,** 397 (1894).
222. Tipson, R. S., "Crystallization and Recrystallization," in A. Weissberger, ed., *Technique of Organic Chemistry,* Vol. III, Interscience, New York, 1950, Chap. VI.
223. Tizard, H. T., and A. G. Marshall, *J. Soc. Chem. Ind.,* **40,** 20 (1921).
224. Vlugter, J. C., H. I. Waterman, and H. A. van Westen, *J. Inst. Petrol. Technol.,* **21,** 661 (1935).
225. Von Fuchs, G. H., and A. P. Anderson, *Ind. Eng. Chem.,* **29,** 319 (1937).
226. Ward, A. L., and S. S. Kurtz, Jr., *Ind. Eng. Chem., Anal. Ed.,* **10,** 559 (1938).
227. Ward, S. H., and M. van Winkle, *Ind. Eng. Chem.,* **46,** 338 (1954).
228. Waterman, H. I., C. Boelhouwer, and J. Cornelissen, *Correlation Between Physical Constants and Chemical Structure,* Elsevier, New York, 1958.
229. *Ibid.,* pp. 61–68.
230. *Ibid.,* p. 91.

231. *Ibid.,* p. 98.
232. Waterman, H. I., H. F. Westerdijk, H. F. O. Span, H. Booij, and K. van Nes, *J. Inst. Petrol.,* **36,** 281 (1950).
233. Watson, K. M., "Correlation of the Physical Properties of Petroleum," in A. E. Dunstan, ed., *The Science of Petroleum,* Vol. II, Oxford University Press, New York, 1938, pp. 1377–1386.
234. Watson, K. M., and E. F. Nelson, *Ind. Eng. Chem.,* **25,** 880 (1933).
235. Watson, K. M., E. F. Nelson, and G. B. Murphy, *Ind. Eng. Chem.,* **27,** 1460 (1935).
236. Wibaut, J. P., and S. L. Langedijk, *Rec. Trav. Chim.,* **59,** 1220 (1940).
237. Wiener, H., *J. Am. Chem. Soc.,* **69,** 2636 (1947).
238. Wiener, H., *J. Phys. Colloid Chem.,* **52,** 1082 (1948).
239. Wilson, A., L. Bini, and R. Hofstader, *Anal. Chem.,* **33,** 135 (1961).
240. Wilson, O. G., Jr., *Ind. Eng. Chem.,* **20,** 1363 (1928).
241. Winterstein, A., and K. Schön, *Z. Physiol. Chem.,* **230,** 139, 146 (1934).
242. Wood, J. C. S., *Anal. Chem.,* **30,** 372 (1958).
243. Wright, W. A., *ASTM Bull.,* July 1956, p. 84.
244. Young, S., *J. Chem. Soc.,* **73,** 905 (1898).
245. Young, S., *Proc. Roy. Irish Acad.,* **B38,** 65 (1928).
246. Zelinsky, N., *Ber.,* **57B,** 42 (1924).
247. Zijll Langhout, W. C. van, thesis, Delft, 1955.
248. Zimmerschied, W. J., R. A. Dinerstein, A. W. Weitkamp, and R. F. Marschner, *Ind. Eng. Chem.,* **42,** 1300 (1950).

Chapter 100

TRACE ANALYSIS: ESSENTIAL ASPECTS

BY R. K. SKOGERBOE* AND G. H. MORRISON, *Department of Chemistry, Cornell University, Ithaca, New York*

Contents

* Present address: Department of Chemistry, Colorado State University, Fort Collins, Colorado.

I. INTRODUCTION

The important role of very small amounts of elements in physical, chemical, and biological systems has emerged as methods of analysis have increased in sensitivity. Much of this progress may be attributed to the demands of materials and biological research and to the availability of modern instrumentation. Although a large number of different physical techniques have evolved, they all have a common goal—the measurement of the elemental composition at the parts per million or parts per billion level—so that the distinct field of trace analysis has emerged.

This chapter is concerned with the essential aspects of the subject; its purpose is to demonstrate the basis of the various methods, the means for standardization and intercomparison, and, finally, to provide a survey of the present capabilities of a number of the most frequently used trace techniques. The subject has been treated in detail in a book edited by G. H. Morrison (16), so that only the highlights of the subject, as well as some more recent developments, are treated here.

At the outset, it should be noted that the term trace, as used in this chapter, refers to constituents present in samples below an upper limit of 100 parts per million; this limit is, of course, an arbitrary one. Included in the goals of trace analysis are the determination of the bulk concentrations of the trace elements, the preconcentration of microconstituent into a small, essentially matrix-free sample to improve detectability and/or remove matrix interferences, the determination of local concentrations by probe techniques in order to establish the topographical distribution of trace elements in solid samples, and the determination of major and minor species in a minute initial sample. In the past much emphasis has been placed on analyses for bulk concentration, but recently there has been increased interest in the determination of local concentrations (12).

Among the many criteria used in the selection of an appropriate trace analytical method; sensitivity, accuracy and precision, scope, selectivity, and sampling and standards requirements are paramount. These aspects will be treated in the appropriate places throughout the chapter.

II. THE ANALYTICAL PROCESS

All trace analytical methods can be divided into three component steps: sampling, chemical and/or physical pretreatment, and instrumental measurement (2). On the basis of the type of information desired in an analysis and the requirements of sensitivity, precision, etc., an appropriate instrumental technique is selected. Therefore, it is essential to know the capabilities and limitations of the various methods of instrumental measurement. Once they are known, appropriate steps can be taken in the sampling and pretreatment steps to provide sufficient microconstituent that is free of interferences and is in the appropriate form for the final measurement.

It must be noted, however, that a strict division of all analytical methods into these phases or steps is not inviolate. Thus, in electron microprobe analysis the sampling and the measurement occur simultaneously. In spite of possible deviations, the important fact is that these steps are interrelated and require different degrees of emphasis, depending on the individual analytical situation. In all cases, the analyst is interested in finally measuring a signal which can be related to the concentration of a particular species in the original material. Therefore, in order to properly evaluate the requirements of pretreatment (preconcentration and/or separation) and sampling, this chapter will discuss the three phases in reverse order of the actual analytical sequence, starting with an examination of the culminating step of analysis, the instrumental measurement.

A. INSTRUMENTAL MEASUREMENT

1. Analytical Sensitivity

The application of any analysis technique to trace characterization is governed to a large extent by the analytical sensitivity that can be achieved for the species of interest in a given material. As used herein, *sensitivity* refers to the ability to discern a small change in concentration or amount of the species of interest. *Detection limit* is a closely related but distinct term, defined here as the lowest concentration or amount that can be determined with a specific degree of confidence. Because the magnitude and the reproducibility of the measurement signal are implicit in the definitions of these pertinent terms, it is necessary to consider the measurement step in greater detail.

All analytical measurements can be categorized into two general classes. When signal reduction methods such as colorimetry are used, the signal is set at a particular level (the 100% level, for example) by means of a blank, and the amount of the species sought is inferred from the reduction in signal due to its presence. Signal increase methods such as emission spectrography deduce the amount of the analyte from the magnitude of the signal emitted by the species. For either category, the final result depends on the ability to measure reliably the difference between the analytical signal and the blank, background, or electronic noise signal. The simultaneous fluctuations in the extraneous and analysis signals determine the reproducibility with which the signal changes can be measured. At the same time, the relationship between the signal intensity and the amount of the analyte must be considered. It is intuitive that greater sensitivity will result from a larger signal change per unit concentration, provided that the measurement precision does not become proportionately worse. Mandel and Stiehler (13) and, more recently, Skogerboe et al. (21) have proposed a mathematical definition of sensitivity based on this concept. The sensitivity, γ, is defined by

$$\gamma = \frac{dI}{dC} \frac{1}{s} \tag{1}$$

where I = the intensity of the signal measured, C = the concentration or mass of the analyte, and s = the standard deviation of the signal measurement. Because I and s have the same units, γ is derived in reciprocal concentration or mass units. Figure 100.1 indicates the dependency of the sensitivity on the slope of the analytical curve and the measurement precision as derived from equation (1) and is obviously consistent with the intuitive concept previously mentioned.

By utilizing this definition of sensitivity, a mathematical definition of detection limit can be derived which is also consistent with the verbal definition given above. At or near the detection limit, the magnitude of the analytical signal, I, and its reproducibility approach those of the background, noise, or blank signal, I_b. The minimum signal change that can be differentiated is given by

$$dI_m = I - I_b \equiv s \tag{2}$$

where s is the standard deviation of measurement for that level. To make the determination of dI_m objective, a confidence or tolerance level must be introduced; this is accomplished by using the t statistic, that is,

$$dI_m = st_{(n-1, 1-\alpha)} \tag{3}$$

where the tabular value of t depends on the number of measurements, n, and the percentage confidence level required, $100(1 - \alpha)\%$. The t sta-

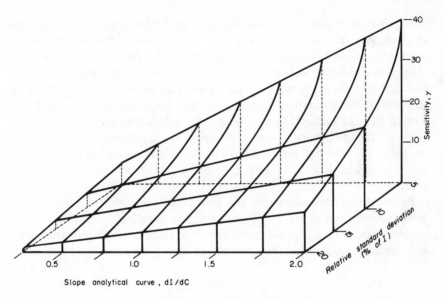

Fig. 100.1. Analytical sensitivity as a function of calibration curve slope and measurement precision.

tistic is specifically useful for small numbers of measurements (<20). Where many measurements are available, the confidence constants associated with the normal error distribution may replace t (11).

The concentration required to produce the minimum discernible signal change defined by equation (3) is readily determined from the equation for the analytical curve. If the curve is linear, for example, then

$$I = mC + I_b \tag{4}$$

where C is the concentration and m is the slope. By substituting equation (3) into (4) and rearranging, an expression for the minimum detectable concentration, C_m, is obtained:

$$C_m = \frac{dI_m}{m} = \frac{st_{(n-1,1-\alpha)}}{m} \tag{5}$$

Note, however, that rearrangement of equation (1) and substitution into (5) produces

$$C_m = \frac{t_{(n-1,1-\alpha)}}{\gamma} \tag{6}$$

Thus, the detection limit is determined for the confidence level required in a manner which is consistent with the verbal definition previously presented. In addition, equation (6) defines the minimum determinable change in concentration at any level if the slope and the standard deviation for this level are used to determine the sensitivity, γ. It must be emphasized that parameters defined in the above manner, although objective, depend on both the mode of data acquisition and the characteristics of the instrument used. Consequently, maximum objectivity will be obtained only by inclusion of complete and exact specifications of the experimental method and the instrumental conditions subject to control.

By means of this approach, the sensitivities of different methods can be objectively compared. If, for instance, the sensitivities of two methods are determined in a consistent manner with regard to concentration level and method of obtaining the data, the following expression may be used:

$$K = \frac{\gamma_1 t_{(n_1-1, 1-\alpha/2)}}{\gamma_2 t_{(n_2-1, 1-\alpha/2)}} \tag{7}$$

The subscripts refer to methods 1 and 2, respectively, and the n_i designate the number of measurements used in the determination of the respective sensitivity parameters. Deviation of K from unity specifies that one method is more sensitive than the other at the $100(1 - \alpha)\%$ confidence level, that is, for $K > 1$, method 1 is preferred.

2. Survey of Detection Limits

A survey of the analytical literature indicates clearly that there is a lack of consistency in definition and/or specification of the detection limits associated with the various analytical methods. Consequently, it is difficult to make critical comparisons between methods for a particular entity. Because of the necessity of knowing the detection capabilities of the measurement steps of the various techniques, the only alternative at this time is to summarize the experimental values that have been published, realizing that large inconsistencies exist. It is hoped that future investigators will adopt the more objective approach.

Table 100.I presents such a summary, based on data in the recent literature reported by authorities for the respective techniques. The values represent limits that have been experimentally achieved under conditions of little or no interference when such limits were available. Absence of values for specific elements by a particular technique does not imply that the determination of said elements by that technique is impossible, but only that data are limited. The lower limit of detection can be expressed in either of two ways. The *absolute limit* is the smallest detectable weight of the substance expressed in micrograms, nanograms, etc. The *relative limit* is the lowest detectable concentration expressed as a percentage, parts per million (on a weight or atomic basis), micrograms per gram, micrograms per milliliter, etc. In order to avoid ambiguity and to permit easy comparison of methods of measurement involving solutions and solids, all limits have been converted to absolute units. The limits, therefore, indicate the smallest weight of an element that must be present in the measurement step to produce a measurable signal (disregarding the level of confidence at which the measurement was made). These absolute limits will assist in indicating the degree of preconcentration required when such an approach is necessary, and the size of sample that must be used.

a. SOLUTION TECHNIQUES

Limits of detection for techniques which utilize solutions in the measurement step are most frequently stated in units of micrograms per milliliter. Multiplication of the relative detection limit by the minimum sample solution volume required to obtain a measurement produces the limit in absolute terms, that is, the smallest weight of the analyte which can be detected.

(1) Flame and Atomic Absorption Spectrophotometry

The best absolute limit is obtained with a measurement made rapidly with sufficient allowance for the instrumental response time. A sample volume of 0.05 ml per measurement, as utilized by Slavin et al. (24), was chosen for these conversions. At a sample consumption rate of 1.5 ml/min, this volume allows 2 sec for the response time (approximately twice that of the average instrument). The values listed were derived from the data by Fassel and his associates (5,6) and by Skogerboe et al. (21) for flame emission, and from those by Slavin (23) for atomic absorption.

(2) Absorption Spectrophotometry

The values listed were determined on the basis of a 1-cm path length and a minimum volume of 0.1 ml. These parameters are consistent with practical cell designs currently in use (14). The limits given by Sandell (20), Boltz (1), and Pinta (19) were used and do not necessarily imply adherence to Beer's law at these levels.

TABLE 100.I
Absolute Detection Limits in Nanograms

Element	Absorption spectrophotometry	UV-visible fluorescence	Atomic absorption	Flame spectrophotometry	Electrochemistry	Emission spectrography				X-ray fluorescence	Neutron activation	Solids mass spectroscopy
						Solution methods	Solid (spark)	De arc solids	Residue methods			
Ag	0.5	0.05	0.5	15		2	3	0.03	0.03	1,200	0.01	0.2
Al	0.05		5	10		10	5	3	0.5	5,000	1	0.02
As	1		25	100		300	100	3	50		0.1	0.06
Au	0.5		5	50	400	2,000	100	8	20		0.05	0.2
B	5	4	800	1,000		5	2	0.2	0.3			0.01
Ba	10		5	2		5	20	25	10	120	5	0.2
Be	0.8	0.1	0.2	50		0.3	10	0.2	0.03			0.008
Bi	60		1	150	10,000		10	0.8	5	600	50	0.2
Br	1					100				330	0.5	0.1
C					5							0.01
Ca	10	0.3	0.2	0.3		1	4	0.3	0.5	140	100	0.03
Cd	0.3	200	0.5	100	1	20	2	0.2	20	770	5	0.3
Ce	4			500		300	300	130	30	170	10	0.1
Cl	10				100						1	0.04
Co	0.3	2	0.4	50	5,000	50	4	0.1	0.3	50	0.5	0.05
Cr	0.7		0.3	5	200	10	10	15	1	220		0.05
Cs			3	0.3				50		150	100	0.1
Cu	0.2	3	0.3	5	1	1,500	1	3	0.03	110	50	0.08
Dy			10	5		5	1,000	20	20		0.0001	0.5
Er			10	5		200	1,000	50	10		0.1	0.5
Eu			10	2	2,500	200	1,000	50	2		0.0005	0.2
F	400	0.1			500	50		2	10		100	0.02
Fe	5	2	0.5	35	5	20	10	2,500	0.3	90	5,000	0.05
Ga	20	0.1	4	4	10	50	20	0.2	0.03		1	0.09
Gd	4		200	100		50	1,000	3	10		0.5	0.5
Ge	400	200	100	30	1,000		10	15	50		100	0.4
Hf	2	10	800	3,800		400	700	3	10		1	0.6
Hg			10	35	2,000		30	25	20		1	0.1
Ho	0.5		20	5		1,000	1,000	8		70		0.1
I					30	50		50			0.5	0.1
In	1	4	3	2	50	300	100	3	10		0.005	0.1
Ir	5		200	5,000		1,000		100	50		0.01	0.3
K	0.6		0.3	0.2		20,000		25	5	120	5	0.03
La	20	20	4,000	50		30	100	15	2		0.1	0.1
Li	400		0.3	0.0002		10		15	0.1			0.006
Lu	0.3		2,500	10		50	1,000	3	1		0.005	0.1

Element												
Mg	0.2	0.02	0.03	10		3	4	0.3	0.03	60	50	0.03
Mn	0.5		0.3	5		2	1	0.8	0.03	400	0.005	0.05
Mo	1		5	2	1,000	30	30	25	2		10	0.3
N	800			0.005	10,000							0.01
Na	0.5	10	0.3	50		50	3	15			0.5	0.02
Nb	400		1,000	50	1	200	600	130	20		0.5	0.08
Nd	0.4		100	30	2,000	500	10	130	20	60	10	0.4
Ni		0.2	0.5		1,000	50	10	3	1		5	0.07
Np												
O	5			500		1,500	100	1,300	10	1,100	5	0.01
Os	10		5	50		100	10	4	1	400	50	0.4
P	0.6		25	50	1	400		0.5	5		1,000	0.03
Pb	1		500	50		200	500	25	20	760	0.05	0.3
Pd	400		25	100		200		130	2		0.05	0.3
Pr	1			200	3,000	100		100			5	0.1
Pt												0.5
Pu	20	0.2	0.3	0.1		500	100	25	20	600	5	0.1
Rb	5		80	50	50	70	10	250	200		0.05	0.2
Re	10	100	0.2	15		200		15	30		0.05	0.09
Rh	10	20	20	15				10	10		1	0.03
Ru	0.4	10			2,500	200		8			500	0.03
S	0.8		10	10	10,000	200	100	3	3	280	0.5	0.03
Sb	20	10	10	4		5	8	15			1	0.04
Sc	10	0.2	25	150	5,000	7,000	500	500			500	0.1
Se	100	8	10	250		100	10	10	1		5	0.03
Si	6	10	250	30		300	500	75	20		0.05	0.5
Sm	1		5	25	500	200	10	3	5		50	0.3
Sn	4		0.5	0.2		6	40		5	280	0.5	0.3
Sr	100	300		900	10,000	200		750	10		5	0.09
Ta	40		300	50		300	1,000	50	100		5	0.2
Tb	5	100	100	50		1,000	400	250	50		5	0.1
Te	1	20	20	30	10,000	300	500	250	20		5	0.3
Th	2				50,000	200	30	25	2	80		0.2
Ti		5		7,500	1,500	10	10	0.2	5		1	0.05
Tl	30	10	5	25		300	100	15	100	700	0.5	0.2
Tm	1	5	10	5	250	200	20	2,500	1	45	0.1	0.1
U	3	2		10	200	300	500	250	10	220	0.1	0.2
V		0.1	1,500	500		10	80	50	1			0.04
W	100	100	5	200	2,500	300	100	15	3	70	0.1	0.5
Y		2	150	15	1	30	50	3	3		10	0.07
Yb	10		20	50		30	60	250			100	0.5
Zn	5	100	2			20						0.1
Zr		2	250	2,500								0.1

(3) Ultraviolet–Visible Fluorescence Spectrophotometry

The tabulated values were derived from those given by Weissler and White (28), Pinta (19), and Sandell (20) and were calculated on the basis of a 0.1-ml volume.

(4) Electrochemistry

The data given represent either coulometric or polarographic determinations as given by Taylor et al. (26). The technique producing the best limit was selected in each case. The sample sizes indicated for the respective elements by the above authors were used for the required conversion.

(5) Emission Spectrography

The data given are for methods that utilize solutions in the excitation process as presented by DeKalb et al. (5) and were converted using a 0.1-ml sample volume. In each case the mode of presentation of the solution to the excitation discharge, for example, porous cup or rotating disk, producing the best limit was selected.

b. SOLID TECHNIQUES

When solid samples are used in the estimation step, detection limits may be stated in absolute terms as for neutron activation or copper spark spectrography, or in relative terms (micrograms per gram) as for dc-arc spectrography. In the latter case, the product of the sample weight and relative detection limit produces the absolute value.

(1) Emission Spectrography

The tabular data are categorized according to three modes of solid sample excitation as obtained from the survey of DeKalb et al. (5). The data designated as residue refer primarily to excitation of micro deposits on some substrate and include copper and graphite spark techniques and hollow-cathode-type discharges. The solid-spark data refer to excitation of solids of bulk nature and were calculated on the basis of a 10-mg sample. The term spark, as used here, encompasses the excitation methods for which there is alternation of the electrode polarity. The arc spectrography data are referred to 25-mg solid samples and include methods using constant polarity and continuous discharge, such as the dc arc.

(2) Neutron Activation

The values listed are those defined by Guinn and Lukens (9) for a thermal-neutron flux of 10^{13} n/cm²-sec, using the most sensitive mode of decay (beta or gamma).

(3) Spark Source Mass Spectrography

The limits given were calculated using the equation given by Morrison and Skogerboe (18) and assuming that 5×10^{11} ions are required for photographic detection. Values obtained in this manner are expected to be accurate to an order of magnitude.

(4) X-ray Spectroscopy

The values are from essentially matrix-free residues contained on ion-exchange resin-loaded filters. Hubbard and Green (10) and Campbell and his associates (3,4) determined these quantities, using fluorescence measurements.

(5) Other Techniques

All techniques are not included in Table 100.I. The exclusion of any one technique does not imply that its use in trace analysis is limited, but rather may suggest that either the number of data available is limited or that it is difficult to ascertain a general set of operative conditions.

c. INTERCOMPARISON

To permit a generalized comparison of the data in Table 100.I, they are summarized graphically in Fig. 100.2. The absolute limits are plotted against the number of elements that can be detected at a particular level. The intersection with the ordinate indicates the total number of elements for which data were available and does not necessarily mean that more elements could not be determined. The techniques have been grouped into two categories according to the form of the sample normally used in the measurement step. Nearly all techniques are generally applicable down to 100 ng. In the 0.01–100 ng range several of the techniques are applicable to a more limited number of elements, but only neutron activation and solids mass spectroscopy are generally usable below this.

It must be emphasized that these data are idealized to a degree and are generally representative of conditions of little or no interference by the matrix or other sample constituents. Failure to list a limit for a particular element–technique combination does not imply that the ele-

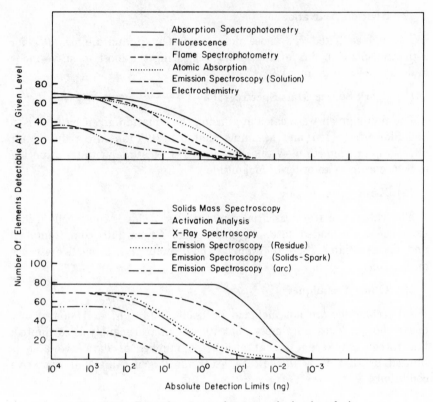

Fig. 100 2.   Detection scope of trace methods of analysis.

ment cannot be determined by that technique. Rather, such omissions are usually the result of a paucity of data for this purpose.

3. Measurement Errors

The description or evaluation of any phase of an analytical procedure is incomplete without reference to the error involved. Particular emphasis must be placed on measurement errors because it becomes increasingly difficult to adequately specify the errors arising from the pretreatment or sampling stages of the analysis when the variability associated with the actual measurement is large.

At the outset, two terms implicit in any discussion of errors must be defined. The term *accuracy* refers to the reliability of the result, while *precision* refers to the repeatability of the result. It should also be understood that the elimination of inaccuracies is primarily an analytical problem, whereas the assessment of precision is, at least partially, within the realm of statistics.

a. Types of Errors

Errors affecting an experimental result may be classed as either systematic or random (2,29). The former, occasionally referred to as determinate errors, are due to causes over which the analyst has control, and their undesirable effects on the accuracy of the result can be avoided or corrected. Random errors are not subject to control and are manifested, even in the absence of systematic errors, by variations in the result. Although the magnitude of random variations can be reduced by carefully keeping all operations identical, they are never eliminated entirely. The precision of the result thus depends on the random errors. In this situation the distinction between precision and accuracy becomes less apparent, that is, the *precision defines the accuracy that can be achieved in the absence of systematic errors*.

b. Additivity of Variances

As previously noted, the analytical process can be subdivided into three stages. This breakdown and precision estimates associated with each stage are presented schematically in Fig. 100.3. The average result, $\bar{x}$, is the sample estimate of the true population mean, μ, and the overall or total standard deviation, s_t, estimates the true standard deviation of the population, σ. The value of s_t is determined by the distribution of errors among the several stages through the additivity of variances, that is,

$$s_t^2 = s_s^2 + s_p^2 + s_m^2 \tag{8}$$

where the subscripts t, s, p, and m refer to the total, sampling, pretreatment, and measurement phases, respectively (2). Because the relative standard deviation is given by $C = 100s/\bar{x}$, equation (8) could be written as

$$C_t^2 = C_s^2 + C_p^2 + C_m^2 \tag{9}$$

Since we are presently concerned with the measurement step, the other sources of error will be considered where appropriate in the succeeding discussion.

c. Isolation of Random Measurement Errors

As indicated by the schematic, the instrumental measurement precision is estimated by assaying k (homogeneous) aliquots of one analysis sample or by computing a pooled estimate (30) from aliquots of several

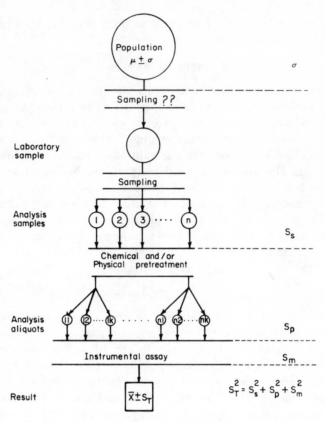

Fig. 100.3. Schematic of the analytical process.

samples. By replication, an improvement in the precision associated with any stage of the analysis and consequently a reduction of the contribution of this stage to s_t can be obtained through the use of the relation

$$s_{\bar{x}} = \frac{s}{(n)^{1/2}} \tag{10}$$

where $s_{\bar{x}}$ designates the standard deviation of the mean of n replicate measurements, and s denotes the standard deviation of a single measurement (2). For example, equation (8) becomes

$$s_t{}^2 = s_s{}^2 + s_p{}^2 + s_m{}^2/n \tag{11}$$

when n measurements are made on an analysis sample. Because of the $(n)^{1/2}$ relationship, the effect of replication on the standard deviation rapidly approaches a point of diminishing gain. In addition, the advisability of replicating any one step, such as the measurement step, is highly dependent on the relative magnitudes of the error terms comprising the overall precision estimates. For a heterogeneous population it would be more advantageous to replicate the sampling step or to utilize an homogenizing pretreatment process. In view of these factors, the argument for a definitive evaluation of the distribution of errors among the analysis stages is obvious. Failure to recognize or consider the distribution of errors has led to a misunderstanding among analysts regarding the use of measurement precision as an estimate of the overall precision of the analysis, for example, counting precision is often used to estimate the reproducibility of radiochemical methods without proving that the sampling and pretreatment errors are negligible.

d. Systematic Measurement Errors

Although the preceding discussion has been specifically concerned with random errors, some comment regarding systematic errors associated with the assay step is appropriate. Failure to resolve the analytical signal from some extraneous signal originating from the sample and failure to make adequate blank corrections are two common errors of this type. As suggested above, these are actually analytical problems related to calibration, choice of standards, and pretreatment and will be considered in these sections.

B. PHYSICAL AND CHEMICAL PRETREATMENT

Although much has been written about the capabilities of the various instrumental techniques for trace analysis, their potentialities for the compositional characterization of complex materials can, in many cases, be fully realized only by coupling the techniques with preconcentrations and separations to remove or minimize the ever-present possibility of interferences or to provide adequate microconstituent for the measurement to be employed. These analytical operations constitute the pretreatment step and may be accomplished by a variety of physical or chemical methods. The specific requirements for preconcentration and separation are intimately related to the sample under consideration and the technique of measurement employed. Many measurement methods accomplish some degree of physical resolution which minimizes the effects of interferences.

In addition to preconcentrations and separations, sample pretreatment may be required to convert a sample to a form more appropriate for the measurement step. Thus, in colorimetric estimation of trace elements in solid samples, the sample must be dissolved, the oxidation state of the ion modified, and a color-producing reagent added before the actual spectral measurement is made. In the determination of the bulk concentration of trace elements in solids the pretreatment step may involve some form of homogenization such as dissolution or grinding and mixing.

Extreme care must be taken in the pretreatment step to avoid contamination or loss of trace elements, particularly at the nanogram level. A number of good treatments of this subject have been published (8,15,17,25,27). If possible, the minimum of pretreatment is advisable, and direct methods of trace analysis are preferable if they are applicable to the problem.

Many good methods of chemical separation are available in trace analysis, the more generally applicable ones being solvent extraction, chromatography, ion exchange, precipitation and coprecipitation, electrodeposition, and volatilization. Ashing is used for the preconcentration of trace elements in biological materials, and the recently developed technique of low-temperature dry ashing looks particularly promising (7). Popular physical methods include selective evaporation and zone melting for solid samples.

1. Recovery and Separation Factors

As stated earlier, the specific requirements for preconcentration and separation are intimately related to the sample under consideration and the technique of measurement employed. Details of separation procedures may be found in the literature (17,25), but two important terms applicable to all methods of separation should be defined at this point. One is the recovery or yield, R_A, of a desired trace element A, and the other is the separation factor, $S_{B/A}$, for the undesired constituent B with respect to a desired trace element A. The definitions are as follows:

$$R_A = \frac{Q_A}{Q_A{}^0} \times 100(\%) \tag{12}$$

$$S_{B/A} = \frac{(Q_B/Q_A)}{Q_B{}^0/Q_A{}^0} \tag{13}$$

where $Q_A{}^0$ is the quantity of A in the sample, $Q_B{}^0$ is the quantity of B in the sample, Q_A is the quantity of A after separation, and Q_B is the quantity of B after separation (20).

The reciprocal of $S_{B/A}$ is called the concentration factor or the enrichment factor of A when B is the matrix. The ideal separation has a recovery of 100% and a separation factor of zero. Actually, these two factors depend on the nature of the sample as well as the method of separation. Knowing the limits of detection of the various methods of measurement, one can then select the sample size and the appropriate method of separation and/or preconcentration to accomplish the trace analysis as outlined in Fig. 100.4. When a direct analysis is possible, time and effort can usually be conserved by avoiding an alternative method which requires separation or preconcentration.

The analytical interrelationship between the measurement and pretreatment steps, as presented in Fig. 100.4, is determined by the solutions to equations (6), (12), and (13). In essence, when the amount of the analyte in the sample is less than the detection (or determination) limit, preconcentration is required and the size of the sample to be used is determined by the values of R_A and $1/S_{B/A}$ for the system chosen to effect the preconcentration. If a physical or chemical interference precludes direct analysis, separation is indicated and the method to be utilized is selected on the basis of the respective recovery and separation

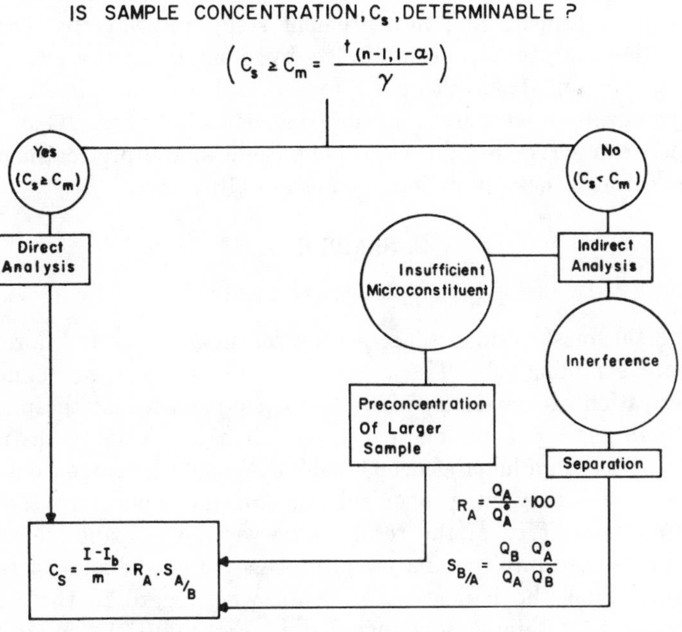

IS SAMPLE CONCENTRATION, C_s, DETERMINABLE ?

$$\left(C_s \geq C_m = \frac{t_{(n-1,\,1-\alpha)}}{\gamma} \right)$$

Fig. 100.4. Interrelation between detection limit and pretreatment.

factors. In either case, the final result is corrected by the yield and concentration factors $(S_{B/A})$ for the systems used, the values of these being unity for direct analysis.

2. Pretreatment Errors

The pretreatment step is subject to a variety of errors which, when combined with those from the sampling and measurement steps, make up the total error of analysis as given in equation (8). Typical sources of error include poor reproducibility of the separation process and inability to reliably assess the yield of the separation. Standards are normally processed along with the sample to quantitate these effects. Contamination is another source of error, and blanks are run to determine the contribution to the trace element of interest by the reagents, container material, etc.

Since the measurement step is required to assess the errors of the pretreatment step, the two combined can be considered as indicative of the analysis error.

$$s_a{}^2 = s_p{}^2 + s_m{}^2 \qquad (14)$$

where s^2 is the variance and the subscripts a, p, and m refer to the analysis, pretreatment, and measurement steps, respectively. The variance for the analysis step is obtained by eliminating the effect of $s_s{}^2$ and computing $s_a{}^2$ from the data for several similar aliquots of the same homogenized laboratory sample as defined in Fig. 100.3. Again, replication can serve to reduce the magnitude of the pretreatment contribution through the application of equation (10).

C. SAMPLING

1. Sample Heterogeneity

One of the most serious sources of error in analysis is improper or inadequate sampling (2). The situation is particularly acute in trace characterization of materials, where the trace elements at the parts per million and parts per billion levels are almost invariably distributed in solid matrices in a heterogeneous fashion. When a heterogeneous population is to be analyzed, proper selection of the laboratory sample is critical to the validity of the result; otherwise, significant inaccuracies may be introduced into the analytical process before the analyst receives the sample. Thus the indication of doubt with regard to the selection of the laboratory sample was included in Fig. 100.3. If heterogeneity is recognized in the laboratory sample, it is certain that the population is also nonuniform, and appropriate sampling techniques should be applied to ensure obtaining an adequate representation of the population. It is not the analyst's fault that the material is heterogeneous but, rather, his responsibility to point out this fact.

Another important factor that influences the sampling is the type of information required for the characterization, that is, bulk, surface, or local concentrations of the impurities. In the past emphasis was almost exclusively on the determination of bulk or average composition—information which was necessary in improving purification methods. Currently there is considerable interest in determining the local concentrations in order to establish the topographic distribution of impurities and their relation to the various physical, biological, and chemical properties being investigated (12). It is only necessary to note the growing interest in probe techniques to verify this statement. The sampling requirements and the data interpretation depend on the type of information required.

As indicated above, the combined variance terms, $s_p^2 + s_m^2$, are indicative of the analysis error and thus determine the reliability of the result for a local analysis of a sample. The deviation between local analysis results determines the sampling variance, which is the measure of the homogeneity of the material. There can be no sampling error for a homogeneous material, and the local results should agree within the limits imposed by the analysis error.

A *homogeneous material* is defined herein as one for which the unit determining heterogeneity (segregation) is considerably smaller than the size of the analysis sample. For instance, the grain and the particle may be units which define segregation between the trace element and the matrix in solid and particulate materials, respectively, whereas the atom, ion, or molecule is the defining unit in solutions. When the probability of including the same number of trace element units in a particular sample size closely approximates unity, the material appears homogeneous. For samples having small units of definition, such as solutions, is is relatively easy to obtain the required probability for nearly any sample size. As the defining unit size increases, however, the sample size necessary to maintain unit probability also increases. As a result the degree of heterogeneity indicated becomes greater as sample size decreases. Skogerboe and Morrison (22) have presented data which support this expectation and which indicate an inverse relationship between the sampling variance and the sample weight, as would also be expected. Hence, when a measure of homogeneity is given, the sample size used to determine this parameter must be stated in the specification.

2. Sampling Errors

When a bulk analysis is required, the reliability of the result is determined by s_t, of which the sampling error may be the most significant part. Three common approaches used to minimize sampling errors are (1) selection of a large sample to ensure adequate representation of the population; (2) induction of uniformity by grinding, mixing, dissolution, etc.; and (3) increasing the number of samples analyzed to obtain the reduction predicted by equation (10). In any case, an evaluation of the sampling error is a prerequisite for achievement of the purpose.

Highly sensitive techniques of analysis often require a small sample and, as a result, suffer in their ability to estimate average population concentrations if heterogeneity prevails. This situation is further complicated by the fact that many techniques of high sensitivity are direct methods of analysis which are not readily amenable to prehomogeneizing techniques. In spite of this limitation, however, most of these techniques are capable of producing more reliable local analyses and consequently can be used with advantage to provide this type of information and to indicate the existence of concentration variations at the micro and submicro levels.

In view of the many factors influencing sampling errors, the difficulties of interpretation associated with this phase of analysis are understandable. To elucidate the capabilities of the different physical methods, Table 100.II presents generalized data which are pertinent to the discussion on sampling. The data may be regarded as typical since they represent the most common practice, usage, and results as determined from the general literature.

From Table 100.II it can be seen that the combined pretreatment and measurement errors for all methods listed are in the same range of 1–5%, depending on the care taken in the analysis. Higher values are possible, of course, if involved pretreatment or separation procedures are required. All of these techniques involve either electronic or photographic detection in the measurement step, and relative standard deviations of 1–5% are reasonable. Thus in spectrography the measurement errors are determined by the ability to reproduce intensity ratios between two lines of the same element.

With regard to the total error involved in the various methods, it is readily apparent that techniques utilizing solutions consistently result in an overall error which is less than that observed for direct techniques. This is expected when it is realized that solution techniques normally utilize relatively large samples, with a resultant reduction in the observed

TABLE 100.II

Distribution of Errors of Trace Methods

Technique	Typical sample size, mg	Most common mode of sample pretreatment	Typical relative standard deviation, %	
			Pretreatment and measurement	Total
Spectrophotometry				
Absorption	100–1000	Homogenization by dissolution	1–5	1–5
Fluorescence	100–1000	Homogenization by dissolution	1–5	1–5
Flame	100–1000	Homogenization by dissolution	1–5	1–5
Atomic absorption	100–1000	Homogenization by dissolution	1–5	1–5
Neutron activation	10–1000	Direct or homogenization by dissolution	1–5	1–15
Emission spectrography	1–50	Direct or homogenization by dissolution	1–5	5–30
Solids mass spectroscopy	0.1–10	Direct	1–5	5–30
Microprobe	0.01–1	Direct	1–5	5–30
X-ray fluorescence	10–1000	Direct or homogenization by dissolution or fusion	1–5	5–30

5855

sampling error. Unfortunately it is also common practice to make several measurements on the same homogeneous solution and to use these data to obtain a precision estimate. This method fails to include the pretreatment and sampling contributions and hence tends to place the technique in a more favorable, but false, position when compared with other methods.

A similarly misleading practice is encountered in publications involving the nondestructive techniques, that is, neutron activation and X-ray fluorescence. Either the square root of the observed count or the reproducibility of subsequent analyses on the same samples is frequently taken as representative of the total error. Again, these estimates are indicative only of the measurement error.

From the foregoing discussion, it should be apparent that comparison of the indicated reliabilities of the analytical results obtained by independent methods can be invalidated by differences in the capabilities of the methods with regard to the interference of heterogeneity and/or the misinterpretation of the total error involved.

III. TOTAL ANALYTICAL METHOD

A. ACCURACY AND PRECISION

When two or more methods of adequate sensitivity for the analysis in question are available, consideration of the overall accuracies of the methods with regard to the analytical purpose is appropriate. Several concepts relating to this in the individual analytical steps have been presented, but a unified treatment is in order.

As noted, precision implies accuracy in the absence of systematic errors. The precision of a bulk analysis is determined by the contributions from all phases as defined by equation (8), whereas the reproducibility of a local analysis depends on the distribution of random errors between the last two analytical steps as specified by equation (14). Since precision varies greatly, depending on how it is determined, the assumption of accuracy–precision equality is acceptable when the precision is obtained through the use of many samples rather than repeated analyses of one or a few samples. This qualification compensates for differences in sample homogeneity, long-term variations in the method, environmental effects, and similar factors. Certainly it is pointless to attempt to maintain a precision (and presumably an accuracy) of $\pm 1\%$ when the sample-to-sample variability is actually $\pm 5\%$.

A useful measure of precision for the satisfaction of this criterion can be obtained by combining the data for a number of sets of analyses to calculate a pooled estimate of the standard deviation (30). This may be done, for example, by treating the data for each concentration or each sample as a separate set and computing the sum of the squared deviations, S_i, from the average for each set. The pooled estimate is then given by

$$s = \sqrt{\frac{\sum_i S_i}{n - k}} \tag{15}$$

where n is the total number of measurements included in the k sets used. This value could be representative of the entire sample concentration range or of different concentration classes chosen so the concentrations included in each grouping are nearly equivalent. Standard deviation values obtained by grouping in this manner can subsequently be used to deduce confidence limits for the respective concentration levels.

Equating precision and accuracy forces a burden of proof on the analyst—the problem of proving that systematic errors are truly absent. Because instrumental trace techniques are calibrated with either synthetic standards or standards analyzed by independent methods, evaluations of systematic errors or corrections for them can be, but frequently are not, obtained. Obviously the concentrations of the standards must be reliably known; and, when chemical interferences are operative, the preparative histories of both samples and standards must be similar or both histories must be reduced to a common state. It is a truism that the accuracy of an analysis depends ultimately on the accuracy of the standards.

The preparation of highly reliable standards, particularly solid-state standards, at the trace and ultratrace levels is a singularly difficult task. The lack of trace standards is recognized as a prime limiting factor in the field and is receiving considerable attention from interested parties and agencies. Calibration against independent methods is often fallacious because the independent methods are not calibrated any better. Participation in well-planned and well-regulated interlaboratory comparisons can be fruitful in establishing accuracies of methods and in providing good standards (29,30).

A survey of the general literature indicates a paucity of uniformity with regard to statements concerning accuracy and precision. A statement such as "The relative standard deviation was found to be ±5%" is meaningless unless the method of determining this value is completely elaborated. The use of a uniform and definitive method of obtaining and stating measures of precision, taking the contribution from all three phases of the analysis into account, will be beneficial both in the reduction of confusion and in current or eventual evaluations of accuracy.

When an adequate evaluation of the random errors is available, some objective statement about the degree of reliability to be associated with the analysis is often required. The confidence level to be used may be determined by the purpose of the analysis, or its selection may be optional. In choosing the confidence level the analyst must decide whether the statement should be conservative with little risk of error or should be more liberal and consequently subject to a higher probability of error. Several factors may influence this choice; at any rate, it is more important that the level chosen and the method of determination be specified.

B. SCOPE AND SELECTIVITY

Although the sensitivity, accuracy, and precision are prime factors considered in the selection or comparison of trace analytical methods, other characteristics, such as scope, selectivity, cost, and time, must also be considered.

There are several facets to the concept of analytical scope, and most of them have a direct or indirect influence on the other characteristics. A method of high scope allows the simultaneous determination of a large number of species; is both sensitive and accurate; is free of interference (selective); requires a minimum of pretreatment operations (direct analysis); utilizes standards which are easy to obtain or prepare; can handle samples of varying shape, form, and composition; provides results which are simple to obtain and interpret; and minimizes the cost–time function.

Several examples can be given to demonstrate the inseparability of these characteristics. For instance, the greater scope of a multielement technique may offer large savings in terms of time and cost but may render the technique less selective than a single-element one. A nonselective method subject to chemical and/or physical interference problems usually necessitates some degree of pretreatment with attendant possibilities of loss and contamination, and this may affect the time–cost gains which accrue from the multielement aspect. A direct method of analysis implies high chemical and physical selectivity but also generally imposes more stringent requirements on the choice of standards should interference exist. On occasion, sensitivity and/or accuracy may be sacrificed in favor of a gain in another characteristic that falls within the aspect of the term scope.

In view of the expense incurred in the purchase and maintenance of modern trace analytical instrumentation, the ability to determine a wide variety of elements in a range of materials is a desirable feature. No one technique will solve all trace problems, but a method offering considerable versatility is generally preferred. In essence, the choice of a trace technique involves a series of compromises which are governed by the requirements of the analysis, the ultimate use of the results, and the various practical considerations mentioned.

Finally, some brief comments with regard to apparent trends in the field of trace analysis are in order. There is a continuing interest in improving the detection capabilities of methods in general, which is a direct result of improvements in the preparation of pure materials and the increasing realization that many elements at the trace and ultratrace levels exert prominent effects on chemical, physical, and biological properties.

Integration techniques designed to improve measurement precision and increase the ability to discern smaller differences between analysis and extraneous signals are receiving attention in a number of instrumental areas. For the same purpose, greater emphasis is being placed on the selection of optimum analytical conditions and on developing better methods of controlling the parameters of the system responsible for the production of the analysis signal. Time-resolved spectroscopy, servo-controlled excitation systems, ultrastable light sources and electronics, phase-sensitive amplifiers, and fast-response electronic integrators are a few examples of developments that can be included in these categories.

An increasing interest in the topographic distribution of trace impurities in host materials is stimulating the development of highly sensitive micro- and milli-sampling techniques which permit analyses *in situ*. The electron microprobe, generally capable of a spatial resolution of 1 μ^2, is proving particularly advantageous for studies of this type. The use of this technique, together with selective etching methods, often makes possible the location and identification of segregates or inclusions and in some cases indicates the chemical form of the species involved. The use of lasers to vaporize, excite, and/or ionize micro samples *in situ* also shows high promise, at least for semiquantitative analyses. Emission spectroscopy and spark-source mass spectrographic techniques are also capable of providing distribution data at a spatial resolution of roughly 1 mm but with greater sensitivity than the electron microprobe. All of these probe techniques have developed rapidly in the past few years. As instrumentation improves further and the demand for this type of analysis increases, the capabilities of the techniques will progress to an even higher status of refinement.

One emerging trend in trace analysis relates to a broadening of the scope of the techniques employed. The increasing complexity of instrumentation required for the types of trace analysis information currently requested necessitates a practical knowledge of chemistry, physics, optics, electronics, mechanics, instrumental design, and statistics, among other subjects.

It is obvious that no one technique can solve the many different types of analytical problems encountered today. Thus it becomes essential that the trace analyst have available a wide variety of the techniques mentioned in this chapter. More detailed information concerning specific techniques and their applications is given in the appropriate chapters in this treatise.

REFERENCES

1. Boltz, D. F., "Spectrophotometric Methods of Analysis," in L. Meites, ed., *Handbook of Analytical Chemistry*, McGraw-Hill, New York, 1963.
2. Calder, A. B., *Anal. Chem.*, **36**, No. 9, 25A (1964).
3. Campbell, W. J., E. F. Spano, and T. E. Green, *Norelco Reptr.*, **XIII, 77** (1966).
4. Campbell, W. J., National Bureau of Standards Symposium on Trace Characterization, October 1966.
5. DeKalb, E. D., R. N. Kniseley, and V. A. Fassel, *Ann. N.Y. Acad. Sci.*, **137**, 235 (1966).
6. Fassel, V. A., and D. W. Golightly, National Bureau of Standards Symposium on Trace Characterization, October 1966.
7. Gleit, C. E., *Am. J. Med. Electron.*, **2**, 112 (1963).
8. Gorsuch, T. T., *Analyst*, **84**, 135 (1959).
9. Guinn, V. P., and H. R. Lukens, Jr., "Nuclear Methods," in G. H. Morrison, ed., *Trace Analysis: Physical Methods*, Interscience, New York, 1965.
10. Hubbard, G. L., and T. E. Green, *Norelco Reptr.*, **XIII, 73** (1966).
11. Kaiser, H., and H. Specker, *Z. Anal. Chem.*, **149**, 46 (1955).
12. Kane, P. F., *Anal. Chem.*, **38**, No. 3, 27A (1966).
13. Mandel, J., and R. D. Stiehler, *J. Res. Natl. Bur. Std.*, **A53**, 1955 (1954).
14. Menis, O., ed., *Natl. Bur. Std. Tech. Note* No. 275 (1965).
15. Mizuike, A., "Separations and Preconcentrations," in G. H. Morrison, ed., *Trace Analysis: Physical Methods*, Interscience, New York, 1965.
16. Morrison, G. H., *Trace Analysis: Physical Methods*, Interscience, New York, 1965.
17. Morrison, G. H., and H. Freiser, *Solvent Extraction in Analytical Chemistry*, Wiley, New York, 1957.
18. Morrison, G. H., and R. K. Skogerboe, "General Aspects of Trace Analysis," in G. H. Morrison, ed., *Trace Analysis: Physical Methods*, Interscience, New York, 1965.
19. Pinta, M., *Recherche et Dosage des Elements Traces*, Dunod, Paris, 1962.
20. Sandell, E. B., *Colorimetric Determination of Trace Metals*, Vol. 3: *Chemical Analysis*, Interscience, New York, 1959.

21. Skogerboe, R. K., Ann T. Heybey, and G. H. Morrison, *Anal. Chem.*, **38**, 1821 (1966).
22. Skogerboe, R. K., and G. H. Morrison, National Bureau of Standards Symposium on Trace Characterization, October 1966.
23. Slavin, W., *Appl. Spectry.*, **20**, 281 (1966).
24. Slavin, W., S. Sprague, and D. C. Manning, *Atomic Absorption Newsletter,* No. 18, Perkin-Elmer Corp., Norwalk, Conn., February 1964.
25. Stary, J., *Solvent Extraction of Metal Chelates,* Pergamon, New York, 1965.
26. Taylor, J. K., E. S. Maienthal, and G. Marinenko, "Electrochemical Methods," in G. H. Morrison, ed., *Trace Analysis: Physical Methods,* Interscience, New York, 1965.
27. *U.S. At. Energy Comm. Repts., Nucl. Sci. Ser.,* NAS-NS, U.S. Department of Commerce, Washington, D.C.
28. Weissler, A., and C. E. White, "Fluorescence Analysis," in L. Meites, ed., *Handbook of Analytical Chemistry,* McGraw-Hill, New York, 1963.
29. Youden, W. J., *Statistical Methods for Chemists,* Wiley, New York, 1951.
30. Youden, W. J., "Statistics in Chemical Analysis," in D. Meites, ed., *Handbook of Analytical Chemistry,* McGraw-Hill, New York, 1963.

DETERMINATION OF PURITY

By F. H. Stross and J. H. Badley, *Shell Development Company, Emeryville, California*

Contents

I. INTRODUCTION

A. SCOPE AND PURPOSE OF THE CHAPTER

The treatment of the subject of purity can be considered from three aspects: the defining, the teleological, and the procedural. These points of view are reflected, in a general way, in the subdivision of our topic as it appears in this chapter.

Many attempts have been made to define the concept of purity rigorously. Eyring, Wichers, and others (39,171) have discussed this problem and have concluded that a practical as well as rigorous definition probably cannot be written. At the heart of the difficulty is the fact that molecules which are otherwise identical can exist in different energy states in which their physical properties and chemical reactivity are different. Furthermore, many substances exist which, from a rigorous point of view, are mixtures, for example, naturally occurring elements are usually mixtures of isotopes, but from a practical point of view are considered pure substances because the components are not perceptibly different when used in ordinary processes. A concept of impurity that has become important in the field of semiconductors is that of features foreign to an idealized perfect crystal lattice. These can include imperfections intrinsic to the material, that is, imperfections or dislocations in the crystal.

It is useful, nevertheless, to think of a pure substance as one made up entirely of molecules of the same atomic composition and spatial configuration, or of one kind of atom. We are concerned in this chapter with methods for testing materials which approximate this condition, particularly those in which the major component comprises 95% or more of the substance. By "determination of purity" we mean the estimation of the fraction of the sample made up of a major component. Methods for this determination which do not require known synthetic mixtures for calibration are called absolute methods, with the understanding, of course, that certain more or less restrictive conditions must be met in these tests if they are to give accurate results. We are also concerned with the measurement of physical properties of substances when these are used to judge the purity of the sample by comparison with the properties of an authentically pure specimen. Under certain conditions, this can be an absolute determination of purity, but usually it is employed in a qualitative manner only. Allied to the determination of purity is the analysis of nearly pure substances for minor components. When all of the minor components are known, the purity can be calculated by difference. In many practical situations, however, the minor components are not all of equal interest, and the determination of purity is replaced by methods for estimating specific impurities. These methods are often much more sensitive and frequently less laborious than purity determination.

The second aspect involves the purpose inherent in the determination. Information on the application intended for the substance under study very often will tell us, or will imply, the probable range of concentrations and the nature of the impurities of interest. Thus chemicals that are sold as solvents, for example, for paints, may be adequately defined by specifying purities of 95–99%. Chemicals used in the manufacture of nylon must contain not more than parts per million of some impurities (such as iron) but have much higher tolerances for others (such as water). Consideration of this aspect may actually invert specifications for certain substances when intended for different applications. For instance, impurity specifications for conductivity water may have to be set lower than those for drinking water by factors of the order of a thousand; yet the latter would not be "pure" if it contained as little as 10 parts per trillion of *Bacillus coli*, whereas conductivity water could tolerate much larger concentrations of this impurity. This aspect, then, is of importance in setting specifications, which in turn largely govern the choice of the methods to be used for determining the purity.

A description of the operations performed, during preparation and purification, on the substance under test normally gives a clue as to the nature of the impurities to be expected. In the manufacture of absolute alcohol, the water is commonly removed by use of the azeotrope with benzene. Hence we expect small quantities of benzene in absolute alcohol obtained by this process, and we will certainly include an analysis for benzene in our purity determinations on this product. The manufacture of sugar yields a product likely to contain other carbohydrates as impurities, but hardly aromatic hydrocarbons or fatty acids. Thus knowledge of the method of preparation of the test material contributes to the selection of the proper method for determining its purity.

The intent of this survey is to aid in dealing with several types of problems: (*1*) to list current sources of information on standards of purity and the methods available for its determination for specific materials, (*2*) to provide a basis for selection among alternative test methods, (*3*) to provide a general background for the development or applicability of test methods for new materials, and (*4*) to help in setting up realistic criteria for the purity of compounds.

In any case, we must not only know the nature and chemistry of the analysand, but also develop a familiarity with the techniques available for making the required determinations. The latter task should be made easier by means of this chapter. The following paragraphs and the Guide in Section I.B contain suggestions on how to use the chapter to the best advantage.

1. To deal with the first problem, the compilations of standards and methods for purity determination most widely used for the purpose and issued at various intervals are the following: the American Chemical Society monograph series *Reagent Chemicals* (111), the *U.S. Pharmacopeia* (153), and the *National Formulary* (152). In addition, pertinent methods and specifications for a number of compounds can be found in the standards published annually by the American Society for Testing and Materials (3). Some of the methods will be listed individually under the appropriate techniques in the following sections.

2. Selection between alternative methods normally is made by comparing them with respect to their important features as outlined in the Guide, Section I.B. This Guide should aid in finding which methods are applicable in a specific case and the descriptions of the methods themselves will offer more detailed information to help in making such comparison possible.

3. Finding methods suitable for use in the case of new products will require a general knowledge not only of the methods available, but also, as far as possible, of the chemical behavior of the product, including the stability of the major component, as well as of the nature, concentration, and chemical behavior of the expected minor components.

4. Criteria for purity normally are established in terms of the results of specific test methods. This is often done by governmental agencies or standardizing bodies such as the American Society for Testing and Materials; sometimes specifications are agreed upon by direct negotiation between producer and consumer.

The sensitivity of available methods and the purity required in the end use of the test substance are important considerations in developing workable purity criteria. As mentioned previously, a knowledge of the impurities arising from the manufacturing process is helpful in this development.

B. GUIDE TO THE USE OF THE CHAPTER

When several methods are applicable to particular problems, the preferred method is usually the one which gives adequate precision and accuracy with a minimum investment in equipment and operator time. These factors are discussed later for many of the individual methods. The paragraphs below indicate the broad categories into which purity methods fall and outline the kinds of materials to which they apply.

Separation techniques represent a direct approach to the evaluation of impurities and give results in terms of concentration (molar, volume, weight, or other) that depend on the response of the detector used. Their effectiveness is a function of the efficiency and selectivity of the separation as well as of the sensitivity of the detector. Among these techniques, gas chromatography has acquired enormous popularity because of its convenience, sensitivity, and wide range of applicability. Any compound that can be volatilized and separated without decomposition under the conditions of analysis can, in principle, be analyzed by this technique. Gas chromatography equipment is available that operates up to 500°C; even petroleum residues such as asphaltic components and waxes have been analyzed by means of such instruments. Gases and liquids are analyzed by standard procedures, and many solids can be handled without much difficulty. The major uncertainty in using gas chromatography for the determination of purity, as defined above, involves the adequacy of the separation; an impurity peak can easily be masked by the major component peak if the retention times are close to each other.

Substances that are not sufficiently volatile, or that decompose under the conditions of experiment, often can be analyzed by liquid chromatography; here the sample is injected as a liquid or in solution. This technique is used in research involving drugs or insecticides, in biochemical work, and in processes falling largely under the heading of the preceding chapter.

Physical properties are often measured for comparison with measurements on a sample set up as a standard. Refractive index and density are most commonly used in this connection. Unless the nature and effect of the impurity are precisely known, these determinations are qualitative and are useful for general characterization only. Nevertheless, they are widely employed for control purposes to detect departures of a product from the norm, and to confirm purity by compliance with suitable standards rather than to measure impurity. Physical properties are also widely used in establishing specification limits in the production and sale of organic and inorganic compounds.

It is difficult to obtain an absolute estimate of the total impurities without recourse to standard substances for comparison. In the past, wide use has been made, for this purpose, of a thermal analysis method, which, within certain limitations discussed in Section III.G, gives such an estimation in terms of mole fraction or mole per cent impurity. This method, which is based on the heat of fusion and the melting point, is well developed for organic compounds; its use for inorganic compounds is limited by temperature considerations, alloying problems involving the container, and other difficulties. Generally speaking, solids melting below 250°C that are better than 95% pure and that do not form solid solutions with their contaminants can be analyzed by this method. The higher the purity of the compound, the greater is the precision of the results obtained by this method.

Chemical and electrochemical methods retain application both in assay (estimation of the major components) and in the determination of impurities, for both inorganic and organic reagents. With the increase in knowledge in the field of solid-state physics, and the growth of radiochemistry, the importance of extremely small amounts of impurity in some substances has become apparent. Impurity concentrations of the order of 1 part in 100,000,000 have significant effects on the properties of materials used in transistors and on the properties of alloys of iron, copper, and aluminum. For the analysis of such materials, polarography, colorimetry, X-ray fluorescence, mass and emission spectroscopy, and activation analysis are widely used.

The discussion of the individual methods should help in relating their application to the different groups of chemicals. Some methods estimate individual elements only, in which case one must know how they are chemically combined before the results of such an analysis can be used for a purity estimate. Some methods determine functional groups; others, the individual compounds. For methods that are very specific and are based on comparison standards, there will not be much doubt as to their scopes. When the methods are very general, such as separation methods, they may be applicable to elements and to both organic and inorganic compounds.

II. SEPARATION TECHNIQUES

A general discussion of separation techniques is given in Section C, "Separation: Principles and Techniques," Part I, Vols. 2 and 3, of this Treatise. The present discussion will deal only with what is considered necessary in the context of this chapter.

The analytical separations that we seek are based on differences in volatility, solubility, or sorbability of the components, and usually employ some kind of multistage treatment to split the sample into a number of fractions. Either these fractions can be retained and estimated by weighing or by volumetric methods, or they may pass through a detector measuring changes in concentration and be discarded without further analysis. Although it is often possible to use either of the two approaches, the latter offers savings in time and sample requirements, and thus has come to be very widely applied in the analytical field. The two approaches are illustrated by preparative chromatographic practice on the one hand, where individual fractions are collected separately for further use, and analytical chromatography on the other, where the main object is to obtain a chromatogram, the corresponding fractions being represented only graphically. In solid–liquid separations, the analogous procedures are represented by fractional crystallization on one hand, and a technique such as thermal analysis on the other. Since our objectives here are predominantly analytical, the far less cumbersome and more widely used procedures of the second type will be given more emphasis in our treatment.

A limitation common to the use of separation techniques for the determination of purity is the tendency of binary or multiple combinations of substances to remain together even though a cursory knowledge of their individual physical properties would give reason for expecting separation. In distillation, such combinations are called azeotropes; the components boil together unchanged, even though they may have different boiling points when heated separately. In crystallization, corresponding behavior is shown by substances forming solid solutions. In chromatographic processes, analogous phenomena involving interaction on adsorbing surfaces and in the liquid phases of gas-liquid chromatography have been designated "asorbotropes" (64) to designate combinations of compounds not separating under conditions that normally would be expected to give resolution of the mixture.

To avoid difficulties from this source in testing unfamiliar materials, it is good practice to apply two or more separation techniques. If possible they should employ different separating principles; at the least, they should operate under a set of conditions different enough to influence the relative order of separation of the components.

Another observation may be of use at this point. Not uncommonly, the sensitivity of a method is confused with its sample requirement. Thus, for instance, it is known that paper chromatography can be carried out on very small samples. Often, however, small samples are submitted for analysis by this method, in which trace components ranging in parts-per-million concentrations are to be determined. The sensitivity of this technique is only about 0.1–1%, and trace determination is not possible directly. To obtain the desired sensitivities, preconcentration will be necessary. Sometimes this can be done by scaled-up versions of the particular separation technique, but often other separation processes will be more advantageous, especially in the light of the preceding paragraph. In that case, a sample of appropriately larger size must be available. Techniques capable of determining components in the parts-per-million range directly (e.g., neutron activation) are discussed elsewhere in this chapter.

A. CHROMATOGRAPHY

Chromatography is a physical method of separation, in which components to be separated are distributed between two phases, one of the phases constituting a stationary bed of large surface area, the other being a fluid that percolates through the stationary bed (61). This definition, which is essentially a pragmatic one, is applicable to the different techniques to be discussed in the following sections, where the stationary phase can be solid or liquid, and the mobile phase can be liquid or gaseous. These techniques are distinguished from other separation processes by their utilization of a stationary phase as a separating medium; the countercurrent principle used in distillation, for instance, is not involved. Partition takes place many times as the moving phase carries the sample through the column.* If the sample components differ in their partition behavior and the column is efficient enough, they will

---

* The term column is used here generically to include paper strips and sheets.

be completely separated. They can be detected by observing their concentration in the eluent stream by means of a suitable detector or indicator, or they can be collected and measured cumulatively by an appropriate device.

Chromatographic operations can be carried out in different ways, of which elution development is the most useful for the purpose at hand. In elution development, a sample of analysand is introduced at the inlet of the column. The sample may be gaseous or liquid, but must be relatively small in order to achieve efficient operation. It is swept over or through a stationary phase by an inert, flowing phase called the carrier. The stationary phase may be liquid, in which case it is usually supported as a thin film on the surface of a suitable solid, or it may be a solid. The partition depends on the preferential solution or sorption of the components of the sample by the stationary phase.

In gas chromatography, the moving phase is a gas, and the stationary phase is a liquid (gas–liquid chromatography, GLC) or a solid (gas–solid chromatography, GSC). In liquid chromatography, the moving phase is a liquid, and the stationary phase is a liquid (liquid–liquid chromatography, LLC) or a solid (liquid–solid chromatography, LSC). If the stationary phase is liquid (GLC or LLC), partition takes place by difference in solution properties; if it is solid (GSC or LSC), by sorption. It is possible, however, to combine these effects, for example, by distributing a relatively small amount of liquid phase on a relatively active solid (liquid–modified solid adsorbent).

Chromatographic methods are highly suitable for component analysis. However, as our interest is restricted to the determination of purity, our discussion will concern itself mainly with their application to this objective. For a more general discussion, a wide variety of books and review articles is available (12,49,60,80,110,116,120,138).

Chromatography is a very effective and relatively economical technique for carrying out separations. When it is used in the determination of purity, we generally have a very large indication of component concentration (such as peak or spot) and one or more relatively small ones. To make sure that a peak of interest is not hidden under a large peak, especially the peak representing the major component, it is highly desirable to repeat a given determination under different conditions, especially by using a stationary phase of very different polarity. One should also try to make sure that the column is not overloaded, and that the detector is sensitive to the minor components sought. The application of these general rules to specific examples requires further acquaintance with the technique and with the particular problem.

1. Gas Chromatography

Experiments in gas chromatography (GC) were made as early as 1930 (24); in 1941 Martin and Synge (87) suggested the principle of GLC. In the years that followed, a number of papers describing the analytical applications of GSC appeared and the technique of GLC was reduced to practice by Martin himself, in collaboration with James, in the years 1949–1952 (58). The years since then have seen a phenomenal rise in the application of gas chromatography, especially of GLC.

Any substance whose components have appreciable vapor pressure at the temperature of the chromatographic column and which are stable under the conditions of the experiment can be analyzed by GC. This in practice allows application to components with boiling points at least up to 400°C.

The sample is vaporized in a stream of carrier gas and swept through a packed column. The composition of the emerging gas, expressed as per cent carrier, is monitored continuously. If a suitable column is used, separate bell-shaped peaks are produced in the composition–gas volume curve for each component of the sample.

The quantitative information is obtained primarily from the peaks of the chromatogram. These peaks reflect the progress of the compounds through the detector; the peak areas, when obtained under proper operating conditions, are proportional to the amount of sample component, and by applying the appropriate response or calibration factor and dividing by the sum of the adjusted areas of all components, the concentration of the individual component is calculated, that is,

$$\% \text{ Component} = 100 \left(\frac{A_c F_c}{\Sigma A_i F_i} \right) \tag{1}$$

where A_c and A_i are the peak areas of the component in question and of any component, respectively, and F_c and F_i are the corresponding response factors.

At this point, it is important to consider the nature of the analysand, as it will affect our method of determining its purity. The sample to be analyzed usually falls into one of the following categories. (*1*) The sample and its impurities are well known, that is, their identities and their response factors are established. This is often the case with routine analyses of refinery streams for control purposes. Then a comparison of the areas of the peaks representing the impurities should be sufficient to give the additional data needed to use equation (1). (*2*) The identities of the major and the minor components are generally known, but the analysis is not an established one and the factors are not known. In this case, it is sometimes possible to assume or to calculate response factors, without having to determine them experimentally. (3) The major component is known, but the minor components are not. This is the most frequent case in research laboratories. It is then necessary to establish the identities of the components and then the response factors, and this problem can be approached in several ways.

One may try to deduce the identity from general chemical considerations, and to verify the conjecture by adding the suspected compound in pure form and looking for a corresponding increase in the size of the peak under scrutiny. Alternatively, one may calculate the specific retention volumes from the retention times observed and from the geometry of the system and look for compounds with the corresponding parameters in tables of absolute or relative retention volumes. Such tables are published in the literature from time to time, especially in the *Journal of Chromatography*. The second approach may be necessary when other approaches fail, but will usually be found to be quite laborious.

Direct identification can sometimes be made by collecting the separated substances after detection in suitable freeze-out traps and subjecting them to qualitative analyses. If additional equipment is available, the task of identification can be greatly facilitated. In some assemblies the gas chromatography is followed by a mass spectrometer, which thus acts as a qualitative detector. In some cases, infrared or ultraviolet spectrometers also are used for this purpose, functioning as detectors either directly or by analyzing the individual components after trapping as suggested, preferably by a suitable microtechnique.

Response factors must be found to relate the amounts of the compounds under study to the areas of the peaks representing them. There is some literature on the calculation of such factors for members of homologous series, for particular detectors. However, in view of the wide variety of detectors and other relevant conditions, it is usually far quicker to determine these factors experimentally in the manner indicated above. Known mixtures containing the compound under study in different concentrations are analyzed by GC, and the areas related to the concentration; attention should be paid particularly to departures from linearity in this relation. Such effects indicate either nonlinearity of the detector or interaction with other compounds present.

The evaluation of the peak areas is sometimes carried out by means of a planimeter or other manual method. The use of mechanical or electronic integrators, especially in conjunction with completely auto-matic systems, however, is becoming increasingly widespread.

Gas–liquid chromatography has been used for a wide variety of mix-tures and systems. Hydrocarbons ranging in volatility from methane to waxes; alkyl and aryl amines, ammonia, and pyridine homologs; fatty acids, sulfur compounds, boron and silicon hydrides, and halogen-ated hydrocarbons; hydrogen isotopes; and trace impurities in a variety of gases have all been studied. Other compounds and systems are being added to this list almost daily.

Gas–solid chromatography, the less flexible of the two techniques, is used chiefly when GLC is inapplicable for some reason. This is the case with the "fixed" gases, whose retention volumes in GLC are normally too low to yield accurate analytical data. Charcoal has been used for the separation of hydrogen, oxygen, nitrogen, and rare gases, as well as carbon monoxide, the nitrogen oxides, and methane; molecular sieve packing is said to give better resolution for the same type of separation.

A wide selection of instruments is available commercially; it is still possible, however, to assemble a serviceable apparatus from components that can be bought at low cost (137). The essential requirements for such an instrument are a source of carrier gas (usually helium in the United States), pressure- and flow-regulating devices, a sample injector, a column, and a detector. The degree of control of the operating vari-ables, such as temperature, and flow rate and pressure of the gas, that is necessary for a given problem will depend on the requirements of the particular analysis. If high temperatures, unusual sensitivity, high separation efficiency, or other extraordinary conditions are to be attained, the investment required will be correspondingly greater than if these conditions are simple.

Determinations by GC can sometimes be made in seconds, but may take hours, depending on the volatility of the compounds and the com-plexity of the mixture. Impurities can normally be determined in concen-trations of around 0.1% of the total directly, but sometimes one can detect impurities of the order of parts per million, as in the coulometric determination of chlorine-containing compounds in hydrocarbon gases.

2. Liquid Chromatography

Liquid chromatography is widely used in the determination of purity, especially when GC is inapplicable. This is the case if the sample is nonvolatile or if it is unstable under the conditions of GC analysis. In such cases LC can give excellent results. Determination by LC usually take hours to complete.

A number of different methods have been developed within the classification of LC. Although the various methods overlap in certain areas, they generally can be grouped as follows:

> Liquid–solid chromatography
> > Adsorption column
> > Ion exchange
>
> Liquid–liquid chromatography
> > Partition column
> > Paper
> > Thin-layer

Liquid chromatography shares with other chromatographic methods the ability under suitable conditions to find all impurities, thus making a determination of purity quantitatively possible. The conditions are as follows: the major component may not swamp out any of the minor components; the minor components may not interfere with one another; and the areas must be quantitative representations of the components.

a. LIQUID–SOLID CHROMATOGRAPHY

(1) Adsorption Column

Liquid–solid chromatography using adsorption columns was the form in which chromatography was first developed. One of the earliest practitioners, the Russian chemist Tswett, named the technique; he used it to separate colored compounds in plant extracts by elution.

In the simplest form of the technique, a solid of suitable particle size and porosity is placed in a column or bed, the sample is placed on top of the column, and a liquid eluent is introduced, which slowly desorbs the sample and transports the individual components down the column in inverse order of their energy of adsorption by the column solids. The eluent, in this method, is more weakly adsorbed than the sample components; this makes it possible to achieve a clean separation of the components from one another. If the eluent can be separated from the sample components easily, the latter can be obtained in the pure form, although this is not always necessary.

The adsorbent usually is a granular and porous solid of relatively large surface area, which adsorbs the sample components with different strengths. The strengths of various adsorbents can be compared by comparing the relative rate at which an adsorbate is eluted under some standardized conditions. Among the generally weak adsorbents are starch, sucrose, and talc; intermediate adsorbents, calcium carbonate, magnesia, calcium phosphate, and calcium oxide; strong adsorbents, alumina and silica gel. The strong absorbents are often intentionally made less active by allowing some water to be adsorbed.

The eluted sample components can be detected by various means. They can be collected, or merely detected, in the effluent (elution analysis) ; or the elution can be stopped before any of the sample has emerged from the column, and the components can be "developed" on the column (development analysis). The latter is the process used by Tswett. He extruded the column from its container, separated the sections from one another by cutting, and extracted them to recover the components. Although this process may not be important to the present objective in this form, the principle of development analysis is used in paper and thin-layer chromatography and will be referred to again under those headings.

In the usual elution analysis, the effluent is collected or detected, and the fractions are collected manually or automatically and analyzed by one of many well-known methods. If one wishes only to obtain the chromatogram, the change in concentration can be detected and recorded automatically in a manner similar to that used in the GC process. Among the most versatile detectors for the purpose are continuous refractometers, with sensitivities limited only by the limits imposed by the best temperature controllers available (about 10^{-7} r.i. units). In certain cases, ultraviolet detectors can be used. Where applicable, electrometric detection methods also are very useful.

The technique is discussed thoroughly by Mair (83). The selection of the various components and conditions of analysis and the application to specific problems are given there. Additional literature references will also be found in the valuable general reference works by Cassidy (21) and Lederer (72).

(2) Ion Exchange

Ionic impurities often can be determined effectively by ion-exchange chromatography (114). To illustrate the action of ion exchangers, we may consider a modern organic cation-exchange resin. Beads of styrene polymer, suitably cross-linked during polymerization and swollen by treating with an organic solvent, are sulfonated. The structure of the polymer leaves a high proportion of the sulfonic acid groups accessible. When the resin is brought into contact with water, some water penetrates into the resin, and the hydrogen atoms associated with the sulfonic acid ionize. The protons then can be replaced by equivalent quantities of other cations.

A number of cation and anion exchangers are available commercially. Widely used cation exchangers include Dowex, Amberlite, and Zeo-Karb products; anion exchangers include resins of the Duolite line.

Ion exchange is especially valuable in auxiliary operations such as removal of interfering components or preconcentration of trace constituents. For example, iron, aluminum, and calcium can be effectively removed by means of cation-exchange resin to prevent their interfering in the determination of phosphorus, and similar application of this operation might facilitate assay of certain other substances.

Ion exchange is useful for the concentration of very dilute constituents of a solution, when they are the impurities to be determined. In determining the impurities of fresh water, for instance, both cations and anions can be retained very effectively by a pair of ion-exchange columns. Then the actual determination can be made by replacing the components now held on the columns, and determining them by usual procedures. A striking example of this kind is the determination of cesium in water. A cation-exchange resin was used for concentrating the cesium ion; its concentration in the original sea water was found to be $4 \times 10^{-9} M$ (129).

Trace concentration by means of ion-exchange resins can also be carried out in nonaqueous media. By using 2-propanol as solvent and concentrating by means of a sulfonated polystyrene column, trace amounts of copper have been determined in lubricating oil (18).

b. Liquid–Liquid Chromatography

(1) Partition Column

In this method of separation, the material is distributed between a moving liquid phase and a liquid phase immobilized on a solid support. The stationary liquid phase is substantially immiscible with the moving liquid phase. Martin and Synge (87), the inventors of this technique, which engendered most of the other chromatographic techniques in wide use today, first used it to separate acetyl amino acids.

Although many modifications of this type of LLC are possible, the one most commonly used involves detection or collection of the effluent material. The solid support usually displays residual adsorption of undefined magnitude and therefore may take a part in the separation. The stationary liquid often is aqueous, but it is possible, by using appropriate solid supports or by treating them with suitable wetting agents, to immobilize nonpolar organic liquids on them.

It is not always easy to find a pair of nearly immiscible liquids which have comparable solvent power for the sample.

Liquid–liquid chromatography columns find more application for purification and preparative purposes than for determining purity. In comparison with adsorption columns, they offer the advantage that partitioning by differential solubility is often a milder process than partitioning by differential adsorption, and therefore there is less chance in LLC for undesirable side reactions to occur.

(2) Paper

Paper chromatography (PC) and, more recently, thin-layer chromatography (TLC) are outgrowths from Martin's work on partition chromatography. They are used very widely in the qualitative and quantitative determination of impurities. Detailed treatment is given by Pazdera and McMullen (101).

In PC, the solid support is the paper, which normally carries on it adsorbed water or an aqueous phase; partitioning is accomplished either by distribution of the sample between the moving liquid and the stationary liquid phases, or by the combined solvent–sorbent action of both the liquid and the paper (73).

The following features have helped PC achieve its outstanding popularity and usefulness. Sample size requirements are small, compared to column chromatography; interesting results can be obtained even with inexpensive equipment; many modifications are possible to fit particular problems; it is easy to multiply or diversify the partitioning action by permitting the progress of different solvents from different directions, or by imposing other driving forces, such as an electrical potential, on the system; often the separated components can be detected by development and inspection without expensive instrumentation.

Martin and his co-workers (28) published the first paper on the subject from which the following illustration of PC is taken. "To a strip of Whatman No. 1 filter paper a mixture of amino acids (glycine, alanine, valine, and leucine) is applied at a spot marked with pencil, near one end. The paper represents the column in ordinary chromatography; it is hung in a chamber, with the upper end held in a trough. At the bottom of the chamber there is a dish for holding the conditioning liquid."

Two partially or sparingly miscible liquids are mutually saturated; in the example at hand, they are phenol and water. The aqueous phase is placed in the conditioning dish so that it will saturate the air in the chamber, and the paper hanging in it. Sorbed on the paper, it provides the stationary phase, which is active in partitioning the sample. The phenol-rich phase is placed in the trough in which the upper end of the paper is held; after the chamber is resealed to permit establishment of equilibrium conditions, this "developer" moves into the paper by capillarity and gravity. The developer dissolves the sample and carries it over the cellulose saturated with the aqueous phase, causing separation just as in a chromatographic column. In the latter, the outer limit is set by the walls of the tube or column; in the case of the paper technique, it is set by the surface forces of the developer.

After the front of the developer has moved close to the lower edge of the paper, the latter is removed, the position of the front marked immediately, and the paper then allowed to dry. The position of the separated amino acids is brought out by spraying the paper with ninhydrin, and the distances from the center of the initial sample zone to the developer front, and to the center of density of each spot, are measured. The ratio of these distances, which is characteristic of the compound, is designated R_F; thus

$$R_F = \frac{\text{Distance from starting line to center of zone}}{\text{Distance from starting line to solvent front}}$$

Therefore substances that are effectively retarded will show smaller R_F values than those that move faster.

To obtain the desired separations, the following conditions should be approached. (1) The zones of the mixture should move at different and reproducible velocities in a straight line in the direction of motion of the developer. (2) The zones should retain a regular (circular or elliptical) form. (3) It must be possible to detect the zone in question. (4) It must be possible to detect the developer front, or at least the zone of a reference substance, so that one can determine the R_F or a relative R_F. (5) The developer should flow along the paper at a reasonable velocity.

The actual procedures used in this technique are manifold; the paper columns have taken the form of strips, sheets, cylinders, and circles, with corresponding change and complication of the flow of the developer. Several different types of sample development have been used. Reversed-phase application of this technique also can be made by filling, coating, or chemically modifying the paper fiber structure in some way with a substance so as to produce a lipophilic surface, and using an aqueous or at least highly polar mobile phase as the developer. Change in the direction of the gravitational field and application of electric fields and of centrifugal force have been used for improved separation and speed.

Quantitative evaluation of the chromatogram also has been carried out in a large number of ways. Most recently, spot area estimation, reflectance measurements, transmission photometry, photoelectric determination of fluorescent materials, autoradiography when radioactive materials are used, elution analysis of excised spots, and many other methods have been used to make the quantitative estimates required. The most complete periodical surveys and bibliographies of recent activity in this field, as in most others in analytical chemistry, will be found in the corresponding articles of the Annual Review (April issue) of *Analytical Chemistry.*

Paper chromatography is applicable to a very wide range of substances, inorganic and organic, and is particularly useful where sample size is very restricted, or where the sample does not crystallize easily, forms solid solutions, or is prone to decompose at melting or vaporizing temperatures. Biochemical materials, carbohydrates, steroids, and alkaloids are typically analyzed by PC, and this technique has played a vital part in the insight into the process of photosynthesis acquired in recent years; in the study of metabolic processes in animal life; in the control of enzymatic processes such as those encountered in making cheese and beer; in forensic chemistry; and in many other fields. Application of PC to purity problems has included its use in research and control of food products, pharmaceutical products, and vat dyes. Many other examples of applications in organic and inorganic chemistry can be found, especially in such specialized periodcials as the *Journal of Chromatography.*

A more detailed discussion can be found in the appropriate chapter of this Treatise (101) and in the many excellent books dealing with this topic (13).

(3) Thin-Layer

Thin-layer chromatography is often considered merely a modification of PC for improved convenience and speed. The advantage usually stressed, greatly increased rapidity, is indeed most valuable. Another very desirable feature of the technique, however, is that it can more cleanly provide separation by sorption (LSC) and by partition (LLC) by choice of the appropriate stationary phases. Because of the greater diversity and latitude in the structure of the solid support, greater resolution and better separations than are possible with PC have been achieved in many cases.

In TLC practice widely used today, reproducible, uniform layers of constant thickness of the sorbent, held together by a suitable bonding agent, are applied on an inert support such as glass or enamelware. The layers are typically 0.25 mm thick, but may range from 0.01 to 0.5 mm and more. Silica gel and diatomaceous earth, alumina, and magnesium silicate powders are much used as sorbents, with plaster of Paris as the bonding agent. Cellulose powder also is useful, the advantage over the corresponding PC support being provided by the difference in geometric structure and pretreatment. Starch is still frequently used as the bonding agent in applications where its organic nature does not introduce problems. A completely inorganic layer, such as silica gel bonded by plaster of Paris, allows separation by sorption and is much more resistant to acids and heating than the paper strips used in PC.

The solvent is selected to correspond to the requirements of the separation, as in the other techniques of chromatography. Some details are given by Stahl (134) on this subject.

The technique of TLC was introduced by Kirchner (63); it permits the separation of water- or organosoluble substances and the determination of the minor components with great rapidity and selectivity, and a sensitivity of detection that is often greater than in the case of PC. Equipment is commercially available, simple, and inexpensive, and typical analysis times range around a half-hour.

Detailed discussions are presented periodically in *Chromatographic Reviews* (34).

Application of TLC in general covers a range similar to that for PC and will be found discussed in the references cited. The following example illustrates such application within the context of our objective.

Tschesche and Sen Gupta (159) have used TLC in examining the purity of esters of two triterpenic acids. Tschesche and his colleagues (160) continued their studies and reached conclusions which included the following: TLC is a technique permitting the rapid and economical analysis of highly complex mixtures, as well as purity determinations; it is capable of very high sensitivity (limit of detection of oleanolic acid = 0.02 μg) and in some cases has made possible the resolution of mixtures (triterpenic acids) that could not be analyzed by PC.

B. LIQUID–LIQUID EXTRACTION

Liquid–liquid extraction (LLE) can parallel liquid–liquid chromatography in providing distribution of the analysand between two liquid phases; in the former, the most popular installations have employed discrete equilibrium stages, while in LLC the partition can proceed in a continuous manner through the length of the column. In preparative and plant-scale operations, however, truly continuous LLE is commonly practiced.

Liquid–liquid extraction has been used for assay and the determination of purity. Thorough discussions of this technique, including this application, are given by Craig and Craig (30) and by Irving and Williams (57). It is very useful for purification of nonvolatile or heat-sensitive substances such as polypeptides. However, as equipment for liquid chromatography is now commercially available, chromatographic techniques are effectively replacing the more complex, bulky, time-consuming, less efficient LLE operation for the determination of purity.

C. DISTILLATION

In ordinary batch distillation, the sample is evaporated slowly and the vapors are condensed, either with or without intermediate rectification. One or more properties of small samples of the distillate are measured as the evaporation proceeds. Usually the condensation temperature of the vapor is the property measured, although refractive index, density, dielectric constant, and many others have also been employed. The property is plotted against the fraction of the sample collected as distillate to give a distillation curve. If the property remains constant, the material is said to be pure with respect to distillation. In principle, one can determine the amount of impurity on an absolute basis by noting the fraction distilled between discontinuities or inflection points in the distillation curve. In practice, however, the inflections are not very sharp, and except in unusual cases impurities corresponding to less than a few per cent of the sample cannot be reliably estimated by this means.

Fractional distillation is sometimes used to concentrate impurities for determination by additional tests. This is particularly convenient when the impurities are volatile compared to the major component. Then enrichment factors of 10–100 to 1 can be achieved under favorable circumstances. When the impurities are less volatile, column holdup limits the enrichment factor obtainable in practical procedures to 5–10 or so.

The conditions under which separation can be achieved and the apparatus and procedure necessary to achieve it are well known. Carney (20) and Zuiderweg (175) give practical instructions for fractional distillation. A useful summary is given by Rose (115). Perhaps the most complete treatments of the theory and practice of laboratory distillation are given by Weissberger (169) and Krell (67). Sample sizes for fractional distillation typically vary from 10 to 1000 ml. Distillation times range from 2–3 hr for a low-efficiency separation to a week or more for the highest, with 8–12 hr perhaps the most practical compromise. Most of the applications of fractional distillation for purely analytical purposes have been superseded by more versatile, sensitive, and convenient gas chromatographic procedures. This is particularly true for the determination of purity.

Distillation is widely used in specification testing. While the purity, that is, the fraction of the major component, of the sample is not measured, it is inferred qualitatively from the temperature difference corresponding to two widely separated points on the distillation curve. The apparatus used in a simple batch still is of closely specified construction. The simplicity of the apparatus and procedure and a long-standing tradition combine to make this test a part of the specifications for practically all volatile liquids sold in commerce. The methods used in this country are generally those described by the American Society for Testing and Materials (4). Several methods, which differ in significant details, are available for materials boiling at atmospheric pressure. The method intended for petroleum product distillation uses a total immersion thermometer without stem corrections and can give temperatures as much as 5°C different from the true value. For nearly pure materials, the method intended for lacquer solvents and diluents should be used. Most specification distillations require 100 ml of sample and about 1 hr for a test.

D. ELECTROPHORESIS

Electrophoresis is a method of particular value in the isolation, purification, and analysis of substances of biochemical interest, such as proteins, enzymes, and many other macromolecular materials. Electrophoretic analysis has the advantage of being quite specific in regard to information on chemical individuality, since two substances rarely appear to have the same mobility throughout the available ranges of pH and ionic strength of the medium.

In electrophoretic methods, one may use boundary or zone separations. Boundary electrophoresis gives high resolution, but only of the boundaries of the components being separated, and no real isolation of these components can be achieved, except perhaps for the slowest and fastest substances. By the zone methods, a nearly or completely quantitative recovery is possible for all components. Zone electrophoresis requires a stationary support or bed in order to keep the zones separated against gravity, and thus acquires some of the features of a chromatographic technique. The stationary phase may be a powder—cellulose or glass—a gel, or a strip of filter paper (paper electrophoresis).

The introduction of zone methods has extended the applicability of electrophoretic analysis to substances of low and intermediate molecular weight. Amino acids, peptides, nucleotides, and even simple inorganic ions can be studied by such methods (158).

E. PHASE SOLUBILITY ANALYSIS

At a given temperature and pressure, the solubility of a pure substance is independent of the ratio of the quantities of solid and solution phases, provided that both are present. Conversely, the variation of solubility with phase ratio can be used to estimate the impurity content of impure substances, provided that the following conditions are met: (1) the impurities form distinct phases in the original sample; (2) the impurities and the major component are not present in the ratios of their solubilities; (3) the solubility of the major component is not affected by the nature or amount of the impurities; and (4) equilibrium is reached in the determination of solubility.

Mader (82) discusses the theory and practice of phase solubility analysis. Data are obtained by mixing sample and solvent in several different ratios, using a different container for each mixture; stirring at a carefully controlled temperature until equilibrium is reached; and measuring the total dissolved solids content of the solution. The variation of the dissolved solids concentration with sample-to-solvent ratio indicates the purity of the sample. The apparatus required is relatively simple. It consists of a number of stoppered bottles, a good thermostatted bath, a shaker, and a means for analysis of the solutions. The elapsed time to reach equilibrium can be rather long, in extreme cases a week or more, and the separation of the solution from the solids requires careful technique. Nevertheless, the actual work involved is not extensive.

The interpretation of the data is based on a plot of dissolved solids versus sample-to-solvent ratio. The simplest case is illustrated in Fig. 101.1. As the sample-to-solvent ratio increases, the dissolved solids concentration also increases regularly. At the solubility limit of the major component, and at higher ratios, the major component concentration in the solution remains constant. However, as long as the solubility of the impurities is not exceeded, the total dissolved solids content will continue to increase. The rate of increase after the major component limit is reached is proportional to the amount of impurity present in the original sample.

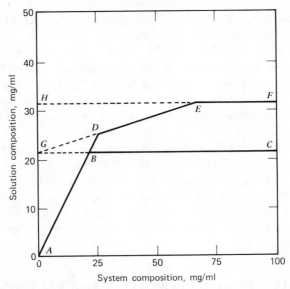

Fig. 101.1. Types of solubility curves (30°C): (*1*) *ABC*, pure DL-isoleucine in water; (*2*) *ADEF*, 85% DL-isoleucine and 15% DL (+)-glutamic acid in water. Reproduced with permission from reference 82.

Insofar as interferences are absent and the solubility can be measured with the required precision, this is an absolute method for the determination of purity in the same sense as are the colligative property methods discussed in Section III.G. In fact, the analogy between the two types of methods is quite close. In both, the solid–liquid phase ratio is varied while the composition of the liquid phase is measured. However, in phase solubility analysis, the quantity of interest usually appears as the difference between two relatively large numbers, namely, the total solids content of solutions prepared at two different sample-to-solvent ratios. Consequently, the sensitivity of the method is limited by the precision of the solubility determination. This is in the region of 0.1–0.5% by weight of the sample in favorable cases.

Although, in the simple case discussed above, the determination of purity by phase solubility analysis is straightforward, there may be further complications in practice. For example, if the solubility limit of all or part of the impurities is reached before that of the major component, the presence of some of the impurities may be concealed by the experimental errors of the solubility measurement, even though they would be detected if they were more soluble. This comes about because the range of sample-to-solvent ratio available to evaluate the effect of the impurity is short. Consequently, experimental errors have a large, and often fatal, effect on the property related to the impurity content, namely, the slope of the solubility versus sample-to-solvent ratio line. Another complication can arise from a change in solubility of the major component caused by the impurity. Such nonideal interactions are not uncommon; in fact, a good deal of effort has gone into discovering additives (impurities) which increse the solubility of pharmaceuticals.

Phase solubility analysis has been used mainly by biochemists who deal with materials of the order of 90–98% pure. Subsequent to Mader's thorough review, very few papers discussing this method have appeared in the literature. The reasons may be the relatively low sensitivity achieved by analyses of this type and the more detailed information provided by other techniques, such as liquid chromatography and electrochromatography (158).

III. PHYSICAL PROPERTY MEASUREMENTS

The use of physical properties in the determination of purity is very old. Archimedes measured the specific volume to test the purity of the gold in Hieron's crown over 2000 years ago. All chemists are familiar with extensive tabulations of easily measured properties which can be used to judge the purity of organic or inorganic compounds. However, the measurement of a physical property does not, in itself, give a definitive estimate of purity, even if a reference value for the property is available, because different impurities can have different effects on the property measured. For example, compared to water at 4°C, the specific gravity of water at 20°C is 0.9982 and mixtures of 1% ethanol and glycerol have specific gravities of 0.9964 and 1.0006, respectively. The effect of the glycerol is of different sign and is ⅓ that of the ethanol. Furthermore, if several impurities are present, and their composition can vary in an unknown way, a reliable estimate of purity cannot be obtained from the measurement of a single physical property. A mixture of 1% ethanol and ¾% glycerol in water probably could not be distinguished from pure water by the determination of specific gravity.

Although there are many instances of this sort in which the measurement of a particular property gives little or no information concerning the purity of a sample, there are three situations in which reliable purity estimates can be obtained. In all of them the effect of the impurity on the property must be known and different from zero. The situations are as follows: (1) the impurity is a single component; (2) all impurities have the same effect; and *(3)* the effect of impurities of the particular composition in the sample is known. When the effect of the impurity on the property can be determined *a priori*, that is, by some means other than calibration with known mixtures, the estimation of purity by measurement of the property becomes an absolute method.

The need for knowledge of the nature of the impurities means that purity determination by physical property measurement is most useful in testing a number of samples from the same source, or samples prepared in a way which leads to the same kind and composition of impurities in all of them. Then separate calibration experiments serve to establish an empirical relation between the measured property and sample purity. Because physical property measurements can be made quickly and often automatically, they are widely used in the manufacturing control of the purity of chemical products.

The most commonly used properties are density and refractive index. Other properties sometimes used are specific electrical conductivity, optical rotation, dielectric constant, and viscosity. The melting and boiling points, also used for this purpose, differ from the other properties because, under certain conditions, all impurities have the same molal effect on the measured property, which can be found without calibration with known mixtures. The melting and boiling points can thus be used in absolute purity methods. Cryometric and ebulliometric methods are discussed in more detail in Sections III.G.1 and III.G.2. Although spectrophotometric methods are, strictly speaking, based on physical property measurements, they are also treated separately, in Section VII.

A. DENSITY

Density is widely used for purity testing. The applications and techniques of density measurement are discussed in several reviews (9,112,150). Gases, liquids, and solids can be tested. The test is less ueful for solids because representative samples free of voids and inclusions are sometimes hard to obtain. The density of nearly pure gases and liquids varies almost linearly with impurity content. Consequently, when any of the three listed conditions is met, calculation of the purity from the density of the mixture is straightforward. For binary mixtures of organic compounds, an empirical determination of the effect of the impurity may not be required if its identity is known and its density can be found in the literature (55,68–70,156,168).

The sensitivity of the measurement of purity depends on the difference in density between the major component and the impurities, as well

TABLE 101.I
Range of Density, Refractive Index, and
Molar Refraction of Organic Liquids

Compound	Temp., °C	d^t, g/ml	n_D	$[R]$
Isopentane	20	.620	1.3537	25.28
Silicon tetramethyl	18.7	.648	1.3591	38.44
n-Heptane	20	.684	1.3876	34.54
Ether	24.8	.708	1.3497	22.51
Cyclopentane	20	.745	1.4065	23.15
Acetone	20	.791	1.3591	16.17
Propionaldehyde	20	.807	1.3636	16.02
n-Amyl alcohol	20	.817	1.4101	26.73
Benzene	20	.879	1.5011	26.18
Methyl propionate	18.9	.917	1.3770	22.10
Adiponitrile	20	.950	1.4597	31.16
Pyridine	20	.982	1.509	24.05
Diethyl adipate	20	1.009	1.4281	51.62
Aniline	20	1.022	1.5863	30.59
Fluorobenzene	20	1.024	1.4677	26.07
Thiophene	19.7	1.065	1.5287	24.35
Chlorobenzene	20	1.107	1.5251	31.16
Nitromethane	20	1.139	1.3935	12.80
o-Dimethyl phthalate	20.8	1.192	1.5155	49.16
n-Amyl bromide	20	1.218	1.4444	32.97
Glycerol	20	1.261	1.4729	20.48
Carbon disulfide	20	1.263	1.6276	21.38
Chloroform	19	1.490	1.4457	21.35
Bromobenzene	20	1.495	1.5604	33.98
Carbon tetrachloride	15	1.604	1.4630	26.42
Lead tetraethyl	20	1.653	1.5198	59.47
o-Dibromobenzene	17.4	1.964	1.6117	41.74
Ethylene bromide	20	2.182	1.5379	26.93
Bromoform	19	2.892	1.5980	29.82

as on the precision of the density determination. The range of densities encountered in organic liquids is shown in Table 101.I. In practice, a difference of more than 0.2 seems rather unlikely. The precision of the measurement of density can vary from 0.1 to 0.001%, depending on the apparatus and technique. Liquid densities repeatable to 2 or 3 in the fourth decimal place can be measured without undue difficulty, for example, with a Westphal balance. With a density difference between major component and impurity of 0.2, about 0.3–0.5% impurity can be detected reliably. Usually, however, the sensitivity is lower.

B. REFRACTIVE INDEX

The refractive index of organic liquids is widely used for purity testing because relatively inexpensive instruments are available for making quick, precise measurements. Furthermore, extensive tables of the refractive indices of various compounds are available (55,68–70,156,168). Representative values are shown in Table 101.I. The techniques and applications of refractometry are reviewed elsewhere (10,75,155). In purity testing, refractometry is most useful for liquids. Tests on gases require relatively complex apparatus, and tests on solids are difficult to interpret because most crystalline solids are optically anisotropic, that is, their refractive index varies with the direction of light passed through the crystal. By using the "ordinary" Abbe refractometer with suitable temperature control, refractive indices precise to 1 or 2 in the fourth decimal place are easily obtained in a few minutes. With a precision refractometer, refractive indices precise to 0.00002 can be obtained. When the ordinary instrument is used in a purity test, the maximum sensitivity is of the order of 0.2%. This is realized, however, only for rather special systems, such as o-dibromobenzene in n-heptane. Usually the sensitivity will be lower. The amount of sample required for a test using an Abbe instrument is a few tenths of a milliliter or less. As in the density measurement, knowledge of the identity of both components of a binary mixture often permits estimation of the approximate purity of the sample by use of published values for the properties of the pure components.

1. Molar Refraction

Much attention has been given in the literature to the molar refraction (10). One definition of this is:

$$[R] = \frac{n^2 - 1}{n^2 + 2} \cdot \frac{M}{d}$$

where $[R]$ = molar refraction, n = refractive index, d = density, and M = molecular weight.

The molar refraction of a mixture is an additive property of the components. Furthermore, it is an approximately additive property of the atomic constituents of each molecule. Consequently, when published data are not available for the minor component, its molar refraction can be estimated from its formula for use in an approximate calculation of the purity of the sample. Representative values of the molar refraction are shown in Table 101.I.

2. Refractive Dispersion

A property associated with the refractive index is the refractive dispersion. This is the difference in refractive index measured at two different wavelengths of light. Because it is based on a difference measurement, it is less precise than the refractive index alone. Furthermore, the variation of refractive dispersion is normally not very large from one substance to the next. Usually other methods can be found which are more sensitive and require a less exacting technique. This property and its applications are discussed by Bauer et al. (10).

C. SPECIFIC ELECTRICAL CONDUCTIVITY

The classic application of electrical conductivity measurements to the determination of purity is the testing of conductivity water. This is not a purity determination in the strict sense because a number is not given to the amount of impurity; rather a limit is set on the permissible conductivity of the water. Many other examples of a similar sort can be found; among them is the testing of pure metals and semiconductor materials. Because conductivity measurements can be made with high precision, and because the specific effect of the impurity is often large, quite sensitive tests can be devised. Parts per billion of certain electrically active impurities in germanium can be detected, although electrically inactive impurities may be as high as 1 part in 100,000 (56). Electrical conductivity measurements on liquids are discussed by Loveland (81) and on metals and semiconductors by Bardeen (8).

D. OPTICAL ROTATION

The rotation of the plane of incident polarized light, and the variation of the rotation with the wavelength, are useful in the investigation of the structure of complex molecules. A large literature exists on the measurement and interpretation of optical rotation. It is probably most useful for the determination of purity when the major component is inactive and the impurities are optically active. Thorough reviews are available (50,139).

E. DIELECTRIC CONSTANT

The measurement and applications of the dielectric constant are discussed by Powles and Smyth (109) and Thomas and Pertel (154). As a purity test, measurement of this property is most useful for nonpolar liquids, such as hydrocarbons, in which the principal impurities are polar, for example, water or oxygenated compounds.

F. VISCOSITY

The viscosity of a liquid could be used to estimate its purity if one of the three conditions listed in the introduction to this section were met. However, examples in the literature are few. Reviews of the technique of measurement and the applications of viscosity are given by Lyons (130), Swindells et al. (146), and Van Wazer et al. (165). Because it appears that the reciprocal of the viscosity is an additive property (47), the determination of viscosity should be suited to the detection of a small amount of a mobile liquid in a viscous one, for example, water in glycerin.

G. COLLIGATIVE PROPERTIES

The colligative property of solutions most commonly used in the determination of purity is the freezing point depression. The theory and practice of this measurement are well known and are discussed by Glasgow and Ross elsewhere in this Treatise (42). However, it has limitations; for example, it is applicable only to crystallizable samples and must be carried out at a particular temperature, the melting point of the sample. It is interesting to generalize the steps of the cryoscopic determination of purity and to examine analogous methods for the determination of purity based on other colligative properties.

The colligative solution properties are changes in the property of a substance which depend quantitatively on the molal concentration of the minor components. Classically, they are freezing point depression, boiling point elevation, vapor pressure lowering, and osmotic pressure. All of these properties are related to the partial molal free energy of mixing of the solvent (major component); and, for a given mixture, if one is known, together with the properties of the solvent, the others can be calculated. The classical colligative properties are measured by establishing equilibrium between two phases and observing a temperature or pressure. In the determination of purity, the sample is manipulated until it has a suitable fraction in each of two phases. When the two phases are different in composition, the observed equilibrium property varies with the fraction of the sample in one of the phases. If one of the phases is pure, the impurity concentration in the other is directly related to the phase fraction; and it is thus possible to extrapolate the observed property to zero impurity concentration. The slope of the extrapolation line depends on the molal concentration of impurity in the entire sample and on certain properties of the major component. Because these properties can be measured independently, this procedure can give an absolute estimate of the purity of the sample. The estimate is absolute in the sense that neither a reference material of certified purity nor the specific impurities need be available or known.

The need for a pure phase in the determination of purity by the use of colligative properties is vital. When it is not met, for example, when solid solutions are formed or the impurities vaporize or permeate the membrane, the determination loses its generality because specific impurities affect the measured property differently. Furthermore, the experimental work required to define the system becomes much greater, often prohibitively so. Consequently, although both theory and experimental technique have been devised to cope with these systems in cryoscopy (7,88,166), their application is difficult and comparatively unreliable (90).

From a practical point of view, the difference in the experimental conditions required for the use of the four colligative properties is interesting. The cryoscopic and ebullioscopic method must work at the freezing point or boiling point of the major component, whereas the vapor pressure lowering and osmotic pressure methods can be used at any arbitrary temperature at which the sample is liquid. The first two require the measurement of precise temperature differences; the others, measurement of differential pressures. Although not all compounds can be obtained in crystalline form, the cryoscopic method, when applicable, is least likely to suffer contamination of the pure phase (by solid solution formation). On the other hand, volatile impurities are quite common in ebullioscopic and vapor pressure measurements. The osmotic pressure method is not practical in general because there appears to be no semipermeable membrane suitable for osmotic pressure measurement on anything except high-molecular-weight impurities. If high-molecular-weight impurities are of interest, the technique used for the determination of osmotic pressure (167) may be useful. This property, although attractive in principle, will not be considered further here.

A colligative property determination of purity can be based on the following steps:

1. Measurement of a property (freezing point, boiling point, or vapor pressure) of the sample.

2. Separation of part of the major component in pure form, leaving all of the impurity concentrated in the residue.

3. Measurement of the property of the residue.

4. Estimation of the dependence of the change in property on the concentration of impurity.

Steps *1*, *2*, and *3* require the usual apparatus and techniques for the indicated physical property measurements or separations and need not be discussed in detail here (42,74a,174). Step *4* requires further explanation.

In sufficiently dilute solution, the relations for all three properties to concentration are of the form:

$$q = q_0 + \frac{X_2}{A} \tag{2}$$

where q = property of the solution;

q_0 = property of the pure major component;

X_2 = concentration of impurity in solution, mole fraction;

$A = -\Delta H_f/RT_{f0}^2$ for the cryoscopic method;

$\quad = \Delta H_v/RT_{b0}^2$ for the ebullioscopic method;

$\quad = -1/P_0$ for the vapor pressure lowering method;

P_0 = vapor pressure of the major component;

ΔH_f = molal heat of fusion of the major component, calories per mole;

ΔH_v = molal heat of vaporization of the major component, calories per mole;

R = gas constant = 1.987 cal/°C-mole;

T_{f0} = absolute temperature of the freezing point of the pure major component;

T_{b0} = absolute temperature of the boiling point of the pure major component.

It is clear that, if fraction $(1 - Y)$ of the sample is separated in pure form, leaving all of the impurity in fraction Y,

$$X_0 = X_2 Y \tag{3}$$

where X_0 = the impurity concentration in the entire sample, mole fraction. When the equations are combined,

$$q = q_0 + \frac{X_0}{AY} \tag{4}$$

Thus measurement of q and Y at two values of Y, say $Y = 1$ and $Y = Y_s$, permits one to solve equation (4) for X_0/A and q_0:

$$\frac{X_0}{A} = \frac{(q_s - q_1)Y_s}{1 - Y_s}$$

$$q_0 = q_1 - \frac{X_0}{A} \tag{5}$$

Alternatively, one might measure q at a number of values of Y, plot q versus $1/Y$, and determine the slope and intercept. Values for the parameter A can be obtained in three ways:

1. By calculation from published physical properties (2,36,37,55,68,69, 70,102,113,156,168).

2. By direct measurement of the physical properties. Equipment has been described for the measurement of heat of fusion (86), heat of vaporization (86), and vapor pressure (174). Usually the heat of vaporization can be calculated with sufficient accuracy from vapor pressure data.

3. By measurement of the change in the measured property on addition of known amounts of impurity. This is the most general means, and the most practical one for novel compounds, if the sample is not too valuable to prohibit contamination.

The procedure described is reliable provided that (*1*) the impurity concentration in steps *1* and *3* is low enough so that the laws of dilute solution are valid; (*2*) a reasonably large fraction of the sample can be removed in pure form; (*3*) temperatures or pressures can be measured with the required precision; and (*4*) no interferences occur in the property measurement.

The impurities in most nearly pure materials are sufficiently dilute so that condition *1* is met. However, certain kinds of impurities can be under- or overestimated even in quite small concentration, for example, polymeric (high-molecular-weight) or dissociating impurities. Extension of the theory to more concentrated solutions in cryoscopy has been discussed by several authors (7,84,121,151).

Recrystallization of the sample from a solvent (157) is usually the best way to separate a pure portion of the sample from the rest. However, the necessary complete removal of the solvent from the residue can be difficult, and a recrystallization from the melt or zone melting may be preferred (100,103). The latter techniques are more subject to difficulties arising from solid solution formation. Liquid extraction (57), fractional distillation (115), or chromatography (116) can be considered for use with samples which cannot be recrystallized in a practical fashion. Separation of a "pure" phase from an already highly purified substance may require special precautions to avoid contamination, both in the separation step and in handling the fractions (43).

Samples of high purity, above 99.9 mole %, require very sensitive temperature- or pressure-measuring devices. Except when it is desired to measure the physical properties of the major component in the same apparatus, they need be only differential instruments. Temperature differences of the order of a few hundredths of a degree Celsius must be measured reliably. The corresponding pressure sensitivity is of the order of one thousandth of the vapor pressure of the test substance.

Interferences in the property measurement can be encountered through such phenomena as solid solution formation, volatility of the impurity, and instability of the sample.

The preceding discussion has been limited to an oversimplified method for the determination of purity in order to show the parallelism of the cryoscopic, ebullioscopic, and vapor pressure lowering methods. In practice, only the cryoscopic method has been used in the manner discussed here, and then usually in a form in which all four steps are carried out in one series of measurements. However, this broader analysis of the purity determination problem may be of value in devising new methods for testing particular substances or in interpreting the results obtained by older methods.

1. Cryometry

The freezing point of a sample can be used to estimate its purity, provided that the cryoscopic constant and the freezing point of the perfectly pure major component are known. However, in this section, we are concerned with the cryoscopic determination of purity when these data are not available. The elements of the cryoscopic purity method are those described for colligative properties in the preceding section. Experimentally, the equilibrium temperature is measured as the sample is melted or frozen. As the phase change progresses, the temperature also changes as it follows the change in composition of the liquid phase.

The special feature of a cryoscopic purity method, compared to a freezing point determination, is the measurement of the phase ratio during the course of the experiment. In the literature, three ways of following the phase ratio change have been described. These serve as a convenient basis for classifying the cryoscopic purity methods discussed below. Reviews of the literature are given by Glasgow and Ross (42), Skau et al. (125), Sturtevant (141), Mathieu (89), and Cines (23).

a. Calorimetric Purity Method

In the calorimetric purity method, the heat content of the sample is used to measure the phase ratio. The sample is placed in an apparatus which prevents all unmeasured heat interchange between the sample and its surroundings. Starting at a temperature at which the sample is solid, heat is added in measured amounts. While the sample remains solid, the temperature increases at a rate proportional to the heat capacity of the solid sample and its container. When the sample begins to melt, the temperature increment for a given increment of heat diminishes. For a hypothetical, perfectly pure substance, the temperature remains constant until an amount of heat has been added equal to the latent heat of fusion of the sample. For real samples of varying degrees of purity, curves of the sort shown in Fig. 101.2 are obtained.

The data can be reduced to calculate the sample purity in a number of ways (7). Conventionally, the heat content–temperature curve for the solid is extrapolated to the temperature corresponding to the freezing point of the sample. The difference in the heat content of the sample at this point and at the corresponding point at which the sample is all liquid is the heat required to melt the sample. Similar extrapolations at intermediate phase ratios give the amount of heat absorbed in melting part of the sample. The fraction melted is calculated by simple proportionality. When the sample is quite impure, for example, less than 99.5% purity, it is difficult to establish the curve for the solid, and extrapolation to the freezing point to find the heat content of the solid may be uncertain. In these cases, the conventional data reduction methods can give ambiguous results, and a method which does not require the explicit evaluation of the heat of fusion is preferred (7).

A basic experimental limitation of the calorimetric purity method is the tendency of the liquid phase to develop composition gradients during melting (90). These come about because the liquid formed from the melting of the solid is purer than that already present. The equalizing of the composition by diffusion is slow compared to the equalizing of the temperature throughout the sample. Data obtained for nonuniform

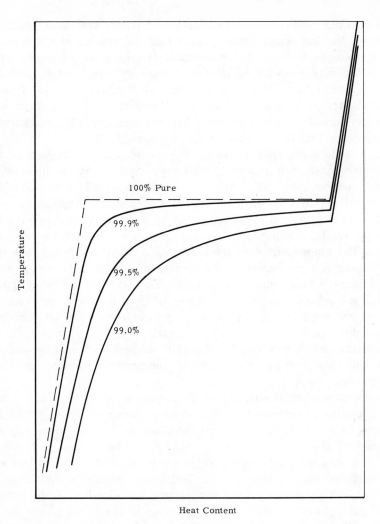

Fig. 101.2. Curves obtained by the calorimetric purity method for real samples of varying purities.

phase systems are hard to interpret, and the amount of impurity can be either over- or underestimated.

The classical calorimetric method for the determination of purity has been employed chiefly as an adjunct to the study of the low-temperature specific heats of very pure substances (99.9–99.995%). In these studies, a very precise calorimeter is used, and the heat content–temperature measurements required for determination of purity are easily obtained. However, the low-temperature specific heat calorimeter is too complex and delicate an instrument for practical determination of purity when this is the only property of interest. Relatively large samples (30–100 ml) are used, and tests may require several days to complete. Typical materials tested in the classical low-temperature specific heat calorimeters include highly purified n-heptane, benzene (41), substituted naphthalenes (91), and various sulfur compounds (92).

For the determination of purity by cryoscopy, an estimate of the cryoscopic constant is required. Most cryoscopic purity methods, when applied to new materials, require a test on a sample with an added amount of impurity to evaluate this property. When large amounts of sample are required, and the sample to be tested is very expensive because it has been highly purified, this is not a very satisfactory procedure. The unique virtue of the calorimetric purity method is the fact that the cryoscopic constant can be measured with little extra work and without contaminating the sample. This has led to the design of several specialized calorimeters (6,104,161,162) in which precision of heat content measurement and operability at extremely low temperatures are sacrificed in order to simplify the construction and operation. These instruments are, however, still relatively complex and delicate and require the services of a competent instrument builder for their construction.

The calorimeters described by Tunnicliff et al. (161, 162) represent a marked departure from the classical instruments, yet retain the features essential for the determination of purity. The more recently designed of the two is shown in Fig. 101.3. The sample (0.7 ml) is contained in the annular space between the wall of a cylindrical cup and that enclosing a thermometer and heater assembly, as shown in Fig. 101.4. This design makes temperature equalization of the sample rapid and reduces the distance for diffusion in the liquid phase during melting. The sample cup is suspended in a shield automatically maintained at a temperature which prevents uncontrolled heat transfer to the sample. The instrument is equipped with a platinum resistance thermometer which is sensitive to 0.001°C and accurate to about 0.02°C. Samples of purity in the range 95–99.95% and melting in the range −130 to +200°C can be tested.

The small sample size and easy cell loading and cleaning, together with a relatively high temperature range, are features of this calorimeter which make it useful for testing a wide variety of materials. However, because the sample is not isolated from the atmosphere, analysis of

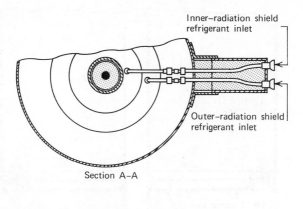

Inner–radiation shield
refrigerant inlet

Outer–radiation shield
refrigerant inlet

Section A–A

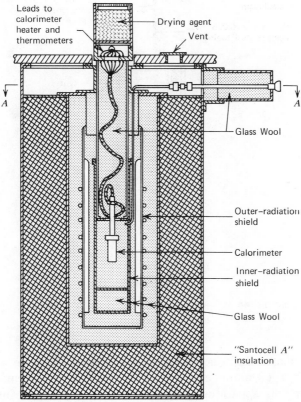

Leads to
calorimeter
heater and
thermometers

Drying agent

Vent

Glass Wool

Outer–radiation
shield

Calorimeter

Inner–radiation
shield

Glass Wool

"Santocell A"
insulation

Fig. 101.3.   Calorimeter and radiation shield assembly. Reproduced with permission
from reference 162.

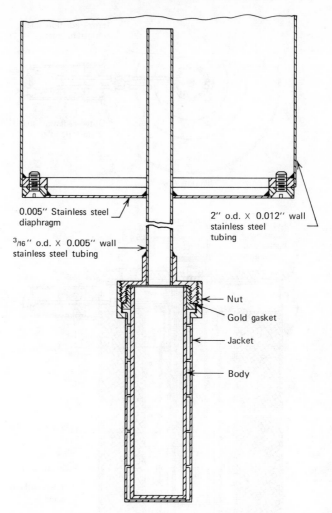

0.005" Stainless steel
diaphragm

$^3/_{16}$" o.d. × 0.005" wall
stainless steel tubing

2" o.d. × 0.012" wall
stainless steel
tubing

Nut

Gold gasket

Jacket

Body

Fig. 101.4. Calorimeter and support. Reproduced with permission from reference
162.

very pure materials is not possible since they cannot be loaded into
the cell without contamination. Also, triple-point measurements cannot
be made in this apparatus as they can in more elaborate instruments.
A test requires from 4 to 8 hr to complete. Typical materials tested include *n*-heptane, benzene, and naphthalene.

A commercial instrument called the Differential Scanning Calorimeter
(DSC) can be used for the calorimetric determination of purity. This

instrument resembles a differential thermal analyzer in that a small sample and a reference specimen are heated in it at a constant rate and the temperature difference between the two is monitored. In the DSC, electronic circuitry keeps this difference small by varying the power supplied to the sample cell. A record of this power difference as a function of time forms the data of the experiment.

After suitable corrections are made for thermal lag, portions of the area under the corrected curve are proportional to the fraction melted of the sample at corresponding temperatures. Values of reciprocal fraction melted are plotted against the temperature. The slope of the line is proportional to the impurity content, and the intercept is related to the freezing point of the pure compound. Instrumental problems make the absolute value of this property accurate only to about 2°C. The DSC is normally calibrated so that the area under the differential power curve can be used to calculate the heat of fusion of the sample. Thus the DSC experiment gives all the information needed to calculate the purity of the sample by the cryometric method.

Temperature differences usually cannot be measured to better than 0.005°C. This limits the maximum purity measurable with reasonable accuracy to about 99.9 mole %. The DSC test uses a remarkably small sample; only 2–5 mg are needed. Measurements can be made over the range from −90 to 400°C. A test takes from ½ to 2 hr to perform, depending on the amount of information available about the sample and the amount of pretreatment necessary to ensure proper crystallization. The apparatus is commercially available but is comparatively expensive. Plato and Glasgow (105) have applied the method to 95 materials and describe a procedure which they feel will give reliable results for about 75% of all crystalline or crystallizable organic compounds.

b. Time–Temperature Method

In the time–temperature method for the determination of purity, the sample is placed in an environment such that it loses or gains heat from its surroundings at a small and constant rate. Thus the heat content change of the sample is proportional to the time after the start of the experiment. The curves resemble those obtained with the calorimetric method and can be used to estimate the fraction melted in the same way. Two approaches have been used to reduce the nonuniformity of the liquid phase. In the first, the solid–liquid mixture is stirred. Although this is the most positive way of reaching the desired end, it has certain limitations. When mechanical stirrers are used, the increasing mass of crystals formed as freezing progresses interferes with the stirring until the stirrer can no longer mix the slurry adequately. Sometimes the amount of mechanical work done on the sample is an appreciable and varying proportion of the energy transferred, and in most cases measurements can be made over only 50–60% of the melting curve. In the second approach, the sample is distributed in a thin film so that temperature gradients and diffusion distances are small.

(1) Stirred-Sample Apparatus

Stirred-sample time–temperature apparatus has been described by a number of workers. The most widely used apparatus for nearly pure materials appears to be that of Rossini and his co-workers (5,44–46,84). It consists of a sample tube, jacket, cooling bath, stirrer, platinum thermometer, and resistance bridge. The apparatus is shown in Figs. 101.5, 101.6, and 101.7 and is discussed in greater detail by Glasgow and Ross (42). It has been used to test a large number of nearly pure hydrocarbons prepared as standards by American Petroleum Institute Research Project 6. The method requires 40–50 ml of sample, and a test can be completed in 2–3 hr. Other workers describe apparatus with reduced sample size (19,51) and different types of stirrers (19,51), as well as with automatic recording of the time-temperature data (140,172). All of the instruments mentioned above are primarily devices for the precise measurement of freezing point. They are most reliable for the determination of purity when the freezing point and cryoscopic constant of the pure major component are known. Taylor and Rossini (151) discuss the conditions under which the freezing point of the pure major component can be calculated from time–temperature data. Mair et al. (84) suggest that, to measure the cryoscopic constant, the apparatus be calibrated as a heat leak calorimeter, using a reference substance of known latent heat of fusion.

(2) Thin-Film Apparatus

Thin-film time–temperature apparatus has been employed by a number of investigators also (126–128,131). Skau's apparatus, shown in Fig. 101.8, illustrates the principal features. The sample (0.5 ml) is held in a sealed thin-wall glass ampoule equipped with a central well. The temperature of the sample is measured by a copper–constantan thermocouple in this well. The ampoule and the thermocouple are suspended within an electrically heated copper shield, thermally insulated from the cooling bath in which it is placed. The temperature of the shield, also measured by a thermocouple, is adjusted manually to warm or

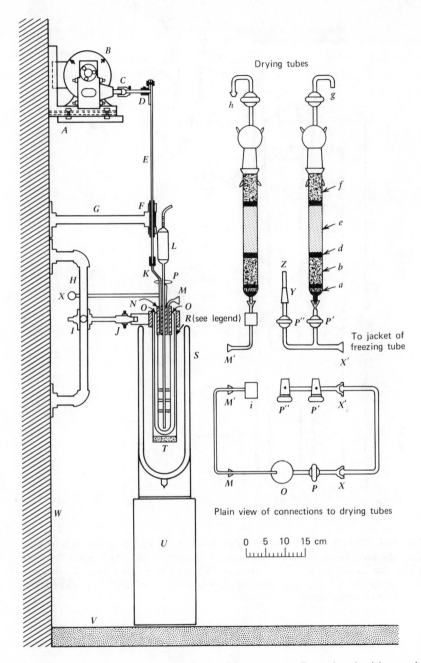

Fig. 101.5. Assembly of the freezing point apparatus. Reproduced with permission from reference 5.

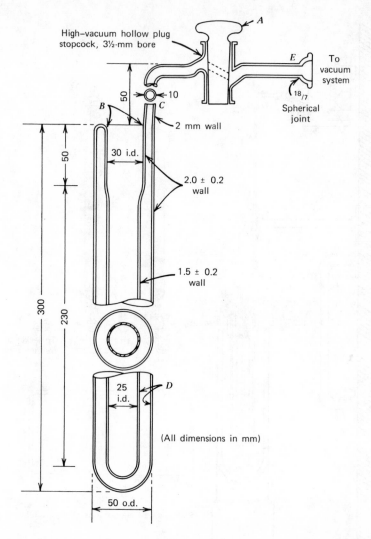

Fig. 101.6. Details of the freezing tube. A—High-vacuum stopcock, hollow plug, oblique 3½-mm bore. B—Inside opening of freezing tube, which must have no bulge at this point. C—Slanted connection to jacket of freezing tube. D—Internal walls of jacket of freezing tube, silvered. E—Spherical joint, 18/7. Reproduced with permission from reference 5.

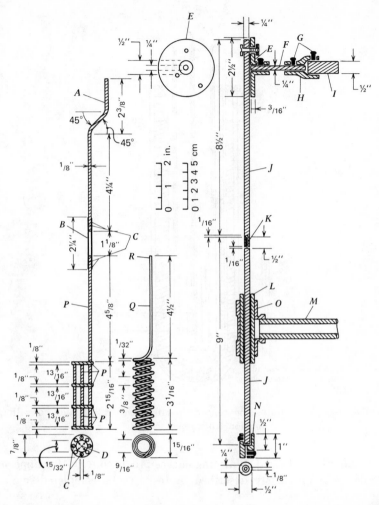

Fig. 101.7. Details of the stirring assembly and supports. *A*—Stainless steel rod, round. *B*—German-silver tube. *C*—Pins. *D*—Holes, ⅛ in. in diameter. *E*—Brass wheel, with three holes; tapped for machine screws, spaced ½, ¾, and 1 in. from center; normal position is ¾ in. from center. *F*—Steel rod. *G*—Set screws. *H*—Brass coupling. *I*—Steel shaft. *J*—Steel rod, round. *J'*—Steel rod, square. *K*—Connecting pin. *L*—Brass sleeve bearing. *M*—Steel pipe, ½ in. nominal size. *N*—Brass coupling. *O*—Brass tee. *P*—Aluminum. *Q*—Double helical stirrer, made by winding 1/16 in. diameter nichrome wire downwards on a cylinder 9/16 in. in outside diameter to form the inner helix, and then upwards over a cylinder 13/16 in. in outside diameter to form the outer helix, with the two ends silver soldered together. *R*—Place where shaft of the double helical stirrer is joined to the stirrer shaft. Reproduced with permission from reference 5.

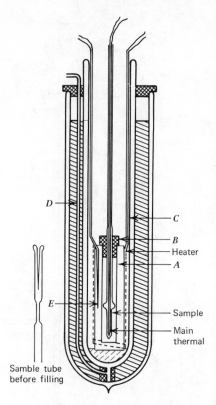

Fig. 101.8. Skau's apparatus for time–temperature curves. Reproduced with permission from reference 141.

cool the sample as desired. Warming curves produced with this apparatus resemble those previously described. A test can be completed in about 1 hr.

The virtues of this apparatus are its relatively simple construction and the small sample size and the short testing time required. However, temperature gradients in the sample can be a problem (131), and the small sample size makes difficult the calibration of the apparatus as a heat leak calorimeter for the measurement of the cryoscopic constant. Consequently, the reliability of this kind of instrument for the absolute determination of purity is lower than that of the other apparatus described above. It should be quite useful, however, for comparative measurements on different samples of the same type of material. Materials tested include naphthalene, benzoic acid, and tin, as well as compounds melting below room temperature.

c. Dilatometric Method

In the dilatometric method for the determination of purity (89,106–108,124,142,145), the volume change of the sample on melting or freezing is used to estimate the fraction melted. The sample is contained in a glass cell with a re-entrant thermometer well. The cell is immersed in a controlled-temperature bath. The temperature of the sample is measured by a suitable thermometer in the well of the cell. Starting at a temperature at which the sample is all liquid, the bath temperature is lowered in steps and maintained at each level until the volume of the sample becomes constant. From the initial volume of the liquid sample and the final volume corresponding to the solid sample, the fraction melted at intermediate temperatures is calculated by interpolation. The apparatus required for this determination is quite simple. The sensitivity obtainable depends on that of the thermometer used and on the stability of the controlled temperature. Samples can be handled in a sealed system and can easily be loaded into the cell without exposure to air. The duration of the test is variable. In contrast to the time–temperature methods, samples which exhibit slow changes in phase can be treated without difficulty other than the increased duration of the experiment.

There are two inherent experimental problems in the dilatometric procedure. The sample in the cell must not contain dissolved gas, as this would be expelled on freezing and lead to erroneous indications of volume change. Also, it is sometimes difficult to secure a reliable final volume of the solid sample because the liquid freezes in the measuring capillary before all of the sample in the cell is frozen.

The cell recommended by Swietoslawski (142) contains about 100 ml. However, there appears to be no reason why small cells cannot be used if the amount of sample available is limited and results of lower sensitivity are acceptable. The original static measurement described by Swietoslawski can give unreliable results because of the formation of a nonuniform liquid phase as discussed above. Plebanski (106) avoids this difficulty by using a cell equipped with a stirrer.

Usually the dilatometric technique measures the triple point of the system. Consequently, in measurements on pure samples for which the freezing point depression is small, the calculation of purity based on freezing points for the perfectly pure compound obtained by the time–temperature method, stirred-sample technique, can be quite erroneous, because the latter values correspond to the air-saturated liquid phase. It is not possible to measure the cryoscopic constant with this technique except by tests on a sample containing a known amount of added impurity.

2. Ebulliometry

The technique of precise determination of boiling and condensation temperatures by measurements on boiling systems constitutes the subject matter of ebulliometry (143). Although ebulliometric measurements have many applications, we are concerned here with their use in estimating the purity of volatile liquids. The difference between boiling and condensing temperatures is used as a qualitative indication of the impurity content, although the sensitivity of the test varies with the nature of the impurities. Swietoslawski and Anderson (144) discuss the problems in ebulliometry and describe apparatus for measuring the ebulliometric degree of purity of liquids. Leslie and Kuehner (74a) describes apparatus and procedures for ebulliometric studies. Lehrle (74) presents an extensive bibliography on ebulliometric apparatus, some of which can be adapted to purity testing, although Lehrle is concerned primarily with the determination of molecular weight.

In an earlier section of this chapter it was pointed out that, in principle, the boiling point elevation can be used for the absolute determination of purity in an analogue of the cryoscopic determination. This method gives unambiguous results only in systems in which none of the impurities is volatile. Such systems are rare in purity determination by ebulliometry because volatile liquids are usually purified by distillation and the remaining impurities are perforce volatile also. When the impurities are volatile, two types of difficulty interfere with securing absolute purity estimates: (*1*) as the boiling point of the impurity approaches that of the major component, the sensitivity of the method decreases; and (*2*) when there is more than one impurity, and the impurities have different boiling points, their specific effects on the boiling point of the mixture are different. The second difficulty is the more serious one and, in effect, makes the absolute determination of purity by ebulliometry impractical.

If sufficient thermometric sensitivity can be obtained, ebulliometry can be used for comparative or qualitative purity testing. The ebulliometer recommended by Swietoslawski for this purpose is shown in Fig. 101.9. This is a differential device which measures both the boiling (liquid) and condensing (vapor) temperatures of the sample. Required auxiliary equipment includes sensitive thermometers and a pressure-controlling system. The ebulliometer must be insulated in some manner to eliminate the effects of drafts and room temperature changes. Mate-

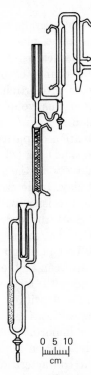

Fig. 101.9. Differential ebulliometer for degree-of-purity studies and boiling point determination. Reproduced with permission from reference 144.

rials which have been tested with this apparatus include benzene and isopropyl alcohol. Differences of 0.000–0.003°C are observed between the boiling and condensing temperatures of many carefully purified materials.

Differential ebulliometry is not widely used in industry for purity testing. The need for the kind of information it supplies is usually met by some sort of distillation procedure or by gas chromatography.

IV. CHEMICAL METHODS

A. TITRIMETRY

Titrimetry is a technique basic to the procedures of quantitative chemical analysis. There is a detailed discussion in Vol. 11, Part I, of this Treatise. A known weight of sample in solution or suspension is caused to react with a liquid reagent of known composition (titrant) by addition of the latter from a graduated dispenser. When the reaction is completed, a change takes place in the color, turbidity, electrical properties, or other characteristics of the solution, and the volume of titrant at which this change occurs is recorded. If the titrant and the sample react in known proportions, the amount of analysand present can be calculated from this volume, and the purity of many substances can thus be evaluated.

In speaking of the determination of purity, we usually think in terms of (1) assay of the major component, or (2) evaluation of the impurities, or minor components. It is largely in the first field that titration procedures are still widely used today. In determining the major component, the analysis is limited by the precision with which the volume can be read; generally speaking, it is of the order of 1 in 1000. In making assays by titration, then, it is the volumetric precision that is to be maximized in order to get the best performance of the method. In many instances, assays by titration still can be the most precise of the methods available, even though instrumental methods often can give answers more quickly, on smaller samples, or with less interference. In the second area, the determination of impurities, the precise measurement of these components usually is not as important as the ability to detect and estimate them semiquantitatively even when they are present in very small amounts. It is here that characterization by physical means will often show a distinct advantage. Titration methods have the great advantage, however, of low capital investment and simplicity of operation, and a small laboratory will often resort to this classical method for the determination of purity for these reasons.

The chemical criteria for evaluation of this method for our use are (1) extent of reaction, (2) specificity of reaction, and (3) rate of reaction. The extent of the titration reaction should be complete, or at least reproducible, and known. The titration should be specific for the compound under test, or interfering substances must be known to be absent. Finally, the reaction should be reasonably rapid, not only for economy in analysis time, but also to avoid or minimize possible side reactions, decomposition, or fading of the end point.

Titrimetry is so widely used in all fields of chemistry that often its applicability in the context of our objective is most practically sought by starting with the substance to be analyzed rather than with the method. If the analysand is an element, Part II of this Treatise, "Analytical Chemistry of the Elements," is a practical source of pertinent information. Application to specific compounds is discussed in general reference works cited in Section I of this chapter and also in the Annual Reviews of *Analytical Chemistry*.

The three large fields in titrimetry are acid–base (31,136), oxidation–reduction (38), and precipitation (15) and complexometric titrations (59). In all of these, detection of end point can be accomplished in various ways; achievement by electrical methods is discussed under separate headings. In any case we need a noninterfering indicator, which signals the end of the titration reaction. Some less common methods of end-point detection, such as thermometric and chemiluminescent detection, will not be discussed here as their usefulness to the objective at hand appears to be quite limited at present.

1. Acid–Base Titrations

For commercial acids and bases, assay by titration is still largely used as a purity specification. As an example of the determination of the minor components, on the other hand, the acid content can be a rather sensitive indication of purity, as in fats, oils, or inhibitors, where acids may appear as a result of a deterioration process.

2. Oxidation–Reduction Titrations

In this method we use the ability of substances to be oxidized or reduced by solutions that can be standardized and stored over reasonable periods of time. A considerable number of metals and nonmetals exist in various states of oxidation and can be determined by tritration with suitable oxidizing or reducing solutions. The choice among competing methods often is very difficult and will depend on the nature and the amount of the impurities and other factors. The purity of peroxides and of many organic oxidizing and reducing agents can be determined by this technique. Details are given by Stenger (38).

3. Precipitation and Complexometric Titrations

In this technique, reactions which give rise to precipitation or complex formation are utilized. Precipitations still are used as the most precise assay methods, in some cases, if carried out with appropriate care. The assay of silver (Volhard) is an example in this category, and it has maintained its usefulness for a long time.

B. GRAVIMETRY

The remarks made in regard to titrimetry, with respect to its utility in assay and its simplicity in equipment requirements, apply also to gravimetry. The problem here is to find a reaction which will yield an insoluble precipitate with the constituent sought without precipitating the minor components. The physical character of the precipitate must then be such that it can be washed free of other ions without significant solubility loss. The chemical properties of the precipitate must allow its conversion to a stable compound of constant composition by drying or ignition at some specified temperature.

An extensive literature exists describing the procedures to be followed, ways to avoid errors, and examples of application; a review and bibliography are given by Rulfs (164).

V. ELECTROCHEMICAL METHODS

Electrochemical methods (22; see also Section D-2, "Electrical Methods of Analysis," in Vol. 4, Part I, of this Treatise) are used when determination of the composition of mixtures, assay of major components, or estimation of minor, including trace, components is the objective. The second and third applications are of interest here, and they are feasible with liquids and solids soluble in the many solvents that can be employed in this type of analysis.

One can subdivide the field into methods that are based on weighing the products of electrochemical separation, without requiring a precise estimate of the electrical quantities, and methods in which the latter are measured exactly and are indicative of the concentration or amount of analysand. Electrogravimetric analysis is the only method of interest here that falls within the first classification, that is, is capable of yielding quantitative results in the ranges under discussion. The second group includes potentiometric, conductometric, polarographic, amperometric, and coulometric analyses. These methods, because of better understanding of their underlying principles and of the instrumentation involved, today show much overlap, but a classification such as that ordinarily given is used for organization of the material. The classification follows that made by Lingane (78).

A. ELECTROGRAVIMETRY

Electrogravimetry (149) is in use for the determination of a number of metals from aqueous solution. Direct current is passed through the solution at controlled pH, and the metal, deposited on a suitable electrode, is weighed after washing and drying. Such metals as copper, bismuth, lead, tin, nickel cadmium, silver, and mercury (and practically all other common metals) are easily determined. A metal can be determined as the major component, for purposes of assay, or in small concentration if present as an impurity. A typical example of the first class is the determination of copper, where a standard deviation of 0.03–0.05% can be expected; the determination of lead can represent the second case, where 5–100 mg of lead can be determined from aqueous solution to about 0.25% of the value.

The electrode potential can be kept constant automatically with good precision (1–5 mV) by means of a potentiostat. More recent equipment is fully electronic and consequently operates almost without inertia.

B. POTENTIOMETRY

Potentiometry (40,78) can be used for the assay and determination of acids and bases, strong and weak, in aqueous and in nonaqueous media. Widely used electrodes are the hydrogen, glass, and quinhydrone electrodes.

Potentiometric precipitation titrations also fall into the realm of determination of composition rather than of purity. The most important analyses of this type are those involving insoluble silver and mercury salts, and also salts of copper, zinc, and lead. A considerable number of anions can be determined by this technique; they include halide ions, sulfate, sulfide, ferrocyanide, cyanide, and thiocyanate ions, and ions of various organic acids. Potentiometric oxidation–reduction titrations fall in this general category, but as their application is related even less to our present purpose, the reader is referred to the discussion of "Electrical Methods of Analysis" in Part I, Vol. 4, Section D-2, of this Treatise.

C. CONDUCTOMETRY

The electrical conductance of a solution is a summation of contributions from all ions present, and hence is not a specific property of any particular ionic species. Nevertheless, conductometry (81) is useful in our context in some cases. The solubility of many inorganic salts, or the degree of dissociation of weak acids in some water-miscible solvents such as acetone or alcohol, is greatly increased by small amounts of water (65). The conductivity of alcohol, when shaken with calcium sulfate, increases with increasing small amounts of water, and the concentration of water can therefore be determined by this method; it is necessary to establish calibration curves as the relation between conductance and water content is not linear.

D. POLAROGRAPHY

Polarography (96) is a technique of very wide application. It is useful in the determination of composition, the state of oxidation, and the chemistry of redox couples, and because of its versatility it can help in solving many problems concerned with the determination of purity.

This type of analysis is based on the properties of current–potential curves obtained with a dropping mercury electrode (53) or other suitable microelectrode. Under certain conditions the current–potential curve will exhibit a limiting current, whose magnitude is directly dependent on the concentration of the electroactive species, and as the decomposition potential of a solution is characteristic of the particular electrode reaction, the value of the potential can be used to identify the reaction. As a result polarography is capable of simultaneous qualitative and quantitative analysis of aqueous and nonaqueous solutions of any substance that can be reduced by sodium or correspondingly oxidized. This includes such a wide range of compounds that reference must be made to other works (17,66,95,96). Because of its flexibility and wide range of application, this method is extensively used, and many types of apparatus, ranging from simple constructions put together in the laboratory to highly versatile and fully automatic equipment, are available.

E. AMPEROMETRY

In amperometry, the concentration of an electroactive substance is measured by the current resulting from its reaction at an electrode. In amperometric titrations, we thus can detect equivalence points by the abrupt change in current that occurs at this point in a titration.

The precision and accuracy of amperometric end-point detection are excellent, and titrations of small concentrations can be carried out by means of this method. Solutions of lead ion as dilute as $0.001M$ can be titrated with an accuracy of $\pm 0.3\%$ with dichromate ion, and $+3$ arsenic in concentrations of $0.001M$ and above can be titrated to $\pm 0.1\%$ with bromine. Many other applications are known and can be found in the references on electrochemistry already cited. An important use is the amperometric determination of very small amounts of water in a great variety of organic and inorganic compounds by the Karl Fischer method. A thorough discussion of this technique, including other methods of detecting the end point, is presented in the monograph by Mitchell and Smith (97). This very sensitive method is often used as a criterion of purity in cases where contaminants other than water are known to be absent or are of no importance.

F. COULOMETRY

In coulometry (33), a substance in solution is determined quantitatively by measuring the quantity of electricity required to bring about complete reaction in an electrolysis cell. Of the two major divisions—the amperostatic titration, with constant current, and the potentiostatic, which operates with controlled potential at the working electrode—the former was developed mainly from the work of Szebelledy and Somogyi (147,148), while Hickling (54) and later Lingane (79) did the early work in the second category.

This technique is attractive because it lends itself well to the automatic performance of routine analysis, because it is useful in continuous control analysis, and because exceptionally great sensitivity and accuracy can be obtained with it. Coulometry is, generally speaking, an absolute method, that is, the target information can be computed from the electrical quantities measured and the stoichiometry of the substances involved. Only in potential scanning coulometry, used for the case of extremely small samples, is this advantage lost.

Coulometry is very widely applicable to the determination, over a wide range of composition, of many metals adn of many organic compounds. The applicability of the method to particular cases is discussed in the references previously cited and in a useful review by Kies (62).

Among the most interesting applications of this principle is its use in the determination of water in hydrocarbons, both gaseous and liquid. Trace amounts of moisture are determined by means of the Keidel cell. The gas or vaporized liquid is passed through a coulometric cell consisting of an anhydrous phosphorus matrix between platinum electrodes. Accurate analyses can be attained at moisture levels down to 1 ppm and less, as the moisture is absorbed and hydrolyzed simultaneously and quantitatively and the current required is measured.

Other interesting uses include the control of mustard gas in air, and of hydrogen sulfide, thiophene, and mercaptans in 0.5 ppm concentration. Borane, ozone, chlorine, halide ions, copper, and bases have been determined continuously. Coulometric detectors have been used in conjunction with gas chromatographic analysis in the determination of minute quantities of halogen in sulfur (29).

G. ANODIC STRIPPING VOLTAMMETRY

Anodic stripping methods (123) offer great sensitivity with good precision for metals which will form amalgams with mercury. The technique involves a pre-electrolysis at controlled potential while using a stationary micro-mercury electrode. Metal ions are reduced in the form of an amalgam at the surface of the electrode. The plating time is varied according to the concentration of the solution and usually is in the range of 1–30 min. After electrolysis, the potential is scanned rapidly in a positive direction and the current–voltage curve is recorded. The technique has been used in the $10^{-9}M$ range, and the reported precision is about 3% at $10^{-8}M$. Simultaneous metal determinations may be made in the usual polarographic way. The simplicity of the method is one of its most attractive features. By reversing the direction of the current during the electrolytic process, anions such as halide can be determined with excellent sensitivity.

This technique is very useful for determining impurities in reagent-grade chemicals such as potassium chloride.

VI. COLORIMETRY

In the field of analytical chemistry, the word colorimetry is understood to mean the determination of a chemical element or compound by measurement of the color of a solution. In addition to the measurement of color itself, a colorimetric method includes procedures for dissolving the sample, removing interfering substances, and developing the color to be measured. A great many more or less specific reagents have been found for forming colored substances with particular inorganic or organic compounds. Consequently, the range of application of colorimetry is very broad. Because very low concentrations of colored substances can be determined, colorimetry is used widely for trace analysis.

The apparatus ranges in complexity from a visual comparator for a solution of the sample and a standard, through filter photoelectric photometers, to spectrophotometers which measure absorption as a function of the wavelength of the incident light. Much useful work has been done with inexpensive apparatus and relatively simple procedures.

The sensitivity of colorimetric methods depends to some extent on the means used to measure the color and more especially on the effectiveness of the procedures for the removal of interfering substances. Sensitivities ranging from 0.1 to 100 ppm are commonly obtained. The precision of measurement is normally for 2 to 5% of the amount present in the sample.

Sandell (118) describes methods for the colorimetric determination of traces of metals, while Boltz (14) supplements this work by providing methods for nonmetallic elements. Snell and Snell (133) and Allport (1) give methods for many inorganic and organic compounds.

VII. SPECTROSCOPIC METHODS

The study of the interaction of electromagnetic radiation with atoms or molecules has found wide application in the determination of the structure of single compounds as well as of the composition of mixtures (see Section D-3, "Optical Methiods," in Part I, Vols, 5 and 6, of this Treatise). Since in general one can expect a precision of about 1% in spectroscopic comparison of samples with a standard, spectroscopic techniques are not very useful for assay tests. On the other hand, these techniques can be extremely sensitive and often can serve to determine minute concentrations of impurities. In fact, the most valuable aspect of these techniques in the determination of purity is their ability to make positive statements regarding the *absence* of impurities. If absorption at a given wave number is not observed, the upper limit to the concentration of the corresponding material can be stated. As a result, spectroscopic methods are very useful for following the steps during purification of a substance. As the impurities are removed, the spectrum of the substance becomes simpler, until it agrees with that of the pure compound. This is in interesting contrast with methods in which one must guard against the tendency of the major components to obscure the minor or trace components and thus to keep the latter from being observed.

The different techniques involving either emission or absorption spectroscopy can be grouped conveniently according to the frequency range of the radiation with which they are associated. The methods interesting in the determination of purity are enumerated in a sequence of decreasing radiation energy in the following paragraphs.

A. X-RAY FLUORESCENCE

When a sample is bombarded with X-rays, electrons in the inner shells of the atoms are excited into orbitals of higher energy. As the electrons return to their ground states, X-rays are emitted. The frequency of these fluorescence X-rays depends on the atom itself, not on the state of chemical binding. This process therefore can be used for qualitative and quantitative estimation of the elements present in a sample, but not directly for the determination of the molecular composition of a mixture (77).

Often it is desirable to know specific contaminants of a substance if other contaminants are not of interest, are known to be absent, or are already known. Then this technique can be of great use, as some elements can be determined very sensitively and rapidly. The arsenic content in certain catalysts, for instance, is a factor of importance; it can be determined satisfactorily by X-ray fluorescence. The purity of organic substances can be evaluated by this method if all the impurities can be expressed by the concentration of elements that are accessible to X-ray fluorescence analysis. It is often desirable to know, for example, before a polymer product is released for shipping, how free it is from catalyst residue. The polymer can be ashed in a container, which is then subjected to X-radiation that gives rise to the fluorescence. Some elements, particularly metals, can be determined in concentration ranges of parts per billion.

The simplicity and speed of the analysis are the most attractive features of this technique. Presence or absence of half a dozen elements can be checked in as many minutes. Quantitative results require calibration and a more involved computation procedure, and each analysis may take as long as 1–2 hr. In the case of samples having bulk, the surface only is observed. Thus the method would not be suitable for metal objects whose surfaces might be susceptible to corrosion, or can be enriched in another component by mechanisms other than corrosion. The method is applicable to elements having an atomic number higher than magnesium, but the sensitivity is not uniformly high in all cases.

The instrument has been commercially available for some time.

B. EMISSION SPECTROSCOPY

When a substance is excited by strong heating or is subjected to the influence of an electric arc or spark, electrons in the outer shells of the atoms can be excited, and the radiation emitted by the electrons returning to their normal state can again serve to identify the element present. The emission spectrum in the ultraviolet and the visible is nearly independent of the chemical state, too. This technique (122) therefore is useful in determining elements rather than compounds in the analysand.

So far, it has not been possible to maintain the conditions of temperature and concentration constant for more than very brief periods of time. Consequently, to obtain quantitative results, the approach has been to integrate the total light emission at each wavelength of interest, and to compare the integrated values with those of an internal standard wavelength similarly integrated. By this process it is possible to compare precisely the unknown and the standard, thus eliminating the problem of rapid fluctuation of the energy source.

Modern instruments use all-electronic means for integration and comparison of the light flux at predetermined wavelengths. Considerable progress has been made recently in lowering the limits of detection of impurities in metals. In direct analyses of nickel samples, concentrations of impurities as low as 0.1 ppm have been reported (117). In the analysis of silicon carbide, under conditions which suppressed volatilization of the SiC, sensitivity limits of the impurities of 10–300 ppb were observed (98). The purity of reactor graphite also has been the subject of study, and boron at concentrations to 3 ppb was determined by the arc method (16).

This technique has also been used in conjunction with preseparation procedures for sensitive purity determination. Nachtrieb (99) combined a cupferron procedure to extract 39 impurity elements from plutonium with determinations by the copper spark method of emission spectroscopy.

The purity criteria for metals, especially in the semiconductor field, have grown to such an extent that the gas content of metals has assumed great importance. Much work has been reported in the determination of oxygen, nitrogen, and hydrogen in metals, and recent literature gives many examples of the application of emission spectrometry for this purpose.

Sample requirements are small, and usually 1–10 mg is sufficient for an analysis.

Instruments of varying complexity are commercially available.

C. MASS SPECTROSCOPY

Mass spectroscopy can logically be classified as an emission spectroscopic technique because, under the action of ionizing radiation, a new observable, the mass spectrum, is generated. The mass spectrum, that is, the distribution of the positive ions formed, is characteristic of the substance ionized and hence is useful for analytical purposes.

The sample vapor is bombarded by electrons of sufficient energy to ionize the molecules. Some molecule ions are formed with enough excess internal energy to lead to their unimolecular dissociation with the formation of fragment ions. The ions are accelerated into a magnetic field, which deflects them into different paths according to their ratio of mass to charge, m/s. The resulting ion beams usually are detected by sweeping them past a slit in front of a collector plate. The collected ion currents are amplified and recorded. "Sweeping" is accomplished by varying either the magnetic field or the ion-accelerating voltage. The relative abundance of the ions is proportional to the relative ion beam intensities, or the peak heights of the record.

The peak heights of the ion fragments at the different mass/charge ratios of the fragments give us the mass spectrum, which is characteristic of each compound. Such a spectrum thus provides us with a qualitative identification of the compound. Since fragments of different classes of compounds will tend to differ from one another to a greater degree than fragments deriving from homologous series, it will be generally easier to resolve a mixture of the former than of the latter. The mass spectrometer therefore can give a good idea of what types of contaminants are present in the compound whose purity is to be examined. In the quantitative computation of the minor components, the sensitivity ranges around 0.02% but can be increased by suitable preconcentration.

Mass spectroscopy is applicable to any substance that is volatile enough to be vaporized into the evacuated ionization chamber of the instrument. Applications involving more vigorous excitation methods, such as the arc and the spark, include the analysis of metals and metal compounds. Mass spectrometry of inorganic materials is thus assuming an important role in analysis.

An interesting application of the mass spectrometer to the study of purity can be described as follows. The leak in the standard inlet system of a mass spectrometer is designed to give molecular flow rates. Therefore, the rate of depletion of a particular compound from the reservoir will be a function of its molecular weight. For the determination of the molecular weight, a smaller reservoir than usual is employed, and the leak size is calibrated with a known standard. This device can be used to check a peak suspected of being caused by an impurity in the sample. If the relative height of the peak changes with the depletion of the sample, at least a part of the peak must be due to an impurity of another molecular weight. The molecular weight of the impurity can also be calculated if the total peak is due to the impurity. Polar compounds, however, tend to associate in the gas phase, so that this method requires a general knowledge of the chemistry of the system to be analyzed, in order to prevent misleading interpretation of the results.

Many discussions of this subject are available (25,93,135); see also Section D-1, "Magnetic Field Methods of Analysis," in Part I, Vol. 4, of this Treatise).

D. FLAME PHOTOMETRY

Flame photometry is a technique of emission spectroscopic analysis, which utilizes a flame excitation source. It is used in the determination of many elements with excitation potentials low enough for flame excitation.

The sample in solution is atomized into a flame, and elements in the sample are vaporized and excited. The light emitted by the excited atoms is filtered or dispersed, so that eventually a narrow portion of the spectrum emitted is allowed to fall on a detector needed to convert the light intensity into a galvanometer deflection or strip-chart record. The fuels used for the flame include air–natural gas, cyanogen–oxygen, and hydrogen–oxygen. The last of these provides a flame temperature high enough to excite many elements (about 2900°K), yet gives a relatively low background.

Quantitative determinations with the flame photometer are very rapid and precise. Single measurements can be made in 10 sec; the precision can be as good as 0.2% of the emission. It is necessary to make calibration curves, as these methods are empirical.

The most frequent applications are in the determination of alkali- and alkaline-earth metals, boron, chromium, cobalt, copper, iron, lead manganese, nickel, and silver. The method is being used also, however, for other elements.

In connection with the determination of purity, flame photometry, in spite of the widening range of application, is still used mostly for the determination of alkali metals in various essentially pure metals, such as sodium or potassium in lithium, aluminum, lead, or tungsten.

Flame photometry is valuable when a simple, reasonably accurate, and rapid method of analysis is required. Flames excite fewer lines of each element than an arc or a spark source, and the conditions of procedure are less difficult to establish than with the other emission methods. Flame methods should be particularly interesting to small foundries and control laboratories which do not have spectroscopic apparatus.

There are many books, chapters, and reviews on the subject which should be consulted for more detail (25,32,52,85,163).

E. FLUORESCENCE AND PHOSPHORESCENCE

When substances are irradiated with light in the ultraviolet or visible range, the molecules absorb most of the radiation. Sometimes, however, it is re-emitted at characteristic frequencies, either immediately or some time later. These processes are known, respectively, as fluorescence and phosphorescence (27). Although the application of these processes to the estimation of purity does not cover a very broad range, the intense fluorescence of many compounds, for example, vitamins, steroids, catecholamines, and polycyclic hydrocarbons, has made techniques based on this property very useful in determining traces of such compounds in solutions and liquids of biological interest. Small amounts of polynuclear hydrocarbons can be estimated in liquid alkanes or other organic solvents that do not themselves fluoresce.

Related compounds sometimes exhibit interesting differences in fluorescence. *Trans*-stilbene is strongly fluorescent, whereas the *cis* isomer is not (76). Donough et al. have shown that "pure" tetraphenylporphyrin consists of two isomers (35). The substance was excited with light at two different wavelengths within the main absorption band, and two different fluorescence spectra were obtained. The technique in general requires care in the interpretation of the spectra, because traces of strongly fluorescing impurities may be thought to belong to the major component. Thus five samples of "pure" cholesterol exhibited different fluorescence spectra (173).

Under certain conditions, fluorescence phenomena can be used in the determination of purity in the inorganic domain. Because of the lack of specificity and the need for rather detailed knowledge of the systems studied, however, this application has not as yet proved of great importance. Details will be found in *Analytical Reviews* (170).

F. ABSORPTION SPECTROSCOPY

Most of the lower-energy radiation falling on a substance is absorbed and converted into thermal energy rather than re-emitted. Ultraviolet radiation is absorbed by the valence electrons and converted into vibrational, rotational, and translational energy. Infrared radiation is not energetic enough to excite the electrons, but it can cause the molecules to vibrate. The lower frequencies, in the far infrared and microwave region, cause rotation of the molecules.

Spectroscopic analysis (11), with the exception of nuclear magnetic resonance, is basically empirical in nature. The absorptivity of a molecule cannot be calculated to the accuracy with which it can be measured. Any procedure that gives accurate results, as determined by independent analytical methods, is valid but should be accepted as valid only for the particular system involved. The law that applies for the ideal case is Beer's law, which states that for a given sample thickness the transmittance depends exponentially on the concentration of the absorbant, c. The absorbance is defined as the logarithm of the ratio of the incident intensity, I_0 (the intensity of the light entering the sample), to the intensity, I, of the light that has passed through the sample of thickness b. The proportionality constant, a, is the absorption constant characteristic of the sample and the frequency of the radiation:

$$\log I_0/I = A = abc$$

An extension of this law, the principle of additivity, is of great importance to us in the context of the objective of this chapter and was alluded to in the introductory paragraph of this discussion of spectroscopy. It states that the absorption of radiation by one species will be unaffected by the presence of other materials, whether they absorb or not. When this principle is not valid, quantitative determinations may still be possible, but they will be more difficult.

1. Ultraviolet–Visible

Ultraviolet–visible absorption (119) is particularly sensitive for unsaturated organic compounds, especially those containing aromatic or carbonyl groups. Many substances that do not normally absorb in a convenient region can be made to do so by forming suitable derivatives, for instance, by combining an organic compound with a metallic ion. For determining low concentrations of contaminants, this is the most sensitive of the spectroscopic methods. Under favorable conditions, between 10^{-3} and 10^{-8} mole of minor component can be detected per mole of major component.

2. Infrared

Infrared absorption (132) is the most nearly universal method of analysis, although it is less sensitive than ultraviolet. All organic compounds and most inorganic compounds absorb in the range of commercial spectrometers. However, a thorough familiarity with the actual appearance of the bands and additional information about the sample are required to make the most of the data obtained by the spectroscope. This information would include the separation methods used, the physical and chemical properties, and the empirical formula of the sample. Interaction of relatively polar compounds with reactive solvents interferes with the absorption bands in this region, and quantitative analysis is generally not satisfactory for such systems. Where applicable, the sensitivity of infrared analysis for a minor component, on the same basis as before, is on the order of 10^{-2} to 10^{-4} mole/mole.

G. NUCLEAR MAGNETIC RESONANCE

Nuclear magnetic resonance (48) is particularly useful for hydrogen and fluorine compounds, although a number of other atoms also can be examined. The spectrum can be predicted *a priori,* and therefore a reference spectrum is not needed. The spectrum appears as series of peaks representing hydrogen (or whatever magnetic nucleus corresponds to the particular frequency used), and the areas of the peaks correspond exactly to the amount of hydrogen in a particular chemical or magnetic environment. The areas of the pure components therefore are in integral proportion to one another, and any departure from this relationship, or the existence of additional peaks, indicates impurities. Normally these impurities can be identified from their position by theoretical considerations. In general, the sensitivity of this technique is not particularly good. About 0.2% hydrogen extraneous to that calculated for the pure compound can be detected; this must be converted to mole per cent impurity by the proper calculation. If the impurity contains a relatively large amount of hydrogen, the sensitivity in terms of mole per cent of the component is good; if a small amount, it is poor. In general the spectrum of the major component should not interfere with that of a minor component, unless they are very closely related.

The installation is a massive and expensive one.

The sample size requirements depend on the hydrogen (or fluorine) content but range around 20 mg; a microcell can handle samples as small as 1–5 mg. Analysis time is 15–20 min; the calculation of the results depends on the complexity of sample and may vary from a few minutes to several hours.

VIII. RADIOCHEMICAL METHODS—ACTIVATION ANALYSIS

Activation analysis (94) is finding its way into an increasing number of research and control laboratories. The neutron sources required for the analysis, although still expensive, are coming within reach of at least the large organizations that they might benefit.

The sample substance is bombarded by a flux of neutrons and captures some of them. Isotopes of the capturing atoms are formed in this process; many of these are unstable and decay to stable states or substances, while giving off characteristic radiation. The gamma radiation, which is a part of the total radiation emitted, can be analyzed spectrally at any convenient time after irradiation. In addition, the decay of radiation of any chosen wavelength can be followed with time, and additional complementary information is obtained in this way. The information so obtained can then be interpreted to give a qualitative and quantitative estimate of the various atomic components. The method is suitable for the determination of minor components rather than for assay, as the precision is on the order of 2–5% of the value over the whole range of sensitivity. As in cases discussed previously, the method can be used as an estimate of purity only if certain conditions are met, that is, if the component under study is an element or if the nature of its chemical combination is known, and if the other minor components are similarly knowable, are of no interest, or are known to be absent.

The sensitivity of the technique depends largely on the flux available from the installation. Large nuclear reactors such as those at Brookhaven, Oak Ridge, and Harwell, which yield fluxes in the range of 10^{11}–10^{13} neturons/cm^2-sec, in many cases can detect elements in concentrations of parts per million or better, and, with radiochemical separation, of parts per billion. Examples are the determination of zinc, copper, gallium, manganese, chromium, scandium, and hafnium in high-purity iron and aluminum; antimony, copper, strontium, arsenic, silver, phosphorus, and gadolinium in silicon; selenium in sulfur; trace elements in silicon carbide; vanadium and aluminum in graphite; sodium in telluric acid; and trace elements in commercial toluene. These and many other examples are discussed in books and review articles (71). The sensitivity of detection that can be reached in some cases is far beyond the limit achievable by any other analytical method, and this is one of the outstanding features of this type of neutron activation analysis.

More recently, smaller neutron sources yielding fluxes of the order of 10^7–10^8 n/cm^2-sec have become available; these are more modestly priced and their sensitivities are correspondingly lower. Among these, the accelerators are characterized by their ability to produce high-energy or "fast" neutrons, which, however, can be slowed by suitable moderators to energies corresponding to the average energy of neutrons typically produced by reactors. An example of a commercial source is the Texas nuclear accelerator, which produces 14-MeV neutrons from the deuterium–tritium reaction.

With the fast neutrons it is possible to include lower elements among those that can be determined. The analysis of outstanding interest is that for oxygen; it is possible to determine oxygen as an impurity far more rapidly and precisely in this way than by any other method, in concentrations from 0.01% up and in virtually any kind of substance.

The sensitivity of the determination for a given flux varies over a wide range for different elements, even if they are chemically related. In a particular installation, for instance, the limit of detection for cobalt is 0.6 μg, whereas it is 470 μg for nickel. However, the availability of neutron sources of different energies largely overcomes this difficulty.

A number of other features are associated with this technique. A great range of concentrations can be handled with little or no change in procedure. Neutron activation provides a nondestructive analysis, except where radiochemical separation precedes the analysis. Volatile, flammable, toxic, and unstable samples can be handled, as the specimens are placed without preparation in relatively large vials (e.g., 30 cc), which are then sealed and never opened thereafter. Quite heterogeneous samples, therefore, are suitable for analysis without the need for homogenizing. Although smaller samples can be handled when conditions warrant, the ability to accommodate such large samples provides a very high concentration sensitivity and reduces the sampling error.

Instrumental activation analysis generally permits short activation and counting periods, often measured in minutes or even seconds, so that the total analysis time can be extremely brief. Thus the method, on one hand, is advantageous for the control of processes involving large production, where much time can be saved by replacing techniques involving a longer period for analysis, and, on the other hand, can prove highly useful in the research laboratory, where nonroutine samples of a volatile, toxic, or reactive nature can be analyzed without the usual extraordinarily time-consuming precautions.

REFERENCES

1. Allport, N. L., *Colorimetric Analysis,* Chapman and Hall, London, 1947.
2. American Petroleum Institute, *Selected Values of Properties of Hydrocarbons and Related Compounds,* API Project 44, College Station, Pa.
3. American Society for Testing and Materials, *ASTM Standards,* Parts 1 to 11, Philadelphia, Pa.
4. ASTM Committee D-2, *ASTM Standards on Petroleum Products and Lubricants,* American Society for Testing and Materials, Philadelphia, Pa., 1962.
5. *Ibid.,* D-1015-55.
6. Aston, J. G., H. L. Fink, J. W. Tooke, and M. R. Cines, *Anal. Chem.,* **19,** 218 (1947).

7. Badley, J. H., *J. Phys. Chem.*, **63**, 1991 (1959).
8. Bardeen, J., "Conduction, Metals and Semiconductors," in E. U. Condon and H. Odishaw, eds., *Handbook of Physics*, McGraw-Hill, New York, 1958, Part 4, p. 73.
9. Bauer, N., and S. Z. Lewin, "Determination of Density," in A. Weissberger, ed., *Technique of Organic Chemistry* 3rd Ed., Vol. 1, Part 1, Interscience, New York, 1959, Chap. 4.
10. Bauer, N., K. Fajans, and S. Z. Lewin, "Refractometry," in A. Weissberger, ed., *Techniques of Organic Chemistry*, 3rd Ed., Vol. 1, Part 2, Interscience, New York, 1960, Chap. 18.
11. Bauman, R. P., *Absorption Spectroscopy*, Wiley, New York, 1962.
12. Bennett, C. E., S. Dal Nogare, and L. W. Safranski, "Chromatography: Gas," this Treatise, Part I, Vol. 3, Chap. 37.
13. Block, R. J., E. L. Durrum, and G. Zweig, *A Manual of Paper Chromatography and Paper Electrophoresis*, 2nd Ed., Academic, New York, 1958.
14. Boltz, D. F., *Colorimetric Determination of Nonmetals (Chemical Analysis*, Vol. 8), Interscience, New York, 1958.
15. Kolthoff, I. M., "Titrimetry: Precipitation Titrations," this Treatise, Part I, Vol. 11.
16. Brandenstein, M., I. Janda, and E. Schroll, *Microchim Acta*, 1960, 935.
17. Brauer, G. M., "Polarography," in G. M. Kline, ed., *Analytical Chemistry of Polymers*, Part II, Chap. XI, Interscience, New York, 1962.
18. Buchwald, H., and L. G. Wood, *Anal. Chem.*, **25**, 664 (1953).
19. Campbell, A. N., and L. A. Prodan, *J. Am. Chem. Soc.*, **70**, 553 (1948).
20. Carney, T. P., *Laboratory Fractional Distillation*, Macmillan, New York, 1949.
21. Cassidy, H. G., *Fundamentals of Chromatography (Technique of Organic Chemistry*, Vol. 10), Interscience, New York, 1957.
22. Charlot, G., J. Badoz-Lambling, and B. Tremillon, *Electrochemical Reactions*, Elsevier, Amsterdam, 1962.
23. Cines, M. R., "Solid-Liquid Equilibria of Hydrocarbons," in A. Farkas, ed., *Physical Chemistry of the Hydrocarbons*, Academic, New York, 1950, Chap. 8.
24. Claesson, S., *Arkiv Kemi Mineral. Geol.*, **23A**, No. 1, 1 (1946).
25. Clark, G. L., *The Encyclopedia of Spectroscopy*, Reinhold, New York, 1960.
26. *Ibid.*, pp. 582–647.
27. Conrad, A. L., "Fluorimetry," this Treatise, Part 1, Vol. 5, Chap. 59.
28. Consden, R., A. H. Gordon, and A. J. P. Martin, *Biochem. J.*, **38**, 224 (1944).
29. Coulson, D. M., L. A. Cavanaugh, J. E. De Vries, and B. Walther, *J. Agr. Food Chem.*, **8**, 399 (1960).
30. Craig, L. C., and D. Craig, "Laboratory Extraction and Countercurrent Distribution," in A. Weissberger, ed., *Technique of Organic Chemistry*, 2nd Ed., Vol. 3, Part 1, Interscience, New York, 1956, Chap. 2.
31. Beukenkamp, J., and W. Rieman, "Titrimetry: Acid-Base Titrations in Aqueous Solution," this Treatise, Part I, Vol. 11.
32. Dean, J. A., *Flame Photometry*, McGraw-Hill, New York, 1960.
33. Deford, D. D., and J. W. Miller, "Coulometric Analysis," this Treatise, Part I, Vol. 4, Chap. 49.
34. Demole, E., "Recent Progress in Thin-Layer Chromatography," in M. Lederer, ed., *Chromatographic Reviews*, Vol. 4, Elsevier, Amsterdam, 1962, p. 26.

35. Dorough, G. D., and K. T. Shen, *J. Am. Chem. Soc.,* **72,** 3939 (1950).
36. Dreisbach, R. R., *Physical Properties of Chemical Compounds* (*Advan. Chem. Ser.,* Nos. 15, 22, and 29), Vols. 1, 2, and 3, American Chemical Society, Washington, D.C., 1955, 1959, and 1961.
37. Egloff, G., *Physical Constants of Hydrocarbons,* Vols. 1, 2, 3, 4, and 5, Reinhold, New York, 1939, 1940, 1946, 1947, and 1953.
38. Stenger, V. A., "Titrimetry: Oxidation-Reduction Titration," this Treatise, Part I, Vol. 11.
39. Eyring, H., *Anal. Chem.,* **20,** 98 (1948).
40. Furman, N. H., "Potentiometry," this Treatise, Part I, Vol. 4, Chap. 45.
41. Ginnings, D. C., and G. T. Furukawa, *J. Am. Chem. Soc.,* **75,** 522 (1953).
42. Glasgow, A. R., and G. Ross, "Cryoscopy," this Treatise, Part I, Vol. 8, Chap. 88.
43. Glasgow, A. R., Jr., G. S. Ross, A. T. Horton, D. Enagonio, H. D. Dixon, C. P. Saylor, G. T. Furukawa, M. L. Reilly, and J. M. Henning, *Anal. Chim. Acta,* **17,** 54 (1957).
44. Glasgow, A. R., Jr., A. J. Streiff, and F. D. Rossini, *J. Res. Natl. Bur. Std.,* **35,** 355 (1945).
45. Glasgow, A. R., Jr., N. C. Krouskop, J. Beadle, G. D. Axilrod, and F. D. Rossini *Anal. Chem.,* **20,** 410 (1948).
46. Glasgow, A. R., Jr., N. C. Krouskop, and F. D. Rossini, *Anal. Chem.,* **22,** 1521 (1950).
47. Glasstone, S., *Textook of Physical Chemistry,* 2nd Ed., Van Nostrand, Princeton, N.J., 1946, p. 500.
48. Goldstein, J. H., "Microwave Spectroscopy," this Treatise, Part I, Vol. 5, Chap. 62.
49. Hardy, C. J., and F. H. Pollard, *J. Chromatog.,* **2,** 1 (1959).
50. Heller, W., and D. D. Fitts, "Polarimetry," in A. Weissberger, ed., *Technique of Organic Chemistry,* 3rd Ed., Vol. 1, Part 3, Interscience, New York, 1960, Chap. 33.
51. Herington, E. F. G., and R. Handley, *J. Chem. Soc.,* **1950,** 199.
52. Herrmann, P., and C. T. J. Alkemade, *Flammenphotometric,* 2nd Ed., Springer-Verlag, Berlin, 1960.
53. Heyrovsky, J., *Chem. Listy,* **16,** 256 (1922).
54. Hickling, A., *Trans. Faraday Soc.,* **38,** 27 (1942).
55. Hodgman, C. D., *Handbook of Chemistry and Physics,* 44th Ed., Chemical Rubber Publishing Co., Cleveland, Ohio, 1962.
56. Hunter, L. P., *Handbook of Semiconductor Electronics,* 2nd Ed., McGraw-Hill, New York, 1962, Sect. 6, p. 13.
57. Irving, H., and R. J. P. Williams, "Liquid-Liquid Extraction," this Treatise, Part I, Vol. 3, Chap. 31.
58. James, A. T., and A. J. P. Martin, *Biochem. J.,* **50,** 679 (1952).
59. Johansson, A., and E. Wanninen, "Titrimetry: Complexation Titrations," this Treatise, Part I, Vol. 11.
60. Keulemans, A. I. M., *Gas Chromatography,* 2nd Ed., Reinhold, New York, 1959.
61. *Ibid.,* p. 2.
62. Kies, H. L., *J. Electroanal. Chem.,* **4,** 257 (1962).
63. Kirchner, J. G., J. M. Miller, and G. J. Keller, *Anal. Chem.,* **23,** 420 (1951).

64. Knight, H. S., and S. Groennings, *Anal. Chem.*, **26**, 1549 (1954).
65. Kolthoff, I. M., *Rec. Trav. Chim.*, **39**, 126 (1920).
66. Kolthoff, I. M., and J. J. Lingane, *Polarography*, 2nd Ed., Interscience, New York, 1952.
67. Krell, E., *Handbook of Laboratory Distillation*, Elsevier, Amsterdam, 1963.
68. Landolt-Börnstein, *Physikalisch-Chemische Tabellen*, 5th Ed., 1st, 2nd, and 3rd Suppl., Springer-Verlag, Berlin, 1923, 1927, 1931, and 1935.
69. Landolt-Börnstein, *Zahlenwerte und Functionen*, 6th Ed., Springer-Verlag, Berlin, 1950–62.
70. Lange, N. A., *Handbook of Chemistry*, 10th Ed., McGraw-Hill, New York, 1961.
71. Leddicotte, G. W., "Nucleonics," Annual Reviews, *Anal. Chem.*, **34**, 155R (1962).
72. Lederer, E., and M. Lederer, *Chromatography*, 2nd Ed., Elsevier, Amsterdam, 1957.
73. *Ibid.*, p. 115.
74. Lehrle, R. S., "Ebulliometry Applied to Polymer Solutions," in J. C. Robb and F. W. Peaker, eds., *Progress in High Polymers*, Vol. 1, Heywood, London, 1961, p. 37.
74a. Leslie, R. T., and E. C. Kuehner, "Ebulliometry," this Treatise, Part I, Vol. 8, Chap. 89.
75. Lewin, S., and N. Bauer, "Refractometry and Dispersometry," this Treatise, Part I, Vol. 6, Chap. 70.
76. Lewis, G. N., T. T. Magel, and D. Lipkin, *J. Am. Chem. Soc.*, **62**, 2973 (1940).
77. Liebhafsky, H. A., H. G. Pfeiffer, and E. H. Winslow, "X-ray Methods: Absorption, Diffraction, Emission," this Treatise, Part I, Vol. 5, Chap. 60.
78. Lingane, J. J., *Electroanalytical Chemistry*, 2nd Ed., Interscience, New York, 1958.
79. Lingane, J. J., *J. Am. Chem. Soc.*, **67**, 1916 (1945).
80. Littlewood, A. B., *Gas Chromatography*, Academic, New York, 1962.
81. Loveland, J. W., "Conductometry and Oscillometry," this Treatise, Part I, Vol. 4, Chap. 51.
82. Mader, W. J., "Phase Solubility Analysis," in J. Mitchell, Jr., I. M. Kolthoff, E. S. Proskauer, and A. Weissberger, eds., *Organic Analysis*, Vol. 2, Interscience, New York, 1954, p. 253.
83. Mair, B. J., "Chromatography: Columnar Liquid-Solid Adsorption Processes," this Treatise, Part I, Vol. 3, Chap. 34.
84. Mair, B. J., A. R. Glasgow, Jr., and F. D. Rossini, *J. Res. Natl. Bur. Std.*, **26**, 591 (1941).
85. Margoshes, M., "Emission Flame Photometry," Annual Reviews, *Anal. Chem.*, **34**, 221R (1962).
86. Kybett, B. D., T. V. Charlu, A. Chaudhuri, Jennifer Jones, and J. L. Margrave, "Calorimetry," this Treatise, Part I, Vol. 8, Chap. 90.
87. Martin, A. J. P., and R. L. M. Synge, *Biochem. J.*, **35**, 1358 (1941).
88. Mastrangelo, S. V. R., and R. W. Dornte, *J. Am. Chem. Soc.*, **77**, 6200 (1955).
89. Mathieu, M. P., *Acad. Roy. Belg. Classe Sci. Mem.*, No. 1639 (1953).
90. McCullough, J. P., and G. Waddington, *Anal. Chim. Acta*, **17**, 80 (1957).
91. McCullough, J. P., H. L. Finke, J. F. Messerly, S. S. Todd, T. C. Kincheloe, and G. Waddington, *J. Phys. Chem.*, **61**, 1105 (1957).

92. McCullough, J. P., H. L. Finke, W. N. Hubbard, S. S. Todd, J. F. Messerly, D. R. Douslin, and G. Waddington, *J. Phys. Chem.,* **65,** 784 (1961).
93. McLafferty, F. W., "Mass Spectrometry," in F. C. Nachod and W. D. Phillips, eds., *Determination of Organic Structures by Physical Methods,* Vol. 2, Academic, New York, 1962, Chap. 2.
94. Guinn, V. P., "Activation Analysis," this Treatise, Part I, Vol. 9, Chap. 98.
95. Meites, L., *Polarographic Techniques,* Interscience, New York, 1955.
96. Meites, L., "Voltammetry at the Dropping Mercury Electrode (Polargraphy)," this Treatise, Part I, Vol. 4, Chap. 46.
97. Mitchell, J., Jr., and D. M. Smith, *Aquametry (Chemical Analysis,* Vol. 5), Interscience, New York, 1948.
98. Morrison, G. H., R. L. Rupp, and G. L. Klecak, *Anal. Chem.,* **32,** 933 (1960).
99. Nachtrieb, N. H., H. A. Potratz, O. R. Simi, S. Wexler, and B. S. Wildi, *U.S. At. Energy Comm.* LA-387, 1957.
100. Parr, N. L., *Zone Refining and Allied Techniques,* George Newnes, London, 1960.
101. Pazdera, H. J., and W. H. McMullen, "Chromatography: Paper," this Treatise, Part I, Vol. 3, Chap. 36.
102. Perry, J. H., *Chemical Engineers Handbook,* 3rd Ed., McGraw-Hill, New York, 1950.
103. Pfann, W. G., *Zone Melting,* Wiley, New York, 1958.
104. Pilcher, G., *Anal. Chim. Acta,* **17,** 144 (1957).
105. Plato, C., and A. R. Glasgow, Jr., *Anal. Chem.,* **41,** 330 (1969).
106. Plebanski, T., *Bull. Acad. Polon. Sci., Ser. Sci. Chim.,* **8,** 23 (1960).
107. Plebanski, T., *Bull. Acad. Polon. Sci., Ser. Sci. Chim.,* **8,** 117 (1960).
108. Plebanski, T., *Bull. Acad. Polon. Sci., Ser. Sci. Chim.,* **8,** 125 (1960).
109. Powles, J. G., and C. P. Smyth, "Measurement of Dielectric Constant and Loss," in A. Weissberger, ed., *Technique of Organic Chemistry,* 3rd ed., Vol. 1, Part 3, Interscience New York, 1960, Chap. 38.
110. Purnell, H., *Gas Chromatography,* Wiley, New York, 1962.
111. *Reagent Chemicals,* Committee on Analytical Reagents, American Chemical Society, Washington, D.C.
112. Reilly, J., and W. N. Rae, *Physico-Chemical Methods,* 5th Ed., Vol. 1, Van Nostrand, Princeton, N.J., 1953, pp. 577–629.
113. Riddick, J. A., and E. E. Toops, *Organic Solvents* (Technique of Organic Chemistry, Vol. 7), 2nd Ed., Interscience, New York, 1958.
114. Rieman, W., and A. C. Breyer, "Chromatography: Columnar Liquid-Solid Ion-Exchange Processes," this Treatise, Part I, Vol. 3, Chap. 35.
115. Rose, A., "Distillation," this Treatise, Part I, Vol. 2, Chap. 29.
116. Rosenthal, I., A. R. Weiss, and V. R. Usdin, "Chromatography: General Principles," this Treatise, Part I, Vol. 3, Chap. 33.
117. Rupp, R. L., G. L. Klecak, and G. H. Morrison, *Anal. Chem.,* **32,** 931 (1960).
118. Sandell, E. B., *Colorimetric Determination of Traces of Metals (Chemical Analysis,* Vol. 3), 2nd Ed., Interscience, New York, 1950.
119. Schilt, A. A., and B. Jaselskis, "Ultraviolet and Visible Spectrophotometry," this Treatise, Part I, Vol. 5, Chap. 58.
120. Schulz, H., "Continuous Counter-Current Separation under Conditions of Elution Gas Chromatography," in M. Van Swaay, ed., *Gas Chromatography,* Butterworths, London, 1962, pp. 225–233.

121. Schwab, F. W., and E. Wichers, "Precise Measurement of the Freezing Range as a Means of Determining the Purity of a Substance," in American Institute of Physics Symposium, *Temperature, Its Measurement and Control in Science and Industry,* Reinhold, New York, 1941.

122. Scribner, B. F., and M. Margoshes, "Emission Spectroscopy," this Treatise, Part I, Vol. 6, Chap. 64.

123. Shain, I., "Stripping Analysis," this Treatise, Part I, Vol. 4, Chap. 50.

124. Simonelli, A. P., and T. Highuchi, *J. Pharm. Sci.,* **50,** 861 (1961).

125. Skau, E. L., J. C. Arthur, Jr., and H. Wakeham, "Determination of Melting and Freezing Temperatures," in A. Weissberger, ed., *Technique of Organic Chemistry,* 3rd Ed., Vol. 1, Part 1, Interscience, New York, 1959, Chap. 7.

126. Skau, E. L., *Proc. Am. Acad. Arts Sci.,* **67,** 551 (1933).

127. Skau, E. L., *J. Chim. Phys.,* **31,** 366 (1934).

128. Skau, E. L., *Bull. Soc. Chim. Belg.,* **43,** 287 (1934).

129. Smales, A. A., and L. Salmon, *Analys,* **80,** 37 (1955).

130. Lyons, J. W., "Measurement of Viscosity," this Treatise, Part I, Vol. 7, Chap. 83.

131. Smit, W. M., *Anal. Chim. Acta,* **17,** 23 (1957).

132. Smith, A. L., "Infrared Spectroscopy," this Treatise, Part I, Vol. 6, Chap. 66.

133. Snell, F. D., and C. T. Snell, *Colorimetric Methods of Analysis,* 3rd Ed., Vols. 1, 2, 3, 4, and 2A, Van Nostrand, Princeton, N.J., 1948, 1949, 1953, 1954, and 1959.

134. Stahl, E., *Chemiker Ztg.,* **82,** 323 (1958).

135. Stewart, D. M., "Mass Spectrometry," in A. Weissberger, *Technique of Organic Chemistry,* Vol. 1, Part 4, Interscience, New York, 1960, Chap. 51.

136. Streuli, C. A., "Titrimetry: Acid-Base Titrations in Nonaqueous Solution," this Treatise, Part I, Vol. 11, Chap. 117.

137. Stross, F. H., and H. W. Johnson, Jr., "Gas Chromatography," in R. E. Kirk and D. F. Othmer, *Encyclopedia of Chemical Technology,* 2nd Suppl., Interscience, New York, 1960, p. 380.

138. *Ibid.,* p. 377.

139. Struck, W. S., and E. C. Olson, "Optical Rotation: Polarimetry," this Treatise, Part I, Vol. 6, Chap. 71.

140. Stull, D. R., *Ind. Eng. Chem., Anal. Ed.,* **18,** 234 (1946).

141. Sturtevant, J. M., "Calorimetry," in A. Weissberger, *Technique of Organic Chemistry,* 3rd Ed., Vol. 1, Part 1, Interscience, New York, 1959, Chap. 10.

142. Swietoslawski, W., *Bull. Intern. Acad. Polon. Classe Sci. Math. Nat.,* **10A,** 113 (1949).

143. Swietoslawski, W., *Ebulliometric Measurements,* Reinhold, New York, 1945.

144. Swietoslawski, W., and J. R. Anderson, "Determination of Boiling and Condensation Temperatures," in A. Weissberger, ed., *Technique of Organic Chemistry,* 3rd Ed., Vol. 1, Part 1, Interscience New York, 1959, Chap. 8.

145. Swietoslawski, W., *Roczniki Chem.,* **21,** 94 (1947).

146. Swindells, J. F., R. Ullman, and H. Mark, "Determination of Viscosity," in A. Weissberger, ed., *Technique of Organic Chemistry,* 3rd Ed., Vol. 1, Part 1, Interscience, New York, 1959, Chap. 12.

147. Szebelledy, L., and Z. Somogyi, *Z. Anal. Chem.,* **112,** 385 (1938).

148. Szebelledy, L., and Z. Somogyi, *Z. Anal. Chem.,* **112,** 313 (1938).

149. Tanaka, N., "Electrodeposition," this Treatise, Part I, Vol. 4, Chap. 48.
150. Taylor, J. K., "Measurement of Density and Specific Gravity, this Treatise, Part I, Vol. 7, Chap. 81.
151. Taylor, W. J., and F. D. Rossini, *J. Res. Natl. Bur. Std.,* **32,** 197 (1944).
152. *The National Formulary,* American Pharmaceutical Association, Washington, D.C.
153. *The Pharmacopeia of the United States of America,* the United States Pharmacopeial Convention, Inc., Washington, D.C.
154. Thomas, B. W., and R. Pertel, "Measurement of Capacity: Analytical Uses of Dielectric Constants," this Treatise, Part I, Vol. 4, Chap. 52.
155. Tilton, L. W., and J. K. Taylor, "Refractive Index Measurement," in W. G. Berl, ed., *Physical Methods of Chemical Analysis,* 2nd Ed., Vol. 1, Academic, New York, 1960, p. 411.
156. Timmermans, J., *Physico-chemical Constants of Pure Organic Compounds,* Elsevier, New York, 1950.
157. Tipson, R. S., "Crystallization and Recrystallization," in A. Weissberger, ed., *Separation and Purification (Technique of Organic Chemistry,* Vol. 3, Part 1) 2nd Ed., Interscience, New York, 1956, pp. 395–562.
158. Tiselius, A., "Purity and Purification of Chemical Substances" *Roy. Inst. Chem. (London), Lectures, Monographs, Rept.* 2 (1958).
159. Tschesche, R., and A. K. Sen Gupta, *Chem. Ber.,* **93,** 1903 (1960).
160. Tschesche, R., F. Lampert, and G. Snatzke, *J. Chromatog.,* **5,** 217 (1961).
161. Tunnicliff, D. D., and H. Stone, *Anal. Chem.,* **27,** 73 (1955).
162. Tunnicliff, D. D., and J. H. Badley, *Rev. Sci. Instr.,* **31,** 953 (1960).
163. Vallee, B. L., and R. E. Thiers, "Flame Photometry," this Treatise, Part I, Vol. 6, Chap. 65.
164. Rulfs, C. L., "Gravimetric Analysis," this Treatise, Part I, Vol. 11.
165. Van Wazer, J. R., J. W. Lyons, K. Y. Kim, and R. E. Colwell, *Viscosity and Flow Measurement,* Wiley, New York, 1963.
166. Van Wijk, H. F., and W. M. Smit, *Anal. Chim. Acta,* **23,** 545 (1960); **24,** 41 (1961).
167. Wagner, R. H., and L. D. Moore, Jr., "Determination of Osmotic Pressure," in A. Weissberger, ed., *Technique of Organic Chemistry,* 3rd Ed., Vol. 1, Part 1, Interscience, New York, 1959, Chap. 15.
168. Washburn, E. W., *International Critical Tables,* McGraw-Hill, New York, 1926–33.
169. Weissberger, A., ed., *Distillation (Technique of Organic Chemistry,* Vol. 4), Interscience, New York, 1951.
170. White, C. E., "Fluorometric Analysis," Annual Reviews, *Anal. Chem.,* **34,** 81R (1962).
171. Wichers, E., "Materials Research and Standards," *ASTM Bull.,* **1,** 314 (1961).
172. Witschonke, C. R., *Anal. Chem.,* **24,** 350 (1952).
173. Wotherspoon, N., and G. Oster, "Determination of Fluorescence and Phosphorescence," in A. Weissberger, ed., *Technique of Organic Chemistry,* 3rd Ed., Vol. 1, Part 3, Interscience, New York, London, 1960, Chap. 31.
174. Young, J. A., "Measurement of Vapor Pressure," this Treatise, Part I, Vol. 7, Chap. 78.
175. Zuiderweg, F. J., *Laboratory Manual of Batch Distillation,* Interscience, New York, 1957.

SUBJECT INDEX

A

Accelerator, activation analysis, thermal
    neutrons, 5602-5603
    14-MeV neutrons, 5601-5602
    sources, 14-MeV neutrons, 5598-5601
Activation analysis, 5387, 5422-5423,
    5583-5641, 5924-5926
    accelerator-produced neutrons, 5598-
      5603
    accuracy and precision, 5632-5640
    applications, 5639-5640
    charged-particle, 5607-5609
    comparator technique, 5588-5590
    definition of, 5584
    errors, gamma-ray self-absorption,
      5634
      geometrical factors, 5632-5633
      thermal-neutron self-shielding, 5633-
      5634
    forms of, 5609-5625
    high-flux thermal neutron, 5625-5630
    irradiation time, 5586-5587
    materials of high molecular weight,
      physical methods, 5746-5799
      viscous petroleum fractions, 5746-
      5781
    moderate-flux 14-MeV neutron, 5630-
      5632
    photonuclear, 5603-5607
    precision, effect of counting statistics,
      5635-5639
    radioactive decay, 5587-5588
    radiochemical separation form, 5621-5625
    research reactors, 5590-5595
    sensitivity, 5625-5632
    sources of error, 5632-5640
    theory of, 5585-5590
    thermal neutron and fast neutron, 5590-
      5598
Activation instrumental analysis, 5609-5621

    calculations, 5619-5621
    decay schemes, 5609-5611
    gamma-ray spectrometry, 5614-5619
    scintillation counter, 5611-5614
Activation reactor, fast neutrons, 5596-
    5598
    thermal neutrons, 5595-5596
Analysis, of complex mixtures, 6553-6555
    of simple systems, 5646-5653
    with radioactive reagents, *see* Radiometric
      analysis,
Aromatic ethers, analysis of 5789-5791

B

Back scatter, 5386

C

Cadmium, separation from contaminants,
    table, 5494
Carrier-force, 5470
Carrier-free separations, 5476-5487
    coprecipitation, 5477-5478
    ion exchange, 5481-5485
    paper chromatography, 5481-5485
    solvent extraction, 5478-5481
    volatilization, 5485-5487
Cerium, determination of, table, 5556
Characterization, uniform fractions, 5805-
    5808
Characterization and analysis, physical and
    chemical properties for, 5643-5833
Coal, analysis of, 5782-5788
Cockcroft-Walton generator, figure, 5416
Coulomb corrections, 5438
Cyclic acids, analysis of, 5792-5794,
    5824-5826
Cyclotron, principle of, figure, 5418

5933

54548